AMERICANIZATION
AND ITS LIMITS

Americanization and Its Limits

Reworking US Technology and Management in Post-War Europe and Japan

Edited by
Jonathan Zeitlin and Gary Herrigel

OXFORD
UNIVERSITY PRESS

OXFORD

UNIVERSITY PRESS

Great Clarendon Street, Oxford OX2 6DP

Oxford University Press is a department of the University of Oxford.
It furthers the University's objective of excellence in research, scholarship,
and education by publishing worldwide in

Oxford New York

Auckland Bangkok Buenos Aires Cape Town Chennai
Dar es Salaam Delhi Hong Kong Istanbul Karachi Kolkata
Kuala Lumpur Madrid Melbourne Mexico City Mumbai Nairobi
São Paulo Shanghai Taipei Tokyo Toronto

Oxford is a registered trade mark of Oxford University Press
in the UK and in certain other countries

Published in the United States
by Oxford University Press Inc., New York

© Jonathan Zeitlin and Gary Herrigel 2000

The moral rights of the author have been asserted

Database right Oxford University Press (maker)

First published 2000
First published in paperback 2004

British Library Cataloguing in Publication Data

Data available

Library of Congress Cataloging in Publication Data

Americanization and its limits: reworking US technology and management in post-war
Europe and Japan / edited by Jonathan Zeitlin and Gary Herrigel.
p. cm.
Selected and revised proceedings of two workshops.
Includes bibliographical references.
1. Industrial management—United States—Congresses 2. Industrial
management—Europe—Congresses. 3. Industrial management—Japan—Congresses.
4. Comparative management—Congresses. 5. Reconstruction (1939–1951)—Congresses.
I. Zeitlin, Jonathan. II. Herrigel, Gary.
HD70.U5 A562 2000 658—dc21 99-059049
ISBN 0–19–829555–3 (hbk.)
ISBN 0–19–926904–1 (pbk.)

1 3 5 7 9 10 8 6 4 2

Typeset by Hope Services (Abingdon) Ltd.
Printed in Great Britain
on acid-free paper by
Biddles Ltd.,
King's Lynn, Norfolk

In memory of
Duccio Bigazzi
(1947–1999)
Indefatigable scholar, exemplary colleague,
warm friend.

Preface

This book draws on the selected and revised proceedings of two workshops held at the University of Wisconsin-Madison. We are pleased to acknowledge the generous funding for these workshops provided by the Japan Foundation, the Joint Committees on Japan and Western Europe of the Social Science Research Council and the American Council of Learned Societies, and the Anonymous Fund and the Global Studies Program of the University of Wisconsin-Madison. Nor would these workshops have been possible without the exemplary moral, material, and administrative support of the Office of International Studies and Programs of the University of Wisconsin-Madison, especially Dean David Trubek, Associate Dean Cathy Meschievitz, and Cynthia Williams. We would also like to thank the other workshop participants whose fine papers could not be included in this volume: Patrick Fridenson, Takahiro Fujimoto, Peter Kjaer, Michael Mende, Philip Scranton, Volker Wellhöner, Håkon With Andersen, and Seiichiro Yonekura; Jonathan Zeitlin's former colleague T. J. Pempel, now at the University of Washington, for his helpful advice and assistance in raising the initial funding; and our editor David Musson, for his patient encouragement and support of this project from the very outset. Jonathan Zeitlin would also like to acknowledge the European University Institute in Florence, and particularly the Robert Schuman Center and the Department of History and Civilization, for providing a splendid collegial and physical setting in which to complete the preparation of the manuscript. Finally, we dedicate this book to the memory of our friend and fellow contributor Duccio Bigazzi, whose fertile and productive scholarly career was tragically cut short by cancer in April 1999 just as this book was going to press.

Contents

List of Figures

List of Tables

List of Abbreviations

AACP	Anglo-American Council on Productivity
AC	Alternating current
ACS	Archivio centrale dello Stato
AEI	Associated Electrical Industries
AEU	Amalgamated Engineering Union [UK]
AGIP	Azienda generale italiana petroli
AID	Agency for International Development
AMG	Allied Military Government
AMI	Associazone meccanica italiana
ARMCO	American Rolling Mill Corporation
ASAR	Archivio storico Alfa Romeo
ASEA	Allmänna Svenska Electriska Atiebolaget
ASF	Archivio storico Fiat
ATH	August Thyssen Hütte AG
ATM	Automatic Telephone Manufacturing [UK, a firm]
BICC	British Insulated and Callender's Cables [a firm]
BIOS	British Intelligence Objectives Sub-Committee
BMC	British Motor Corporation
BOF	Basic oxygen furnace [technology]
BPC	British Productivity Council
BSI	British Standards Institution
C&W	Cable & Wireless [a firm]
CA	Continental Archive [German tyre company]
CCS	Civil Communications Section [Allied Powers]
CED	Committee for Economic Development [USA]
CEO	Chief executive officer
CGIL	Confederazione generale italiana del lavoro
CGP	Commissariat général du plan
CIOS	Combined Intelligence Objectives Sub-Committee
CIR	Comitato interministeriale per la ricostruzione
CISL	Confederazione italiana dei sindicati lavoratori
CKD	Completely knocked down
CMI	Cantieri Metallurgici Italiani
CNAM	Conservatoire National des Arts et Métiers
CNPF	Conseil national du patronat français
COLIME	Comité de liaison des industries métalliques européennes
CPS	Comptoir des produits sidérurgiques

CSEU	Confederation of Shipbuilding and Engineering Unions [UK]
CSSF	Chambre syndicale de la sidérurgie française
DAI	Divisione affari internazionali
DHHU	Dortmund-Hörde Hütten Union
EAC	Eastern and Associated Companies
EAC	Engineering Advisory Council [UK]
ECA	Economic Cooperation Administration [USA]
ECA	European Cooperation Administration
ECSC	European Coal and Steel Community
EEC	European Economic Community
EEF	Engineering Employers' Federation [UK]
EMD	Edelmetall-Motor-Drehwähler
ENIOS	Ente nazionale italiano per l'organizzazione scientifica del lavoro
EPA	European Productivity Agency
EPAB	European Production Assistance Board
ERP	European Recovery Programme
EU	European Union
FBI	Federation of British Industries
FEA	Foreign Economic Administration [USA]
FIA	Ford Industrial Archives
FIAT	Field Information Agency, Technical
FIM	Féderation des Industries Mécaniques
FOA	Foreign Operations Administration [USA]
FSSP	Foreign Student Summer Project [USA, MIT]
GEC	General Electric Company [UK]
GHQ	General Headquarters [Allied Powers]
GM	General Motors
GMOO	General Motors Overseas Operations
HCLC	Holding Company Liquidation Commission (Japan)
HFM	Henry Ford Museum
HGW	Hermann Göring Werke
HOAG	Hüttenwerk Oberhausen AG
HVDC	High voltage direct current
IBRD	International Bank for Reconstruction and Development
ICA	International Cooperation Administration; International Cooperation Agency [USA]
IHK	Industrie- und Handelskammer
IMF	International Monetary Fund
INCIO	Italian National Craftsman Industries Organization
IP	Industrial products
IPEJ	Institute of Production Engineers Journal
IRI	Instituto per la ricostruzione industriale
IRSID	Institut de recherches de la sidérurgie

ISA	Industrie siderurgiche associate
JMA	Japan Management Association
JPC	Japan Productivity Council
JPC	Joint Productivity Council [UK]
JUSE	Japanese Union of Scientists and Engineers
LO	Landsorganisationen i Sverige
MDAA	Military Defence Assistance Act 1949 [USA]
MERB	Mitsubishi Economic Research Bureau
MIP	Management incentive plans
MIT	Massachusetts Institute of Technology
MITI	Ministry of International Trade and Industry [Japan]
MoS	Ministry of Supply [UK]
MRC	Modern Records Centre [University of Warwick, UK]
MSA	Mutual Security Agency [USA]
MTM	Methods-Time Management
MTP	Management Training Programme [of Japanese Ministry of Trade and Industry]
MW	Megawatt
NACMMI	National Advisory Council for the Motor Manufacturing Industry [UK]
NAM	National Association of Manufacturers [USA]
NARA	National Archives and Records Administration [USA]
NATO	North Atlantic Treaty Organization
NEJTM	National Engineering Joint Trades Movement [UK]
NFEA	Nya Förenade Electriska AB [Swedish transport firm]
NFTC	National Foreign Trade Council [USA]
NICB	National Industrial Conference Board [USA]
NMC	National Management Council [USA]
NPC	National Productivity Centres
NPI	Norwegian Productivity Institute
NTT	Nippon Telephone and Telegraph
OECD	Organization for Economic Cooperation and Development
OEEC	Organization of European Economic Cooperation
OMO	[Olivetti's engineering division]
ORGALIME	Organisme de liaison des industries métalliques européennes
OSP	Off-shore procurement
PCM	Pulse-code modulation
PEP	Political and Economic Planning
PRA	[Committee for] Productivity and Applied Research [Europe]
PRO	Public Record Office [UK]
R&D	Research and Development
Rheinstahl	Rheinische Stahlwerke AG
RM	Rehn–Meidner [model]
SAC	Société Anonyme Française

SCAP	Supreme Command of the Allied Powers [Japan]
SEEA	Swedish Engineering Employers Association
SGIMTM	Syndicat général des industries mécaniques et transformatrices des métaux [later, Féderation]
SMWU	Swedish Metalworkers Union
Sollac	Société lorraine de laminage continu
SQC	Statistical quality control
SSPB	Swedish State Power Board
STC	Standard Telephone and Cables [a firm]
TAT	Trans-Atlantic Telephone
TMM	Textile Machinery Manufacturers Ltd [UK firm]
TMSC	Time and Motion Study Committee
TMU	Time measurement units
TQC	Total Quality Control
TUC	Trades Union Congress [UK]
TWI	Training Within Industry
Usinor	Union sidérurgique de Nord de la France
USTA&P	United States Technical Assistance and Productivity Programme
VCA	Verbali del Consiglio di amministrazione
VCD	Verbali del Comitato direttivo
Vestag	Vereinigte Stahlwerke AG
WNRC	Washington National Record Centre
WPB	War Production Board [USA]
WTO	World Trade Organization

List of Contributors

Duccio Bigazzi was Associate Professor of Modern History at the University of Milan, Italy, until his death in April 1999.

Paul Erker is Assistant Professor at the Institute of Economic History, Free University of Berlin, Germany.

Henrik Glimstedt is Research Fellow at the Institute of International Business, Stockholm School of Economics, Sweden.

Gary Herrigel is Associate Professor of Political Science and the College at the University of Chicago, USA.

Matthias Kipping is Associate Professor of Economics and Business, Universitat Pompeu Fabra, Barcelona.

Kenneth Lipartito is Professor of History at Florida International University, Miami, USA.

Jacqueline McGlade is Associate Dean of the Graduate College and Associate Professor of History, University of Nothern Iowa.

Ruggero Ranieri is Senior Lecturer and Jean Monnet Professor of History, University of Manchester.

Takao Shiba is Professor of Business History at Kyoto Sangyo University, Japan.

Steven Tolliday is Professor of Economic History at the University of Leeds, UK.

Kazuo Wada is Associate Professor of Business History at the University of Tokyo, Japan.

Jonathan Zeitlin is Professor of History, Sociology, and Industrial Relations and Director of the European Union Center at the University of Wisconsin-Madison.

Chapter 1

Introduction: Americanization and Its Limits: Reworking US Technology and Management in Post-War Europe and Japan

JONATHAN ZEITLIN

A conspicuous feature of the development of the modern world economy has been the emergence of new models of productive efficiency and their attempted diffusion across national boundaries. Britain in the late eighteenth and early nineteenth centuries; the United States from the late nineteenth century to the end of the 1960s, and once again perhaps in the 1990s; Japan in the 1970s and 1980s: each of these countries generated innovations in technology and business organization which were widely believed to define a transnational standard of productive efficiency. In each case, foreign observers flocked to the rising industrial power of the day to determine the secrets of its success, while business people and government officials sought through a variety of means to transplant the new methods into their own domestic soil. In each case, moreover, such experiments touched off far-reaching debates about the essential features of the new production paradigm; its economic, social, cultural, institutional, and political preconditions; and its transferability across national borders.[1]

Although responses have varied widely across firms, countries, and periods—and any definitive judgement would be premature in the most recent cases—historical experience suggests that wholesale imitation of foreign 'best practice' has typically proved less common than piecemeal borrowing and selective

For incisive comments, criticisms, and suggestions on earlier drafts of this introduction—not all of which I have been able to incorporate in the text—I am very grateful to Gary Herrigel, Patrick Fridenson, Henrik Glimstedt, Steve Tolliday, Kazuo Wada, Peter Kjaer, Matthias Kipping, Ruggero Ranieri, Chuck Sabel, and Sue Helper. The usual caveats *a fortiori* apply.

[1] The literature on the international diffusion of technology, organization, and management models is too vast to cite here. For a representative collection of historical studies covering a wide range of countries over the past three centuries, see David J. Jeremy (ed.), *International Technology Transfer: Europe, Japan, and the USA, 1700–1914* (Aldershot: Edward Elgar, 1991); *The Transfer of International Technology: Europe, Japan, and the USA in the Twentieth Century* (Aldershot: Edward Elgar, 1992). For a broad overview of social-scientific diffusion theory and research, see Everett M. Rogers, *Diffusion of Innovations*, 4th edn. (New York: Free Press, 1995).

adaptation to suit the divergent requirements of local economic and institutional circumstances. Often, too, such incremental modifications of the internationally dominant model have given rise to innovative hybrids which became sources of competitive advantage in their own right—as in the case of Japanese manufacturers' post-war transformation of US mass-production techniques.[2] In theory as in history, moreover, there are strong grounds for believing that 'any successful imitation of foreign organizational patterns requires innovation', and that 'the process of transfer and adaptation of a productive model from a parent context to another site will always lead to the hybridization of the logic and elements of the productive organization' because of inevitable differences between the original and new environments. Modification and hybridization of imported technology and management methods, on this view, should not be understood as a negative expression of domestic resistance to the transfer process, nor even as a regrettable if perhaps necessary consequence of compromises in adapting a foreign 'best practice' model to fit local constraints, but rather as a positive opportunity for decentralized innovation and learning by self-reflective actors.[3]

This book focuses on the largest and to date most significant example of this global phenomenon: that of Americanization after the Second World War. For a central problem confronting Western European countries and Japan alike was how far their domestic industries should be reshaped in the image of the United States, unquestionably the dominant economic and military power of the post-war world. To contemporaries on both sides of the Atlantic and Pacific, the 'American model' meant above all mass production—the high-volume manufacture of standardized goods using special-purpose machinery and predominantly unskilled labour—together with the host of 'systematic' management techniques, organizational structures, and research and marketing services

---

[2] A standard history of European industrialization emphasizing the diffusion of best-practice technology and productive organization from leader to follower countries is David S. Landes, *The Unbound Prometheus: Technological Change and Industrial Development in Western Europe from 1750 to the Present* (Cambridge: Cambridge University Press, 1969). For an alternative view which highlights the critical reception of foreign models and their selective appropriation and transformation by local actors, see Charles F. Sabel and Jonathan Zeitlin (eds.), *World of Possibilities: Flexibility and Mass Production in Western Industrialization* (Cambridge: Cambridge University Press, 1997); and for a national study in a similar spirit, see Gary Herrigel, *Industrial Constructions: The Sources of German Industrial Power* (Cambridge: Cambridge University Press, 1996). On the pivotal Japanese case, see for example, Michael A. Cusumano, *The Japanese Automobile Industry: Technology and Management at Nissan and Toyota* (Cambridge, MA: Harvard University Press, 1985); Koichi Shimokawa, 'From the Ford System to the Just-in-Time Production System: A Historical Study of International Shifts in Automobile Production Systems, their Connection, and their Transformation', *Japanese Yearbook on Business History* 10 (1993), 84–105; Steven Tolliday, 'The Diffusion and Transformation of Fordism: Britain and Japan Compared', in Robert Boyer, Elsie Charron, Ulrich Jurgens, and Steven Tolliday (eds.), *Between Imitation and Innovation: The Transfer and Hybridization of Productive Models in the International Automobile Industry* (Oxford: Oxford University Press, 1998), 57–95.

[3] See D. Eleanor Westney, *Imitation and Innovation: The Transfer of Western Organizational Patterns to Meiji Japan* (Cambridge, MA: Harvard University Press, 1987), 6; Boyer *et al.*, *Between Imitation and Innovation*, esp. chs. 1, 2, and 16 (quotation, 4).

developed for its efficient administration and effective exploitation. Beyond the intrinsic appeal of such methods to nations aspiring to emulate American productivity and abundance, US policy makers actively sought to promote their diffusion through the technical assistance programmes and counterpart funds associated with the Marshall Plan in Europe and on a more modest scale with the military occupation and procurement authorities in Japan. At a deeper level, finally, US officials and business leaders aimed to recast European and Japanese patterns of corporate organization and competitive order through assertive support for antitrust, decartellization, and deconcentration policies, together with international market integration and trade liberalization.[4]

Much of the historical literature on post-war Americanization has tended to assume without extensive supporting evidence that this process proceeded relatively smoothly and rapidly, at least in its narrowly economic and technological dimensions. The real barriers to Americanization, from this perspective, lay rather in the social, cultural, institutional, and political spheres, where established élites and popular classes each proved reluctant, to varying degrees and for different reasons, to embrace transatlantic models of labour-management relations, mass consumption, and macroeconomic management. Western Europe, as one influential formulation puts it, was only 'half-Americanized' during the post-war reconstruction period; but productive organization and techniques in such accounts are squarely allocated to the 'Americanized' half.[5] Even where the limits of industrial Americanization are recognized, as in recent studies of post-war Britain, the persistence of 'pre-Fordist' production methods is typically taken as a sign of backwardness and complacency, an avatar of and contributory factor in the subsequent decline of domestic manufacturing.[6] Only

[4] For recent overviews on the European side, see John Killick, *The United States and European Reconstruction, 1945–1960* (Edinburgh: Keele University Press, 1997); David W. Ellwood, *Rebuilding Europe: Western Europe, America and Post-War Reconstruction* (Harlow: Longman, 1992); René Girault and Maurice Lévy-Leboyer (eds.), *Le Plan Marshall et le relèvement économique de l'Europe* (Paris: Comité pour l'histoire économique et financière de la France, 1993); Jacqueline McGlade, 'The Illusion of Consensus: American Business, Cold War Aid, and the Reconstruction of Western Europe', unpublished Ph.D. dissertation (George Washington University, 1995). On the Japanese side, see Andrew Gordon (ed.), *Postwar Japan as History* (Berkeley: University of California Press, 1993); Haruhito Shiomi (ed.), 'Postwar Revival and Americanization', special issue of the *Japanese Yearbook on Business History* 12 (1995). From 1953 onwards, first the Foreign Operations Administration and then the International Cooperation Administration took over responsibility for US technical assistance and the promotion of productivity activities in Japan as in Western Europe.

[5] For this formulation, see Michael J. Hogan, *The Marshall Plan: America, Britain, and the Reconstruction of Western Europe, 1947–1952* (Cambridge: Cambridge University Press, 1987), 436, drawing on Pier Paolo d'Attorre, 'ERP Aid and the Politics of Productivity in Italy during the 1950s', *European University Institute Working Paper* 85/159 (Florence: European University Institute, 1985), Italian version in *Quaderni storici*, n.s., 58 (1985), 55–93. For a fuller discussion of their argument and its relationship to the broader historiography of post-war Americanization, see section 1.1 of this chapter.

[6] See, for example, N. F. R. Crafts, ' "You've Never Had It So Good?": British Economic Policy and Performance, 1945–60', in Barry Eichengreen (ed.), *Europe's Post-War Recovery* (Cambridge: Cambridge University Press, 1995), 246–70; S. N. Broadberry and N. F. R. Crafts, 'British Economic Policy and Industrial Performance in the Early Post-War Period', *Business History* 38, 4 (1996),

in the Japanese case has there been much explicit discussion of possible efficiency gains obtained by modifying the American model to suit local circumstances; and even there, the reconstruction of the post-war workplace is often none the less assimilated to the broader triumph of a transnational 'politics of productivity' exported from the United States.[7] Yet in an era when American manufacturers have themselves struggled to respond to the challenges of new competitive strategies based on greater product diversity and productive flexibility, there can be little justification for considering mass production and systematic management as they were practised in the United States during the 1940s and 1950s as a universal model of industrial efficiency which other nations failed to embrace at their peril.[8]

Based on a richly detailed set of empirical studies by an international group of leading scholars, this book seeks to develop a new comparative analysis of industrial Americanization in post-war Europe and Japan aimed at overcoming the conceptual limitations of the existing literature. First, the essays in this volume closely examine European and Japanese responses to post-war efforts at promoting the transfer and diffusion of US management methods and manufacturing practice. Paying particular attention to issues of impact and implementation at the level of individual sectors and firms, the authors emphasize the autonomous and creative role of local actors in the reception—both positive and negative—of American techniques and methods, above and beyond the influence of US government agencies, Marshall Plan institutions, or national productivity councils. Second, the contributors look carefully not only at *what* the historical actors did, but also at *why* they did it: at the processes of reflection

65–91; Nick Tiratsoo and Jim Tomlinson, 'Exporting the "Gospel of Productivity": United States Technical Assistance and British Industry, 1945–1960', *Business History Review* 71 (1997), 41–81; Jim Tomlinson and Nick Tiratsoo, 'Americanisation Beyond the Mass Production Paradigm: The Case of British Industry', in Matthias Kipping and Ove Bjarnar (eds.), *The Americanisation of European Business, 1948–1960: The Marshall Plan and the Transfer of US Management Models* (London: Routledge, 1998), 115–32; Paddy Maguire, 'Designs on Reconstruction: British Business, Market Structures and the Role of Design in Post-War Recovery', *Journal of Design History* 4 (1991), 15–30.

[7] See Cusumano, *Japanese Automobile Industry*; Tolliday, 'Diffusion and Transformation of Fordism'; Shimokawa, 'From the Ford System to the Just-in-Time Production System'; Shiomi, 'Postwar Revival and Americanization'; Andrew Gordon, 'Contests for the Workplace' in id., *Postwar Japan as History*, 373–94, esp. pp. 375–8. For the original formulation of the 'politics of productivity', see Charles S. Maier, 'The Politics of Productivity: Foundations of American International Economic Policy after World War II', in id., *In Search of Stability: Explorations in Historical Political Economy* (Cambridge: Cambridge University Press, 1987), 121–52, and the fuller discussion in section 1.1 of this chapter.

[8] For a sample of the critical literature linking post-war mass production and systematic management to the competitive difficulties experienced by US manufacturers during the 1970s and 1980s, see William J. Abernathy, Kim B. Clark, and Alan M. Kantrow, *Industrial Renaissance: Producing a Competitive Future for America* (New York: Basic Books, 1983); Michael J. Piore and Charles F. Sabel, *The Second Industrial Divide: Possibilities for Prosperity* (New York: Basic Books, 1984); H. Thomas Johnson and Robert S. Kaplan, *Relevance Lost: The Rise and Fall of Management Accounting* (Boston: Harvard Business School Press, 1987); Michael L. Dertouzos, Richard K. Lester, Robert M. Solow, and The MIT Commission on Industrial Productivity, *Made in America: Regaining the Competitive Edge* (Cambridge, MA: MIT Press/New York: Harper Perennial, 1989).

and debate, both public and private, which underlay their strategies and choices. Historical actors, like contemporary historians, disagreed sharply about the possibilities and limitations of post-war Americanization in different national and sectoral contexts, while the ensuing debates, as these essays demonstrate, often exercised a decisive influence on the decisions taken, and thus on the trajectory of economic development in the broadest sense. Contemporary objections to the American model, as we shall see, were not purely sociocultural, nor can they easily be dismissed as blinkered conservatism even in hindsight: on the contrary, their economic and technological reservations foreshadowed much of the subsequent critique of US manufacturing practice in the face of the Japanese challenge. Third, the contributors treat established market and industrial structures not simply as objective parameters for entrepreneurial decision making, but rather as contested terrains whose contours were shaped by rival visions—both foreign and domestic—of the bases for competitive order, technical efficiency, and democratic stability in a modern economy. Fourth, rather than posing the problem in terms of wholesale acceptance or rejection of the American model, the essays in this volume underscore instead the importance of selective adaptation to fit the demands of domestic markets and institutions, giving rise to a multiplicity of hybrid forms of productive organization, some of which would eventually develop into significant innovations in their own right. Such creative modifications of US practice, as the authors show, could be observed not only among outspoken critics of Americanization, but also paradoxically among many of its most ardent European and Japanese admirers. For all these reasons, finally, this book argues for a shift in analytical perspective from the transfer and diffusion of US technology and management to their active reworking in post-war Europe and Japan, while the contributors prefer in the end to speak not so much of 'Americanization', or even of its limits, but rather of 'American engagements', with all its multiple, ambivalent, and actively charged connotations.

The balance of this introductory chapter is divided into three main sections. Section 1.1 re-examines the historiography of post-war Americanization, highlighting the theoretical assumptions underlying contending perspectives in order to bring out the distinctive features of the conceptual approach developed in this book. Only by substantially modifying or discarding altogether a series of widely held assumptions about the nature and transferability of productive models, it argues, can the pervasive evidence of selective adaptation and innovative modification of US techniques and methods uncovered by the studies in this volume be convincingly accommodated. The second section (1.2) draws together the empirical findings of the individual chapters to sketch out a complex, multi-level comparative analysis of similarities and variations in post-war European and Japanese engagements with the American model across firms, sectors, and national economies, stressing the creativity and reflexivity of local actors together with the resulting proliferation of hybrid forms and practices. The third and final section of the chapter (1.3) considers the implications of the book's interpretation of post-war Americanization for current debates on the transfer and diffusion of foreign productive models across national borders, underlining the historical grounds for scepticism about the likelihood and desirability of international convergence around any single 'best practice' model of economic and technological efficiency, whether Japanese or Anglo-American.

1.1   Post-war Americanization: Contending Perspectives and Theoretical Assumptions

Few historiographical propositions are more deeply entrenched than the claim that the transfer of US technology and managerial know-how lay at the heart of the extraordinary economic growth experienced by Western Europe and Japan during the 'golden age' of the long post-war boom. This view, which originated in the self-presentation of the Marshall Plan institutions and their contemporary supporters, has been reinvigorated over the past decade and a half by the burgeoning economic literature on international catch-up and convergence of productivity. 'The spread of best practice American technologies and systems of work organization throughout Western Europe and Japan', write Andrew Glyn, Alan Hughes, Alain Lipietz, and Ajit Singh in a widely cited essay on 'The Rise and Fall of the Golden Age', 'was reflected at the macroeconomic level in the slow process of "catch-up" of average productivity levels . . . Common to all [countries] were productivity missions sent to the US to bring back the message as to how American prosperity could be emulated.' 'High [post-war] growth', Nicholas Crafts and Gianni Toniolo likewise observe in an overview of current historical-economic research, 'was made possible by the gains deriving from the transfer of the (Taylorist) mass production technology in a receptive (socially capable) environment stabilized by a strong American leadership'.[9]

Many post-war historians carry this interpretation further in stressing not only the transfer of American production techniques and management methods to Western Europe and Japan, but also the realignment of economic structures, institutions, and practices in the latter countries with those of the United States. Thus as John Killick contends in a recent synthetic text on the United States and European reconstruction:

Since 1945, the European economy has developed many characteristically American features. For instance, huge increases in intra-European trade, encouraged by improved transport and EC legislation, have produced large-scale industrial restructuring and many firms now operate throughout Europe. These new corporations are organised

---

[9] Andrew Glyn et al., 'The Rise and Fall of the Golden Age', in Stephen A. Marglin and Juliet B. Schor (eds.), The Golden Age of Capitalism: Reinterpreting the Postwar Experience (Oxford: Oxford University Press, 1990), 56; Nicholas Crafts and Gianni Toniolo, 'Postwar Growth: An Overview', in eid. (eds.), Economic Growth in Europe since 1945 (Cambridge: Cambridge University Press, 1996), 25; cf. also Moses Abramovitz, 'Catch-up and Convergence in the Postwar Growth Boom and After', in William J. Baumol, Richard R. Nelson, and Edward N. Wolff (eds.), Convergence of Productivity: Cross-National Studies and Historical Evidence (Oxford: Oxford University Press, 1994), 100–1.

more like American oligopolies than traditional British or German firms: their managers use American methods, often learned in American-style management schools; their products and services are advertised in American-style media and are marketed in American-type stores. This market is kept closer to full employment than in the 1930s by the use of relatively active and coordinated fiscal and monetary policies—which were developed, in key respects, in the USA. The market is policed by European adaptations of American anti-trust legislation and regulatory agencies.[10]

Much writing in this vein similarly emphasizes the more or less transformative influence on West European and Japanese society resulting from the post-war diffusion of American models of mass consumption, commercial culture, industrial relations, and the displacement of distributive conflict by an ideological consensus around the pursuit of economic growth—what Charles Maier has influentially termed 'the politics of productivity'. Radio, television, advertising, and above all Hollywood cinema, according to this view, worked alongside Marshall Plan propaganda to diffuse seductive images of the 'American way of life', driving 'the demand side of the economic and social transformation, speeding and channelling the changes in mentality and behaviour' towards an Americanized 'era of high mass consumption'.[11]

For other post-war historians, however, the European and Japanese adoption of US production techniques and management methods was not matched, at least in the short and medium term, by a parallel embrace of the social, cultural, institutional, and political components of the American model. Business and political élites in many countries, on this view, long remained highly sceptical of, if not actively hostile to, the New Deal-inspired dimensions of the Marshall Plan programme such as high wages, domestic mass consumption, co-operative union–management relations, public welfare expenditure, decartellization, and Keynesian macroeconomic management, as well as to US proposals for international market integration and the liberalization of trade and payments. Important and in some cases dominant sections of the labour movement likewise rejected the US-sponsored vision of a productivity partnership between

[10] Killick, *United States and European Reconstruction*, 155.

[11] Maier, 'Politics of Productivity'; Ellwood, *Rebuilding Europe*, ch. 12 (quotation, 227). See also Anthony Carew, *Labour under the Marshall Plan: The Politics of Productivity and the Marketing of Management Science* (Manchester: Manchester University Press, 1987); and the literature discussed in Killick, *United States and European Reconstruction*, ch. 14. An influential national study along these lines is Volker Berghahn, *The Americanisation of West German Industry, 1945–1973* (Leamington Spa: Berg, 1986); for extensions of his argument, see also id., 'Technology and the Export of Industrial Culture: Problems of the German-American Relationship, 1900–1960', in Peter Mathias and John A. Davis (eds.), *Innovation and Technology in Europe: From the Eighteenth Century to the Present Day* (Oxford: Blackwell, 1991), 142–61; id., 'Resisting the Pax Americana? West German Industry and the United States, 1945–55', in Michael Ermath (ed.), *America and the Shaping of German Society, 1945–1955* (Oxford: Berg, 1993), 85–100; id., 'West German Reconstruction and American Industrial Culture, 1945–1960', in Reiner Pommerin (ed.), *The American Impact on Postwar Germany* (Providence, RI: Berghahn Books, 1995), 65–82. For 'the postwar triumph of the "politics of productivity"' and the 'Westernization of consumption patterns' in Japan, see Gordon, 'Struggles for the Workplace', 375–8; and Charles Yuji Horioka, 'Consuming and Saving', in Gordon, *Postwar Japan as History*, 259–92, esp. 273–9.

depoliticized unions and progressive managements based on plant-level contractual bargaining. For all these reasons, US diplomatic historian Michael Hogan concludes, borrowing a phrase from Pier Paolo d'Attorre's analysis of Italy, 'In the end . . . Western Europe was only "half-Americanized" '; whereas 'the Marshall Plan had aimed to remake Europe in an American mode . . . America was made the European way'—a judgement which could readily be extended with appropriate modifications to the Japanese case.[12]

In response to the conflicting evidence thrown up by the opposed positions in this debate, some recent accounts of post-war Americanization accordingly emphasize the coexistence of trends towards international convergence of productive systems resulting from the attempted diffusion of the US model with the continuing persistence of national differences. Post-war Americanization, on this view, involved not only a transfer of US production techniques and management methods to Western Europe and Japan, but also a partial transformation of economic structures, institutions, and sociocultural practices. The extent and forms of this latter transformation, however, varied across countries depending on pre-existing features of their domestic environment, together with the opportunities these created for local resistance to the adoption of the American model. Thus as David Ellwood writes in his broad synthetic text, *Rebuilding Europe*:

In historical terms Americanisation appears as a particularly distinctive form of modernisation, superimposed with great political, economic and cultural force . . . on each European country's own variant . . . Every nation arrived at its own synthesis of production and consumption, of collective and individual spending, of traditional ways and new practices directed to growth.[13]

---

[12] Hogan, *Marshall Plan*, 436, 445; d'Attorre, 'ERP Aid and the Politics of Productivity in Italy', 37–41. Cf. also Federico Romero, *The United States and the European Trade Union Movement, 1944–1951*, trans. Harvey Ferguson II (Chapel Hill, NC: University of North Carolina Press, 1992); Richard Pells, *Not Like Us: How Europeans Have Loved, Hated, and Transformed American Culture since World War II* (New York: Basic Books, 1997). Both Harioka and Gordon also emphasize the social, cultural, institutional, and political limits to Japanese convergence on Western models of mass consumption, mass culture, and labour-management relations: see Harioka, 'Consuming and Saving', 290–1; Gordon, 'Contests for the Workplace', esp. 377–8, 385–7, 391–4; and id., 'Conclusion', in *Postwar Japan as History*, 454–6. In the French case, similarly, Richard F. Kuisel highlights the strength of domestic resistance to American models of management, industrial relations, mass consumption, and mass culture despite industrialists' willingness to borrow US manufacturing techniques and the post-war development of a consumer society: see *Seducing the French: The Dilemma of Americanization* (Berkeley, CA: University of California Press, 1993); and id., 'The Marshall Plan in Action: Politics, Labor, Industry and the Program of Technical Assistance', in Girault and Lévy-Leboyer, *Le Plan Marshall*, 335–57. In Italy, conversely, Vera Zamagni argues that American consumption models found easier post-war acceptance than did US production processes and methods of industrial organization, which required more substantial adaptation to domestic market conditions: see her 'The Italian "Economic Miracle" Revisited: New Markets and American Technology', in Ennio di Nolfo, *Power in Europe? II: Great Britain, France, Germany and Italy and the Origins of the EEC, 1952–1957* (Berlin: de Gruyter, 1992), 197–226, esp. 209.

[13] Ellwood, *Rebuilding Europe*, 236; cf. also id. *et al.*, 'Questions of Cultural Exchange: The NIAS Statement on the European Reception of American Mass Culture', in Rob Kroes, Robert W. Rydell, and Doeko F. J. Bosscher (eds.), *Cultural Transmissions and Receptions: American Mass Culture in Europe* (Amsterdam: VU University Press, 1993), 321–33.

Or as Marie-Laure Djelic puts it more theoretically in her comparative sociological study, *Exporting the American Model*:

[T]he American model was not accepted nor adopted to the same extent in all Western European economies. National peculiarities remained and they were more or less significant in each case. Indeed, for each country, the transfer process was embedded in different economic, political, cultural, and institutional environments. In turn, those national differences had an impact not only on transfer mechanisms and their efficiency but also on the nature and degree of resistance and opposition that was to emerge, nationally, to the cross-national transfer process.[14]

Whereas Djelic and others influenced by the new sociological institutionalism adopt a self-consciously agnostic stance towards the efficiency or performance consequences of such national differences in the reception and transfer of the American model, other writers—above all historical economists—have no such reservations. For the latter school, the effective absorption of US mass-production technology, Taylorist or Fordist work organization, and systematic management methods—regarded as the key to productivity catch-up—depended in turn on complementary shifts in socio-economic institutions and practices, from wage bargaining and union structure to corporate organization, market regulation, and macroeconomic management. National institutional environments across Western Europe (and Japan) varied in their compatibility with a growth model based on the importation of American productive techniques, and the resulting differences are assigned a key part in explaining cross-national variations in economic performance during the post-war 'golden age'. 'In order to take full advantage of the adaptation of American technologies to European conditions', Crafts and Toniolo insist, 'business and trade union practices had to be adjusted accordingly'. Even if 'the spread of the new productivity ideology . . . was universal', they observe, 'the speed and lasting impact of adaptation varied from country to country . . .'.[15] For Eichengreen, similarly,

Institutions were not equally well adapted to the needs of growth in all European countries. Some, notably the UK and Ireland, failed to develop the relevant domestic

---

[14] Marie-Laure Djelic, *Exporting the American Model: The Post-War Transformation of European Business* (Oxford: Oxford University Press, 1998), 2–3; cf. Ove Bjarnar and Matthias Kipping, 'The Marshall Plan and the Transfer of US Management Models to Europe: An Introductory Framework', in Kipping and Bjarnar, *Americanisation of European Business*, 6. Both Djelic and Bjarnar and Kipping draw explicitly on the work of Mauro F. Guillén, who argues that 'Nation-states are . . . the structured setting in which a variety of institutional patterns, as well as economic and technological factors, affect the adoption of different models of management': see his *Models of Management: Work, Authority, and Organization in a Comparative Perspective* (Chicago: University of Chicago Press, 1994), 6–7.

[15] Crafts and Toniolo, 'Postwar Growth', 23; contrast Djelic, *Exporting the American Model*, 10–11, 274. For the new sociological institutionalism and its rejection of efficiency-driven explanations of changes in business organization, see Walter Powell and Paul DiMaggio (eds.), *The New Institutionalism in Organizational Analysis* (Chicago: University of Chicago Press, 1991); Neil Fligstein, *The Transformation of Corporate Control* (Cambridge, MA: Harvard University Press, 1990); William G. Roy, *Socializing Capital: The Rise of the Large Industrial Corporation in America* (Princeton: Princeton University Press, 1997).

institutions. Others, such as France and Italy, managed to do so only with delay.
. . . These different institutional responses go a fair way towards accounting for variations across countries and over time in European growth performance.[16]

Superimposed on these conflicting views of the extent and consequences of
post-war Americanization is a cross-cutting debate about the role of the United
States and the Marshall Plan in the transfer and diffusion of productive models
and techniques. For many authors, predominantly but by no means exclusively
American, US initiatives such as the Marshall Plan, the technical assistance programmes, and the policies of the German and Japanese occupation authorities
were crucial in transferring managerial expertise and know-how, financing
investment in mass production technologies, reshaping institutions, and creating a consensus around productivity and economic growth across the Atlantic
and the Pacific. For others, notably but not solely non-American, the primary
impetus behind post-war reorganization of production and the introduction of
new techniques and methods—including those borrowed from the United
States—came rather from the European and Japanese themselves, deriving its
real power from national policies and domestic institutions, with the Marshall
Plan and other US programmes significant mainly at the margin.[17] Here, too,
much of the recent literature has sought to steer a middle course between these
polarities, presenting the Marshall Plan and other US policies as an 'important
catalyst' rather than the driving force behind the post-war economic transformation of Western Europe and Japan; 'not the fuel' but instead 'the lubricant in
[the] engine', according to a slightly different metaphor. 'Transfers of complex
models', argue Ove Bjarnar and Matthias Kipping in their introduction to a
recent edited volume on *The Americanization of European Business*, 'are likely
to take place more effectively when an active exporter is faced with an active
importer'; while for Djelic, similarly, the post-war success of the US authorities
in exporting the American productive model depended on their relative ability

[16] Barry Eichengreen, 'Institutions and Economic Growth: Europe after World War II', in Crafts
and Toniolo, *Economic Growth in Europe*, 38–72 (quotation, 41). The French 'Regulation School'
similarly emphasizes the importance of domestic institutional forms—notably the 'capital–labour
nexus'—in shaping the post-war diffusion of 'Fordist' production methods and consumption models from the United States to Western Europe and Japan, together with the impact of the resulting
differences on national growth trajectories: for a recent restatement of this approach, see Robert
Boyer, 'Capital–Labour Relations in OECD Countries: From the Fordist Golden Age to Contrasted
National Trajectories', in Juliet Schor and Jong-Il You (eds.), *Capital, the State and Labour: A
Global Perspective* (Aldershot: Edward Elgar, 1995), 18–69.

[17] The debate has been particularly fierce in the German case: see Charles Maier and Günther
Bischof (eds.), *The Marshall Plan and Germany* (Oxford: Berg, 1991). Werner Abelshauser has been
the principal spokesman for the autochthonous view of the sources of post-war German growth
which downplays the impact of US assistance; for an English-language statement of his position, see
'American Aid and West German Economic Recovery: A Macroeconomic Perspective', in Maier
and Bischoff, *Germany and the Marshall Plan*, 367–409. At a pan-European level, Alan Milward has
argued most forcefully for the primacy of endogenous forces in post-war economic recovery and
the marginal contribution of the Marshall Plan: see his *The Reconstruction of Western Europe,
1945–51* (London: Methuen, 1984); *The European Rescue of the Nation-State* (London: Routledge,
1992); and for a particularly trenchant statement, 'Was the Marshall Plan Necessary?', *Diplomatic
History* 13, 2 (1989), 231–53.

in different countries to collaborate with institutionally powerful local élites in a cross-national modernization network.[18] Most of the contributors to this book, however, go significantly further in emphasizing the active reworking and transformation of the American model in post-war European and Japanese industry. Modification and hybridization of US technology and management practices, on this view, should not be interpreted as a negative phenomenon, an index of domestic resistance to the transfer process, nor even as a sign of unavoidable compromises in adapting or 'translating' the American model to fit local constraints, but rather as a positive source of experimentation, innovation, and learning for European and Japanese firms during the post-war era.[19]

Beneath this welter of contrasting historiographical interpretations, however, can be discerned a deeper set of theoretical differences about the nature of the 'American model' and its transferability to other national settings. Although these differences will be examined here in the context of the literature on post-war Americanization, very similar theoretical oppositions and assumptions, as we shall see in section 1.3, inform current debates about the international transfer and diffusion of productive models, whether from Japan or once again from the United States. With appropriate adjustments, therefore, the conceptual approach and historiographical critique developed here should prove more widely applicable.

A first axis of disagreement concerns whether the United States should be considered a unitary or heterogeneous model for European and Japanese industry. For most writers on post-war Americanization, there can be little doubt about the essential features of the US model during this period: large, hierarchically managed corporations, using mass production and distribution techniques to compete in oligopolistic markets policed by antitrust regulation. In so far as internal diversity within the US economy is acknowledged at all in such accounts, it is typically assimilated to the persistence of small and medium-sized firms in labour-intensive industries peripheral to what Alfred Chandler has termed 'competitive managerial capitalism'.[20] At the opposite pole are historians such as Alf Lüdtke, Inge Marssolek, and Adelheid von Saldern, who emphasize instead the 'polymorphous' multiplicity of competing representations of US

[18] Killick, *United States and European Reconstruction*, 184; Charles Maier, 'The Two Postwar Eras and the Conditions for Stability in Twentieth-Century Western Europe', in id., *In Search of Stability*, 153–86 (quotation, 173); Bjarnar and Kipping, 'The Marshall Plan and the Transfer of US Management Models to Europe', 12; Djelic, *Exporting the American Model*, 275.

[19] For a convergent view in the context of more recent debates over Japanization, see Boyer *et al.*, *Between Imitation and Innovation*. Contrast Djelic, *Exporting the American Model*, esp. 272; Bjarnar and Kipping, 'The Marshall Plan and the Transfer of US Management Models to Europe', 11–14. For similar positions in historical debates over the reception of American mass culture in Europe, see Kroes *et al.*, *Cultural Transmissions and Receptions*, especially Rob Kroes, 'Americanization: What Are We Talking About?', 302–18; Pells, *Not Like Us*, esp. ch. 9.

[20] See, for example, Djelic, *Exporting the American Model*, ch. 1; Alfred D. Chandler, Jr., *Scale and Scope: The Dynamics of Industrial Capitalism* (Cambridge, MA: Harvard University Press, 1990); id., 'The United States: Engines of Economic Growth in the Capital-Intensive and Knowledge-Intensive Industries', in id., Franco Amatori, and Takashi Hikino (eds.), *Big Business and the Wealth of Nations* (Cambridge: Cambridge University Press, 1997), 63–101.

economy and society which allowed European and Japanese observers to inter-
pret and appropriate contemporary discourses of 'Americanization' in contradictory ways according to their own subjective experiences, interests, and
desires.[21] Between these two extremes stand those, like the contributors to this
volume, who accept the idea of an 'American model' with certain core characteristics as a contemporary historical construct, while at the same time calling
attention to significant ambiguities, undercurrents, and disparities within US
industrial practice, from which foreign visitors could accordingly draw a variety of lessons.

A second and closely related polarity regards the relationship between the
constituent elements of the American model. For many writers on post-war
Americanization, as we have seen, there was a close linkage not only among
mass-production techniques and systematic management methods, but also
between these and US forms of corporate organization, industrial relations, and
market regulation. Though it could be applied to varying degrees in different
settings, the American model, on this view, can best be understood as a coherent package of tightly coupled elements characterized by a high degree of
mutual complementarity. Another group of historians maintains by contrast
that even the Americans themselves did not seek to market a single self-
contained productive model, but instead offered a wide array of discrete stand-
alone techniques from which European and Japanese industrialists could select
the most useful and cost effective, as if in a 'sort of department store', to cite
Luciano Segreto's evocative if exaggerated phrase.[22] Here again, the authors in
this volume pursue an alternative path, highlighting both the deconstructibility
of the American model and the interdependencies among its elements. Thus as
a number of the chapters in this book demonstrate, high-volume US manufacturing techniques and management tools such as standardization, automation,
flow-line layout, mechanized materials handling, job evaluation, statistical
quality control, training within industry, or time-and-motion study could be
successfully introduced by European and Japanese firms making a more diverse
range of products in smaller quantities, but only through careful adaptation and

---

[21] Alf Lüdtke, Inge Marssolek, and Adelheid von Saldern, 'Einleitung', in eid. (eds.),
Amerikanisierung: Traum und Alptraum im Deutschland des 20. Jahrhunderts (Stuttgart: Franz
Steiner, 1996), 7–33, esp. 14–15.

[22] For this second view, see Bjarnar and Kipping, 'Marshall Plan and the Transfer of US
Management Models to Europe', 6 ('Actors were not compelled to "buy" one big American model.
Instead, it seems that they tried to pick up specific techniques from an ideologically defined
"American" set of models and attempted to mould them into existing institutions and business prac-
tices.'); Luciano Segreto, 'Americanizzare o modernizzare l'economia? Progetti americani e risposte
italiane negli anni Cinquanta e Sessanta', *Passato e presente* 14, 37 (1996), 55–86 (quotation, 57–8).
See also the essays in Kipping and Bjarnar, *Americanisation of European Business* by Bent Boel, 'The
European Productivity Agency: A Faithful Prophet of the American Model?', esp. 43; Rolv-Petter
Amdam and Ove Bjarnar, 'The Regional Dissemination of US Productivity Models in Norway in
the 1950s and 60s', esp. 103; and Tomlinson and Tiratsoo, 'Americanisation Beyond the Mass
Production Paradigm', esp. 115: 'what the Americans wanted to supply was not a self-contained and
entire "model" of production'.

modification to fit with the other elements of their own production systems as well as the external environment.

A third contested issue involves the degree of universality or context-dependence of the American model. In a simple neoclassical world characterized by perfect information, competitive markets, and uniform factor supplies, a more efficient new technique will be rapidly adopted by all producers through-out the economy. Yet few if any historical commentators subscribe in practice to such a naïve view. Thus Moses Abramovitz, one of the founders of the mod-ern convergence approach, insists that productivity catch-up is not an automatic process, but depends on supplementary conditions such as 'social capability' (a loosely defined complex of national attributes, attitudes, and institutions favourable to the absorption of foreign innovations), natural resource endow-ments, and 'technological congruence': 'the relevance or usefulness to less advanced countries of the techniques and forms of organization that character-ize the frontiers of productivity in a leading economy'. Given the distinctive developmental path followed by the United States during the nineteenth and early twentieth centuries, Abramovitz suggests, 'countries less well endowed with natural resources and with smaller domestic markets could not easily adopt and exploit American technology'; only as resource disparities became less important, incomes rose, and markets became more integrated through foreign trade could Europe and Japan begin to catch up by emulating US pro-duction methods.[23] Other writers in this tradition, as we have seen, place sub-stantially greater emphasis on the economic, organizational, and institutional requirements for successful adoption and exploitation of American techniques and methods. Most of the contributors to this book would push this line of argu-ment further to question how far market structures and institutions in Western Europe and Japan did in fact converge with those of the United States, and thus to what extent the conditions for technological congruence across these economies have ever fully obtained. In so far as economic and institutional con-ditions in Europe and Japan continued to diverge from those in the USA, local adaptations, alterations, and hybridization of the American model, on this view, remained both necessary and desirable in theory as well as in practice. Only by being substantially modified or even transformed to fit a broader range of local circumstances, paradoxically, could this putatively universal productive model be widely diffused beyond its original economic and institutional con-text.

A fourth opposition centres on the extent of institutional plasticity or path dependence in the receiving countries. For some writers in the catch-up and con-vergence tradition, the institutional environments of post-war Western Europe

---

[23] Abramovitz, 'Catch-up and Convergence', 88, 96; cf. also id., 'Catching up, Forging Ahead, and Falling Behind', in his *Thinking About Growth* (Cambridge: Cambridge University Press, 1989), 220–42; and Richard Nelson and Gavin Wright, 'The Erosion of US Technological Leadership as a Factor in Postwar Economic Convergence', in Baumol *et al.*, *Convergence of Productivity*, 129–63, esp. 148.

and Japan were sufficiently plastic to permit extensive transfer of US techniques and management methods even under conditions of high context dependence and tight coupling among their constituent elements. For others, by contrast, institutional environments, even more than technologies themselves, are highly path dependent, creating substantial barriers to the adoption of new production and growth models outside of extraordinary historical moments. 'Socio-economic institutions', Eichengreen argues,

necessarily displayed considerable inertia. Their function, in part, being to serve as coordinating mechanisms, their very nature created coordination problems for altering them. Institutions function as standards, giving rise to network externalities that tend to lock in their operation. The exceptional circumstances of war and reconstruction provided singular opportunities for coordinating wholesale adjustments in institutional arrangements. Even under these extraordinary conditions, however, radical changes in coordinating institutions were necessarily difficult to organize. Inevitably, the important institutional changes of the postwar period were only marginal adaptations. They were feasible only where considerable progress had already been made in developing the institutional structures required for growth after World War II.[24]

More typically, however, as we have seen, recent scholars have viewed the institutional environments of post-war Western Europe and Japan as both plastic enough to allow a significant transfer of US techniques and methods, and at the same time sufficiently path dependent to inhibit full convergence on the American model. Many of the contributors to this volume would contest this polarity altogether, emphasizing the ways in which even quite stable institutional arrangements, like technologies and production models, may be reconfigured through apparently marginal modifications to operate quite differently under new environmental conditions. Thus continuing relationships or network ties between institutions may belie a deep transformation in the ways actors conceive of themselves, their mission, and their strategic possibilities, as we shall see, for example, in the case of the German and Japanese steel industries before and after the Second World War. History, on this view, surely, matters, as in the path-dependency story; but its consequences may often be to facilitate rather than to obstruct economic adjustment by serving as a cognitive and practical resource for self-reflective actors in responding to external challenges—without, however, leading to convergence around a single set of institutions, techniques, or practices.[25]

A fifth and perhaps most crucial line of theoretical cleavage concerns the underlying efficiency characteristics of the US model. For most writers on post-

---

[24] Eichengreen , 'Institutions and Economic Growth', 41–3.

[25] I am indebted to Gary Herrigel for this formulation. For a broader critique of the path-dependency literature along these lines, see Charles F. Sabel, 'Intelligible Differences: On Deliberate Strategy and the Exploration of Possibility in Economic Life', paper presented to the Meeting of the Italian Economics Association, Florence, 20–21 October 1995, available from his website (http://www.law.columbia.edu/sabel/). This paper draws on arguments from our joint introduction to *World of Possibilities*, 'Stories, Strategies, Structures: Rethinking Historical Alternatives to Mass Production', 1–33.

war Americanization, as we have seen, the US model of mass production and systematic management unambiguously represented a global productivity frontier, deviations from which, for whatever reason, would give rise to inferior economic performance, as most notably in the case of Britain.[26] For most of the authors in this book, by contrast, US manufacturing techniques and management practices represented at best a more or less effective response to a historically specific set of environmental conditions, outside of which there could be no presumption they would prove equally successful. In a longer historical perspective, such as that adopted in this volume, many core features of the American model would widely come to be seen as liabilities rather than assets as the international environment became increasingly volatile from the 1970s onwards, while the individual chapters, as we shall see, offer striking examples of the pitfalls resulting from excessive emulation of post-war US practice—both technological and managerial—under rapidly changing competitive circumstances. Deviations from the American model, on this view, need not result in inferior economic performance, but could instead give rise to incremental innovations which enhanced productivity and competitiveness, as is broadly acknowledged in the case of Japan.[27] Synchronically, too, European and Japanese manufacturers might develop alternative technologies and production methods better adapted to their own circumstances but functionally equivalent or even superior in performance to that of their American counterparts, as for example in the case of Michelin and the radial tyre revolution discussed in Paul Erker's contribution to this volume.

On the basis of various combinations of these underlying theoretical oppositions, the salient differences among the contending historiographical perspectives on post-war Americanization can be represented schematically as in Table 1.1. The remainder of this section draws together and elaborates the conceptual approach to post-war Americanization pursued by the editors of this volume and shared to varying degrees by the other contributors. In each case, as can be seen from Table 1.1, the approach pursued in this book either reverses the dominant position within the historiography or rejects the theoretical opposition on which it is founded. Only by substantially modifying or abandoning altogether these widely held theoretical assumptions about the nature and transferability of productive models, we contend, can the pervasive evidence of selective

---

[26] See the references cited in note 6 above.

[27] See the references cited in notes 7 and 8 above. Some catch-up and convergence theorists recognize that diffusion of new technologies to different physical and market environments may give rise to product or process improvements which are then transferred back to the country where the initial innovation occurred: see William J. Baumol, 'Multivariate Growth Patterns: Contagion and Common Forces as Possible Sources of Convergence', in id. *et al.*, *Convergence of Productivity*, 62–85, esp. 76. Steven Broadberry argues that US-style mass-production methods yielded superior manufacturing productivity performance to that of the flexible production methods predominantly employed by British industry from the late nineteenth century until the 1980s, when the balance of competitive advantage swung back in the opposite direction: see *The Productivity Race: British Manufacturing in International Perspective, 1850–1990* (Cambridge: Cambridge University Press, 1997).

Table 1.1. Theoretical Approaches to Americanization

Unitary or heterogenous model?	Tightly or loosely coupled elements?	Universally applicable or context dependent?	Global or local efficiency advantage?	Institutional plasticity or path dependency?	Theoretical approach
+	+	+	+	+	Naïve convergence theory (pure neo-classical economics)
+	+	−	+	+	Mainstream catch-up and convergence theory: transfer of best-practice techniques dependent on technological congruence, resource endowments, and social capability (Abramovitz)
+	+	−	+	−	Institutional lock-in/lock-out of best-practice techniques as an explanation of national differences in economic performance (Eichengreen, Crafts, Toniolo)
+	−	+/−	+	−	'Half-Americanization' of European and Japanese societies (Hogan, d'Attorre)

+	+	−	+/−	Transfer process embedded in and shaped by national institutional environments (Djelic, Ellwood)
−	−	+	+/−	'American model' as a divisible set of globally efficient and transferable stand-alone techniques (Bjarnar/Kipping, Segreto)
x	−	−	x	'American model' as a locally effective ensemble of interdependent elements, which could be deconstructed, modified, and recombined to suit foreign circumstances by self-reflective actors (Zeitlin/Herrigel)

Key: + = first alternative; − = second alternative; +/− = intermediate position; x = rejected polarity.

adaptation, creative modification, and innovative hybridization of US technology and management in post-war Europe and Japan uncovered by the studies
in this volume be convincingly accommodated.

'Americanization', in our view, should be understood not as a neutral analytical concept but rather as a contested historical project, referring to the putative
diffusion of an ensemble of interdependent characteristics, techniques, and
practices—from mass production and systematic management to corporate
structure, oligopolistic competition, and antitrust policy—which domestic and
foreign commentators alike took to be distinctive features of mid-twentieth-
century US industry.[28] Such contemporary definitions of an 'American model'
were far from arbitrary, often reflecting, as the essays in this book show, long
acquaintance with and keen observation of US industrial reality stretching back
well before the Second World War. These contemporary definitions, however,
were by no means univocal, nor could they be, given the internal diversity and
heterogeneity of industrial practice across the American economy at the time,
which remained visible to attentive foreign visitors—as to latter-day historians—not only in the surviving redoubts of speciality manufacture, but even in
core mass-production sectors like motor vehicles.[29] US industry, moreover, was
by its very nature a moving target, as newer high-profile practices such as the
creation of centralized corporate R&D laboratories and the application of
government-sponsored 'big science' to technology increasingly caught the eye of
foreign observers, while public antitrust policy slalomed through a dizzying
series of twists and turns from the 1910s to the 1950s in its stance towards competition, co-operation, and merger among rival firms.[30] Different observers
therefore interpreted the 'American model' in varying ways, depending on their
own perceptions, experiences, frames of reference, and not least their political
and ideological relationship to the United States, drawing divergent conclusions
about its substantive merits, ease of exportation, and applicability to European
or Japanese circumstances.

[28] For a similar argument against the use of Americanization as an analytical category rather than
a historical concept rooted in the perceptions of contemporary actors, see Paul Erker,
'"Amerikanisierung" der westdeutschen Wirtschaft?', in Konrad Jarausch and Hannes Siegrist
(eds.), Amerikanisierung und Sowjetisierung in Deutschland, 1945–1970 (Frankfurt: Campus,
1997), 137–45, esp. 145.

[29] For modern historical studies documenting diversity within US industry, see Philip Scranton,
Endless Novelty: Specialty Production and American Industrialization, 1865–1925 (Princeton:
Princeton University Press, 1997), esp. ch. 13; Michael Schwartz, 'Markets, Networks, and the Rise
of Chrysler in Old Detroit', forthcoming in Enterprise and Society 1 (2000); id. and Andrew Fish,
'Just In Time Inventories in Old Detroit', Business History 40 (1998), 48–71.

[30] For a historical overview of the development of R&D in American industry, see David
Hounshell, 'The Evolution of Industrial Research in the United States', in Richard S. Rosenbloom
and William J. Spencer (eds.), Engines of Innovation: U.S. Industrial Research at the End of an Era
(Boston: Harvard Business School Press, 1996), 13–85. On the metamorphoses and contending posi-
tions within US antitrust policy, see Rudolph Peritz, Competition Policy in America, 1888–1992
(Oxford: Oxford University Press, 1996), chs. 2–4; Gerald Berk, 'Neither Competition Nor
Administration: Brandeis and the Antitrust Reforms of 1914', Studies in American Political
Development 9 (1994), 24–59; id., 'Communities of Competitors: Open Price Associations and the
American State, 1911–1929', Social Science History 20, 3 (1996), 375–400.

Many contemporary actors, as several of the chapters in this volume underline, thought of themselves as 'modernizers' rather than 'Americanizers', or understood the latter in terms of the former. But others, as a number of the chapters also document, contested this universal vision of industrial modernity in the name of economic and technological objections which cannot easily be discounted in retrospect: indeed, their warnings about the inflexibility of special-purpose equipment, the high overhead costs of bureaucratic management, the wasteful accumulation of buffer stocks and work-in-progress, and the restrictive impact of standardization on product innovation anticipated much of the critique of US manufacturing practice in comparison to that of the Japanese which became commonplace during the 1970s and 1980s. Post-war strategic debates and decision-making processes, it cannot be too strongly emphasized, were conducted under conditions of radical uncertainty: about the size, structure, and stability of demand in markets at home and abroad; about the trajectory of technological development for particular products and entire industries; about the institutional framework for business activity across individual nations, regional trading blocs, and the international economy as a whole. Even where visionary entrepreneurs and policy makers apparently succeeded in reshaping industries and markets through 'self-fulfilling' bets on American-style mass-production technologies and supranational commercial integration—as Matthias Kipping for instance suggests in his chapter on French steel users and producers—the outcome does not thereby 'prove right' their position against the doubts raised by contemporary critics. For the realization of the optimistic projections of rapid and steady expansion of demand which underlay these investment strategies depended in part on contingent developments beyond the full control of the relevant actors themselves, from the Korean war boom to the completion of the Treaty of Rome. Under somewhat different macroeconomic circumstances, growth of demand for steel and metalworking products might easily have proved subject to sharper cyclical fluctuations, leading to substantial underutilization of costly investments in high-throughput, high minimum-efficient-scale plant, as European and Japanese firms following similar strategies discovered to their chagrin during the 1970s.

For just such reasons, as many of the essays in this book demonstrate, enthusiastic supporters and sceptical critics of the American model alike often sought to hedge against uncertainty and improve its fit with domestic markets and institutions through selective adaptation and modification of US techniques and methods, thereby giving rise to innovative hybrid forms of productive organization rooted in indigenous as well as imported practice. Crucial to such creative hybridization, paradoxically, was contemporary actors' attentiveness to the close linkages between productive models and particular economic and institutional contexts, since as we shall see it was precisely those European and Japanese industrialists most acutely conscious of the distance separating domestic from US conditions who proved most aggressive and adept at deconstructing, modifying, and recombining elements of the American model for their own purposes. But this hybridization process, finally, was no one-way street, for in reworking US technology and management to suit local circumstances, post-war European and Japanese actors at the same time reinterpreted and reshaped their own practices and institutions in ways which theorists of path dependency would scarcely have imagined possible.

1.2   American Engagements: Beyond Transfer and Diffusion

The essays in this volume are divided into two main parts. Part One examines post-war efforts at exporting US industrial practice by public agencies and private enterprises respectively, highlighting the internal tensions and diversity among the would-be prophets of the American model. Part Two, the core of the book, then goes on to examine the variety of European and Japanese engagements with US technology and management through a series of country chapters focusing on key metalworking industries such as steel, motor vehicles, mechanical engineering, and electrical equipment, together with closely allied sectors like rubber tyres, electronics, and telecommunications. These sectors, it is widely agreed, lay at the heart of post-war reconstruction and economic growth in Western Europe and Japan, as well as of contemporary debates and struggles over industrial Americanization. The construction of US-designed wide-strip mills, for example, was among the very largest new investment projects of the late 1940s and early 1950s, while the French and Italian steel and automobile industries, as the chapters by Kipping, Ranieri, and Bigazzi show, absorbed a high proportion of Marshall Aid directed to these countries, and indeed to Western Europe as a whole. A more comprehensive survey of post-war Americanization would doubtless devote greater space and attention to light consumer goods like clothing and furniture, process industries like oil or chemicals, or services like finance and retailing, though individual chapters do consider European interactions with the US model in related sectors such as rubber products (Erker), telecommunications (Lipartito), and electricity supply (Glimstedt, Zeitlin), as well as artisanal firms and regional networks of small and medium-sized enterprises (McGlade). By focusing on a critical group of related sectors, however, the book more than makes up in analytical depth for any sacrifice in empirical breadth through the resulting opportunities for complex, multi-level comparisons and contrasts across industries, firms, and national economies.[31]

In chronological terms, too, the volume's coverage radiates outwards from a central core. Thus all of the essays focus on the key reconstruction years

---

[31] For sectorally broader but chronologically narrower surveys of post-war Americanization, see Girault and Lévy-Leboyer, *Le Plan Marshall et le relèvement économique de l'Europe*; and Dominique Barjot, John Gillingham, and Terushi Hara (eds.), *The Productivity Missions and the Diffusion of American Economic and Technological Influence after the Second World War* (papers presented to the Caen preconference, 18–20 September 1997, to be published by Presses de l'Université de Paris-Sorbonne, 2000).

(1945–60), when European and Japanese engagements with the American model were at their most intense, though many go further in following the trajectory of enterprises, sectors, and national economies through the heyday of the long post-war boom which ended with the first oil shock of 1973–4. In nearly all cases, however, the authors also find it necessary to look back in greater or lesser detail at the pre-war period in order to understand how far national and company responses to Americanization were inspired by earlier engagements with Fordism, Taylorism, and US business more generally, as well as by the historical contingencies of the post-war era itself. Many of the chapters, finally, glance forward, if only briefly, to the crisis of mass production and the resurgence of flexibility since the mid-1970s as a means of contextualizing the distinctive environment of the post-war years and assessing its influence on the strategic choices of the historical actors.

Such multiple, overlapping, and cross-cutting chronological perspectives—together with associated techniques such as flashbacks, anticipations, and epilogues—can also be understood as narrative strategies for challenging and subverting the unilinear, teleological presentation of post-war Americanization as progressive modernization which remains predominant in the existing literature. Like synchronic comparisons across firms, sectors, and countries, or the polyphony of multiple voices within a single text, the plurality of temporalities in this book contributes to what literary theorists Gary Saul Morson and Michael André Bernstein have termed 'sideshadowing': the narrative representation of action as a process of deliberative choice among an open (though not of course infinite) set of alternative possibilities, more than one of which might in fact have been realized. Morson and Bernstein counterpose sideshadowing narratives to those based on 'foreshadowing' and 'backshadowing': the abuse of hindsight to recount events as if their outcome were predetermined and could be used to judge the choices of historical actors irrespective of what the latter could realistically have been expected to know at the time. Only by rigorously eschewing such fore- and backshadowing narratives, we argue, can studies such as those collected in this volume hope to recover the decision-making horizon of contemporary actors and thereby arrive at genuinely *historical* accounts of post-war European and Japanese engagements with the American model.[32]

EXPORTING THE AMERICAN MODEL?

One obligatory point of departure for historical analysis of post-war Americanization is the perspective of the Americans themselves. Many accounts of post-war reconstruction, as we have seen, depict the US authorities as

---

[32] On sideshadowing, foreshadowing, and backshadowing, see Gary Saul Morson, *Narrative and Freedom: The Shadows of Time* (New Haven, CT: Yale University Press, 1994); Michael André Bernstein, *Foregone Conclusions: Against Apocalyptic History* (Berkeley, CA: University of California Press, 1994). For a fuller discussion of narrative strategies and the representation of strategic action in industrial history, see Sabel and Zeitlin, 'Stories, Strategies, Structures'.

seeking to export a unitary and coherent 'American model' to Western Europe and Japan. Yet such interpretations, as Jacqueline McGlade's chapter demonstrates, gloss over the deep internal divisions, conflicting objectives, and shifting priorities within the US camp itself which magnified the ambiguities surrounding the American model and inhibited its transnational projection. US policy makers during the late 1940s and 1950s, McGlade argues, were divided between liberal and conservative internationalists whose 'global developmentalist' and 'strategic security' agendas coincided only briefly during the high tide of the Marshall Plan and the creation of NATO. The business community, too, was deeply split between liberal reformers associated with the Committee for Economic Development who played a leading role in the European Recovery Programme (ERP) on the one hand, and more conservative trade protectionists associated with organizations such as the National Association of Manufacturers and the National Industrial Conference Board who were also hostile to New Deal domestic policies on the other. The US occupation authorities in Germany and Japan, as Gary Herrigel observes, were similarly riven by internecine struggles between radical progressive trustbusters, often linked to the Republican Party, and 'New Dealer' advocates of American-style oligopolistic big business and mass production, as well as between both of these groups and conservative stabilizers more concerned with economic revival and internal order than with democratic institutional reform.

Not only were the would-be exporters of the American model themselves deeply divided, but the international priorities pursued by US policy makers also shifted repeatedly in reaction to changing domestic political alignments and external challenges. Thus the US Technical Assistance and Productivity Programme (USTA&P), as McGlade shows, was initially created in response to a conservative backlash against the Marshall Planners' attempts to encourage broad-based recovery, industrial modernization, and democratic reform of West European economies in the context of mounting cold war tensions. As military preparedness, rearmament, and containing Communism became increasingly prominent among US foreign economic policy goals during the 1950s, the USTA&P's original aim to boost European industrial output through a massive transfer of American business practices and production methods gradually gave way to more narrowly targeted and less directive assistance to defence manufacturers holding off-shore procurement contracts.

Alongside these internal divisions and shifting priorities ran parallel spatial and temporal variations in the scope, modalities, and effectiveness of American influence. Only in directly occupied territories like West Germany and Japan, as Herrigel notes, could the USA deploy coercive power to reshape industrial structure, economic institutions, and market order in conformity with American normative ideals. But even there, as he also demonstrates, the externally imposed reforms were progressively modified in key details by domestic authorities, and exerted their deepest long-term impact by creating conditions within which local actors could revise their own self-understanding and practices in ways that ultimately diverged significantly from both indigenous traditions and the American model.

Elsewhere in Western Europe, despite the evident power conferred by the United States' enormous military and economic resources, American influence on post-war reconstruction could be exercised principally through varying combinations of negotiation, persuasion, and voluntary emulation. US leverage might appear to have been greatest over financially weak countries like France and Italy which were particularly dependent on American aid. But as Chiarella Esposito has documented elsewhere, US authorities' reluctance to bring down fragile centrist coalitions in these strategically critical nations by withholding conditional aid drastically curtailed their ability to insist on policies opposed by domestic governments, such as macroeconomic expansion in Italy or increased expenditure on low-cost housing and financial stabilization to contain inflation in France.[33] American support and Marshall Aid were most effective in providing approved domestic actors with the additional financial and political resources needed to implement their own strategic visions, as Matthias Kipping argues in the case of Jean Monnet, René Damien (Usinor), Pierre Lefaucheux (Renault), the French continuous strip mill installations, and the opening of the domestic steel market to competition through the ratification of the Schuman Plan. Even in such instances, however, as Ruggero Ranieri shows in a marvellous piece of historical detective work, there was often a plurality of conflicting views among US advisers and policy makers at different levels, so that only by mobilizing an extraordinary coalition including American engineering consultants, steelmakers, and equipment suppliers as well as Fiat could Finsider's Oscar Sinigalia succeed in overcoming the combined opposition of the World Bank, local Marshall Plan officials, Italian private steel companies, and other prominent US industrialists and consultants to the construction of a new integrated wide-strip mill complex at Cornigliano on the Genoese coast.

American negotiating leverage over domestic economic and industrial policies during the Marshall Plan era was weakest, finally, in the case of more politically and financially independent Social Democratic nations such as Britain, Norway, and Sweden. Thus the British and Norwegian Labour governments insulated their ambitious capital investment policies from US interference by using counterpart funds mainly to retire public debt, while carefully devolving responsibility for collaboration with the American-led productivity drive to bipartite union-management bodies outside direct state control. Sweden, which remained politically neutral, created no counterpart fund at all because her

---

[33] Chiarella Esposito, *America's Feeble Weapon: Funding the Marshall Plan in France and Italy, 1948–1950* (Westport, CT: Greenwood Press, 1994); id., 'Influencing Aid Recipients: Marshall Plan Lessons for Contemporary Aid Donors', in Eichengreen, *Europe's Post-War Recovery*, 68–90. For an unconvincing attempt to substantiate the claim that 'the administrative agencies in charge of managing [the Marshall Plan] . . . developed powerful means of control that reached down to the country and even to the project level', see Djelic, *Exporting the American Model*, esp. ch. 7 (quotation, 275).

Social Democratic government accepted only small loans but not outright grants from the Marshall Plan.[34]

Chronologically, too, US authorities' capacity to promote the transfer and diffusion of the American model to Western Europe and Japan moved through several distinct phases during the post-war reconstruction period. US direct influence over the European and Japanese economies reached its zenith during the immediate post-war years of military occupation and massive foreign aid, but began to fall off by the early 1950s with the onset of large-scale rearmament and the ascendancy of strategic concerns with Communist containment. Overt American hegemony declined still further during the late 1950s with the increasingly autonomous decision-making role assumed by European governments in multilateral institutions such as the Organization for European Economic Cooperation (OECD) and the European Productivity Agency (EPA), and the gradual reintegration of both Japan and West Germany into the international community. During this latter period, as McGlade comments, initiative in the planning and usage of US aid and technical assistance passed increasingly into the hands of European actors, who creatively adapted it for their own reform purposes rather than mechanically seeking to convert domestic manufacturing into a mirror image of the American model; and similar if perhaps less sweeping observations could also be made about the productivity movement in Japan, which has been termed 'America's star pupil'.[35] Few national productivity centres long survived the dissolution of the EPA in 1961, while even those countries such as West Germany, Japan, or Britain, which had enacted antimonopoly legislation under US pressure in the late 1940s, as Herrigel among others shows, deviated significantly from the American antitrust paradigm during the 1950s and 1960s.[36] Attempts to institutionalize the transfer of the American

[34] On the use of Marshall Aid counterpart funds and the creation of the national productivity centres in Britain and Norway, see, in addition to the chapter by Zeitlin in this volume: Killick, *United States and European Reconstruction*, 101–2, 109–10, 128–9; Jim Tomlinson, 'Corelli Barnett's History: The Case of Marshall Aid', *Twentieth-Century British History* 8, 2 (1997), 222–38; id., 'The Failure of the Anglo-American Council on Productivity', *Business History* 33, 1 (1991), 82–92; Gunnar Yttri, 'From a Norwegian Rationalization Law to an American Productivity Institute', *Scandinavian Journal of History* 20 (1995), 231–58; Helge Pharo, 'Norway, the United States and the Marshall Plan, 1947–1952', in Richard T. Griffiths, *Explorations in OEEC History* (Paris: OECD, 1997), 73–85. For the Swedish case, see Leon Dalgas Jensen, 'Denmark and the Marshall Plan, 1947–8: The Decision to Participate', *Scandinavian Journal of History* 14 (1989), 73.

[35] On the Japanese productivity movement during the 1950s and early 1960s, see Shiomi, 'Postwar Revival and Americanization', particularly the essays by Takenori Saito, 'Americanization and Postwar Japanese Management', esp. 21–3; Kinsaburo Sunaga, 'American Technical Assistance Programs and the Productivity Movement in Japan' (quotation, 38); and Satoshi Saito, 'The Emergence of the Productivity Improvement Movement in Postwar Japan and Japanese Productivity Missions Overseas', esp. 66–7.

[36] On the US campaign for foreign antitrust legislation during the 1940s and its practical consequences, see in addition to the chapters by Herrigel and Lipartito, Helen Mercer, 'The Rhetoric and Reality of Anti-Trust Policies: Britain, Germany and Japan and the Effects of US Pressure in the 1940s', in Carlo Morelli (ed.), *Cartels and Market Management in the Post-War World* (London: London School of Economics: Business History Unit Occasional Paper 1997, no. 1); Matthias Kipping, 'Concurrence et compétitivité. Les origines de la législation anti-trust française après 1945',

model in Western Europe and Japan thus proved largely unsuccessful, leaving the enduring influence of US industrial practice dependent primarily on conversion and voluntary emulation, and hence ultimately on its interpretation, adaptation, and modification by local actors.[37]

But public policies such as the Marshall Plan, the USTA&P, and decartellization comprised only one side of contemporary American efforts to transform European and Japanese industry in their own image. Private business enterprises such as capital equipment suppliers, consultancies, and final goods manufacturers also played an important and growing part in the attempted transfer of US techniques and methods through a variety of channels from machinery sales, contractual advice, and patent licensing to technical co-operation agreements, joint ventures, and foreign direct investment. Thus in nearly every case considered in this book, European and Japanese manufacturers acquired American technical know-how in embodied form during the post-war reconstruction years by ordering machinery and capital equipment from US suppliers, often but by no means exclusively financed with American aid. In a few extreme cases, such as Finsider's Cornigliano complex, Simca's Nanterre automobile factory, or Nissan's Oppama works, European and Japanese companies employed US engineering consultants and equipment suppliers to design, lay out, and even install entire plants. Other companies like Volvo brought in American consultants to implement new payment and work-study systems such as MTM (Methods-Time Measurement), while Finsider, as Ranieri details, worked closely with US steel companies and management consultants to import not only technological know-how but also budgeting, planning, and standard costing methods, as well as supervisory training, job analysis, and job evaluation. Many European and Japanese firms in sectors such as steel, rubber, automobiles, and electronics likewise purchased manufacturing licences to patented American products and processes either for their own use or to service a geographically delimited market, and such arrangements typically included fees for technical assistance and information as well as a pure royalty component. Not all technical co-operation agreements, however, involved cash payments, since well-established European companies like Fiat or Continental possessed sufficient research, design, production, and marketing expertise of their own to place them on a more or less equal footing with US partners such as Chrysler, Goodyear, or General Tire. More rarely, as in the case of Phoenix-Gummiwerke and Firestone Tire & Rubber, such transatlantic collaboration could extend as far as the marketing of jointly branded goods and a US equity stake in the

---

Études et Documents VI (1994), 429–55. On the British case more specifically, see Helen Mercer, *Constructing a Competitive Order: The Hidden History of British Antitrust Policies* (Cambridge: Cambridge University Press, 1995); Tony Fryer, *Regulating Big Business: Antitrust in Great Britain and America 1880–1990* (Cambridge: Cambridge University Press, 1992), chs. 7–8.

[37] For a related typology of cross-national transfer mechanisms, which places greater emphasis on the normative institutionalization of the American model in Western Europe, see Djelic, *Exporting the American Model*, esp. 129–33.

partner company, leading to more rapid and intense know-how transfer, but also, as Paul Erker points out, to a risky and eventually dangerous dependence on American products and technology.

A striking feature of industrial Americanization during the early post-war period is the comparatively limited impact of US management consultancies, especially in larger European countries such as Britain, France, and West Germany. In each of these markets, as Matthias Kipping has suggested elsewhere, US consultancy firms made little inroad during the 1940s and 1950s, both because of the prior existence of domestic service providers apparently able to meet the needs of local clients, and because of their own relative failure to adapt the American model to indigenous corporate structures and management styles. It was only during the 1960s that US management consultancies, notably McKinsey & Co., established a major role for their overseas operations in transferring the multidivisional form of corporate organization to large British and to a lesser extent French and West German companies. Even then, however, as a growing historical literature shows, the multidivisional structure was often implemented in incomplete, 'corrupted', or modified form, especially in Britain, while an important group of West German steel, automobile, engineering, and electrical equipment firms either never adopted the multidivisional form at all, or returned to older functional and holding company structures during the 1970s.[38]

In each of these cases, whatever the eventual outcome, US consultants and manufacturers functioned primarily as brokers or intermediaries in the attempted transfer of American techniques and methods, responding to initiatives from European client firms and/or US public authorities. The major exception to this pattern is of course foreign direct investment by US multinational companies, the eye of the resurgent storm over the 'American challenge' to European business during the 1960s.[39] Surely it is to such companies if anywhere that we should look for robust, coherent, and successful efforts at transplanting the American model? Although historical research on this crucial question is still

[38] On the European operations of US management consultancies and their role in the transfer of the multidivisional form, see Matthias Kipping, 'The U.S. Influence on the Evolution of Management Consultancies in Britain, France, and Germany since 1945', *Business and Economic History* 25, 1 (1996), 112–23; Christopher D. McKenna, ' "The American Challenge": McKinsey & Company's Role in the Transfer of Decentralization to Europe, 1957–1975', *Academy of Management Proceedings* (1997), 226–30; Bruce Kogut and David Parkinson, 'The Diffusion of American Organizing Principles to Europe', in Kogut (ed.), *Country Competitiveness: Technology and the Organizing of Work* (Oxford: Oxford University Press, 1993), 179–202; John Cable and Manfred J. Dirrheimer, 'Hierarchies and Markets: An Empirical Test of the Multidivisional Hypothesis in West Germany', *International Journal of Industrial Organization* 1 (1983), 43–62; Herrigel, *Industrial Constructions*, 222, 232. In Japan, too, most leading manufacturing companies during the post-war period either rejected the multidivisional form altogether in favour of elaborated functional organizations or constrained divisional decision-making autonomy to a much greater extent than in the US model: see W. Mark Fruin, *The Japanese Enterprise System: Competitive Strategies and Cooperative Structures* (Oxford: Clarendon Press, 1994), esp. 176–8.

[39] For the *locus classicus* of this debate, see Jean-Jacques Servan-Schreiber, *The American Challenge*, trans. Ronald Steel (New York: Atheneum, 1968).

in its preliminary stages,[40] Steven Tolliday's study of the European operations of US multinational automobile firms yields some surprising answers. Even the 'Big Three' car makers themselves did not fully conform to a single American model: thus the Ford Motor Company itself had to be comprehensively reorganized with the assistance of ex-General Motors managers during the late 1940s to repair the damages inflicted by the latter days of its founder's rule before it could begin to tackle the problems of rebuilding the company's various European subsidiaries. And even once Ford had ostensibly embraced GM's principles of managerial decentralization and divisional autonomy, the former continued to diverge from the latter in its greater functional centralization and stronger orientation towards the pursuit of production economies of scale through automation and plant integration.[41] Nor did it prove an easy task to integrate and co-ordinate Ford's separate British, French, and German companies, each of which had its own distinct products, production facilities, distribution networks, and management strategies, especially since trade barriers, government policies, and variations in competitive conditions continued to fragment European automobile markets even after the creation of the EEC. Under these conditions, as Jonathan Zeitlin also shows in the case of Britain, adjustment of US mass-production methods to the size and structure of domestic markets, including the development of distinctive local designs and the use of intermediate technological solutions for accommodating shorter runs and more frequent changeovers, gradually became a distinctive feature of the post-war European operations of both Ford and GM (with the partial but significant exception of Ford-France, which was sold to Simca in 1954). Even during the late 1950s and early 1960s, proposals by Ford-International to achieve European-wide economies of scale through cross-national harmonization of model planning and commonization of components ran into fierce opposition not only from the management of its British and German companies, but also from its own central marketing staff. In many cases, therefore, as Tolliday observes, through much

---

[40] The classic studies remain those of Mira Wilkins: see her *The Maturing of Multinational Enterprise: American Business Abroad from 1914 to 1970* (Cambridge, MA: Harvard University Press, 1974); ead. and Frank E. Hill, *American Business Abroad: Ford on Six Continents* (Detroit: Wayne State University Press, 1964). On the British case, which has received closest attention, see John H. Dunning, *American Investment in British Manufacturing Industry* (1st edn., London: George Allen & Unwin, 1958; 2nd rev. edn., London: Routledge, 1998); Geoffrey Jones and Frances Bostock, 'US Multinationals in British Manufacturing before 1962', *Business History Review* 70, 2 (1996): 207–56. For a fuller discussion of the literature, see Tolliday's chapter in this volume.

[41] See also the recent work of David Hounshell: 'Assets, Organizations, Strategies, and Traditions: Organizational Capabilities and Constraints in the Remaking of Ford Motor Company, 1946–1962', in Naomi Lamoreaux, Daniel M. G. Raff, and Peter Temin (eds.), *Learning by Doing in Markets, Firms, and Countries* (Chicago: University of Chicago Press, 1999), 185–218; 'Ford Automates: Technology and Organization in Theory and Practice', *Business and Economic History* 24, 1 (1995), 59–71; 'Planning and Executing "Automation" at Ford Motor Company, 1945–65: the Cleveland Engine Plant and Its Consequences', in Haruhito Shiomi and Kazuo Wada (eds.), *Fordism Transformed: The Development of Production Methods in the Automobile Industry* (Oxford: Oxford University Press, 1995), 49–86; 'Automation, Transfer Machinery, and Mass Production in the U.S. Automobile Industry in the Post-World War II Era', forthcoming in *Enterprise and Society* 1 (2000).

of the post-war period US multinational automobile companies often found themselves running third- or fourth-rate European subsidiaries and struggling to establish effective managerial control from Detroit, while European manufacturers such as Volkswagen, Fiat, Renault, and even for a time Simca surpassed the masters in finding ways to adapt, develop, and reinterpret American practice to suit domestic markets and conditions.

Reworking US Technology and Management: National, Sectoral, and Firm-Level Variations

European and Japanese responses to post-war efforts at exporting the American model and their engagements with US technology and management can be analysed at a number of different levels. In broad terms, the countries considered in this book may be grouped into three related pairs, as reflected in the organization of the chapters, based on their degree of political and financial autonomy from the United States, their economic and technological self-confidence, and the enthusiasm of key decision makers in business and government for emulation of the American model. Each of these axes of variation can in turn be further specified. We have already seen how the forms and extent of US influence on domestic policy varied across militarily occupied territories like Japan and West Germany, financially weak but strategically critical countries like France and Italy governed by fragile centrist coalitions, and more politically and economically independent Social Democratic nations like Britain and Sweden. Such differences can be clearly observed, for example, in the relative importance of Marshall Aid: thus in 1948–9, as Milward has shown, net ERP aid as a proportion of national income ranged from 6.5 per cent in France and 5.3 per cent in Italy to 2.9 per cent in Western Germany, 2.4 per cent in Britain, and 0.3 per cent in Sweden.[42]

A key dimension of national economic and technological self-confidence is the degree of industrial backwardness relative to the United States in the minds of domestic engineers, managers, and government officials, which was arguably greatest in the cases of France, Italy, and from some points of view Japan, but least in those of Britain, Sweden, and Germany. Another important element is the intensity and continuity of interactions with American industry before 1945. Here the UK was clearly in a class of its own, as Anglo-American economic co-operation and technological exchange deepened substantially during the war, in

---

[42] Milward, *Reconstruction of Western Europe*, ch. 3, esp. 95–8. The picture changes somewhat in the case of Britain and Western Germany, as Killick shows, if other forms of US assistance in 1946–7 are also taken into account, while Vera Zamagni arrives at a lower figure for Marshall Aid as a proportion of Italian national income by upwardly revising the index of industrial production for the late 1940s: see Killick, *United States and European Reconstruction*, 96–7; Zamagni, 'Betting on the Future. The Reconstruction of Italian Industry, 1946–1952', in Josef Becker and Franz Knipping (eds.), *Power in Europe? Great Britain, France, Italy and Germany in a Postwar World, 1945–1950* (Berlin: de Gruyter, 1986); ead., *The Economic History of Italy, 1860–1990* (Oxford: Clarendon Press, 1993), 332.

contrast to hostile powers such as Germany, Italy, and Japan, which had already moved towards autarky in the late 1930s, or occupied countries like France, whose industrial contacts with the USA were largely cut off after 1940. More elusive but no less significant is the extent of domestic industrialists' faith in the competitive value of indigenous technological styles, organizational models, and innovative capabilities. Here the Japanese, with their long and proud tradition of modifying and improving imported technologies, arguably stood closer to the British, Swedes, and Germans than to the Italians or the French, though even in the latter cases, as the chapters by Bigazzi and Kipping show, the creativity and self-reliance of domestic engineers and managers should not be underestimated.[43]

The intersection of these axes of variation shaped the relative strength of domestic commitment to and enthusiasm for emulation of the American model. Britain and Sweden on the one hand and France and Italy on the other could thus be located at opposite ends of an international spectrum of post-war responses to industrial Americanization. The position of West Germany and Japan on this continuum is complicated by the disjunction between their dependent status as occupied powers, which, unlike Britain and Sweden, permitted the external imposition of far-reaching changes in industrial structure and market order, and their autonomous technological style and innovative capabilities, which made them more critical and selective in the reception of US practice than France and Italy.[44]

Such overarching generalizations must be immediately qualified by a recognition of the multiplicity of actors and their divergent stances towards Americanization within each country itself. In Italy, for example, both the Christian Democratic-dominated government and the peak business association Confindustria were actively hostile to the Marshall Planners' self-consciously progressive efforts to promote mass consumption and union-management co-operation, while both bodies remained lukewarm at best towards the USTA&P

---

[43] On pre-war Japanese modification and improvement of imported products and processes as a source of confidence in domestic technological creativity and innovative capabilities, see in addition to the chapter by Wada and Shiba in this volume: Tetsuro Nakaoka, 'The Role of Domestic Technical Innovation in Foreign Technology Transfer: The Case of the Japanese Cotton Textile Industry', *Osaka City University Economic Review* 18 (1982), 45–62; Tessa Morris-Suzuki, *The Technological Transformation of Japan: From the Seventeenth to the Twenty-first Century* (Cambridge: Cambridge University Press, 1994), chs. 4–6; William Mass and Andrew Robertson, 'From Textiles to Automobiles: Mechanical and Organizational Innovation in the Toyoda Enterprises, 1895–1933', *Business and Economic History* 25, 2 (1996), 1–37; Tolliday, 'Diffusion and Transformation of Fordism', 72–3, 76. On the technological creativity and self-reliance of French engineers and managers, see also Patrick Fridenson, 'L'Industrie automobile française et le plan Marshall', in Girault and Lévy-Leboyer, *Le Plan Marshall*, 283–9; id., 'L'Industrie automobile: la primauté du marché', *Historiens et Géographes* 361 (1998), 227–42.

[44] These country classifications are evidently different from those of other authors operating with alternative interpretative frameworks. In particular, the essays in this volume place greater emphasis than much of the existing literature on the critical and selective reception of post-war Americanization in West Germany, Japan, and Sweden, as well as on the positive engagement with the US model in important sections of Italian industry.

and the National Productivity Centre (Comitato nazionale per la produttività) until the latter's reorientation in the mid-1950s towards the organization of management education courses and the provision of financial assistance to small and medium-sized firms. It was only the alliance between public-sector technocrats from the great state holding companies IRI (Instituto per la ricostruzione industriale) and AGIP (Azienda generale italiana petroli), big private industrial companies such as Fiat, and their supporters within the governing parties, that ensured the use of a substantial proportion of Marshall Aid funds to support large-scale investments in American-style mass-production technologies. But serialization of production and the assimilation of US high-volume methods in Italy were not confined to large public and private enterprises: as Duccio Bigazzi points out, a significant body of smaller firms manufacturing consumer-orientated products like typewriters, sewing machines, motor scooters, and domestic electrical appliances also took great strides in this direction during the 1950s and 1960s.[45]

In most European countries, the post-war productivity drive raised sensitive and contentious questions about the role of the state in private industrial decision making, which aroused mistrust and even hostility among business-interest organizations. Business spokesmen returning from transatlantic study missions expressed open reservations about the domestic applicability of American manufacturing techniques and management methods not only in Britain and Italy but also in France, where their peak association, the Conseil national du patronat français (CNPF), like the Federation of British Industries (FBI), was officially committed to the US-sponsored productivity programme and directly involved in the administration of ERP counterpart funds.[46] In most countries, too, including France, Britain, West Germany, and Japan, both peak and

[45] In addition to the chapters by Ranieri and Bigazzi, see Segreto, 'Americanizzare o modernizzare l'economia'; Zamagni, 'Betting on the Future', and 'Italian "Economic Miracle" Revisited'; Sergio Chillè, 'Il "Productivity and Technical Assistance Program" per l'economia italiana (1949–1954): Accettazione e resistenze ai progetti statunitense di rinnovamento del sistema produttivo nazionale', *Annali Fondazione G. Pastore* 22 (1993), 76–121; d'Attore, 'ERP Aid and the Politics of Productivity in Italy'.

[46] In addition to the chapters by Kipping and Zeitlin, see Kipping, ' "Operation Impact": Converting European Employers to the American Creed', in id. and Bjarnar, *Americanisation of European Business*, 55–73; Vincent Guigueno, 'L'Éclipse de l'atelier. Les missions françaises de productivité aux Etats-Unis dans les années 1950', unpublished mémoire de Diplôme d'Etudes Approfondies (Ecole Nationale des Ponts et Chaussées, Université de Marne-La-Vallée, 1994); Kuisel, 'Marshall Plan in Action' and *Seducing the French*, ch. 4; Nick Tiratsoo and Jim Tomlinson, *Industrial Efficiency and State Intervention: Labour 1939–1951* (London: Routledge, 1993), chs. 4–7. According to an official of the US International Cooperation Agency (ICA), 'The Germans had resisted the sending of teams to the US since they believed that German technology was more advanced than American technology. Nevertheless, the Germans finally agreed to send teams, and upon doing so, they also learned about these attitudes which contribute in the United States to higher productivity' (Richard Goodrich to Frank Turner of the ICA Japan Mission, 15 April 1955, quoted in Sunagawa, 'American Technical Assistance Programs and the Productivity Movement in Japan', 31). Cf. also Henry Wend, ' "But the German Manufacturer Doesn't Want Our Advice": The Limits of American Technical Assistance in West Germany, 1950–1954' (unpublished paper, Boston University, April 1999).

sectoral trade associations staunchly opposed actual or proposed antitrust legislation inspired by the American model, and fought with varying degrees of success during the 1950s against statutory prohibitions of cartel arrangements.[47]

In France and Italy, the trade union movement split after 1947 into a Communist-led majority wing, which was politically and ideologically hostile to Americanization and the official productivity drive (while often sharing a similar underlying vision of technological advance), and a Social Democratic/ Catholic minority wing, which was initially sympathetic to the idea of plant-level co-operation for productivity improvement, but became progressively disillusioned by management's unwillingness to negotiate over the distribution of the ensuing benefits. In Britain, as in West Germany, the unitary trade union confederation actively supported the post-war productivity drive and associated US study missions. So, too, did most of its constituent organizations, despite criticisms from the left-wing Confederation of Shipbuilding and Engineering Unions of the Labour government's unwillingness to make factory-level Joint Production Committees compulsory or to give tripartite consultative bodies an effective role in industrial planning. In Sweden, as Henrik Glimstedt's chapter shows, the Metalworkers' Union, like the Engineering Employers' Association, initially opposed the introduction of MTM at Volvo both as a violation of industry-wide collective bargaining agreements and as an undesirable deviation from domestic standards of good engineering practice, but eventually negotiated a compromise solution over the objections of the Communist-majority factory club in the face of company managers' evident determination to impose the new payment system. In Japan, as Herrigel observes, blue-collar workers, foremen, and enterprise unions during the 1950s pushed steel plant managers to modify imported US job evaluation and statistical quality control methods to balance merit assessment with seniority in wage payment and to diffuse responsibility from centralized engineering bureaux to shop-floor quality circles respectively.[48]

As in the case of the American model itself, the emergence of broad national patterns in post-war European and Japanese engagements with US technology and management could thus go hand in hand with significant interpretative ambiguities and finer-grained variations among domestic actors within each

---

[47] In addition to the chapters by Kipping and Herrigel and the references cited therein, see Mercer, 'Rhetoric and Reality of Anti-Trust Policies'. In *Constructing a Competitive Order*, Mercer argues that the FBI played a positive role in the drafting of the 1956 Restrictive Trade Practices Act, but her own evidence shows that the primary concern of organized British business in this struggle was to eliminate the independent investigatory authority of the Monopolies and Restrictive Practices Commission rather than to pave the way for the sweeping legal assault on cartel arrangements which followed its passage.

[48] In addition to the chapters by Zeitlin, Herrigel, and Glimstedt, see Romero, *United States and the European Trade Union Movement*; Carew, *Labour under the Marshall Plan*; Kuisel, 'Marshall Plan in Action' and *Seducing the French*, ch. 4; Andrew Gordon, *The Wages of Affluence: Labor and Management in Postwar Japan* (Cambridge, MA: Harvard University Press, 1998).

country.[49] The most important such variations for our purposes, on which the essays in this volume largely concentrate, are those between individual sectors and firms.[50] In Britain, for example, as Zeitlin notes, components manufacturers saw greater scope for reduction of product variety than did makers of capital equipment and consumer goods, while several of the most ambitious post-war experiments with standardization and mass production were undertaken by nationalized enterprises such as the railways or the electricity supply authorities. In France and Italy, as Kipping and Ranieri show, a major impetus to the expansion of steelmaking capacity, the importation of American strip-mill technology, and the opening of the domestic market to greater competition came from the user industries, especially but not exclusively automobile manufacturers. In Sweden, as Glimstedt demonstrates, the motor vehicle and electrical engineering industries followed opposite trajectories during the 1950s and 1960s. Thus Volvo moved towards volume production of standardized passenger cars in response to government restrictions on demand for its main heavy truck business and the opening of the domestic automobile market to foreign imports. Allmänna Svenska Elektriska Atiebolaget (ASEA) conversely used the income stream from light, mass-produced, 'bread-and-butter' lines such as meters and small motors to finance a move into more complex and technologically sophisticated heavy equipment such as transformers, circuit breakers, and high voltage direct current distribution systems in collaboration with its major customer, the Swedish State Power Board.[51]

Within a single industrial sector, as the essays in this volume reveal, there could often be sharp differences in strategic approaches to Americanization and mass production among rival enterprises. Thus, as Ranieri describes, the Italian steel industry in the late 1940s and early 1950s was bitterly divided between pub-

[49] For an extended discussion of the problem of national models which argues that these should be understood as complex and contingent historical constructions whose unity and coherence vary empirically depending on their institutional characteristics, see Steven Tolliday and Jonathan Zeitlin, 'National Models and International Variations in Labour Management and Employer Organization', in id. (eds.), *The Power to Manage? Employers and Industrial Relations in Comparative-Historical Perspective* (London: Routledge, 1991), esp. 273–9.

[50] Another significant dimension of subnational variation, which receives relatively little direct attention in this book, is that between regions. Thus responses to post-war Americanization and engagements with US technology and management could be very different in regions dominated by large, hierarchical companies and in those characterized by more diffused networks of small and medium-sized enterprises, such as the Norwegian counties of Møre and Romsdal or the industrial districts of the 'Third Italy'. In addition to the brief discussion in McGlade's chapter, see Amdam and Bjarnar, 'Regional Dissemination of US Productivity Models in Norway'; id., 'Regional Business Networks and the Diffusion of American Management and Organisational Models to Norway, 1945–65', *Business History* 39, 1 (1997), 72–90; Segreto, 'Americanizzare o modernizzare l'economia?'; David Ellwood, 'The Limits of Americanisation and the Emergence of an Alternative Model: The Marshall Plan in Emilia Romagna', in Kipping and Bjarnar, *Americanisation of European Business*, 149–68.

[51] In France, too, as Patrick Fridenson observes in an unpublished paper presented to the March 1997 and June 1998 Madison conferences, the 'American model' meant very different things to automobile and electronics manufacturers during the post-war reconstruction era: standardization, automation, market research, after-sales service, job evaluation, and other systematic management tools to the former; centralized, science-based corporate R&D to the latter.

lic-sector managers led by Sinigalia on the one hand, who wanted to build giant integrated coastal works capable of turning out vast quantities of sheet steel for nascent mass manufacturers of consumer durables, and private-sector industrialists led by Falck on the other, who wanted instead to concentrate investment on more flexible plants capable of serving diversified final markets, even at the cost of importing cheap semi-finished ingots from other European countries. In France, as Kipping likewise details, François de Wendel, the largest Lorraine steel magnate, initially denounced Usinor's continuous strip-mill plans as a 'crazy' venture 'not suitable for the conditions of the French market', and could only be induced to participate in a second such installation under intense pressure not only from the state planning commission (Commissariat général du plan) but also from Renault, which credibly threatened to acquire a semi-continuous mill for its own steelmaking subsidiary. Nor was there any necessary or uniform connection between public enterprise and support for Americanization: as Bigazzi documents, Pasquale Gallo, interim chief executive officer (CEO) of the state-owned Alfa Romeo, argued immediately after the war that the Italian automobile industry could aspire to nothing more than 'organized craftsmanship' based on short-series quality production, while Vittorio Valletta of the privately owned Fiat defended his company's measured but determined drive towards mass production based on expanding volumes of small, cheap cars and the exploitation of scale economies.

In Britain, as Zeitlin points out, pronounced variations in managerial strategy could be observed between a firm like Standard Motors, which staked its post-war fortunes on high-volume output of standardized cars and tractors using a single interchangeable engine for both types of vehicle, and other domestic automotive manufacturers such as Austin, Morris, Leyland, or Rover, which maintained more diversified model ranges while selectively modifying US-style mass-production methods to accommodate greater product variety and smaller quantities. In the West German rubber tyre industry, as we have already seen, Phoenix and Continental pursued contrasting approaches to co-operation with US companies, as the former committed itself to a long-term dependent relationship with Firestone, while the latter remained careful to preserve its technological and commercial autonomy in successive partnerships with General Tire and Goodyear. In Japan, similarly, Nissan and Toyota after as before the war followed opposed routes to the mastery of Western mass-production methods, with the former focusing on imported technology and licensing agreements with foreign manufacturers, while the latter relied instead, as Kazuo Wada and Takao Shiba shows, on independent adaptation of American techniques and borrowing of organizational practices from other domestic industries like textile machinery, aircraft, and shipbuilding.

Even inside a single firm or enterprise group, finally, conflicting attitudes towards the benefits of Americanization could on occasion precipitate fierce struggles between managers and engineers at different levels. Thus within IRI, for example, as Ranieri observes, Ilva managers remained much less enthusiastic than their Finsider counterparts about US technology and management, including wide strip mills, open-hearth converters, standard costing, statistical planning, and budgeting, while the differences between these groups persisted even after the two companies were merged into Italsider at the beginning of the 1960s. Within Volvo, similarly, as Glimstedt shows, car and truck manufacturing continued to be organized along distinct lines throughout the post-war period, while craft-orientated engineers hostile to the wholesale adoption of mass-production methods were forced out of the passenger car division in the late 1950s, only to be recalled during the mid-1960s to assist in overcoming mounting quality problems with its vehicles.

History and Hybridization

A central theme of the essays in this book is proactivity. European and Japanese industrialists did not wait passively after the end of the Second World War for enlightenment about the American efficiency gospel from Marshall Plan evangelists nor for invitations to participate in productivity missions to the USA. More influential and significant in many cases than the public missions organized by national productivity centres were the private American study trips organized by leading firms such as Lucas, Renault, Peugeot, Thomson, Fiat, Alfa Romeo, Finsider, Continental, Phoenix, or Toyota, often involving repeated visits by managers and engineers at different levels over a period of several years. Still more striking are the images of American dreams in occupied France delineated by Kipping, with Renault engineers secretly developing the 4CV and a new type of modular transfer machine, Peugeot managers clandestinely ordering US machine tools, and the Vichy Organization Committees (Comités d'organisation) in steel and motor vehicles formulating plans for post-war installation of high-throughput strip mills on the one hand and a drastic rationalization of models and standardization of components on the other.[52]

European and Japanese manufacturers could be so proactive in engaging the US model precisely because Americanization was far from a new issue for them after 1945. Fordism, Taylorism, and the 'American System of Manufactures' had already begun to attract widespread foreign interest before the First World War, while public debates about them took on new cultural and political as well as economic urgency during the 1920s as US mass-produced goods and investment capital flooded into international markets. In Europe and Japan alike, industrialists and engineers closely monitored the evolution of US mass-production techniques and systematic management methods, heatedly discussed their applicability to domestic economic and institutional conditions, and sought with varying degrees of ambition and success to transplant into their

---

[52] On the automobile industry, see also Fridenson, 'L'Industrie automobile française' and 'L'Industrie automobile'.

home soil elements of the American model.[53] Many European and Japanese manufacturers, as the chapters in this volume document, also had extensive pre-war contacts with US industry through a multiplicity of channels from study visits, equipment purchases, and consultancy contracts to product licensing, technological co-operation agreements, joint ventures, and participation in international cartels.

Nor was the United States the only foreign point of reference for European and Japanese industrialists during the 1940s and 1950s. A significant counter-current to post-war Americanization was the re-evaluation of Germany as an alternative economic and technological as well as political and cultural model. In many cases, not surprisingly, the outcome of the war induced a sharp reaction against German industrial practice even among its most ardent erstwhile admirers. Thus Sinigalia, as Ranieri details, explicitly saw Americanization of the Italian steel industry as a rejection of the pre-war German model based on vertical intrafirm linkages between Thomas converters and small, flexible rolling mills turning out a wide array of low-volume products marketed under cartel agreements. In engineering, similarly, as Bigazzi observes, Italian manufacturers turned back to the USA after 1945 as their main external source of technological and organizational know-how after an intense—and in his view fruitful—period of engagement with large German enterprises during the late 1930s and early 1940s. The depth to which German industrial prestige had sunk in British eyes can be gauged from the plant visit reports of post-war intelligence missions: despite grudging respect for the skills and training of the work-force, UK engineers and managers, as Zeitlin notes, found 'very little to learn' in terms of production methods or internal organization from their major pre-war competitors in many metal-working sectors. In other cases, however, the eclipse of German influence was less extreme. In Japan, for example, as Wada and Shiba demonstrates, the 'takt' or 'rhythm' system and other modified flow-line methods adapted from German aircraft manufacturers during the 1940s were carried into leading post-war metalworking assembly industries such as ship-building, cameras, and automobiles by domestic efficiency engineers with experience in wartime production. In France, too, despite the post-war ascendancy of American technology and management models, Patrick Fridenson has argued that expatriated German engineers and scientists 'played an important part, and

---

[53] For a selective introduction to pre-war European and Japanese debates over and experiments with Fordism, Taylorism, and American mass production, see Steven Tolliday and Jonathan Zeitlin (eds.), *Between Fordism and Flexibility: The Automobile Industry and Its Workers* (2nd edn., Oxford: Berg, 1992; orig. 1986); Jonathan Zeitlin, 'Between Flexibility and Mass Production: Strategic Debate and Selective Adaptation in British Engineering, 1830–1914', in Sabel and Zeitlin, *World of Possibilities*, 241–72; Patrick Fridenson, 'Un Tournant taylorien de la société française (1904–1918)', *Annales ESC* 5 (1987), 1031–60; Aimée Moutet, *Les Logiques de l'entreprise: La rationalisation dans l'industrie française de l'entre-deux-guerres* (Paris: Éditions de l'École des Hautes Études en Sciences Sociales, 1997); Heidrun Homburg, 'Anfänge des Taylorsystems in Deutschland vor dem ersten Weltkrieg', *Geschichte und Gesellschaft* 4, 2 (1978), 170–94; Mary Nolan, *Visions of Modernity: American Business and the Modernization of Germany* (Oxford: Oxford University Press, 1994); Tolliday, 'Diffusion and Transformation of Fordism'.

sometimes a key role, in the modernization of major branches such as the automobile, chemical, aluminium and professional electronics industries'.[54]

The wartime productive achievements and post-war commercial ascendancy of US industry thus rekindled long-standing sparks of attraction to the American model in Europe and Japan, while dimming if not altogether extinguishing interest in alternative national paradigms. Yet internal disparities and ambiguities within American industry meant that the US model did not always offer clear-cut lessons even to its most enthusiastic would-be imitators. British civil servants and telecommunications engineers, as Kenneth Lipartito shows, greatly admired the US Bell System and sought to learn from its achievements in reforming domestic provision at various points during the post-war era. But which features of the US system should the British emulate: centralized corporate research and vertical integration of equipment design and manufacturing; universal service and the rapid expansion of network demand through low-cost flat-rate pricing; or the broader antitrust paradigm of arm's-length competitive relations among suppliers? Each of these features of the 'American model', as Lipartito argues, pointed in a different direction and carried a divergent implication for the restructuring of British telecommunications, while oscillation between efforts at rationalization along Bell lines and contrary attempts to promote increased competition proved, in his judgement, highly damaging.[55]

Visitors from different countries might also draw opposite conclusions from the same observations. Both British and Italian engineers, as Zeitlin and Bigazzi each comment, were surprised by the large quantities of materials and work-in-progress they encountered in post-war US automobile and metalworking factories. But whereas the British and later the Japanese rejected such multiplication of buffer stocks as a wasteful deviation from good manufacturing practice, the Italians accepted it as a necessary sacrifice for the attainment of higher production volumes which was further justified by the greater complexity of product variations handled by US factories. The Japanese in particular often seized upon US practices and ideas which were either neglected or becoming obsolete in their original setting, such as quick die-change equipment for stamping presses, mixed-model assembly, or the preventative and participatory approach to quality control management associated with Deming, Juran, and Feigenbaum.[56]

---

[54] Patrick Fridenson, 'Who is Responsible for the French Economic Miracle (1945–1960)?', in Michael Adcock, Emily Chester, and Jeremy Whiteman (eds.), *Revolution, Society, and the Politics of Memory* (Melbourne: University of Melbourne Press, 1997), 309. Both the British and Italians, as Zeitlin and Bigazzi remark, paid tribute to Switzerland, with its smaller firms, more skilled workforce, and specialized higher-value products, as an attractive alternative model of industrial prosperity more applicable to European conditions than the United States, though there is little evidence of direct attempts in either country to borrow from this example.

[55] Cf. also Tolliday's observations in the concluding section of his chapter in this volume on the pervasive contrasts between GM and Ford's pre- and post-war approaches to product strategy, production, and organization, together with the resulting ambiguities of 'Americanization' for European automotive manufacturing.

[56] On the Japanese case, see, in addition to the chapter by Wada and Shiba, Tolliday, 'Diffusion and Transformation of Fordism'; Shimokawa, 'From the Ford System to the Just-in-Time

Determined efforts to emulate American practice under sharply different product-market and factor-supply conditions could lead European and Japanese in technologically conservative as well as innovative directions. Both Fiat and Volkswagen managers, for example, were extremely cautious in automating their production operations during this period, with the latter in particular waiting until they were sure the market would absorb the additional output before introducing Detroit-style transfer equipment first in body manufacture and then in the mechanical departments.[57] In Japan, too, as Wada and Shiba notes, companies like Canon abandoned their relatively successful 'work centre' systems, an indigenous analogue to cellular manufacturing, in favour of dedicated automation and conveyor assembly during the late 1950s as output volumes reached minimum efficient scale for American-style mass production. In Italy, where the size and composition of future steel demand remained highly uncertain, Finsider hedged its bets by opting for a semi-continuous but upgradeable wide-strip mill at Cornigliano which could also turn out ship plate if sales of auto body sheet proved disappointing, whereas in the larger French market Usinor and Sollac installed more powerful continuous strip mills but with less up-to-date capacity specifications and plant layouts.

In some cases, too, uncritical imitation of foreign practice without sufficient appreciation of its potential limitations in different organizational and environmental contexts could yield highly deleterious results. Thus, as Zeitlin observes, British public-sector attempts during the late 1940s to standardize steam locomotives and turbo-alternators slowed down the introduction of more powerful and efficient alternatives, raising rather than reducing operating costs in the longer term as contemporary critics had forewarned. In the private sector, too, Standard Motors' post-war use of a common engine in its Ferguson tractor and medium-size Vanguard car compromised the latter's export market success as competing models became more widely available. In telecommunications, similarly, Lipartito contends that American-inspired efforts by British Post Office administrators during the 1950s and 1960s to promote both competitive tendering and consolidation among equipment suppliers disrupted established patterns of co-ordination between technology, manufacturing, and the network without putting in place a coherent market alternative, thereby contributing to development delays and the eventual commercial failure of its System X digital switching project. In the Italian steel industry, as Ranieri points out, American management tools such as job analysis, job evaluation, and standard costing,

Production System'; Cusumano, *Japanese Automobile Industry*; Satoshi Sasaki, 'The Development and Results of Japan's Productivity Improvement Movement: Observation Tours by Participants in the Electrical Equipment Industry', paper presented to the Caen Preconference on 'The Productivity Missions and the Diffusion of American Economic and Technological Influence after the Second World War', 18–20 September 1997.

[57] For the case of Volkswagen, see Volker Wellhöner, *Wirtschaftswunder-Weltmarkt-Westdeutscher Fordismus. Der Fall Volkswagen* (Münster: Westfälisches Dampfboot, 1996); Steven Tolliday, 'Enterprise and State in the German *Wirtschaftswunder*: Volkswagen and the Automobile Industry, 1939–1962', *Business History Review* 69 (1995), 273–350.

which had worked reasonably well at Cornigliano, became ineffective and even counterproductive when extended to the much bigger work-force and more diverse plants of Italsider, a vast merged corporation whose centralized functional organization was closely modelled on that of US Steel. The huge new coastal plant complex built at Taranto in southern Italy during the 1960s, like its contemporary French counterpart at Fos-sur-Mer near Marseilles, proved both too large and too inflexible to cope successfully with the turmoil in world markets during the 1970s, though Taranto's internal problems also stemmed from Italsider managers' misreading of the Japanese model, which had begun to displace the US steel industry as an international yardstick of best practice. In the rubber industry, finally, as Erker argues, over-reliance on the technological and commercial judgement of US manufacturers of fabric-breaker tyres inhibited German companies from appreciating and responding to the challenge of innovative steel-belted radial designs developed by Michelin in France, a fatal error in the case of Phoenix and very nearly so in that of Continental as well.

Often, however, European and Japanese industrialists' prior familiarity with American-style practices, whether through direct contact or domestic analogue, enabled them to treat US technology and management not as a unitary model to be imitated wholesale, but rather as a suggestive point of departure for selective adaptation, creative modification, and innovative hybridization. Thus European automobile manufacturers such as Volkswagen, Renault, Citroën, Fiat, and British Motor Corporation (BMC) all developed original small car designs far removed from those of their US counterparts during the 1940s and 1950s in order to reach mass sales in countries whose lower per capita incomes, higher fuel prices, shorter driving distances, more limited highway networks, and differential motoring taxation regimes held down demand for larger cars with powerful engines. Production engineers at Austin/BMC, as Zeitlin shows, developed novel types of transfer equipment based on standard modular units which could be easily reconfigured to cope with periodic design changes or even wholly new models, unlike contemporary 'Detroit automation'.[58] Both British and Japanese motor vehicle firms, as Zeitlin and Wada describe, modified Fordist flow-line methods to handle greater product variety and shorter runs through practices such as quick tooling changes, mixed-model assembly, multi-skilling, and functional flexibility, while likewise introducing innovative systems of production control and just-in-time logistics aimed at minimizing internal buffer stocks and work-in-progress. Japanese metalworking, steel, and engineering companies also transformed US statistical quality control and job modification techniques by diffusing them from supervisors and specialist staff to groups of shop-floor workers.

[58] The Austin/BMC modular transfer machinery drew on earlier unit construction designs developed by Renault during and immediately after the war: see Jonathan Zeitlin, 'Reconciling Automation and Flexibility? Technology and Production in the Postwar British Motor Vehicle Industry', forthcoming in *Enterprise and Society* 1 (2000); Jean-Pierre Poitou, *Le Cerveau de l'usine: Histoire des bureaux d'études Renault de l'origine à 1980* (Aix-en-Provence: Université de Provence, 1988); Fridenson, 'L'Industrie automobile française' and 'L'Industrie automobile'.

In Sweden, as Glimstedt demonstrates, Volvo integrated US volume manufacturing methods such as MTM with indigenous traditions of high value-
added craft production, sociotechnical approaches to work organization, and
pragmatic union–management relations to carve out an international market
niche based on a distinctive new class of vehicle: safe, high-quality, but relatively inexpensive station wagons. Italian manufacturers of sewing machines,
typewriters, motor scooters, and domestic electrical appliances, as Bigazzi
points out, fused indigenous wartime experience of new processes and materials
with imported American techniques and methods to reconcile low-cost
serial production with a wide range of colours, a variety of shapes, and frequent
adjustments in design. In West Germany, Herrigel argues that the revised
antitrust law of 1957, which deviated from the American occupation statute by
permitting rationalization cartels, smoothed the path of steel firms' adjustment
to the emergence of excess capacity in the late 1960s through the formation of
distribution syndicates for the allocation of their finished products, which then
facilitated a recombination of assets and increased specialization within the
industry. At plant level, too, the co-determination system which emerged from
the post-war encounter between American pluralist ideals and German unions'
challenge to capitalist property rights turned out to foster patterns of shop-floor
co-operation between labour and management which gave domestic steel
companies far greater flexibility within high-volume production than their US
counterparts.

Sometimes the innovations resulting from European and Japanese hybridization of US practices in turn became the basis for imitation and reverse learning
by the Americans themselves. During the late 1950s and early 1960s, for example, the 'Big Three' automobile firms brought out 'compact' car models of
their own in a partially successful effort to repel the invasion of their own home
market by smaller European designs. Both Ford and GM likewise introduced
'building block' or 'unitized' transfer lines with a family resemblance to those
developed by French and British manufacturers in response to the 'horsepower
race' of the mid-1950s and the premature obsolescence of tightly integrated
automated facilities like Ford's celebrated Cleveland engine plant.[59] During the
1980s, as Erker observes, Continental used its lead in the assimilation of French
radial tyre technology first to instruct and eventually to take over altogether its
former US partner General Tire. But undoubtedly the most conspicuous example of this phenomenon is the take-up in the United States and other Western
countries over the past two decades of Japanese manufacturing practices such
as quality circles, just-in-time logistics, teamworking, continuous improvement,
and collaborative supplier relations.[60]

---

[59] On the Cleveland engine plant, the 'horsepower race', and the evolution of transfer equipment
design in the US automobile industry, see Hounshell, 'Planning and Executing "Automation"'; id.,
'Automation, Transfer Machinery, and Mass Production'.

[60] For overviews of the voluminous literature on the diffusion of Japanese manufacturing practices, see Boyer et al., Between Imitation and Innovation; Tony Elger and Chris Smith (eds.), Global

Not all modifications of American practice, however, proved equally successful. At Usinor, for example, the productive efficiency of the new continuous strip-mill installation arguably suffered, as Kipping suggests, from the physical separation of the hot- and cold-rolling facilities and the failure to adopt an integrated organization structure with more elaborate budgeting and control procedures, choices motivated at least in part by a desire to maintain the autonomy of the original family owners within the merged company, as well as to avoid excessive centralization and bureaucracy. At Fiat's Mirafiori plant, conversely, as Bigazzi comments, company technicians made far greater use of overhead conveyors than was typical in the US auto industry, permitting a rapid expansion of facilities in response to sudden changes in demand but necessitating a rigid central control and tight integration of the production process which would ultimately exacerbate both the logistical and labour problems of this huge manufacturing complex.

Yet without imposing teleological assumptions about the evolutionary trajectory of industrial and technological development, there can be no *a priori* grounds for determining *which* modifications of foreign practice may ultimately lead to significant innovations in a particular set of economic and institutional conditions, nor any solid theoretical basis for distinguishing between 'creative' or 'progressive' and 'adaptive' or 'regressive' forms of hybridization.[61] Nor could it be specified in advance how far and in what ways the American model could be deconstructed, adapted, and recombined to suit European and Japanese circumstances, since this depended in the end on the imagination, reflexivity, and experimentation of local actors. Yet successful hybridization, as the essays in this volume suggest, also requires careful attention to the independencies among the constituent elements of imported and indigenous productive models. As in the case of modular product architecture, individual components of US mass-production practice could be redesigned or replaced altogether, provided that an appropriate interface could be devised to fit with the other parts of the model.[62] Thus, for example, American-style automated equipment and flow-line production could be reworked to accommodate a more diverse range

Japanization? The Transnational Transformation of the Labour Process (London: Routledge, 1994); Thomas A. Kochan, Russell D. Lansbury, and John Paul Macduffie (eds.), *After Lean Production: Evolving Employment Practices in the World Auto Industry* (Ithaca, NY: ILR Press, 1997); Jeffrey Liker, Mark Fruin, and Paul Adler (eds.), *Remade in America: Transplanting and Transforming Japanese Production Systems* (Oxford: Oxford University Press, 1999).

[61] For an ambitious and well-developed but ultimately unconvincing attempt to contrast 'backward' and 'forward' adaptations of Fordism in the British and Japanese automobile industries respectively on the basis of a product life-cycle model, see Joseph Tidd and Takahiro Fujimoto, 'The UK and Japanese Auto Industry: Adoption and Adaptation of Fordism', paper presented to the conference on 'Entrepreneurial Activities and Corporate Systems', University of Tokyo, 29–31 January 1993; Tidd and Fujimoto, 'Work Organization, Production Technology and Product Strategy of the Japanese and British Automobile Industries', *Current Politics and Economics of Japan* 4, 4 (1995), 241–80. Cf. also the alternative interpretation in Tolliday, 'Diffusion and Transformation of Fordism'.

[62] For the example of modular product architecture, see Karl Ulrich, 'The Role of Product Architecture in the Manufacturing Firm', *Research Policy* 24 (1995), 419–40.

of products in smaller quantities, but only through complementary adjustments in tooling, scheduling, logistics, and job definitions, which in turn implied deeper shifts in management organization, industrial relations, payment systems, training, and supplier relations. To take full advantage of the additional flexibility gained through such modifications of US high-volume production methods, furthermore, European and Japanese manufacturers also needed to integrate them with product and marketing strategies aimed at extracting a commercial premium from the resulting ability to alter their output mix and introduce new models rapidly in response to changing demand. 'Learning', as Boyer *et al.* observe of contemporary experience with the diffusion and transformation of Japanese manufacturing practices, 'may come from multiple, and even incoherent, sources. What matters is the capacity of firms and related institutions to pull such changes together and make their patterns of hybridization coherent'—though any such coherence, like that of the original model itself, should be regarded as at best partial and provisional.[63]

FROM TRANSFER TO CROSS-FERTILIZATION

Compared to much of the literature on post-war industrial Americanization, the essays in this volume tell a substantially different story. In these tales, as we have seen, the 'American model' itself turns out to be riven with internal ambiguities, tensions, and disparities, while direct transfer mechanisms such as Marshall Aid, US technical assistance, transatlantic productivity missions, and even the imposition of antitrust reforms exercised a decidedly limited impact on European and Japanese industry. The crucial initiatives in the post-war assimilation of US technology and management methods came instead from European and Japanese manufacturers, engineers, and public officials, who drew on prior experience with imported mass-production practices or domestic analogues to deconstruct, modify, and recombine elements of the American model to suit local circumstances. In so doing, however, European and Japanese industrial actors at the same time reinterpreted, reshaped, and sometimes transformed their indigenous practices and institutions, while the ensuing process of hybridization not infrequently gave rise to productive innovations significant enough to be re-exported back to the USA itself. For each of these reasons, as we noted at the outset, 'American engagements', with its multiple, ambivalent, and highly charged meanings, rather than 'Americanization', or even its limits, most fully encapsulates the theme of this book.

The findings of this volume point to the need to revise not only the historiography of post-war Americanization, but also the conceptual framework and vocabulary used in wider analyses of the international transfer and diffusion of productive models. The ambiguity and deconstructibility of American

---

[63] Robert Boyer *et al.*, 'Conclusion: Transplants, Hybridization, and Globalization: What Lessons for the Future?', in id., *Between Imitation and Innovation*.

practices; the internal conflicts and diverse objectives of the would-be exporters; the knowledgeability, self-reflectiveness, and proactivity of the receiving agents; the prevalence of selective adaptation and hybridization; the blurred line between imitation and innovation; and the shifting roles and relationships among the participants: all these features of post-war European and Japanese reworking of US technology and management call into question standard conceptions of the 'transfer', 'diffusion', 'dissemination', or 'transplantation' of a single globally efficient model of 'best practice' across national boundaries. Even apparently more interactive categories such as 'transmission' or 'translation' seem inadequate under the circumstances, since they imply that the underlying 'message' of the original productive model remains fundamentally unchanged when it is 'switched up' and 'switched down' between different levels of abstraction, or converted into a foreign linguistic idiom. A more satisfactory metaphor for the processes analysed in this book would thus be 'cross-fertilization', as Giuliana Gemelli proposes in her study of post-war American influence on the development of European management education: an apt term for what Paul Lillrank calls in the contemporary context 'the most promising [form of] organizational transfer': 'an intelligent learning process, where examples from abroad are used as stimulation for one's own thinking'.[64]

1.3   Against Convergence: Contemporary Debates in Historical Perspective

Historical interpretations of post-war Americanization, as we have often had occasion to observe, are closely bound up with wider contemporary debates about the putative transfer and diffusion of productive models across national boundaries. But just as the problematic of this book is deeply informed by current controversies over 'Japanization' and the 'new American challenge', so too its conceptual approach and historical findings offer ample grounds for scepticism about the likelihood and desirability of international convergence around any single 'best-practice' model of economic and technological efficiency, whatever its geographical origins.

In many respects, to be sure, the European and Japanese economies of the post-war reconstruction period constitute a 'world we have lost', more on the whole for the better than for the worse. Inconvertible currencies, capital controls, and scarcity of foreign exchange; import quotas and high tariff barriers; shortages of labour and raw materials; slow recovery of domestic demand and international trade from the ravages of depression and war: these distinctive characteristics, as the essays in this volume emphasize, mark the distance separating the conjuncture of the 1940s, 1950s, and even to some extent the 1960s from that of the 1980s and

[64] Giuliana Gemelli (ed.), *The Ford Foundation and Europe (1950's–1970's): Cross-Fertilization of Learning in Social Science and Management* (Brussels: European Interuniversity Press, 1998; Paul Lillrank, 'The Transfer of Management Innovations from Japan', *Organization Studies* 16, 6 (1995), 988. Both Gemelli and Lillrank, however, also couch their analyses in terms of 'translation' and 'transmission', respectively, of organizational models and innovations from one context to another.

1990s. In international terms, too, this era differed significantly from our own, with countries divided into hostile camps by the overarching bipolar conflict of the cold war, and national welfare states partially—though never entirely—insulated from the destabilizing forces of the world economy by the fragile regime of 'embedded liberalism', which emerged from the post-war policy struggles between the United States, Britain, Western Europe, and Japan.[65]

JAPANIZATION

Despite these evident contrasts, however, there are also remarkable similarities, both theoretical and empirical, between the issues at stake in historical and contemporary debates over the attempted transfer and diffusion of productive models, which can accordingly be examined within a common conceptual framework. Nowhere are the parallels with the historiography of post-war Americanization more striking than in the recent literature on 'Japanization'. Here, too, for example, there are multiple and conflicting interpretations of the original model itself. Is the 'Japanese production system' a sophisticated refinement of Fordism and Taylorism, which appropriates the tacit knowledge of shop-floor workers and suppliers to achieve higher quality and greater flexibility?[66] Or does it instead represent a radically new technological and organizational paradigm, such as 'lean production', which reconciles quality, flexibility, and productivity by systematically eliminating waste and mobilizing the capacities and initiative of front-line workers; or 'innovation-mediated production', which blurs the line between new product development and incremental improvement in manufacturing through the functional integration of intellectual and manual labour?[67] Does the 'Japanese model' entail an empowerment of shop-floor workers through enhanced opportunities for organizational learning

[65] For the original formulations of the emergence and subsequent decay of 'embedded liberalism', see John G. Ruggie, 'International Regimes, Transactions and Change: Embedded Liberalism in the Postwar Economic Order', in Stephen D. Krasner (ed.), *International Regimes* (Ithaca, NY: Cornell University Press, 1983), 195–231; Robert O. Keohane, 'The World Political Economy and the Crisis of Embedded Liberalism', in John H. Goldthorpe (ed.), *Order and Conflict in Contemporary Capitalism: Studies in the Political Economy of Western European Nations* (Oxford: Clarendon Press, 1984), 15–38; Ruggie, 'At Home Abroad, Abroad at Home: International Liberalisation and Domestic Stability in the New World Order', *Millennium* 24, 3 (1994), 507–26. For a persuasive argument that 'embedded liberalism' was a contested outcome of the post-war bargaining processes and shifting US priorities discussed in the previous section rather than an immediate product of the wartime Bretton Woods agreements, see Milward, *Reconstruction of Western Europe* and *European Rescue of the Nation-State*; and for an incisive analysis of the intrinsic instability of the gold-dollar exchange system which underpinned this international economic regime, see Barry Eichengreen, *Globalizing Capital: A History of the International Monetary System* (Princeton: Princeton University Press, 1996), ch. 4.

[66] See, for example, Knuth Dohse, Ulrich Jürgens, and Thomas Malsch, 'From "Fordism" to "Toyotism"? The Social Organization of the Labor Process in the Japanese Automobile Industry', *Politics & Society* 14 (1986), 45–66.

[67] See, respectively, James P. Womack, Daniel T. Jones, and Daniel Roos, *The Machine That Changed the World* (New York: Rawson Associates, 1990); Martin Kenney and Richard Florida, *Beyond Mass Production: The Japanese System and Its Transfer to the U.S.* (Oxford: Oxford University Press, 1993).

and devolved decision making, or does it rather amount to increased exploitation through intensification of work and 'management by stress'?[68] Can a single national model be identified within Japan itself, or are there a multiplicity of industry and company variants, which differ significantly from one another in key respects? Is the 'Japanese system' a bundle of discrete techniques and methods, such as just-in-time logistics, total quality management, teamworking, and continuous improvement? Or is it rather a tightly coupled package of complementary elements, including on some views lifetime employment, seniority wages, enterprise unionism, and *keiretsu* relationships with suppliers and financial institutions? Is the efficiency of the 'Japanese model', however defined and understood, globally and absolutely superior to that of foreign alternatives, or is its performance advantage instead relative and dependent on a particular economic, social, cultural, institutional, and even political context?[69]

As in the case of post-war Americanization, too, there has been fierce controversy over the transferability of the Japanese model to other national settings. For some authors, such as the International Motor Vehicle Project at the Massachusetts Institute of Technology (MIT) or Martin Kenney and Richard Florida, the Japanese system constitutes a universal model of productive efficiency 'applicable anywhere by anyone', whose principles and practices 'can be successfully inserted into another society and then begin to reproduce successfully in the new environment'.[70] Others, by contrast, have questioned how far the 'Japanese model' could—or should—be transferred to other countries with very different economic conditions, social institutions, and cultural values. Adversarial industrial relations, job-control unionism, high rates of worker mobility, and strongly individualist attitudes in Britain and the United States; occupationally based identities and skill profiles, industrial unionism, and statutory co-determination rights in Germany; authoritarian management traditions, limited work-force skills and educational attainment, weakly developed supplier networks, and macroeconomic instability in developing countries like Brazil: all these have been seen as formidable barriers to a full-scale diffusion of the 'Japanese production system', whether through directly-owned transplants or through indirect imitation by indigenous firms.[71] As in the case of post-war Americanization, moreover, contemporary controversies over Japanization

---

[68] For representative statements of the second view, see Steve Babson (ed.), *Lean Work: Empowerment and Exploitation in the Global Auto Industry* (Detroit: Wayne State University Press, 1995).

[69] For a discussion of these polarities in the debate, see Robert Boyer, 'Hybridization and Models of Production: Geography, History, and Theory', in id. *et al.*, *Between Imitation and Innovation*, esp. 23–32.

[70] Womack *et al.*, *Machine That Changed the World*, 9; Kenney and Florida, *Beyond Mass Production*, 8; cf. Tolliday *et al.*, 'Introduction', in Boyer *et al.*, *Between Imitation and Innovation*, 4–5.

[71] On Britain and the USA, see Stephen Ackroyd *et al.*, 'The Japanization of British Industry?', *Industrial Relations Journal* 19, 1 (1988), 11–23; Stephen Wood, 'Japanization and/or Toyotaism?', *Work, Employment & Society* 5, 4 (1991), 567–600; J. Rogers Hollingsworth, 'The Logic of Coordinating American Manufacturing Sectors', in John L. Campbell, J. Rogers Hollingsworth,

also reflect deeper theoretical disagreements about whether productive models can be effectively broken down into their constituent elements or must be implemented as a coherent whole, as well as about the balance between institutional plasticity and path dependency in the receiving societies themselves.

Current debates about Japanization began in the 1970s, and a large body of empirical research has now accumulated on the attempted transfer and diffusion of the 'Japanese model' to different national and industrial environments. Although there is still considerable room for divergent interpretations of this literature, a number of substantive conclusions can none the less be drawn from its findings. First, Japanese manufacturing techniques and management practices have been broadly, though by no means universally, taken up and emulated in an extremely diverse range of contexts, including many which had initially been regarded as highly inauspicious such as the north-east of England, the US rustbelt, eastern Germany, or metropolitan Brazil.[72] Second, however, as in the case of post-war Americanization, the 'Japanese model' has been widely deconstructed into a set of interrelated elements, which have been more or less significantly modified and adapted to suit local circumstances, giving rise to a broad spectrum of national, sectoral, and firm-level variants. Thus, for example, Japanese automobile transplants in the USA and the UK abandoned their highly individualized wage and assessment systems in favour of rigorous initial selection and training of new recruits, while their counterparts in the North American and Australian electronics industries have typically made less use of long-term employment and seniority-based pay than of other practices such as job rotation and internal promotion. Among US and European companies, similarly, Japanese techniques and methods such as flexible job assignments, quality circles, or just-in-time component supply have often been integrated piecemeal into more traditional mass or craft-based production systems; while wide variations can be likewise observed across the international automobile industry in the organization and operation of 'work teams', particularly as regards the role of team leaders or supervisors.[73]

and Leon N. Lindberg (eds.), *Governance of the American Economy* (Cambridge: Cambridge University Press, 1991), esp. 70–1. On Germany, see Wolfgang Streeck, 'Lean Production: A Test Case for Convergence Theory', in Suzanne Berger and Ronald Dore (eds.), *National Diversity and Global Capitalism* (Ithaca, NY: Cornell University Press, 1996), 138–70. On Brazil, see John Humphries, ' "Japanese" Methods and the Changing Position of Direct Production Workers: Evidence from Brazil', and Anne C. Posthuma, 'Japanese Production Techniques in Brazilian Automobile Firms: A Best Practice Model or Basis for Adaptation', both in Elger and Smith, *Global Japanization?*, 327–77.

[72] In addition to the surveys cited in note 54 above, see Charles F. Sabel, 'Bootstrapping Reform: Rebuilding Firms, the Welfare State, and Unions', *Politics & Society* 23, 1 (1995), 5–48; Michael C. Dorf and Charles F. Sabel, 'A Constitution of Democratic Experimentalism', *Columbia Law Review* 98, 2 (1998), esp. 292–314; Raphael Kaplinsky with Anne Posthuma, *Easternisation: The Spread of Japanese Management Techniques to Developing Countries* (London: Frank Cass, 1994).

[73] For these examples, see Boyer *et al.*, *Between Imitation and Innovation*, chs. 1–2; Kenney and Florida, *Beyond Mass Production*, ch. 8; Elger and Smith, *Global Japanization?*, chs. 1, 6, and 7; John Paul MacDuffie and Frits K. Pil, 'Changes in Auto Industry Employment Practices: An International Overview', in Kochan *et al.*, *After Lean Production*, 9–42.

As with post-war Americanization again, deviations from the original 'Japanese model' are increasingly seen not as negative consequences of resistance, retreat, and compromise in the transfer process, but rather as positive opportunities for experimentation, innovation, and learning. Thus Japanese automobile and electronics firms have discovered through modifications of domestic production and employment practices at their North American and European transplants new possibilities—some of which are being re-exported back to Japan itself—for organizing just-in-time logistics over wider geographical distances, reducing dependence on compulsory overtime, revising the division of labour between blue and white-collar employees, or even coping flexibly with a more diverse range of models. Often, too, European and US companies have drawn on a complex mix of indigenous and imported influences from several national traditions, including but not confined to Japan, to develop creative and original hybrid systems, such as Renault's Spanish subsidiary FASA, Volvo's joint venture with Mitsubishi at the formerly Dutch-owned NedCar factory, Opel's Eisenach plant in eastern Germany, or General Motors do Brasil. In some cases, finally, Western manufacturers have used the Japanese example as an indirect inspiration for more radical and potentially innovative departures from established productive models like GM's labour-management partnership at Saturn or Volkswagen's 'modular consortium' of assembler and suppliers at Resende in Brazil.[74]

THE 'NEW AMERICAN CHALLENGE'

But what of the 'new American challenge' of the 1990s? How does the re-emergence of the US economy as a model for foreign emulation fit into our conceptual framework, and what are its implications for the critical assessment of industrial Americanization advanced in this book? Like post-war Americanization itself, current debates on the 'new American challenge' are the product of a very specific historical conjuncture, in which the US industrial resurgence and 'jobs miracle' has coincided with the tarnishing of alternative models resulting from the end of the cold war and the collapse of Communism; slow growth and high unemployment in European countries like Germany and France; prolonged recession and macroeconomic immobility in Japan; and most recently the Asian financial crisis. For many commentators on both sides of the Atlantic and the Pacific, moreover, these developments reflect deeper structural pressures towards international convergence of financial systems, corporate governance, labour markets, and productive organization on an Anglo-American model of 'free market' capitalism: 'globalization' of production,

---

[74] For these examples and the underlying typology of hybridization, see Boyer et al., Between Imitation and Innovation, esp. chs. 1, 2, and 16; and for an additional case of reverse learning outside the automotive sector, see Bill Taylor, Tony Elger, and Peter Fairbrother, 'Transplants and Emulators: The Fate of the Japanese Model in British Electronics', in Elger and Smith, Global Japanization?, esp. 200–1.

investment, and capital markets; the creation of new opportunities for exit by domestic firms from national systems of finance, innovation, skill formation, and industrial relations; the superior glamour and attractiveness of fast-moving American commercial/managerial culture compared to more staid forms of 'Rhine' and Japanese capitalism; the declining policy and regulatory autonomy of national states; and mounting external demands for realignment of domestic institutions and practices with neo-liberal competitive norms from supra-national bodies like the European Union (EU) and multilateral agencies like the World Trade Organization (WTO) or the International Monetary Fund (IMF), as well as from the USA itself.[75]

Such sweeping claims about transnational pressures for institutional convergence on the Anglo-American model have not gone uncontested, however, while a growing body of critical arguments and counter-evidence suggests that contemporary processes of 'globalization' are neither so novel nor so far-reaching as is often contended. Thus international capital, product, and labour markets were in significant respects more integrated before 1914 than they are today, while government expenditure still accounts for a higher proportion of gross domestic product in most advanced economies than even during the 1950s and 1960s. Most so-called multinational enterprises continue to rely heavily on the economic, cultural, institutional, and political resources of their home country; productive capital remains far from fully mobile; and national states may enhance rather than reduce their capacity for effective economic governance through participation in multilateral agreements and supranational bodies.[76]

It is beyond the scope of this introductory essay to adjudicate these debates, which are in any case still at a preliminary stage. But a few observations can none the less be ventured on the basis of the post-war European and Japanese engagements with the 'American model' analysed in this book. A first concerns the prevalence of heterogeneity and hybridization within the contemporary US economy itself. Even more than in the 1940s and 1950s, there is no single, coherent 'American model' of productive organization today, while the resurgence of

---

[75] For introductions to this debate, see Paul Bracken, 'The New American Challenge', *World Policy Journal* 14, 2 (1997), 10–19; Michel Albert, *Capitalism Against Capitalism*, trans. Paul Haviland (London: Whurr Publishers, 1993); Berger and Dore, *National Diversity and Global Capitalism*; Colin Crouch and Wolfgang Streeck (eds.), *Political Economy of Modern Capitalism: Mapping Convergence and Diversity* (London: Sage, 1997).

[76] For such critical arguments and counter-evidence, see Paul Hirst and Grahame Thompson, *Globalization in Question* (Cambridge: Polity Press, 1996); Robert Boyer, 'The Convergence Hypothesis Revisited: Globalization but Still the Century of Nations?', and Robert Wade, 'Globalization and its Limits: Reports of the Death of the National Economy Are Greatly Exaggerated', both in Berger and Dore, *National Diversity and Global Capitalism*, 29–88; Paul Bairoch, 'Globalization Myths and Realities: One Century of External Trade and Foreign Investment', in Robert Boyer and Daniel Drache (eds.), *States Against Markets: The Limits of Globalization* (London: Routledge, 1996), 173–92; John Zysman, 'National Roots of a "Global" Economy', in Lee-Jay Cho and Yoon Hyung Kim (eds.), *Hedging Bets on Growth in a Globalizing Industrial Order: Lessons for the Asian NIEs* (Seoul: Korea Development Institute, 1997), 33–88; Geoffrey Garrett, 'Global Markets and National Politics: Collision Course or Virtuous Circle?', *International Organization* 52, 4 (1998), 787–824.

US industrial competitiveness has taken very different forms across sectors such as automobiles, steel, electronics, telecommunications, and biotechnology. Some elements of the current American industrial revival are rooted in domestic institutions and policies from antitrust, health care, and defence procurement to interorganizational networks between firms, universities, and government agencies. But others are heavily dependent on selective adaptation and modification of Japanese methods of flexible production, rapid product development, and collaborative supplier relations, thereby reinforcing rather than refuting earlier critiques of the limitations of US mass production and systematic management.[77]

A second observation inspired by historical experience with post-war Americanization concerns the practical limitations, political obstacles, and unintended consequences of attempts to reshape foreign institutions and practices along US lines. Even in a post-cold war world with a single superpower, the risks of political destabilization may limit external leverage over strategic countries' domestic policies, as can be seen for example in the modification of IMF adjustment programmes in response to the 1997–8 Asian financial crisis. Liberalizing initiatives enacted in response to transnational pressures such as deregulation and privatization may likewise be reinterpreted and reworked by domestic actors to conform with established national patterns, as in the case of Japanese reform of retail trade regulation or Italian corporate governance arrangements in formerly state-owned firms.[78] In the transitional economies of East-Central Europe, similarly, privatization of state-owned enterprises has given rise to sharply contrasting patterns of ownership networks, with direct interfirm linkages predominant in Hungary and meso-level ties between banks and investment funds more significant in the Czech Republic, differences rooted in the pre-existing organizational structures of the two economies during the Communist era as well as in the contingent institutional choices of reforming governments during the transformation process itself.[79]

---

[77] See, for example, Richard K. Lester, *The Productive Edge: How U.S. Industries Are Pointing the Way to a New Era of Economic Growth* (New York: Norton, 1998); Zysman, 'National Roots of a "Global" Economy'; id., 'Globalization with Borders: The Rise of Wintelism as the Future of Global Competition', *Industry and Innovation* 4, 2 (1997), 141–66; Sabel, 'Bootstrapping Reform'; Dorf and Sabel, 'Constitution of Democratic Experimentalism'.

[78] On Japanese retailing reform, see Frank K. Upham, 'Retail Convergence: The Structural Impediments Initiative', in Berger and Dore, *National Diversity and Global Capitalism*, 263–97. On Italian privatization and corporate governance, see Fabrizio Barca and Sandro Trento, 'State Ownership and the Evolution of Italian Corporate Governance', *Industrial and Corporate Change* 6, 3 (1997), 533–59; Magda Bianco and Sandro Trento, 'Capitalismi a confronto: I modelli di controllo delle imprese', *Stato e mercato* 43 (1995), 65–93.

[79] See David Stark and Lásló Bruszt, *Postsocialist Pathways: Transforming Politics and Property in East Central Europe* (Cambridge: Cambridge University Press, 1997), esp. ch. 5. This outstanding book, whose theoretical perspective is largely convergent with our own, emphasizes the centrality of organizational 'bricolage' and 'recombinatory innovation' by reflexive actors rather than slavish imitation of foreign models or regressive path-dependent involution in shaping the restructuring and transformation of post-socialist economies.

A final observation concerns the likelihood and desirability of selective adaptation, creative modification, and innovative hybridization of imported practices in meeting the 'new American challenge'. European and Japanese automobile manufacturers, as we have already seen, have borrowed liberally from one another as well as the USA to enrich and revise their indigenous forms of productive organization. Across a wide range of sectors from electric power and food processing equipment to pharmaceuticals and electronics, moreover, multinational enterprises increasingly serve as agents of technological and organizational cross-fertilization between national or regional systems of production and innovation with distinctive capabilities.[80] Nor is such hybridization confined to the sphere of production itself. Thus informed commentators discern signs of the emergence of new hybrid forms of financial intermediation in historically bank-based systems such as Germany, in which increased reliance on external securities markets and emphasis on 'shareholder value' goes hand in hand with pursuit of 'learning by monitoring' through deliberative assessment and benchmarking of investment projects.[81] A number of European countries such as Denmark and The Netherlands have reduced unemployment to near-US levels during the mid-1990s without sharp increases in inequality through various combinations of consensual wage restraint, negotiated reform of public social welfare programmes, and intensive training efforts; while others such as

[80] For multinationals as agents of cross-fertilization between distinctive national and regional economies, see Zysman, 'National Roots of a "Global" Economy', esp. 48–50; John Cantwell, 'The Globalisation of Technology: What Remains of the Product Cycle Model?', in Daniele Archibugi and Jonathan Michie (eds.), *Technology, Globalisation and Economic Performance* (Cambridge: Cambridge University Press, 1997), 215–40; Henrik Glimstedt, 'From Country Competitiveness to Techno-Globalism? Rethinking National Path Dependency and Global Convergence', unpublished paper, Institute of International Business, Stockholm School of Economics, 1998. For specific industrial examples, see also Poh-Kam Wong, 'Creation of a Regional Hub for Flexible Production: The Case of the Hard Disk Drive Industry in Singapore', *Industry and Innovation* 4, 2 (1997), 183–206; Steven Casper and Catherine Matraves, 'Corporate Governance and Firm Strategy in the Pharmaceutical Industry', Wissenschaftszentrum Berlin, Research Area Market Processes and Corporate Development, Working Paper FS IV 97-20, 1997; Peer Hull Kristensen, 'Strategies in a Volatile World', *Economy and Society* 23, 3 (1994), 305–34; id. and Jonathan Zeitlin, 'The Local Uses of Global Networks: A Proposal to Study Industrial Restructuring in a Multinational Corporation' (Copenhagen Business School/University of Wisconsin-Madison, February 1996).

[81] For emergent trends towards 'learning by monitoring' in German banking, see Charles F. Sabel, John R. Griffen, and Richard E. Deeg, 'Making Money Talk: Towards a New Debtor-Creditor Relation in German Banking', in John C. Coffee, Ronald J. Gilson, and Louis Lowenstein (eds.), *Relational Investing* (New York: Columbia University School of Law and Economic Studies, Institutional Investor Project, 1993). For other recent discussions of hybridization among national financial systems and corporate governance, see Stephen Prowse, 'Corporate Governance in an International Perspective: A Survey of Corporate Control Mechanisms Among Large Firms in the U.S., U.K., Japan, and Germany', *Financial Markets, Institutions & Instruments* 4, 1 (1995), 1–63; Pieter Moerland, 'Corporate Ownership and Control Structures: An International Comparison', *Review of Industrial Organization* 10 (1995), 443–64; W. Carl Kester, 'American and Japanese Corporate Governance: Convergence to Best Practice?', in Berger and Dore, *National Diversity and Global Capitalism*, 107–37; John R. Griffin, 'Institutional Change as a Collective Learning Process: A US–German Comparison of Corporate Governance Reform', unpublished paper, Department of Political Science, Massachusetts Institute of Technology, 1996; Martin Rhodes and Bastiaan van Apeldoorn, 'Capital Unbound? The Transformation of European Corporate Governance', *Journal of European Public Policy* 5 (1998), 406–27.

Italy, Ireland, Portugal, and Finland have similarly struck tripartite 'social pacts' aimed at balancing macroeconomic stability, increased labour market flexibility, and high levels of social protection.[82] For all these reasons, therefore, today's 'new American challenge', like that of post-war Americanization, seems likely to leave wide scope for strategic choice and creative hybridization by local actors at a variety of levels, leading to continued diversity of productive and institutional models across firms, sectors, and national economies.

[82] For Denmark and The Netherlands, see Jelle Visser and Anton Hemerijck, A 'Dutch Miracle': Job Growth, Welfare Reform, and Corporatism in the Netherlands (Amsterdam: Amsterdam University Press, 1997); Paul Hirst, 'Can the European Welfare State Survive Globalization? Sweden, Denmark, and the Netherlands in Comparative Perspective', European Studies/Global Studies Working Paper (University of Wisconsin-Madison, 1998). On 'social pacts' and the concertative renegotiation of employment legislation, industrial relations, and the welfare state in Europe, see Martin Rhodes, 'Globalisation, Labour Markets and Welfare States: A Future of "Competitive Corporatism"?' in id. and Yves Mény, The Future of European Welfare: A New Social Contract? (London: Macmillan, 1998), 178–203; Marino Regini, 'Still Engaging in Corporatism? Recent Italian Experience in Comparative Perspective', European Journal of Industrial Relations 3, 3 (1997), 259–79; Giuseppe Fajertag and Philippe Pochet (eds.), Social Pacts in Europe (Brussels: European Trade Union Institute/Observatoire Social Européen, 1997).

PART ONE

Exporting the American Model?

Chapter 2

Americanization: Ideology or Process? The Case of the United States Technical Assistance and Productivity Programme

JACQUELINE MCGLADE

The spread of US economic influence overseas, particularly after the Second World War, is a topic that has recently occupied scholars in many fields of historical study. While the phenomenon has been loosely called 'Americanization', it has yielded a wide variety of academic definitions as well as conclusions as to its actualization (or non-actualization) as a trend or 'process' in foreign economic development. To date, diplomatic historians, especially revisionists and corporatists, have provided the greatest bulk of studies on the rise of US economic expansionism as a political ideal and its transplantation into foreign policy.

In general, revisionist and corporatist scholars have held that, by the 1920s and 1930s, business and government groups increasingly shared a common set of views regarding US leadership and overseas economic development which, through consensus-style politics, was institutionalized as policy before the end of the 1940s. The apex of such policy making came with the formation of the Marshall Plan in 1948 which sought to 'reorient' or reform European business and economic activities in line with American practices. Subsequently, the United States has continued to guide, through government influence and business activities, the reshaping or 'Americanization' of foreign economies around the world.

Over the past decade, however, several works have appeared to challenge this interpretation of US global expansionism as achieved through grand political plans and deliberate economic design. Current debates regarding the legacy of the Marshall Plan as an impetus for business change and industrial reform in post-war Western Europe have also generated further scepticism over New Left views. This chapter briefly examines and casts a critical eye upon the New Left argument that the Marshall Plan evolved out of the corporatist framework of American foreign economic policy making set before the Second World War. It also challenges the notion that the economic goals and programmes of the Marshall Plan enhanced and remained compatible with the thrust of cold war strategic defence. As evidence to the contrary, this chapter focuses on one Marshall Aid programme, the US Productivity and Technical Assistance Programme (USTA&P), and its struggle to advance business reform overseas in the face of shifting cold war military objectives and European reactions.

2.1   Ideological Determinism in United States Foreign Policy and Historiography

At first glance, the USTA&P would seem an obvious by-product of what revisionists have called the new 'corporatist' arrangement of post-war American government-business relations—a programme conceived co-operatively by corporate executives and the state in 1948 to promote and advance economic and business reforms overseas. According to such scholars, this intersection of business and government interests had given rise to a kind of 'ideological determinism' by the early twentieth century. Underpinned by the spread of *laissez-faire* business practices, democratic governance, and free markets, the new expansionist ideal, which historian Emily Rosenberg has termed 'liberal-developmentalism', continued to reshape American economic diplomacy into the 1940s.[1]

In his work on the Marshall Plan, Michael Hogan has expanded upon the doctrinal model of American expansionism first presented by Rosenberg to include important new dimensions inspired by government–business experiences during the New Deal and the Second World War. While still committed to liberal developmentalism as an ideal, American policy makers and business leaders were, nevertheless, profoundly affected by the struggles of the nation during the Great Depression which, in turn, 'tempered their faith in free-market forces' and the return of economic prosperity through private business initiatives. Instead, New Dealers and business supporters supplemented their older beliefs in *laissez-faire* with a new advocacy for increased government assistance and 'frameworks of public–private cooperation' as an impetus for national economic recovery and business growth. Subsequently, the Marshall Planners extended or 'supranationalized' these new American creeds of 'networks of corporative collaboration', government aid, and macroeconomic co-ordination in their creation of a programme for European post-war recovery. According to Hogan, the aim of the Marshall Plan to forge 'a viable balance of power among the Western European states' through trade integration and business reform also acted as the 'conceptual link between [cold war] strategic and economic goals . . . to build a bipolar equilibrium'.[2]

---

[1] E. Rosenberg, *Spreading the American Dream: American Economic and Cultural Expansion, 1890–1945* (New York: Hill and Wang, 1982), 7–13.

[2] M. Hogan, *The Marshall Plan: America, Britain and the Reconstruction of Western Europe, 1947–1952* (Cambridge: Cambridge University Press, 1987), 21–2.

Melvyn Leffler has also argued that the post-war model of American foreign expansionism forged by the Marshall Planners, which was essentially rooted in the liberalization of European business and trade practices, was inherently compatible with the constricting thrust of cold war containment policy. Like Marshall Aid reformers, cold warriors also sought to preserve and expand US overseas influence and power by 'refashioning the world in America's image' of 'economic and technological superiority'. Thus, post-war reformers and planners became bound together by the common goal to spread and maintain US 'strategic superiority' in the face of foreign challenge whether economic or military in its form.[3]

Overall, revisionists and corporatists have presented the modern evolution of American foreign expansionism as a cohesive continuum of events marked by an overlapping, complementary pattern of business pursuits and state aims. However, these historians have been less successful in their attempts to explain adequately the subsequent failure of policy makers wholly to preserve and extend US strategic 'superiority' or hegemony beyond the early cold war era. The difficulty that such historians face when charting the ensuing erosion of American hegemony after the Second World War stems from their persistent efforts to equate the less obtrusive, reformist impulse of liberal developmentalist doctrine with the domination and control features of cold war strategic planning.

One can argue that the doctrines of liberal developmentalism and strategic security did indeed intersect initially in the early phase of US post-war foreign policy making, as demonstrated by the almost simultaneous creation of the Marshall Plan and NATO in 1948–9. However, as the cold war wore on, these two policy initiatives came increasingly into conflict as the older reformist ideal of foreign political and economic transformation through governmental self-determination and private business initiative failed to spur immediate increases in European anti-Communist defence and NATO military production. The US imposition of strategic trade controls in 1953 which sought to realign and separate the markets of East and West also severely endangered the achievement of global free trade and the realization of the 'liberal marketplace utopia' imagined by post-war economic reformers. Put simply, global developmentalism and strategic security became increasingly incompatible, if not irreconcilable, as American international aims by the 1950s. Instead of meshing into a 'grand strategy', as claimed by revisionists, these two ideologies or 'models' of global economic leadership were, in fact, in competition and often confuted US foreign policy-making efforts after 1948. Ultimately, the paradoxical nature of post-war US economic diplomacy disabled the efforts of both reformers and planners to achieve hegemonic redevelopment or 'Americanization' as a global 'process'— a feat that would have required the systematic implementation and acceptance of a consistent set of policies and programmes overseas.

[3] M. Leffler, *The Preponderance of Power: National Security, the Truman Administration, and the Cold War* (Stanford, CA: University of California Press, 1993), 3–4, 15–19.

In recent years, academic theories that promote consistency or 'determinism' as the predominant feature of post-1945 US diplomatic policy making have faced serious challenges. Noted diplomatic historian Frank Ninkovich has cautioned scholars against viewing the rise of American geopolitical expansion solely as 'a function of economic imperatives' since the 'role of ideas' has also been an extremely important and volatile factor in the shaping and reshaping of modern foreign policy aims. As a 'political belief system', US expansionism developed in the manner of any other ideology whose 'proponents . . . must also face the challenge of changing the doctrine itself' in order to ensure its longevity. In a recent article on American trade policy in the early cold war, Thomas Zeiler notes that liberal developmentalists often compromised the orthodoxy of free trade economics in order to 'manage' protectionist attacks. As a result, trade liberalism became 'embedded' in political compromise and evolved 'less as a revolution than a revision of protectionist policy'.[4] According to Ninkovich, this paradox of the survival of ideology through contradiction often gives rise to a 'bifurcation . . . of two distinct outlooks' or belief systems which can co-exist, but which are inherently incompatible. As this chapter will show, the ideology of American foreign expansionism did indeed experience such fragmentation as its original tenets rooted in liberal developmentalism were tested and strained by the new priorities of military protectionism and economic containment forged by the cold war.

In his call for the return of complexity over reductionism in American foreign policy studies, Ninkovich also offers scholars the useful advice to revisit the 'actual, as apart from the perceived, relations of economics to politics'.[5] When re-examined in the more pragmatic light of economic realities, the predominant picture of US government–business relations seems far less congenial and co-ordinated than that previously painted by revisionist historians. Indeed, as business historians such as Mira Wilkins and William Becker have frequently pointed out, governmental policies rarely, if ever, coincided with the actual pattern of American multinational development and corporate interests overseas.[6] Economic historian John Braeman also has shown that the governmental impulse towards foreign expansion was framed more by motives of geopolitical leadership than by business interest or pressure. In his study of US overseas trade after the First World War, Braeman concluded that, instead of a growing reliance on foreign markets, the United States demonstrated an obvious lack of dependency on imports, which comprised less than 5.0 per cent of GNP in 1928. The inconsequential state of import markets was matched only by the 'relative

    [4] Thomas W. Zeiler, 'Managing Protectionism: American Trade Policy in the Early Cold War', *Diplomatic History* 22, 3 (Summer 1998), 337–60.
    [5] F. Ninkovich, 'Ideology, the Open Door, and Foreign Policy', *Diplomatic History* (1983), 186–208.
    [6] W. H. Becker, *The Dynamics of Business–Government Relations: Industry & Exports 1893–1921* (Baltimore: Johns Hopkins University Press, 1982); M. Wilkins, *The Maturing of Multinational Enterprise: American Business Abroad from 1914 to 1970* (Cambridge, MA: Harvard University Press, 1974).

unimportance of foreign investment' which stood at only 1.8 per cent of total national wealth by 1929—a figure that would decline even more during the 1930s. On the basis of his analysis, Braeman dismisses the New Left assertion that a growing communality of government–business interests arose in foreign economic policy making as a 'case of ideology imposing an artificial simplicity upon a complex reality'.[7]

US trade figures during the early cold war era also bear out Braeman's assertion that the nation's income, as well as business interests, rested primarily in the activities and expansion of domestic not foreign markets. In the period 1945–1960, the combined income from export and import trading rarely rose annually above 10–11 per cent of total GNP. On the other hand, income from domestic business sectors accounted for 75–80 per cent of the nation's wealth in the same period. Annual trading figures also provide evidence that cold war foreign economic policies hindered, instead of assisted, US overseas business expansion after the Second World War. In 1946, the American business community enjoyed a $11.5 billion surplus in overseas export over import trading. While European recovery problems precipitated a dip in export totals from 1946 to 1947 (from $19.7 billion down to $16.8 billion) this proved a minor bump in light of the plunge that was to come in 1948–9 as a result of the enactment of Economic Co-operation Administration (ECA) trade policies. By 1950, US export markets had slumped to $13.8 billion, a 33 per cent decrease since 1946, and import markets (largely composed of European products) had risen to $12 billion, a 30 per cent increase since 1945. As a result of increased imports, the annual surplus of export trading fell rapidly to only $1.8 billion and did not go above $5.7 billion in any year from 1950 to 1960.[8] On the whole, such data suggest that scholars should continue to analyse American business activities and interests as a separate, sometimes conflictual sphere, instead of a leading dimension, of cold war geopolitics.

European scholars have also questioned New Left assertions that American foreign expansionism arose as a consequence of an aggressive, successful set of state policies. Thus Geir Lundestad has suggested that US global expansionism was not forged by a progressive grand design, but inherited, sporadically and hesitantly, as an 'empire by invitation' gained through the intentional offerings of war-weary and waning European powers.[9] Alan Milward also makes an outstanding case that a reassertion of European political nationalism and economic self-determination shortly after 1945 often stymied the hegemonic aims of American planners bent on guiding the liberalization and integration of overseas financial and trading markets.[10] Among others, Anthony Carew, Richard

[7] J. Braeman, 'The New Left and American Foreign Policy during the Age of Normalcy: A Reexamination', *Business History Review*, LVII, 1 (Spring 1983), 85, 102.

[8] US Bureau of the Census, *Historical Statistics of the United States, Colonial Times to 1970* (Washington, DC, 1970), 228–30.

[9] G. Lundestad, *The American Empire* (Oxford: Oxford University Press, 1990).

[10] A. Milward, *The Reconstruction of Western Europe, 1945–51* (Berkeley, CA, University of California Press, 1984).

Kuisel, and Federico Romero have also demonstrated that strong European resistance existed, especially among labour unions, against the import of American business practices and methods through such Marshall Aid programmes as the USTA&P.[11]

With these criticisms in mind, why then do some historians insist that a consensual arrangement of American geopolitical objectives and business interests arose by the 1940s, was fashioned into compatible 'macro-policies' such as the Marshall Plan and NATO, and then acted to recast the United States as a 'hegemonic'[12] world power after the Second World War? Unfortunately, the theory that 'lines of continuity'[13] marked government policy making and business activity serves as a tidier explanation for the sudden rise of 'Americationization' as a post-war phenomenon. In an attempt to promote ideological determinism as the impetus behind US post-war expansionism, revisionists and corporatists have often missed charting the many complexities and incongruities inherent in cold war foreign relations and policy making. It is only when such policies are examined in a less dogmatic fashion that a different picture of American post-war expansionism emerges—a picture in which the results of bureaucratic actions rarely coincided or even conformed to the original intentions of legislative and political plans. This chapter shows that government officials and business executives instead of exporting a single US 'model' for foreign expansionism, often acted in an independent manner and sometimes enacted contradictory measures which impeded, instead of promoted, the systematic spread of American influence and authority overseas. The picture it paints is also one in which the implementation of US policies is further complicated by the independent (sometimes rival) aims of private businesses and recovering European states. In the case of Western Europe, then, the importing of US business practices and economic models since the Second World War occurred less as a consequence of concerted plans, than as an outcome of a wide spectrum of diffuse, autonomous actions. Once such scholarly assertions of consistency are rejected, US global expansionism appears, not as a synthetic policy event, but as a syndetic array of multiple government strategies, private business initiatives, and foreign responses.

An examination of one foreign programme, the United States Technical Assistance and Productivity Programme (USTA&P), demonstrates quite clearly that a case can be made for the prevalence of 'incongruity' over 'continuity' in the making and administration of American post-war diplomacy. When launched in 1948 as an initiative under the Marshall Plan, the USTA&P was envisaged by its founders as a programme intended to introduce American-style

---

[11] A. Carew, *Labour under the Marshall Plan: The Politics of Productivity and the Marketing of Management Science* (Manchester: Manchester University Press, 1987); R. Kuisel, *Seducing the French: The Dilemma of Americanization* (Berkeley, CA: University of California Press, 1993); F. Romero, *The United States and the European Trade Union Movement, 1944–1951*, trans. Harvey Ferguson II (Chapel Hill, NC: University of North Carolina Press, 1992).

[12] Leffler, *Preponderance of Power*, 3–20.   [13] Hogan, *Marshall Plan*, 21.

business practices into Western Europe. Over the course of its ten-year life, however, the USTA&P experienced many dramatic challenges to its original reformist purpose and, as a result of cold war pressures, was transformed from a small 'business exchange' programme into a massive 'production drive' for European industry. This chapter suggests that factionalism, rather than corporatism, served as the political 'dynamic' or lever that tipped the USTA&P out of its early phase as a co-operative business programme into its final stage as a compulsory drive for cold war military production. The following analysis highlights, rather than one model framed by consensus-style politics, the rise of two divergent post-war models for American overseas leadership and expansionism. The chapter also traces the eventual conflict that policy makers faced as they tried to equate and preserve both models—European self-reform and strategic security—during the administration of the USTA&P. Finally, it briefly examines the resurgence of European nationalism and self-determination as a hindrance to the USTA&P and its mission to advance the hegemonic reform or 'Americanization' of businesses and industries overseas.

2.2   The ERP as a Liberal Business Programme

In 1945, ambivalence, not resolve, marked the mood of the US 80th Congress on the issue of European economic distress. Choosing to concentrate instead on domestic recovery, legislators ignored requests by the State Department in early January for an appropriations plan intended to assist in the 'reconstruction of European economies'.[14] To reinforce America's withdrawal of assistance, President Truman authorized the termination of the Lend-Lease aid programme and its funding agency, the Foreign Economic Administration (FEA) in August 1945. All other aid to Western Europe formally came to an end in September after the United States refused to replenish the rapidly depleting funds of the United Nations Relief and Rehabilitation Administration (UNRRA).[15]

In 1946, however, American policy makers began to re-examine the state of overseas industrial recovery in the light of growing new concerns that Stalin intended to further Soviet power by fostering the 'systematic disintegration' of Western Europe.[16] Previously, US officials had been confident that Europe would recover on its own as overall industrial production in many countries had been restored to 60 per cent of pre-war levels. Such hopes remained buoyant as France, Belgium, and The Netherlands went on to achieve nearly 90 per cent of

[14] J. Blum, *V Was for Victory: Politics and Culture During World War II* (New York: Harcourt Brace, 1976), 310–12.

[15] Prior to the European Recovery Programme (ERP), the USA extended approximately $16.6 billion in overseas recovery aid through UNRRA. From 1944 to 1947, UNRRA aid was distributed in Great Britain, France, Italy, and Germany through direct recovery grants. The rapid exhaustion of UNRRA funding by its recipients led many US policy makers to call for greater spending accountability measures to be built into the ERP: US Department of Commerce, *Foreign Aid by the United States Government* (Washington, DC, 1948), 16–17.

[16] Leffler, *Preponderance of Power*, 102–4.

pre-war production rates, with Great Britain and Norway exceeding such levels by 10–15 per cent by the spring of 1946. However, ongoing interruptions in the British and German coal markets began to severely hamper European steel production so that it fell by over 40 per cent in 1947. Disrupted consumer and financial markets, depleted tax bases, currency shortages, and illegal profiteering continued to constrict business capital and limit the accumulation of state revenues for recovery. Harsh weather conditions in 1946–7 further added to the woes of the European continent as agricultural production experienced a 20–30 per cent drop by the spring of 1947.[17]

As a result, the US Congress began to consider what had seemed an unwarranted, if not unwanted, act in early 1945—to assist European economic recovery. Despite a growing sense of resolve, however, Washington legislators remained deeply divided over remedies to end European economic distress and counter Soviet aggression. In many ways, the nascent stage of American internationalism as a foreign policy doctrine precipitated the split in congressional attitudes in 1947. Before the Second World War, one individual had virtually set the entire ideological and political course for US internationalism—Franklin D. Roosevelt.[18] After Roosevelt's death, American internationalists began to squabble over such issues as European assistance and Soviet relations, and had realigned into two distinct camps by 1947. Led by former New Deal planners, political liberals, and business free traders, 'liberal internationalists' tended to promote developmentalist visions of a permanent world peace achieved through overseas adoption of the US model of democratic capitalism. Conversely, national security advocates, trade protectionists, and anti-Communist supporters demanded a more immediate advance of American hegemony (both military and economic) as a strategic deterrent to Soviet expansion.

While understated, the ideological differences and political disagreements between liberal and conservative internationalists continued to fester without clear resolution in 1946–7. As a result, the fragile association maintained by internationalist groups began to collapse as the Truman administration mounted a forceful push for a European aid package in the summer and autumn of 1947. Although European desperation provided a temporary platform for a brief reconciliation, American internationalists continued to remain divided, sometimes violently, on the aims and administration of the ERP (and foreign aid in general) after 1948. It is important to re-examine the rivalry between liberal and conservative internationalists as it acted to shape the initial direction of the ERP as a business reform programme and spur its later transformation as a military production drive. Ultimately unresolved, this doctrinal split also intro-

---

[17] Summaries of European post-war economic problems in United Nations, *Survey of the Economic Situation and Prospects of Europe*, 1948 and Economic Co-operation Administration, Third Report to Congress, 1950; Hogan, *Marshall Plan*. For an alternative view of European economic conditions from 1945 to 1947, see Milward, *Reconstruction of Western Europe*.

[18] For additional insight into the pre-war origins of US internationalism, see Robert A. Divine, *Second Chance: The Triumph of Internationalism in America During World War II* (New York: Atheneum Press, 1971).

duced a paradoxical set of aims into post-war American foreign policy making—global developmentalism versus Communist containment. As the cold war escalated, ERP administrators would find it increasing difficult to satisfy the independent recovery aims of Western European countries and the push by the Pentagon for an immediate increase in overseas military production for NATO.

The infancy of the US–Soviet conflict, however, worked to the advantage of liberal internationalists who wished to capture control of a European aid programme when it was first proposed in 1947. In the interest of maintaining world peace, influential liberals such as Bernard Baruch, Eleanor Roosevelt, and the president of Studebaker Motorcar Co., Paul Hoffman, lobbied the secretary of state, George C. Marshall, hard against a military agenda for the ERP. Mindful of Baruch's advice to 'let the new global sights guide you',[19] Marshall instructed his policy planning staff in late April 1947 to draft a plan for the 'revival of the Western European economy' which concentrated on the 'deeper meaning' of the 'submerged and protracted causes of Europe's debility' beyond 'the Soviet Menace'.[20] Encouraged by the non-aggressive tone of Marshall's directive, liberal internationalists stepped up their campaign to establish economic reform, not remilitarization, as the central goal of an American aid package to Western Europe. As members of the Harriman Committee, business liberals or 'progressives' played a particularly strong role in setting the economic reform agenda of the ERP.

As an influential group in the American business community, 'progressive' executives first emerged in the late 1930s and 1940s as supporters of Franklin Roosevelt's New Deal Administration.[21] As the leading lobby for business progressives, the Committee for Economic Development (CED) had shunned the isolationist traditions of other business groups and had vigorously pursued a direct, working relationship with government administrators since its founding in 1942. Unlike the National Association of Manufacturers (NAM) and the National Industrial Conference Board (NICB), the CED also had demonstrated a willingness to address national social issues that were often perceived by conservative executives as lying beyond the scope of traditional business concerns.

As a result, the CED was seen as 'moderate, even liberal in its political outlook', with a membership drawn from 'a small business elite ... who shared [a] broad economic vision'.[22] Along with the chairman of the CED, the president of Studebaker Motorcar, Paul Hoffman, the organization included such

---

[19] Letter from Bernard Baruch to George C. Marshall, 22 January 1947, Papers of George C. Marshall, Box 57, Folder 18, George C. Marshall Library (Virginia Military Institute), Lexington, VA.

[20] H. B. Price, *The Marshall Plan and its Meaning* (Ithaca, NY: Cornell University Press, 1955), 71–3.

[21] Further discussion of the rise of business liberalism and its influence in modern US governmental administration may be found in K. McQuaid, *Uneasy Partners: Big Business in American Politics, 1945–1990* (Baltimore, MD: Johns Hopkins University Press, 1993); R. M. Collins, *Business Response to Keynes, 1929–1964* (New York: Columbia University Press, 1981).

[22] W. Sanford, *The American Business Community and the European Recovery Program, 1947–52* (Westport, CT: Greenwood Press, 1987), 73–4.

business leaders as Phillip Reed of General Electric, Marion Folsom of Eastman Kodak, meat-packing mogul Jay Hormel, Harrison Jones of Coca-Cola, Clarence Francis of General Foods, Charles Kettering of General Motors, and Will Clayton, who would later serve as a close aide to the secretary of state, George C. Marshall.[23] As 'broad' economic thinkers, these CED members and other liberal-minded executives had supported the efforts of Roosevelt's New Deal Administration to improve social and business conditions through greater state involvement and economic planning.[24] Reviled by business conservatives, Roosevelt and his New Deal staff welcomed the support offered by business progressives and began appointing CED members and other such executives to government commissions and planning boards.

As co-operative and willing partners with the Roosevelt Administration, many business progressives also captured powerful government positions during the Second World War. By 1945, these business executives had widened their political influence as chief administrators of the War Production Board (WPB) and the Foreign Economic Administration (FEA), and as diplomats and military production co-ordinators overseas. On the basis of their wartime experience and expertise, several top CED leaders and business progressives including Paul Hoffman, Phillip Reed, the president of S.F.K. Industries, William Batt, and pressed-metals producer William Foster, stood in a strong position to influence the direction of post-war foreign economic policy making.

Starting in the summer of 1947, CED leaders and other liberal developmentalists began to push for a European aid programme that would transplant models of American financial investment, corporate practices, and labour management overseas. Recovery would come swiftly, they argued, for those European countries which adopted the American economic strategy of the 'liberalism of abundance'—a strategy whereby national economic growth was to be achieved through an ever-expanding system of mass production fuelled by expanding consumer markets, liberalized trade and financial investments, and state policies which encouraged, not inhibited, the development of private enterprise.[25] The Harriman Committee proved to be an important forum for the airing of the views of business liberals that the ERP should serve as a 'business programme' which would spur an economic overhaul or 'reconstruction' of Western Europe.[26] When proposed in early January 1948, the influence of busi-

[23] CED Board of Trustees Roster, *Community Handbook on the Special Problems of Small Business* (New York, 1944), 1.

[24] For more on the CED and its role in the Marshall Plan, see Hogan, *Marshall Plan*, 12–16.

[25] For more on the foreign policy views of the CED, see the CED Pamphlet Collection (Phillip Reed Papers, Hagley Museum and Library, Wilmington, DE). The triumph of the business progressives' liberalized free trade and economic agenda for the ERP is also chronicled in Hogan, *Marshall Plan*, 95–101; and C. Maier, 'The Politics of Productivity: Foundations of American International Economic Policy After World War II', *International Organization*, 31 (1977), 607–33 and in id., *In Search of Stability: Explorations in Historical Political Economy* (Cambridge: Cambridge University Press, 1987), 121–52.

[26] President Harry S. Truman formed the Harriman Committee in the autumn of 1947 as one of three committees to examine the European relief question and its remedy through an aid package.

ness liberals was clear in the original legislation for the European Recovery Programme (ERP) which outlined provisions for overseas economic and business relief but made no mention of US assistance for military rearmament or Soviet containment in Western Europe.

While concerned over the lack of a defence measure in the ERP, conservative internationalists did not possess the political clout in the spring of 1948 either to amend or to block its passage. The only hope for a stronger military agenda for the ERP was dashed by February 1948 when George C. Marshall announced his support for the ERP as drafted largely by the Harriman Committee. Since early 1947, Marshall had rapidly been assuming the role of 'global leader' vacated by Franklin Roosevelt and had emerged as the new post-war prophet of US internationalist doctrine.[27] Upon Marshall's endorsement of the ERP, conservative internationalists and military hardliners had little choice but to accept his decision, as a former army general and now secretary of state, that, while relatively benign, economic assistance would be enough to deter Communist expansion in Western Europe. Backed by several key Republicans including Arthur Vandenberg, chair of the Senate Foreign Relations Committee, the ERP was finally passed in April 1948.

While consensus had been achieved, it broke down almost immediately as liberal and conservative internationalists began to fight for control over the ERP and its administration. Conflict between the two factions re-emerged over President Truman's choice for director of the ERP's new administrative agency, the Economic Co-operation Administration (ECA). When Truman announced that Dean Acheson, a staunch cold warrior and Democrat, would run the ECA, several key Republicans led by Arthur Vandenberg and business progressives immediately countered by insisting that the ERP administrator come 'from the outside business world with strong industrial credentials and *not* via the State Department'.[28] Siding with business executives and Republicans, Marshall advised Truman to appoint Vandenberg's candidate, the CED Chairman, Paul Hoffman, as the founding administrator of the ERP. The battle over the ECA administrator stands as one of several key events that split liberals and conservatives over the ERP, its aims, and management. While national security advocates and political conservatives, most notably Truman, would have preferred a strong cold warrior bureaucrat like Acheson at the helm of the ERP, they were forced to accept the more pacific Hoffman who, as a business progressive, had helped draft the programme's non-aggressive economic reform agenda. As a result, a cloud of conservative resentment and political antagonism hung over Hoffman and his ECA staff as they struggled to formulate and implement economic reform programmes under the ERP.

[27] See J. McGlade, 'The Illusion of Consensus: American Business, Cold War Aid and the Recovery of Western Europe, 1948–1958', unpublished Ph.D. dissertation (George Washington University, 1995), 115–18.

[28] Letter from Arthur Vandenburg to George C. Marshall, 24 March 1948 (Box 1, Folder 3, Papers of George C. Marshall).

At first, Hoffman remained undaunted in his vision that the ERP serve as a 'business advisory' agency to ailing overseas countries. Intent on organizing the ECA in a 'business-like manner', Hoffman and his aides persuaded hundreds of executives, as well as labour leaders and government bureaucrats, to serve in the agency.[29] But Hoffman's preference for businessmen over political appointees angered Truman, who warned on 9 August 1948 that 'it is customary for the President to have at least [some] representation' in executive agencies such as the ECA.[30] For the next two years, Hoffman would frequently antagonize the Truman White House as his expansive personality and ebullient management style often clashed with standard bureaucratic procedures. Despite such conflicts, Hoffman and his ECA staff went on to create an extraordinary array of programmes aimed at stimulating European financial reform, industrial modernization, and business recovery by the end of 1949. Nevertheless, ECA efforts to spearhead an 'economic reorganization' of Western Europe did little to assuage the concerns of the Truman White House over the lagging state of defence production overseas. While empowered to assist European rearmament through the Military Defense Assistance Act (MDAA) in 1949, Hoffman and his ECA staff also continued to favour the recovery of civilian industries over that of defence firms despite the ensuing new pressure of NATO material quotas.

2.3   The Conservative Backlash Against the European Recovery Programme

Along with the Truman administration, the ECA also clashed with trade protectionists and national security advocates within the American business community. While several leading business associations including the NAM and the NICB had endorsed the ERP, others such as the National Foreign Trade Council (NFTC) had condemned the aid package as a threat to US overseas trade interests.[31] In the end, many business groups had endorsed the ERP, not as a show of support for its progressive economic agenda, but mainly out of a lack of alternatives for the aid programme in 1948.

The ECA's mission to spur the rise of an integrated system of self-sufficient European trading markets had particularly clashed with the sentiments of American business protectionists. Soon many executives who had anticipated lucrative profits flowing out of ERP subsidies also became deeply disturbed as the ECA began to adopt policies that seemed to restrict, not expand, American business and trading opportunities overseas. In particular, protectionist groups protested at ECA restrictions on the export of American pulp and paper, tobacco, tuna, aluminium, hand tools, and livestock products overseas and the placement of a '50–50 quota' on all USA–Europe shipping activities.

[29] Price, *Marshall Plan*, 73.

[30] National Archives and Records Administration [hereafter NARA], Record Group [hereafter RG] 469, Economic Co-operation Administration [hereafter ECA], Office of the Administrator [hereafter OA], Box 3, Memo from Harry S. Truman to Paul G. Hoffman, 9 August 1948.

[31] McGlade, 'Illusion of Consensus', 386–92; Sanford, *American Business Community*, 73–4.

Disgruntled business groups, including some former supporters of the ERP such as the NAM and the NICB, began to accuse the ECA of consciously discriminating against US overseas trading interests. In general, these disputes demonstrated the fact that business executives had never fully understood that the 'ERP was not designed, [nor] . . . administered, as a vehicle for increasing American exports abroad'.[32] By early 1949, the NAM began calling for the reorganization of the ECA in order to stem 'the growth of certain policies and trends in participating countries which threaten not only the success of the recovery program . . . but the creation as well of permanent barriers to an expanding world trade'.[33]

Thus, the deteriorating state of business support for the ERP worked to the advantage of business protectionists and national security pundits who sought to end the programme in favour of stronger US trade incentives and defence measures overseas. As persistent critics of the ERP as run by progressive executives and liberal internationalists, conservative business groups had allied with cold war advocates and began to exert strong pressures against its economic developmentalist agenda for Western Europe. In the spring of 1949, the Truman White House also took steps to amend the non-aggressive direction of USA–Europe foreign policy by championing the formation of NATO and the passage of the MDAA.

Besieged by growing conservative pressures, Paul Hoffman continued to shield the ERP and its original reform agenda from new plans for European remilitarization. As a countermeasure, Hoffman directed his staff to create and enlarge programmes aimed at the recovery and modernization of European civilian business and industry. By doing so, ECA administrators sought to increase private business development and industrial performance overseas through acts of European self-initiative, not through American direction. As liberal developmentalists, ECA officials were still committed to the founding goal of the ERP—to aid Western Europe upon an independent path for economic recovery and reform. Nevertheless, conservative pressures did force changes in the primary focus of the ERP after 1949. While previously preoccupied with setting up civilian relief, monetary reform, and trade export programmes, the ECA began organizing and promoting new programmes intended to raise levels of industrial production in Western Europe immediately. Consequently, several industrial reform programmes, some of which were quite minor in importance prior to 1949, were elevated to the centre of ERP administration.

As one such programme, the USTA&P rapidly assumed responsibility for carrying out the ECA's new charge to spur an immediate expansion of European industrial production. Initially, the USTA&P had been organized by a cooperative committee of American and British business leaders and government

[32] Sanford, *American Business Community*, 131.
[33] NAM, '1948 International Relations Report' (Papers of the National Association of Manufacturers, Box 4, Hagley Museum and Library, Wilmington, DE, 4).

officials known as the Anglo-American Council on Productivity (AACP)[34] to serve as an 'exchange of persons in industry' programme. As first designed by the AACP in conjunction with the ECA, the USTA&P would sponsor visits for teams of European executives and labour leaders who wished to examine American business practices and industrial operations. In early 1949, ECA officials promoted the team visits as an important mechanism to reinvigorate transatlantic business partnerships and stimulate European interest towards industrial reform. By the end of 1949, however, administrators abandoned co-operative exchange as the framework of the USTA&P and refashioned it as a front-line agency in the ECA's new battle to raise levels of European industrial production immediately.

Instead of overseas initiative, ECA officials began to promote the idea that a massive 'transfer' of American business practices and production methods was needed to 'reorient' European managers and workers and foster widespread industrial reform overseas. By adopting 'reorientation' as the USTA&P's new agenda, the ECA formally succumbed to cold war pressures and actively promoted American direction, over European self-initiative, as the impetus for overseas industrial reform. As a result, officials began refashioning the USTA&P and other ERP industrial aid programmes in support of a massive 'production drive' that the ECA would launch into Western Europe in the spring of 1950.[35]

In order to expand its activities, the USTA&P entered into a series of co-operative agreements with American business groups such as the National Management Council (NMC) and the NAM and several major universities including the Massachusetts Institute of Technology (MIT), Cal Tech, Stanford, Columbia, Northwestern, and New York University to act as hosts for the European team visits. Also, USTA&P officials worked closely with the partnership agencies to restructure the team visits as intensive retraining seminars for the Europeans who attended.[36] As a follow-up to the training seminars, the USTA&P arranged for hundreds of private American business consultants and industry advisers to travel overseas to aid European firms which were interested in implementing production reforms. Under the code name, 'Operation Impact', the USTA&P also devised several other programmes including an

[34] For more on the formation of the AACP, see Carew, *Labour under the Marshall Plan*; Nick Tiratsoo and Jim Tomlinson, *Industrial Efficiency and State Intervention: Labour 1939–1951* (London: Routledge, 1993).

[35] McGlade, 'Illusion of Consensus', 386–460.

[36] The initiative taken by the USTA&P of offering higher management training to overseas managers resulted in a dramatic post-war expansion of foreign student education in US universities. By the 1950s, leading centres of US scientific and business education had started exchange programmes such as MIT's Foreign Student Summer Project (FSSP), which helped host the post-war education of thousands of young European scientists and industrialists. See Papers of the FSSP, MIT Archives, Cambridge, MA. For an extended discussion of the impact of US business education and 'management science' on post-war higher education and vocational learning in Great Britain and Western Europe, see R. Locke, *Higher Management Education Since 1940* (Cambridge: Cambridge University Press, 1991).

International Productivity Congress which brought hundreds of European managers to the United States in 1950 and 1951.[37] By the end of the ERP in 1952, over 5,000 managers from a wide variety of overseas firms had received special training at American universities or through USTA&P-sponsored National Productivity Centrers (NPCs) and the Organization of European Economic Cooperation (OEEC) in Western Europe.[38]

Despite the launching of its 'production drive', the ECA continued to serve as a target for conservative attacks and steadily began losing ground in its fight to preserve business reform as the central aim of European aid. After the outbreak of the Korean war, Paul Hoffman and his staff attempted to shore up the business mission of the ERP by backing the passage of the Benton–Moody Amendment. Under this, European governments were required to create 'national productivity centres' (NPCs), which promoted the spread of 'free enterprise' practices, particularly the exclusion of Communist labour unions, in civilian industries or face a forfeit of further ERP funds.[39] Before the new aid restrictions were implemented, however, Paul Hoffman resigned, in August 1950, throwing the ECA administration into a state of disheartened confusion.

Under its new director William C. Foster, the ECA mounted one final push to retain its control over ERP and MDAA assistance programmes in Western Europe. Unlike Hoffman, Foster took several steps to reorient a number of ERP programmes, including the USTA&P, away from European civilian recovery towards NATO defence production. Under Foster's leadership, the ECA also succeeded in implemented the Benton–Moody Amendment by prohibiting aid to overseas firms which maintained Communist unions and aiding 11 of the 16 ERP recipient countries to create NPCs by the end of 1950. In addition to the NPCs, the ECA launched, through the USTA&P, a 'pilot plants' programme, which extended special assistance to European firms willing to implement American labour practices and industrial management methods. In order to more closely control USTA&P activities overseas, ECA and Pentagon officials also reconstituted the membership of the European Production Assistance Board (EPAB). While European officials had previously staffed the EPAB, American officials comprised the majority of the board after 1950. Finally, the USTA&P expanded its management retraining programme once again to accommodate European firms engaged in military as well as civilian production.[40]

[37] NARA, RG 469, ECA, Productivity and Technical Assistance Division [hereafter PTAD], OD, General Subject File [hereafter GSF], Box 4, File: 'National Management Council Proposal', Letter from National Management Council [hereafter NMC] President, Eldridge Haynes to ECA Assistant Administrator, William Joyce, 8 November 1950.

[38] NARA, RG 469, Deputy Director for Technical Services [hereafter DDTS], Office of Industrial Resources [hereafter OIR], Industrial Training Division [hereafter ITD], In-Plant Training Branch [hereafter IPTB], 1954–1958, Boxes 5–6.

[39] NARA, RG 469, ECA, PTAD, OD, Subject File [hereafter SF], Box 1, File: 'Congressional Presentation 1950 Technical Assistance Quarterly Report and Yearly Projection Plan', 24 May 1950.

[40] NARA, RG 469, Mutual Security Agency [hereafter MSA], Office of the Special Representative/Europe [hereafter OSR/E], PTAD, Labor Productivity Branch [hereafter LPB], GSF, Box 2, File: 'PTAD Conference Paris, September, 1953', 43–4.

Despite such efforts, administrators ultimately failed fully to recast the ECA as a cold war agency and assuage its conservative critics. As one ECA official noted, 'After Korea, there was a filtering down of uncertainty from the top' as to the agency's direction.[41] Finally, in late 1950, the ECA lost its battle to remain an independent agency and was formally absorbed by the State Department. Upon its annexation, the agency's mission as a 'business reform bureau' also came to an end. More importantly, the ECA forfeited its role as the sole authority over US production aid and its administration in Western Europe. Starting in the spring of 1951 then, aid programmes such as the USTA&P were managed by Pentagon officials as well as ECA staff members. In addition, the State Department authorized the dismantling of many ERP economic reform programmes prior to their scheduled end in 1952. The USTA&P, however, survived the bureaucratic upheaval that marked the ECA's final days and was transplanted into the newly created Mutual Security Agency (MSA) in 1953. Under the MSA, the USTA&P finally realized its transformation from a business reform programme to a military production drive for Western Europe.

2.4   United States Hegemony versus European Autonomy, 1953–1958

With the outbreak of the Korean war, cold war bureaucrats, backed by conservative internationalists and trade protectionists, had succeeded in gaining control over European aid. Determined to strip the ERP of its liberal developmentalist agenda, cold war planners began to push for an end to assistance for overseas civilian businesses and industries. By the autumn of 1950, however, domestic goods markets and manufacturing experienced a 30 per cent drop in Western Europe as many countries geared up for NATO military production.[42] As a result, cold war conservatives reluctantly realized that, in order to avoid further economic downturns, civilian as well as military production aid must continue to Western Europe beyond the end of the ERP. In late 1951, the US Congress formalized the mixed arrangement of European aid when it created the successor agency to the ECA, the Foreign Operations Administration (FOA), under the administration of the MSA.

As directed by Harold Stassen, an influential Republican and ardent cold warrior, the FOA was forged in 1953 as an 'economic defence' agency and stood in stark contrast to the ECA and its former reformist efforts.[43] Unlike the ECA, the FOA actively sought to implement or make 'operative' US policies aimed at the economic containment of the communist world. Instead of the ECA's old motto of 'co-operation', the FOA followed a new pledge to garner European 'compliance' for cold war containment measures. In addition to the agreements under the Benton–Moody Amendment, the agency took steps to enforce the Battle Act of 1951 that prohibited aid to countries that 'knowingly' shipped any 'strategic

[41] Price, *Marshall Plan and its Meaning*, 226.
[42] MSA, *Worldwide Enforcement of Strategic Trade Controls* (Washington, DC, 1953), 20–1.
[43] MSA, *First FOA Report to Congress* (Washington, DC, September 27, 1953), 1–2.

materials' including 'arms and atomic energy' to the Soviet bloc.[44] Almost immediately, the emerging system of FOA strategic trade controls, which included a growing list of commercial as well as military supplies to Communist countries, acted to accelerate the fledgling split between East and West in business enterprise and trade markets.

Not surprisingly, relations between the USA and Europe soon stiffened in response to the FOA's actions. Overseas governments and firms began to protest at FOA restrictions that had abruptly cut off access to raw materials and civilian trading markets in Eastern Europe and Asia. After only a few short months in office, the Eisenhower administration had to deal with the first of several serious diplomatic crises brought on by the policies of cold war economic containment. Led by France, several European countries insisted that the FOA ease its strategic trade restrictions or compensate for business losses by increasing once again civilian as well as military levels of industrial aid.[45] While Great Britain opted for only increased military aid, other European countries, particularly France, West Germany, and Italy, received $8 million to $10 million in additional civilian production aid per year from 1952 to 1958.[46]

Along with increased levels of aid, European nations gained other advantages as US cold war foreign policy shifted from its developmental or 'fundamental' stage towards its 'operative' or functional phase.[47] As first managed by liberal developmentalists, USA–Europe diplomacy had been primarily reformist and co-operative in its aims and framework. Furthermore, it carried the distinctive mark provided by liberal developmentalists and business progressives, who were determined to raise European industrial performance through ideological reform. As an impetus for reform, these early administrators of European aid tried to mount a wholesale advance or 'transfer' of the American model of economic growth achieved through liberalized trade, rationalized production, and increased consumerism overseas. When the American model of economic and corporate activity was introduced by the USTA&P, however, many European countries viewed it as either reminiscent of or incompatible with their traditional patterns of industrial production and business management. Nevertheless, USTA&P officials persisted in pushing for an ideological or 'fundamental' conversion of Western European business and industry and in doing so raised instead of lowered overseas resistance to change.

The advent of the Korean war also exposed the many strategic weaknesses inherent in an USA–Europe industrial aid policy framed by the aims of ideological reform and political co-operation. As a result, functionalism began to take the ascendancy over fundamentalism in the administration of US aid to Western

---

[44] MSA, *Worldwide Enforcement of Strategic Trade Controls*, 20–1.

[45] McGlade, 'Illusion of Consensus,' 550–60.

[46] International Cooperation Administration [hereafter ICA], *European Productivity: A Summing Up* (Paris, 1958), 43.

[47] For this distinction between 'fundamental' and 'operative' phases of a political ideology see Ninkovich, 'Ideology, the Open Door, and Foreign Policy', 186–208.

Europe, particularly after 1953. Ironically, it was during this functional or 'operational' phase that European nations and companies began to realize the greatest degree of flexibility, creativity, and self-determination in utilizing USTA&P aid. The rise of European autonomy in aid planning and usage can be traced in two areas—nationally through the activities of the NPCs and transnationally through the OEEC.

On a national level, the activities of Norway's NPC stands out as one such case in which European self-determination predominated over American reformism under the USTA&P. In their work on the Norwegian Productivity Institute (NPI), Rolv Petter Amdam and Ove Bjarnar have shown that national administrators used USTA&P support to strengthen and expand existing patterns of regional business activity as a way to achieve greater overall industrial productivity. Instead of promoting widespread adoption of the mass-production model, the NPI sought to 'rationalize' Norwegian industry by encouraging greater co-operation and 'networking' among firms engaged in flexible batch production within a region. As part of its reform strategy, the NPI also selectively promoted and thus 'diffused' certain components of the American mass production model in the Norwegian business community. As Amdam and Bjarner note, these efforts by the NPI did not alter but instead advanced the 'tendency evident from the 1930s to develop tight and flexible co-operation among independent firms'.[48]

By fashioning an alternative productivity strategy, the NPI demonstrated the common tendency among European NPCs after 1953 to urge firms to 'creatively adapt', instead of 'convert' their manufacturing operations along the lines suggested by the American model. By promoting incremental adaptation over wholesale adoption, the NPCs and other European economic recovery agencies were able to defuse business resistance towards change as well as advance the overall US objective of increased production. The activities of Italy's Inter-Ministerial Committee on Reconstruction (CIR) also serves as an illustration of the trend towards 'creative adaptation' as a business reform strategy under the USTA&P. While the central focus of the CIR's 'productivity drive' reflected the American objective to modernize larger industries, it also drafted its own reform initiative for another important sector of the Italian business community—small, handicraft firms.

From autumn 1948, CIR officials along with influential members of the Italian National Craftsman Industries Organization (INCIO) lobbied ECA administrators hard for the inclusion of artisanal firms as recipients of USTA&P aid. Frustrated by the ongoing resistance of large Italian firms to participating in the USTA&P, American officials reluctantly agreed to extend funds to the

---

[48] R. P. Amdam and O. Bjarner, 'Regional Business Networks and the Diffusion of American Management and Organisational Models to Norway, 1945–65', *Business History*, 39, 1 (1997), 73–90. See also G. Yttri, 'From a Norwegian Rationalization Law to an American Productivity Institute', *Scandinavian Journal of History*, 20 (1995), 231–58, for a discussion of Norwegian political self-determination in the creation of the NPI.

CIR to help artisan firms 'considerably reduce costs of production and improve the quality of goods . . . so that foreign trade will develop accordingly'.[49] As a result, the INCIO received $4,625,000 from the ECA to establish a five-year programme under which Italian small firms could acquire American factory machinery, management consultants, and raw materials. The INCIO, in partnership with the CIR and the ECA, also set up a clearing-house 'corporation' called the 'House of Italian Handicrafts' to import large quantities of raw materials and handle the export of finished artisanal products. As a result, Italian artisans gained the ability by May 1949 to acquire imported materials such as hides, semi-precious stones, decorative paints, silver, and tin, and production tools such as electrical and abrasives equipment and automatic grinding machines at a significantly reduced cost. In addition to facilitating the 'bulk buying' of supplies and equipment, the INCIO clearing-house programme aided small producers in the marketing and distribution of their products in the United States as well as continental Europe. Due to its success, the artisanal reform programme was continued after 1953 under the auspices of the OEEC.[50]

Thus, the Italian artisan programme represented another case in which national self-determination won out over American direction in setting a course for business reform. Instead of bowing to the USTA&P plan to modernize large firms, the CIR and the INCIO developed a 'productivity' strategy aimed at the elevation of the weakest, instead of the strongest, sectors of Italian industry. In the end, such support may have provided for a substantial rearrangement and strengthening of the post-war pattern of Italian industrial activity, in which medium-sized companies engaged in mixed production increasingly acted as a bridge between large corporations and smaller, handicraft firms.[51]

European autonomy in relation to US aims can also be traced on a transnational, as well as a national level, through the activities of the OEEC. Since 1949, the OEEC had assisted the USTA&P in organizing an array of business retraining and technical assistance programmes for European civilian companies. While the OEEC had created some programmes for military defence firms, especially those industries involved in the production of electronics and computers, shipbuilding, aircraft, advanced weaponry, and jet engines, it primarily catered to civilian consumer industries even after the end of the ERP. In order to facilitate greater civilian business development and reform, the OEEC had also created a Committee for Productivity and Applied Research (PRA) to assist the NPCs in administering technical assistance projects. On the basis of the

[49] NARA, RG 286, ECA, PTAD, OD, Country Subject File [hereafter CSF], Box 7, File: French Productivity Teams, 'OEEC Italian Delegation Report', 7–8.

[50] NARA, RG 469, ECA, PTAD, OD, CSF, Box 18, File: 'Italy—General—Position of Miscellaneous Industries and the Craftsman Industry', 29 May 1949, 1–10.

[51] Some historians have credited post-war assistance with a role in the rise of a more 'diffused' or decentralized pattern of industrialization in the 'Third Italy', the central and north-eastern regions of the country. In particular, see Luciano Segreto, 'Americanizzare o modernizzare l'economia? Progetti americani e riposte italiane negli anni Cinquanta e Sessanta', Passato e presente, 37 (1996), 55–86.

recommendations forwarded by the PRA, the OEEC moved in March 1953 to create a separate agency—the European Productivity Agency (EPA)—to spearhead an expanded drive for European business reform, work-force retraining, and increased industrial production and modernization.[52] As historian Bent Boel has noted, the EPA played 'a unique and innovative role' in building a greater 'network' between 'numerous national and transnational organizations' involved with productivity reform and was viewed by 'many [European] member countries . . . as an instrument of integration'.[53]

In general, the MSA welcomed OEEC efforts to assume greater authority over a productivity drive for civilian industries and the management of business education and retraining programs. By centralizing the civilian assistance drive under the OEEC, the MSA also rid itself of another wearisome task—the annual negotiation and maintenance of the cumbersome, and often controversial, technical assistance or '115(k)' agreements.[54] Increasingly after 1953, then, the MSA relied on the OEEC to gain overseas compliance for the thornier anti-Communist and 'free enterprise' provisions set for cold war industrial aid under the 1950 Benton–Moody and 1954 Thye amendments.

Upon its assumption of greater authority, the OEEC managed to build an independent drive for civilian management re-education and industrial reform among its member countries. For the next six years, participant nations in the OEEC annually contributed 8 per cent of their generated counterpart funds towards the growth of the reform drive as managed by the EPA. In turn, the EPA was primarily responsible for co-ordinating the expansion of industrial assistance programmes carried out under the NPCs. Starting in 1954, the EPA administered a wide array of productivity programmes with an annual operating budget of $7.5 million. As a further sign of the 'increasing self-reliance of the Europeans',[55] the OEEC voted to continue the EPA for another three years after the end of USTA&P funding in 1957, with member countries contributing $1.7 million towards its annual operating costs. In recognition of continuing European support for the EPA, the United States also bestowed a special grant of between $1 million and $1.5 million annually to the programme until its formal end in 1960.[56]

While supportive of the independent actions of the NPCs and the OEEC, the MSA swiftly turned its attention (and with it a vast tide of US money and

[52] International Cooperation Administration (ICA), *European Productivity and Technical Assistance: A Summing Up, 1948–1958* (Paris: ICA, Technical Cooperation Division, 1958), 47–51.

[53] Bent Boel, 'The European Productivity Agency: A Faithful Prophet of the American Model?', in Matthias Kipping and Ove Bjarnar (eds.), *The Americanisation of European Business, 1948–1960: The Marshall Plan and the Transfer of US Management Models* (London: Routledge, 1998), 41.

[54] The term '115(k)' referred to the specific section of the Benton–Moody Amendment which restricted US technical aid to European socialist governments and labour organizations and required recipients to comply with American 'free enterprise' provisions, as outlined in the USTA&P funding agreements.

[55] Boel, 'European Productivity Agency', 47.

[56] ICA, *European Productivity and Technical Assistance*, 47–51.

support) away from European civilian business reform towards NATO military production and development. From the start, MSA Director Harold Stassen and his staff had reorganized the USTA&P to include the negotiation and administration of lucrative 'off-shore procurement' (OSP) contracts with European industries. As a support for the OSP contracts programme, USTA&P officials began to solicit European firms engaged in military R&D projects and defence production to take part in technical assistance visits and consultancy ventures. By 1954, powerful European companies such as Rolls-Royce, Renault, and Fiat topped the list of OSP contractors. As a result, the focus of USTA&P technical assistance shifted away from, not only the European civilian business sector, but small and medium-sized firms, in favour of larger producers. As part of its military aid programme, the USTA&P distributed over $370 million annually to European firms for the production of jet aircraft, engines and parts, submarines, electronic equipment, field armaments and ammunition, and transport vehicles for home defence initiatives and NATO.[57] By 1958, the USTA&P had let out more than $8 billion in OSP contracts overseas, primarily to large military manufacturers and suppliers.[58]

Under the OSP programme, European governments and firms had benefited once again from the shift that had occurred in US cold war strategic interests and planning. The USTA&P became less concerned after 1953 under its new 'operative' phase, than it had been in its 'fundamental' or reform phase, that European firms adhere to American corporate methods in the carrying out of OSP contract work. Guided by its new functional aim of spurring immediate production, the USTA&P allowed overseas companies to exert greater autonomy in the setting of management objectives and manufacturing practices under the OSP contracts. As a result, the USTA&P finally abandoned many of its regulations that had required European firms to implement and comply with American labour and production standards before receiving technical assistance funding.

While less reformist in its final phase, the USTA&P continued to expose European business owners, managers, and workers to American corporate practices and strategies through the combined experience of OEEC retraining programmes and OSP contract work. It remains a strange irony, however, that the USTA&P lost its mandate and momentum to reform or 'Americanize' European business at a time when overseas managers and firms were becoming increasingly interested in change. To what extent, then, European business executives took advantage of the unique opportunities afforded through the USTA&P to modernize their companies is a question for further study.

---

[57] NARA, RG 469, Agency for International Development [hereafter AID], DDTS, OIR, Industrial Procurement Division, Box 6, File: 'Procurement Programs'.
[58] NARA, RG 469, AID, Office of the Comptroller, Budget Division, Box 1, File: 'Military—Miscellaneous Papers'.

2.5   Conclusion

While designed systematically to reform or 'Americanize' European business, the USTA&P, at its end, remained more a reflection than a realization of its original aims. As this chapter has shown, hegemonic reform or 'Americanization' never fully emerged as a 'process' under the USTA&P. This failure was due to the overall conflicting nature of government strategies, bureaucratic struggles, business concerns, and foreign reactions to the programme's administration. This chapter has also re-examined the question of whether, in fact, government and business groups shared a common sets of beliefs which led to the formation of a single 'model' for US foreign economic leadership and expansion in the early cold war era.

In contrast to the 'continuity' thesis offered by revisionist and corporatist scholars, this chapter has argued that American expansionist aims varied widely in the public and the private sector and thus led to the evolution of a more conflictual, if not incongruous, framework for foreign economic policy making during the cold war. It should also be noted that US economic diplomacy, in its pre-war developmental or 'fundamental' form, was indeed extended into the post-war period through the formation of the Marshall Plan, but ultimately its reformist structure collapsed under escalating cold war military pressures. In order to ensure the rapid remilitarization of Western Europe, US officials forged a more 'operational' or strategic set of goals for the administration of overseas aid programmes, particularly after 1953. As a result, the pre-war orthodoxy of foreign economic reform or 'liberal developmentalism' did not prevail or, for that matter, co-exist comfortably when placed alongside cold war strategic security objectives.

The history of the USTA&P also provides insight into the question of whether a single 'model' of business redevelopment emerged and was advanced under the Marshall Plan. While scholars are right in assuming that the business model of mass production or 'maximized productivity' prevailed in the American debates over the design of the USTA&P, it is not so clear whether or not this concept survived over the course of the programme. As demonstrated in this chapter, the USTA&P did allow European governments and businesses creatively to adapt, rather than adopt, American industrial strategies and practices in a fashion appropriate to their needs, particularly in the period 1953–8. Nevertheless, more case studies are needed on firms that received USTA&P aid and implemented reforms before a final conclusion can be reached on the legacy of the Marshall Plan and the post-war modernization of European business.

Finally, historians should continue to cast a critical eye towards 'Americanization' as a distinct 'process' in the redevelopment of Western European countries since 1945. While this is not fully discussed here, the USTA&P often reached its limits, even if it did not meet its match, when it tried substantially to alter overseas economic, social, and cultural traditions. Thus, 'Americanization' occurred more often as a forceful expression of economic ideology than a realized hegemonic process in post-war Western Europe under such cold war programmes as the USTA&P.

Chapter 3

Transplanting the American Model? US Automobile Companies and the Transfer of Technology and Management to Britain, France, and Germany, 1928–1962

STEVEN TOLLIDAY

3.1 Introduction

Recent analyses of the dynamics of Europe's long boom after the Second World War have generally accorded a leading role to the transfer of American technology and organizational practices within a broad process of 'catch-up and convergence'.[1] Within this process, American multinational companies are generally seen as central actors: 'a powerful and relatively low-cost mode' of diffusing technology and management practices.[2] In the light of this, it is perhaps surprising that the overwhelming focus of studies of the transfer process has been on efforts by public agencies to export US industrial practices, and on the responses of European manufacturers to the example and recommendations of American 'best practice'.[3]

The research for the study reported in this chapter was supported by funding from the Leverhulme Trust (Grant F/122/AN). The main primary sources are the archives of Ford and its European subsidiaries contained in the Ford Industrial Archives (FIA) and the Henry Ford Museum (HFM), both in Dearborn, MI.

[1] For representative accounts, see: Nicholas Crafts and Gianni Toniolo (eds.), *Economic Growth in Europe since 1945* (Cambridge: Cambridge University Press, 1996) esp. ch. 1; Max-Stephan Schulze (ed.), *Western Europe: Economic and Social Change since 1945* (London: Longman, 1999); Moses Abramovitz, 'Catch-up and Convergence in the Postwar Growth Boom and After', in William J. Baumol, Richard R. Nelson, and Edward N. Wolf (eds.), *Convergence of Productivity. Cross-National Studies and Historical Evidence* (Oxford: Oxford University Press, 1994). For a wider review of this literature, see Jonathan Zeitlin, 'Introduction' in this volume.

[2] N. F. R. Crafts, 'The Great Boom', in Schulze, *Western Europe*, 55.

[3] Michael J. Hogan, *The Marshall Plan. America, Britain, and the Reconstruction of Western Europe, 1947–1952* (Cambridge: Cambridge University Press, 1987); David W. Ellwood, *Rebuilding Europe. Western Europe, America and Post-War Reconstruction* (Harlow: Longman, 1992); Matthias Kipping and Ove Bjarnar (eds.), *The Americanisation of European Business, 1948–1960: The Marshall Plan and the Transfer of US Management Methods* (London: Routledge, 1998); Nick

Detailed studies of the actions and operations of American multinationals are relatively scarce. There are some excellent surveys of the scope and dimensions of US foreign investment in this period,[4] and Mira Wilkins' massive overview of the subject remains unsurpassed.[5] But this work has generally not been followed up by detailed studies at company level,[6] and although some material is contained in recent histories of major US corporations, it is often scattered, or treated as incidental to the main themes of the development of the core US operations.[7]

In the absence of such studies, many works on the subject rely on broad generalizations about the functioning and practices of US multinationals. Among recent studies, Bruce Kogut, for example, argues that American companies enjoyed general advantages arising from the transfer of 'national organizing principles of work', embodied in a 'dominant heuristic' of standardization, rationalization, and Taylorism. The ability of US corporations to transfer these principles internally through their subsidiaries was, he argues, the key to 'the erstwhile dominance of the American multinational corporation' in post-war Europe. In a similar vein, Marie-Laure Djelic portrays US companies as one of the important transmission belts for the transfer to Europe of a superior and universal system in the post-war decades.[8] In an earlier generation of work, it is evident that similar observations and images of the large US multinational corporations of the 1950s and 1960s as bearers of advanced and superior forms of technology, organization, and management, played a large role in the development of many of the most formative and influential theories of multinational business.[9]

Tiratsoo and Jim Tomlinson, 'Exporting the "Gospel of Productivity": United States Technical Assistance and British Industry, 1945–1960', *Business History Review* 71, 1 (1997).

[4] For Britain, see: Geoffrey Jones and Frances Bostock, 'US Multinationals in British Manufacturing before 1962', *Business History Review* 70, 2 (1996); Frances Bostock and Geoffrey Jones, 'Foreign Multinationals in British Manufacturing, 1850–1962', *Business History* 36, 1 (1994). These studies and other related work build on the fundamental contemporary contribution by John H. Dunning, *American Investment in British Manufacturing Industry* (1st edn., London: George Allen & Unwin, 1958; rev. 2nd edn., London: Routledge, 1998). For the rest of Europe, see: James W. Vaupel and Joan P. Curhan, *The Making of Multinational Enterprise* (Cambridge, MA: Harvard University Press, 1969); id., *The World's Multinational Corporations* (Cambridge, MA: Harvard University Press, 1974).

[5] Mira Wilkins, *The Maturing of Multinational Enterprise. American Business Abroad from 1914 to 1970* (Cambridge, MA: Harvard University Press, 1974).

[6] The main exception again involves Wilkins: Mira Wilkins and Frank E. Hill, *American Business Abroad: Ford on Six Continents* (Detroit: Wayne State University Press, 1964).

[7] Two recent studies which provide more extended coverage of international operations are: Bettye Pruitt, *Timken. From Missouri to Mars—A Century of Leadership in Manufacturing* (Boston, MA: Harvard Business School Press, 1998); E. N. Brandt, *Growth Company: Dow Chemical's First Century* (East Lansing, MI: Michigan State University Press, 1997).

[8] Bruce Kogut, 'National Organizing Principles of Work and the Erstwhile Dominance of the American Multinational Corporation', *Industrial & Corporate Change* 1, 2 (1992), 288–310; see also Bruce Kogut and David Parkinson, 'The Diffusion of American Organizing Principles to Europe', in Kogut (ed.), *Country Competitiveness: Technology and the Organizing of Work* (Oxford: Oxford University Press, 1993); Marie-Laure Djelic, *Exporting the American Model. The Post-War Transformation of European Business* (Oxford: Oxford University Press, 1998).

[9] For an excellent overview of such theories, see: Geoffrey Jones, *The Evolution of International Business* (London: Routledge, 1996), esp. ch. 1. For a recent restatement of his views on US multinationals in the 1950s, see: John H. Dunning, 'US-Owned Manufacturing Affiliates and the Transfer

Such approaches beg many broader theoretical questions which are explored more fully by other authors in this book. This chapter, however, analyses the history of the major European operations (in Britain, Germany, and France) of the two largest US automobile companies (Ford and General Motors) from the 1920s to the early 1960s, and identifies some deep-seated problems of such broad-brush approaches.

Among other themes, it stresses that the 'American' model (even in firms and a sector often regarded as the *locus classicus* of that model) was often partial, incomplete, or even illusory: the theory and practice of the leading companies diverged widely both from each other, and from a putative American norm. Moreover, the 'model' was continuously changing. Certain elements, once central, were forgotten or repudiated: others, once scorned, were incorporated. Beyond this, the process of transfer was costly and complex. It required a self-knowledge and understanding of the dynamics of their own production systems that the parent companies often lacked, while, even in the most receptive contexts, transfer required innovation and flexibility. Without that, it could degenerate into the repetition of a sterile 'gospel', or lead its subsidiaries into strategic culs-de-sac. Even where transfer successfully married American elements with local knowledge and capabilities to create something strong and new, these new productive structures could develop in unforeseen ways, at times contrary to the intentions of the parent. On many occasions, foreign companies (despite a range of other disadvantages) were able to use their greater commitment to specific markets, their 'embeddedness', and a selective approach to learning from American experience, more effectively than their American-owned rivals.[10]

This chapter does not examine all aspects of the transfer and transplantation of American practice. It focuses on the core elements of strategy, organization, technology, and product policy.[11] It stresses the role of continuity and long-run dynamics in the transfer process (rather than seeing 'Americanization' as an epiphenomenon of post-war reconstruction). And it focuses primarily 'inside the firm' rather than on the discourses of agencies and observers about the transfer and dissemination of broad elements and processes.

---

of Managerial Techniques: The British Case', in Kipping and Bjarnar, *Americanisation of European Business*, 74–90.

[10] Many of these issues are more fully developed elsewhere in a parallel collaborative study of the process of 'Japanization': Robert Boyer, Elsie Charron, Ulrich Jurgens, and Steven Tolliday (eds.), *Between Imitation and Innovation: The Transfer and Hybridization of Productive Models in the International Automobile Industry* (Oxford: Oxford University Press, 1998).

[11] It omits extensive examination of industrial relations. For more detail on these aspects see: Steven Tolliday, 'The Diffusion and Transformation of Fordism: Britain and Japan Compared' in Robert Boyer *et al.*, *Between Imitation and Innovation*, 57–96; Tolliday, 'Ford and "Fordism" in Post-War Britain: Enterprise Management and the Control of Labour, 1937–87', in Steven Tolliday and Jonathan Zeitlin (eds.), *The Power To Manage? Employers and Industrial Relations in Comparative-Historical Perspective* (London: Routledge, 1991), 81–117.

3.2   Ford and General Motors in Europe, 1920–1940

CHANGE AND CONTINUITY IN THE AMERICAN MODEL:
FORD AND GM IN THE USA, 1920–1940

Before the Second World War, Ford and General Motors provided two defining instances of American big business. The first was the rise of mass production, based on economies of scale, co-ordination, and an intensive division of labour, symbolized by the assembly line and the Model T at the Ford Motor Company.[12] The second was the culmination of the managerial revolution, where Alfred Chandler has famously argued that General Motors 'perfected the structure' of the modern business enterprise in the early 1920s through the creation of the decentralized, divisionalized corporation, which co-ordinated flows and integrated production and marketing in revolutionary new ways.[13]

But the almost iconic status of these episodes should not obscure the fact that these developments were not conclusions, but simply moments in a process of continuing change. The subsequent inter-war story of Ford (well known through the vivid account of Nevins and Hill) was one of organizational degeneration, the creation of personal fiefdoms, the rise of a destructive anti-administrative culture, and a growing disorganization and dissipation of a once clear productive model, which took the company close to collapse.[14] In the same period at General Motors, as recent work by Kuhn, Cheape, Freeland, and Gartman has shown, Sloan's celebrated paradigm of managerial decentralization became something of a myth as the company retreated from divisional autonomy to centralization in the 1930s, masking increased standardization and the blurring of divisional organization under superficial styling diversity.[15]

These developments had significant implications for the companies' overseas operations. GM's history of growth through amalgamation and greater organizational decentralization helped to shape its policy of acquiring foreign

---

[12] The best single account is: David Hounshell, *From the American System to Mass Production, 1800–1932* (Baltimore: Johns Hopkins University Press, 1984). For recent debates on the dynamics of the classic Ford model see Karel Williams, Colin Haslam, and John Williams, 'Ford versus Fordism: The Beginnings of Mass Production', *Work, Employment & Society* 6, 4 (1992), reprinted in Steven Tolliday (ed.), *The Rise and Fall of Mass Production* (Cheltenham: Edward Elgar, 1998), which also contains several other important texts on this issue.

[13] Alfred D. Chandler Jr., *The Visible Hand: The Managerial Revolution in American Business* (Cambridge, MA: Harvard University Press, 1977), 456–63; id., *Strategy and Structure: Chapters in the History of American Enterprise* (Cambridge, MA: MIT Press, 1962), 114–63.

[14] Allan Nevins and Frank E. Hill, *Ford: Decline and Rebirth, 1933–1962* (NY: Charles Scribner & Sons, 1962); see also Hounshell, *American System*, chs. 6–7.

[15] Arthur J. Kuhn, *GM Passes Ford, 1918–38: Designing the General Motors Performance-Control System* (University Park, PA: Pennsylvania State University Press, 1986); Charles W. Cheape, *Strictly Business. Walter Carpenter at Du Pont and General Motors* (Baltimore: Johns Hopkins University Press, 1995); Robert F. Freeland, 'The Struggle for Control of the Modern Corporation: Organizational Change at General Motors, 1924–58', *Business and Economic History* 25, 1 (1996); id., 'The Myth of the M-Form? Government, Consent and Organizational Change', *American Journal of Sociology* (September 1996); David Gartman, *Auto Opium. A Social History of American Automobile Design* (London: Routledge, 1994), 100–36.

companies and managing them at relatively arm's length. At Ford, domestic developments had even more specific and important implications for the conduct of overseas operations. First, Ford's organizational commitment to a single model (the Model T followed by the Model A), until 1932, was strictly carried over into its foreign operations with initially positive, but later highly detrimental effects.[16] Subsequently, the dominance of Ford's V8 engine technology in the 1930s played a similar role. Henry Ford's mass-produced V8 engine was his 'last mechanical triumph', and it pulled Ford out of a period of decline.[17] Launched in 1932, the V8 rapidly became Ford's core technology, and by 1934 Dearborn (Ford's US headquarters) wedded itself wholly to the V8 in the USA, resolutely rejecting pressures to produce other engine types.[18] Henry Ford adamantly refused to follow Chevrolet and Plymouth into the production of low-cost six-cylinder vehicles, and Charles Sorensen, a central figure in the management of foreign operations, was perhaps even more ferociously committed to V8 technology than Henry himself.[19]

The V8 sold well in the 1930s, but it could not defend Ford's dominance in the low-price field against Chevrolet and Plymouth. Ford's primary strategy was to sell the V8 in large volumes through low prices, achieved by conservative product changes and heavy investment in modern plant. It also sought, unsuccessfully, to develop a smaller, cheaper V8 which could compete against the six-cylinder Chevrolets and Plymouths. But the quest for a cheap V8 failed, and Ford finally concluded that such a car could not be viably made. Only after this did Ford partially abandon its old policies of low costs and standardization amd follow GM into annual model changes and model diversification.[20] These model policies shaped the heart and soul of the organization. Their sole aim and target was the domestic US market, but, as we shall see, they had profound ramifications for Ford's overseas operations. In contrast, the much more complex and looser model policies of GM carried far fewer direct reflexes abroad.

BRITAIN: AMERICANIZATION OR ADAPTATION, 1920–1939

Ford

Ford made a deep and systematic attempt to transplant its methods and systems to Britain in the inter-war period. In the 1920s, while it produced the Model T, it resisted adaptation and compromise and sought to install its American system

---

[16] Tolliday, 'Diffusion', 60–7.

[17] Charles E. Sorensen, *Forty Years with Ford* (London: Jonathan Cape, 1957).

[18] W. J. Abernathy, *The Productivity Dilemma: Roadblock to Innovation in the Auto Industry* (Baltimore: Johns Hopkins University Press, 1978), 92–5; Nevins and Hill, *Decline and Rebirth*, 64–7; David L. Lewis, Mike McCarville, and Lorin Sorensen, *Ford 1903 to 1984* (Skokie, IL: Publications International Ltd., 1983), 90; Sorensen, *Forty Years*, 225–31.

[19] He was nicknamed 'Cast-Iron Charlie' for his crucial contribution to solving the technical problems of casting the V8 engine.

[20] Lewis et al., *Ford 1903–1984*, 104; Nevins and Hill, *Decline and Rebirth*, 117–18, 75, 64–9, 57–60.

to the maximum, with ultimately disastrous effects. In the 1930s, in response to these failures, its policy became more complex. On the one hand, it sought to develop better conditions for Americanization by introducing a pan-European production and marketing strategy. On the other hand, it moved away from some of the rigidities of the 1920s by developing a small car specifically for the British market and devolving greater responsibility for British management. However, this did not signal an overall strategic shift to decentralization or pluralism. The small car initiative became a neglected, almost an orphan, project; local management carved out a limited and contested autonomy only in the interstices of Dearborn control and supervision; and with the arrival of the V8, many of the features of the centralizing and homogenizing campaigns of Model T days resurfaced in the late 1930s—once again with quite damaging results.

I have described the failure of Ford's Model T policy in Britain more fully elsewhere. After considerable success before the First World War, Model T sales collapsed in the 1920s in face of the challenge from lighter, cheaper, and more modern cars launched by British makers. To the distress of local managers, Ford refused to allow any modifications to the car to adapt it to local tastes and needs. Instead, it sent in a succession of American managers under the strictest instructions to imitate precisely Detroit methods of manufacturing and organization. The inflexibility of these policies resulted in demoralization and the collapse of Ford's market share from 24 per cent in 1913 to 4 per cent by 1929.[21]

In response to these setbacks, Ford launched its 1928 plan to restructure its European operations. The centrepiece of the plan was the construction of a massive new works at Dagenham. Dagenham was designed to be a one-tenth scale version of the River Rouge plant in Detroit. It would be a massive manufacturing centre serving not only the British market but also the smaller European assembly plants. Dagenham would have a similar relation to Europe as River Rouge had to the US branch plants. As a result, it was hoped, Ford could tap economies of scale by integrating its European operations. The plan envisaged a partial retreat from the 'Americanization' of the 1920s, in so far as it allowed for a greater role for European nationals in top management and for some product adaptation for European markets. The Model A, launched in 1928, included a smaller engine for Europe. But Ford cars remained distinctly large and American, and the construction of Dagenham also emphatically reasserted the primacy of American manufacturing methods.[22]

A central tenet of the 1928 plan was that 'the European demand for Ford products must be filled from Dagenham to justify the plant', and in the initial plan, Germany, France, and Italy were to absorb one-third of Dagenham's planned output of 250,000 cars per year.[23] But the Depression and the rise of European protectionism in the 1930s meant that Dagenham was never able to

---

[21] For further details see: Tolliday, 'Diffusion', 60–6.

[22] Tolliday, 'Diffusion', 66–7; Wilkins and Hill, *American Business Abroad*, 288–92.

[23] Percival Perry to Edsel Ford, 8 Aug. 1928 (HFM, Acc. 6, Box 252); 2 Dec. 1931 (HFM, Acc. 572, Box 18).

operate on this scale or fill this role. Dagenham's capacity was cut back to 120,000 and its peak annual output before the Second World War never exceeded 72,000.

Thus Dagenham was built according to a template incorporating the economic and manufacturing logic of Detroit-style production. Planning assumed that economies of scale, standardization, and the rigorous implementation of 'Ford methods' would cut costs and expand markets. But when changing market conditions vitiated such a project almost from the start, Ford's responses were clumsy and confused. Conflicts between local operational needs and the doctrinaire application of 'Ford policies' pervaded all aspects of product policy, engineering, and marketing.

In terms of product policy, Dearborn at first stubbornly resisted the arguments of British management that they needed a new small 8hp car as their primary product for Europe. Instead, it insisted on tooling up Dagenham for the big Model A despite sluggish sales. At the last moment, however, Henry Ford was unexpectedly converted to the idea of building a small car for Europe. At his insistence, a Dearborn design and engineering team carried this out 'pretty darn quick'. Based on intensive study of the best small European cars of the day, a new car (the Model Y) was developed from scratch in a matter of months.[24]

The developmental capacity of Dearborn engineering, even with a type of small lightweight car that was technologically unfamiliar to them, was remarkable. But this achievement was not matched in terms of engineering for manufacture in British conditions. The designers knew little about production and markets in England. But Dagenham was barred from introducing local modifications even to meet the superior equipment of British rivals, or when the costs of locally available materials made minor design changes necessary. Instead Dearborn insisted on absolute conformity to Detroit practice. In the run-up to production, top managers were pulled out of Dagenham and sent to Dearborn for several months to study Ford practice there, rather than being on the shopfloor at Dagenham solving problems. Meanwhile Dagenham failed to train Dagenham engineers in unfamiliar high-volume production techniques involving chrome steel, precipitating a serious crisis of persistent axle failures, and producing delay and disorganization in the launch of the Model Y.[25]

In terms of marketing, Sorensen insisted on putting the Model Y on the market at super-low prices. The Detroit way, he insisted, was that 'our costs have come down to meet our price . . . In setting the price we never gave a thought to what it would cost in the first instance . . . This was exactly the way we do it here at Dearborn, and this same idea must be carried out in England'.[26] The result

[24] Perry to Charles Sorensen, 5 Oct. 1932 (HFM, Acc. 572, Box 18); Dave Turner, *Ford Popular and the Small Sidevalves* (London: Osprey, 1984), 15; Wilkins and Hill, *American Business Abroad*, 237.

[25] Perry to Sorensen, 26 Apr. 1932 (HFM, Acc. 38, Box 71); Perry to Sorensen, 21 Mar. 1933; Sorensen, 'Memo', 10 Apr. 1933 (HFM, Acc. 572, Box 18).

[26] Sorensen to Smith, 11 Aug. 1932 (HFM, Acc. 572, Box 18).

was consternation in England where, as the managing director, Percival Perry, pointed out, there was no basis for the idea that price reductions would increase sales in Europe. In the face of import quotas and sales restrictions in key markets, cuts in prices would simply result in cuts in revenue![27]

Dagenham management chafed at the over-rigid application of Dearborn policy, but they were still enthusiasts for Ford methods properly targeted. While Perry criticized Dearborn's mandated price cuts, or the inappropriate criteria underpinning the formal statistical audits used to measure British cost performance, he wholeheartedly embraced the classic Ford priority of a low (£100) selling price for the Y, as a way of enlarging the market. However, he recognized that achieving this would require a significantly different approach in Britain. While Dearborn focused almost solely on Dagenham's process costs (the legacy of its experience with highly integrated operations in Detroit), Perry and his purchasing manager, Patrick Hennessy, instead focused most intensively on the reduction of the cost of bought-in components which comprised some two-thirds of total manufacturing costs at Dagenham. Through a mixture of advice, technical assistance, and bullying, Hennessy and his staff drove down supplier prices to make it possible for Ford to regain its lead in low prices.[28]

In this case the use of Detroit technical expertise and assistance was effectively channelled and focused on precise problem solving. But inside Dagenham itself there was a lack of systematic technical collaboration. Perry wanted permanent American engineering staff to 'teach men from the outside the proper meaning and values of Ford methods'. But Dearborn engineers came and went, and often ended up as 'regular loose ends', 'under nobody's real control', and outside the local managerial hierarchy. As Perry put it, they saw themselves as 'critics with supervisory functions' and their work occasionally degenerated into embarrassing shouting matches on the shop-floor.[29] In other respects, Dagenham perhaps imitated the disorganized and factional River Rouge plant a little too directly, as visiting staff from Detroit engaged in spying or took sides in factional politics.[30]

In terms of model policy, Henry Ford's interest in smaller cars for Europe was shortlived. Perry hoped to develop quality low-price small cars, possibly relying on 'more labour and less mass production', but Dearborn showed no interest and discontinued development work on the Model Y after 1934, focusing its developmental attention on the larger Model C (a car which could accommodate both

[27] Perry to Sorensen, 30 Aug. 1932, 5 Oct. 1932 (HFM, Acc. 572, Box 18).

[28] Hennessy specifically posed the question: 'Can we apply the methods which you use for your American suppliers to the British equivalents?' (Patrick Hennessy to A. M. Wibel (Dearborn Purchasing Director), 21 Sept. 1934 (HFM, Acc. 157, Box 201)); for an example of this work, see: 'Report on Mr. Siminick's Visit to Briggs Motor Bodies Ltd., Dagenham', 13 Sept. 1937 (HFM, Acc. 38, Box 36).

[29] Perry to Sorensen, 15 May 1929; 5 Dec. 1932; ibid., 13 Dec. 1932 (HFM, Acc. 572, Box 18).

[30] One accountant spiced his reports on time studies and operations with gossip on Perry's alleged narcotics habits and cronyism: 'Mr. Klopsic's Report on Progress at Dagenham', 13 Feb. 1933 (HFM, Acc. 572, Box 18).

four-cylinder and V8 engines).[31] The Model Y (with some reluctance) was allowed to continue to run in parallel, however, and Dagenham engineers devoted their own attention to refining it. By stripping and de-costing it they produced the £100 Popular in October 1935, and then redesigned and improved it to create the 7Y in 1937. Such work was strictly against the rules: when Dagenham first showed its designs for the 7Y to Dearborn, Sorensen 'blew up' and told them to 'take an axe and chop them up'.[32] But, finally, Dagenham was allowed to go ahead and launch the 7Y in August 1937. This car is sometimes described as the first Ford car to be designed and developed in the UK, but it was primarily a reworking and facelift of the original Y, cannibalizing mechanical innovations in brakes, wheels and steering from Dearborn's Model C programme.[33] Yet while sales of the Model C line were moderate, the 7Y proved to be Ford's most successful car of the late 1930s.

By the late 1930s, however, Dearborn's primary interest (in conformity to its US policies) was the transfer of its new V8 technology to Europe. Dearborn wanted Dagenham to move into large-scale production of V8 cars, involving over $1 million of new investment. But Perry saw no substantial market for such a large, powerful car, and he resisted. Ultimately, Perry convinced Dearborn to test the market with imported V8s (sold at a loss at very low prices to attract consumers). Perry's doubts were vindicated. Demand was so weak, even at artificially low prices, that Dearborn realized that the car made no sense for England and abandoned plans to manufacture V8 cars at Dagenham.[34]

However, Dearborn still insisted on shifting its British commercial vehicle programme wholly to V8 engines—with disastrous effects. The V8 was overpowered for UK conditions, and the use of low gears to compensate for this resulted in heavy petrol consumption and bad engine wear. Morris and Bedford swooped on the market. Market share fell from 25 per cent to less than 15 per cent within a year. Dealers were unanimous in clamouring for the return of a four-cylinder engine, and in 1938 Ford bowed to pressure and reintroduced a four-cylinder truck, primarily designed at Dagenham, and discontinued V8 engine production at Dagenham.[35] Reflecting on the disastrous history of the V8 trucks, Perry commented to Sorensen, 'I hope you will not mind me saying that GM seems to have outmanoeuvred us by using their foreign factories for the production of vehicles suitable to the country of origin and not facsimiles of American production'.[36]

---

[31] Perry to Edsel Ford, 21 Nov. 1933 (HFM, Acc. 38, Box 13); Sorensen to Perry, 26 Apr. 1934 (HFM, Acc. 38, Box 21).

[32] Mira Wilkins, Notes of an interview with Sir Patrick Hennessy, 23 March 1960 (HFM).

[33] Turner, Ford Popular, 42.

[34] Sorensen to Perry, 28 June 1934; Perry to Sorensen, 8 May and 27 July 1934 (HFM, Acc. 38, Box 21).

[35] Perry to Sorensen, 26 Nov. 1934 (HFM, Acc. 38, Box 21); Perry to Sorensen, 14 Apr. 1937 (HFM, Acc. 38, Box 36); Turner, Ford Popular, 94.

[36] Perry to Edsel Ford, 21 June 1938 (HFM, Acc. 38, Box 39).

General Motors

Perry was correct that General Motors had pursued a wholly different approach to its European product, production, and organizational policies. While Ford often dogmatically insisted on the maximum direct transfer of US product, production, and organization, GM was flexible and *ad hoc*, prepared to devolve control, to accept local initiative, and to play a role of support and assistance, although at times it also actively injected selected elements of US practice into its relatively autonomous subsidiaries, notably in the form of pioneering practice in unit-construction technology in the 1930s.

In the 1920s, these policies reflected confusion and uncertainty in its overseas policy. GM preferred simply to export or to assemble Completely Knocked Down (CKD) kits in small foreign assembly plants. Despite pressure from the chairman of their Export Companies, James D. Mooney, to add leading European firms to the GM stable, and develop smaller, cheaper cars specifically for European markets, the Executive Committee in the USA remained divided and uncertain and, according to Sloan, 'in search of a policy'. They feared unpredictable governments and worried that they would dilute their already overstretched management resources if they tried to run substantial European operations.[37]

Nevertheless, Mooney persistently championed the issue and came close to clinching the takeover of Citroën in 1919, and Austin in 1925.[38] Shortly after this, Mooney engineered the purchase of Vauxhall, a second-rank, financially troubled, British maker of expensive cars. Mooney bought it very cheaply, but he acted without the full support of his colleagues and was bitterly attacked for buying 'a run-down outfit living on its reputation'. The opposing executives labelled the Vauxhall cars, 'Jim Mooney's hearses'. Although the purchase was allowed to stand, Mooney's related development plans were blocked by Sloan and the GM Executive, and Vauxhall continued to produce large outdated cars and sustain substantial losses for a further five years.[39]

Subsequent development after 1929 was only grudgingly approved by the GM Executive as it became clear to them that 'we either had to build up Vauxhall or else give up on the English market'.[40] The backbone of development in the 1930s was a light truck based on a Chevrolet chassis (the Bedford truck). While Vauxhall's car sales remained sluggish, truck sales prospered. Vauxhall took 27

[37] Alfred P. Sloan, *My Years with General Motors* (New York: Doubleday, 1963; Penguin edn., Harmondsworth, 1986), 316–21.

[38] Sloan, *My Years*, 317–20; P. S. Steenstrup, 'Confidential Notes on the European Car Situation in June 1920' (Pierre S. Du Pont Papers, Box 624, Hagley Museum and Library).

[39] Lawrence Hartnett, *Big Wheels and Little Wheels* (Sydney, Australia: Gold Star Publications, 1965 and 1973), 35–7; Len Holden, 'A History of Vauxhall Motors to 1950: Industry, Development and Local Impact on the Luton Economy', M.Phil. thesis (Open University, 1983), 39; Sloan, *My Years*, 320; Frederic G. Donner, *The Worldwide Industrial Enterprise. Its Challenge and Power* (New York: McGraw-Hill, 1967), 15.

[40] James D. Mooney to Pierre S. Du Pont, 15 May 1929 (Walter Carpenter Papers, GM II.2, Box 820, Hagley Museum and Library).

per cent of the light truck market in 1936 and sold more trucks than cars in the decade. GM provided the initial capital for investment in a factory at Luton but thereafter truck operations were highly profitable and provided a stream of income to fund further investment.[41]

GM made little attempt to introduce American methods to Luton for most of the 1930s. Although assembly-line manufacture was introduced, it was at first a low-volume system, oriented primarily to trucks, which simply switched periodically to produce batches of cars for periods of days at a time.[42] GM also showed no inclination to introduce American management. After briefly, and somewhat unwillingly, putting in an American managing director, Mooney quickly found a British replacement, Charles Bartlett, the former head of GM's British assembly plant. As Len Holden has shown, Bartlett was 'as English as they come', a sound financial manager, and strongly committed to truck development, who developed a thoroughly English and paternal style of plant management in the 1930s, relatively untroubled by Detroit.[43]

On the car side, GM's first attempts to produce a product for Europe were unsatisfactory. A product study group at GM's technical centre hurriedly designed a low-priced six-cylinder car (the rather unsatisfactory Cadet) based on a Chevrolet, which was introduced in 1930 and which was as out of touch with the needs of the European market as the Ford Model A.[44] However, the study group continued its work, resulting in the improved Light Six of 1933 which began Vauxhall's car sales revival, and also providing the basis for a much more far-reaching product innovation: GM's development of the infant technique of unitary construction for its European cars.

Although Ford transferred many American product features and manufacturing practices to Britain, all of them, except the Model Y, were imitations of existing established practice in the USA. No attempt was made to develop innovations that had value in Europe unless they also had clear advantages and applications in the US context. But GM's development of unitary construction for its European cars was a striking contrast, directly applying the innovative capabilities of the US firm to produce technologically innovative solutions to distinctly different market and product problems.

Unit-body construction (a method of design and manufacture in which a single structure serves as both body and chassis of the car) was pioneered in the USA in the early 1930s by American bodymakers when closed steel bodies and related advances in metal stamping and welding made such bodies feasible. These methods had great strength-to-weight advantages, and increased internal space compared to traditional body-on-chassis designs, but US car makers

---

[41] Hartnett, *Big Wheels*, 39–42; Holden, 'Vauxhall Motors', 89–94.

[42] 'The Works of Vauxhall Motors Ltd.', *Automobile Engineer* (August 1930), 284.

[43] For details, see: Holden, 'Vauxhall Motors', 45–7; id., ' "Think of Me Simply as the Skipper": Industrial Relations at Vauxhall, 1920–50', *Oral History* 9 (1984); Maurice Platt, *An Addiction to Automobiles* (London: Warne, 1980), 90–1.

[44] Sloan, *My Years*, 256, 320; Hartnett, *Big Wheels*, 43; Michael Sedgwick calls the Cadet 'cheap and nasty', in his *The Motor Car, 1945–56* (London: Batsford, 1979), 123.

showed limited interest in these techniques. The potential savings in weight and size were of limited importance in American cars, and were offset by high tooling costs, design rigidities, and high costs of incremental styling changes. Consequently, although GM, Chrysler, and Ford all investigated the method in the 1930s, only one significant chassis-less car (the 1937 Lincoln Zephyr) was (briefly) produced in the United States before the war.[45]

In Europe, however, the market was much more sensitive to size and weight, and less attuned to constant style changes, and it was European manufacturers who first seized on the advantages of this form of body, most notably the 1934 Citroën 7CV which was the first mass-produced unit-body passenger car. While Ford had extensive capabilities in this field, it targeted them only at its luxury Lincoln cars for the home market. GM, however, saw the potential rewards of applying and developing the technique in Europe rather than in the USA. In 1935, GM's leading engineer, Earl A. McPherson, designed a small unit-construction car for Chevrolet. GM decided not to build the car in the USA, but adapted it to become the basis of the 1936 Opel Olympia in Germany and the 1938 Vauxhall 10 in the UK.[46] Indeed. by 1940 *all* of GM's European models had dispensed with a separate chassis, and while GM never built a unitary construction car in the USA until the advent of the subcompact models in 1959, it built *only* chassis-less cars in Europe during this period.

Thus Vauxhall (together with GM's Opel subsidiary in Germany), alongside Citroën and Lancia, became the European leaders in unitary construction on the eve of the Second World War. Vauxhall's rich cash flow from its truck sales was channelled into the very large fixed investment needed for the introduction of unitary construction from 1937. Vauxhall at Luton created its *own* innovative body plant for this purpose (a rarity in Britain at this time), the so-called Million Pound Shop. The result was a significant technical lead, and a strong rally in Vauxhall car sales. By the 1950s, unitary construction became the predominant mode for the leading continental European and British automobile companies.[47]

GERMANY: AMERICANIZATION OR NAZIFICATION, 1928–1940

Ford and GM both entered manufacturing in Germany in the late 1920s. Both acted reluctantly in face of protectionist pressures and fears of exclusion. But while GM bought a successful local company and managed it opportunistically

---

[45] Abernathy, *Productivity Dilemma*, 193–5.

[46] T. P. Newcomb and R. T. Spurr, *A Technical History of the Motor Car* (New York: Adam Hilger, 1989), 281–2.

[47] 'Producing the Vauxhall Ten All-Steel Body', *Automobile Engineer* (January 1938); Nick Georgano, Nick Baldwin, Anders Clausager, and Jonathan Wood, *Britain's Motor Industry: The First Hundred Years. The Fluctuating Fortunes of Britain's Car Manufacturers from Queen Victoria's Time to the Present Day* (Sparkford, Somerset: G. T. Foulis & Co., 1995), 103; Michael Sedgwick, *Cars of the Thirties and Forties* (Göteberg: Nordbok, 1979), 69–73; Sedgwick, *Motor Car*, 126–7.

at arm's length, Ford remained deeply committed to the transfer of its own systems and methods even in the most adverse circumstances of the autarkic Nazi economy.

GM entered Germany through the acquisition of Germany's leading car producer, Adam Opel, in 1928. The entry was a reluctant response to protectionism and a bank-led movement to create trusts in the car industry to resist US influence. The GM board members were divided and indecisive over what to do, but ultimately took opportunistic action when GM's own banking connections unexpectedly opened doors for it to acquire Opel.[48]

Opel was a company with its own lengthy traditions of imitating and adapting American methods dating back to the 1890s. Its key model in the 1920s (the 5CV 'tree-frog') was a direct copy of the Citroën 5CV, the car with which André Citroën had consciously sought to Americanize sales and production methods in France, and Opel's version was built largely with second-hand machine tools bought from GM in the USA.[49]

Because Opel was an efficent and effective producer, GM made few changes to products or production in the early 1930s, and relied heavily on Opel's engineering staff. But in the late 1930s GM was able opportunistically to blend increased Americanization with a strong dose of Nazification. GM had no compunction about cultivating close relations with the government and the Nazi Party, and the 'German-ness' of Opel shielded them from adverse reactions in the USA. It carried Opel into truck production for the first time, using an old Buick chassis and powertrain (driveshaft), building the resulting Blitz truck as the vehicle of choice of the German army. GM accepted an invitation from the Nazi government to build a truck plant at Brandenberg to service this military demand, and the substantial profits from the trucks bankrolled the new facilities required to pioneer unitary construction cars (the Olympia) in Germany, to modernize production facilities, and to enlarge market share by aggressive pricing.[50] As the largest car company, it was able to turn many of Hitler's initiatives to its own advantage. Opel placed itself at the centre of Hitler's 'German standardization' drive, espoused nationalist local content programmes, and used its multinational marketing linkages to perform as export champions in Hitler's export drives. Somewhat ironically, GM's *lack* of multinational co-ordination or integration facilitated this, in contrast to Ford. Standardization posed no obstacles to GM because they had no cross-national model policies, and Opel

---

[48] Yuji Nishimuta, 'German Capitalism and the Position of the Automobile Industry between the Two World Wars', *Kyoto University Economic Review* 1, 61, 1 (1991); 2, 61, 2 (1991), 34; Sloan, *My Years*.

[49] Anita Kugler, 'Von der Werkstatt zum Fliessband. Etappen der fruehen Automobilproduktion in Deutschland', *Geschichte und Gesellschaft* 13 (1987), 317–37; see also ead., *Arbeitsorganisation und Produktionstechnologie der Adam Opel Werke (von 1900 bis 1929)* (Berlin: Wissenschaftzentrum, 1985).

[50] Karl Ludvigsen, *Opel. Wheels to the World* (Princeton: Princeton Publishing, 1975), 46–52; Sloan, *My Years*; Kugler, 'Von der Werkstatt', 335–7; Simon Reich, *The Fruits of Fascism. Postwar Prosperity in Historical Perspective* (Ithaca, NY: Cornell University Press, 1990), 111.

exports from Germany did not have to contend with the demands of rival GM subsidiaries for export territories.[51]

Ford followed GM into Germany in 1929, but followed a very different course. Ford's entry reflected its fear of GM snatching an advantageous position, and a desire to retain a foothold in a major market. But the move involved more difficult choices for Ford than for GM. Ford's commitment to Ford methods and Ford products ruled out the acquisition of a local company. But the construction of its own manufacturing facilities in Cologne cut across the international strategic aims of Ford's 1928 plan. The German market was supposed to be an important outlet for Dagenham's immense capacity, but Cologne not only took over this market but also created new capacity that would require its own export outlets to achieve economies of scale.

Ford made large losses in Germany for several years. As in Britain and France, its rigid approach to model policy isolated it from the widest markets. Despite the clamour of local management for a small car (preferably even smaller than the Model Y), Ford insisted on pushing the Model C and its derivatives in Germany, with little success.[52] Ford's partial salvation came, as with GM, through establishing a profitable alliance with the Nazi government. Ford found this more difficult than GM. Its commitment to its own proprietary technology and cross-national model policies hampered its efforts to promote itself as a 'German' firm. It could not embrace 'German standardization' because of its reliance on American designs and on specialized tooling for its own equipment. It only supported Hitler's export drive with great reluctance, since this required it to transfer valuable export markets to Cologne at the expense of an irate Dagenham.[53]

Beyond this, Ford was also less adept at commercial *realpolitik* than GM. Its attempts to curry favour with the Nazi regime were clumsy and half-hearted. In pursuit of political kudos, it embroiled itself in protracted and ultimately futile merger discussions with a near-bankrupt German engineering company. And it sought Party favour by 'Germanizing' its management: but it appointed a German Jew as managing director and then tried to retrieve the position by firing him.[54] But despite this, for a long time, it refused to give the Nazis what they most wanted: a good three-ton truck for the military, such as Opel was enthusiastically producing. Henry Ford's old pacifist principles blocked a military truck for almost two years. But, in face of the displeasure of the regime, the

[51] Heinrich Albert, 'Memorandum on the Situation of Ford AG', Feb. 1936; Albert to Sorensen, 20 June 1936 (HFM, Acc. 38, Box 33); Reich, *Fruits of Fascism*, 110–11; Richard J. Overy, *War and Economy in the Third Reich* (Oxford: Clarendon Press, 1994), 79.

[52] E. Vitger to E. Heine, 4 Apr. 1934 (HFM, Acc. 38, Box 22); Albert, 'Memorandum on the Situation of Ford AG', Feb. 1936; Vitger, 'Popular Car and Sales Possibilities', 20 Feb. 1936 (HFM, Acc. 38, Box 33); Wilkins and Hill, *American Business Abroad*, 206, 233.

[53] Albert to Sorensen, 20 Jan. 1936 (HFM, Acc. 38, Box 33;); 'Sales Meeting, Cologne', 9 May 1935 (HFM, Acc. 507, Box 98); Wilkins and Hill, *American Business Abroad*, 272–80.

[54] Erich Diestel was appointed managing director in 1935. Albert to Sorensen, 20 Jan. 1936 (HFM, Acc. 38, Box 33); V. Y. Tallberg, 'Reminiscences' (unpublished manuscript, HFM).

principles melted away. Ford reversed its position and built the trucks, and, partly as a result, Henry Ford was ceremonially awarded the Grand Order of the German Eagle in 1938.[55] As Wilkins and Hill frankly note, Ford's late rally revolved around close co-operation with Hitler. The road to success 'was one of teaming up with the National Socialists . . . and swallow[ing] any qualms they may have had . . . in a bid for prosperity'.[56]

Ford's rigid attachment to its American model severely handicapped it until this late grab at 'Germanization'. Ford hesitated much more about military contracts, took on far less than Opel, but received much more criticism in the USA. It clung more rigidly to its American methods than GM, yet it was unable to inject as much effective American support into its German organization as its rival. Meanwhile, GM/Opel secured the benefits of being a 'local' company, and evaded significant scrutiny from the American public in its relations with the Nazi regime. Its position as established market leader, its clear primacy within GM's European organization, and its autonomy (supported rather than compromised by its links with the USA) put it in a powerful position.

France: The Gospel of Americanization, 1934–1940

Ford and GM faced much the same dilemma in France as in Germany in the early 1930s. In face of tightening protectionism, they could either accept the loss of an export market, or move from kit assembly to local manufacture. In this case, however, GM showed almost no interest in entry and Ford, therefore, experienced no follow-my-leader pressure from its great rival. Moreover, Ford's leading European managers warned that the French market was too small to be economic, and that additional plant in France would still further compromise the position of Dagenham as a pan-European supplier for Ford.[57] Nevertheless, even though its existing sales in France were less than 3,000 per year, Ford decided to enter manufacture in France in 1934. Reversing its earlier resistance to acquisitions or joint ventures, it formed a joint venture (Matford) with Mathis, the fourth-ranking car producer in France.

This decision is somewhat puzzling. The French managing director, Maurice Dollfus, urged the scheme as a low-cost route to French status and a manufacturing presence in France, taking advantage of Mathis' desperate financial position. But Ford was unimpressed by Mathis' physical assets, and, as we shall see, showed little interest in developing the Mathis range of cars. It seems that the primary motivation was to secure recognition as a 'French' producer. However, in contrast to GM's approach at Opel, Ford did not attempt to develop and build on the local capabilities of its acquisition. Moreover, within months of

---

[55] The award came just after Hitler's annexation of Austria and provoked howls of protest in the United States. One month later Jim Mooney of GM received a similar award with little publicity in the USA (*Washington Post*, 30 Nov. 1998).

[56] Wilkins and Hill, *American Business Abroad*, 284.

[57] Perry to Sorensen, 17 July and 19 Dec. 1931 (HFM, Select File).

completing the Matford agreement, Ford turned its back on an opportunity to acquire a position in the French industry that would have been closely analogous to that of GM in Germany. Citroën, France's leading car company, went bankrupt in 1935, and André Citroën, the leading exponent of the emulation and adaptation of American methods in the French car industry, approached Ford for a merger.[58] The combination of Matford and Citroën would have created by far the largest car company in France, and the cost was not high. But Dearborn turned the proposal down flat. Sorensen told the New York bankers that Citroën engaged as intermediaries that, 'as far as we were concerned, they could not give us the plant!'[59] In the event, Michelin (with the backing of the government) arranged a financial restructuring with the banks and emerged with uncontested control. Citroën swiftly returned to its prime position in France.[60]

Ford believed that its dominating role in its partnership with Mathis was a superior alternative. In particular, the Mathis venture would not compromise the independence of Ford's model policy. Despite some early promises, Ford showed almost no interest in nurturing the Mathis line of four-cylinder cars, and within a year of the merger, Ford dropped all four-cylinder car production, effectively eliminating the Mathis marque from the French market. Instead, Dearborn focused entirely on the V8. As in Britain, Dearborn perceived the V8 as a significant technical advance and a superior product which was bound to attract a large following in Europe, as in the USA, if adequately promoted and marketed. It strongly believed (on very little evidence) that European demand was swinging towards the V8 and decided to concentrate all its energy on this line. Dearborn displayed remarkable optimism about the V8's sales potential in France once the company's 'French' identity broke down chauvinist sales resistance. But, above all, Dearborn believed that sales of V8s would increase as a result of 'the education of the auto buying public to proper appreciation of V8 motors'.[61] The notion of 'educating the public' to the 'V8 idea' or the 'V8 spirit' resonated through the Franco-American correspondence of the time. The key task was to break consumer resistance. Thus, for example, when buyers turned away from Ford's new V8 trucks, Dollfus' response was simply to remove all other Ford trucks from the dealers.[62] Meanwhile, criticisms from Emil Mathis, that it was 'absurd' to 'try to force an 8-cylinder on the public in France at the

[58] Citroën had previously sought mergers with Ford in 1919 and with GM in 1925. On Citroën's methods see Sylvie Schweitzer, Des engrenages à la chaine. Les usines Citroën, 1915–1935 (Lyons: Presses Universitaires de Lyon, 1982).

[59] Sorensen to Dollfus, 25 Jan. 1935; Dollfus to Edsel Ford, 22 Jan. 1935 (HFM, Acc. 38, Box 27). The cost would have been approximately $6.5 million, and Ford spent c.$8m. at Matford between 1935 and 1939.

[60] Hubert Bonin, 'Les Banques ont-elles sauvé Citroën? (1933–5) Reflexion sur la marge d'initiative bancaire', Histoire, Economie et Societé (July–Aug. 1984).

[61] Dollfus, 'Matford Programme', 9 Oct. 1934; Panier and Delamare to Dollfus, 6 July 1934 (HFM, Acc. 507, Box 64); Dollfus to Sorensen, 10 Aug. 1934 (HFM, Acc. 38, Box 22).

[62] Dollfus to Edsel Ford, 14 Jan. 1935 (HFM, Acc. 507, Box 64).

present moment . . . because conditions call for a small very cheap 4-cylinder job',[63] were brushed aside.

Yet the V8 was fundamentally unsuited to generate sales beyond a core of well-off buyers in France. It was a high-price car that was large and costly to run and sales quickly plateaued. From 1936 onwards, V8 sales steadily deteriorated, but Sorensen continued to reject calls to develop a smaller alternative, and insisted that the real problem was poor marketing: 'I cannot understand why (V8) sales in France should not be doubled'.[64] Meanwhile, the problems of inappropriate products were exacerbated by rigid thinking on manufacturing and engineering methods. Selected French engineering staff were taken to Detroit and thoroughly indoctrinated in Ford methods. But this resulted in bitter conflicts with local engineers and managers on their return. In particular, this focused on the use of 'overluxurious' tooling which, Dollfus observed, was 'completely incompatible with our necessarily limited production'.[65]

As with the Model T and Model A strategies in Britain, it was only the demonstrable and lamentable failure of the V8 strategy that finally forced Ford to change course. As the Mathis joint venture became an increasingly troublesome burden, Ford realized that any future in France would require a new modern manufacturing plant and a new product strategy. Within Dearborn, Edsel Ford changed tack and began to argue for the development of a four-cylinder car for France based on Ford's German Eifel/Taunus car. By late 1938 Edsel and the top European managers were finally able to persuade Sorensen that a new plant and a four-cylinder car represented the only way forward. By that time, the company was close to disaster, and it was only able to limp along while the new plant (Poissy) was built by accepting contract work from the government for aero-engines.

Even with the construction of Poissy, however, Ford continued to resist adaptation and selective application of its methods. During the Poissy construction programme Dollfus argued for machinery 'more of the universal type', and wanted to focus on low-cost and flexible plant construction. He resisted sending the architect selected to design Poissy to Dearborn to study the Detroit factories because 'Dearborn construction methods could not be applied here economically'.[66] But Sorensen utterly rejected this: 'Every time we have ever built a plant it has been based on detailed studies of US factories.' The aim had to be to emulate Dearborn practice as far as possible, and 'get it functioning like we have in our plants'.[67]

Ford's Matford episode highlights in extreme form the contradictions involved in Ford's pre-war approach to 'Americanization'. While GM judged that the context was simply inhospitable for it, Ford spent a lot of money on a

[63] As reported in Dollfus to Sorensen, 27 Aug. 1935 (HFM, Acc. 38, Box 27).
[64] Sorensen to Dollfus, 14 Jan. 1938 (HFM, Acc. 38, Box 40).
[65] Dollfus to Sorensen, 11 Dec. 1934 (HFM, Acc. 507, Box 64).
[66] Dollfus to Sorensen, 21 Aug. 1938 (HFM Acc. 38, Box 40).
[67] Sorensen to Dollfus, 22 Aug. 1938 (HFM Acc. 38, Box 40).

second-rate French car company which bogged it down in time and trouble. It spurned an opportunity to take control of the French market leader and opted instead for partnership with a much frailer company in order to ensure complete control. Yet it made no attempt to take advantage of what that company could offer, in making and marketing cars specifically adapted for the French market. It preferred simply to buy a French name to hang over Ford operations. And it then doggedly and evangelistically sought to apply standard methods in often wholly inappropriate conditions and ill-adapted factories, turning its back on 'French' products to try to force pure American V8s on unwilling consumers.

3.3   Ford and General Motors in Europe, 1945–1962

THE POST-WAR AMERICAN MODEL: FORD AND GM IN THE USA, 1945–1962

Before the war, there had been persistent and striking contrasts between the productive and organizational models of Ford and GM in the United States which had deep consequences for the conduct of their overseas operations. After the war, there was some convergence in their organizational structures, production organization, and product strategies, although substantial differences remained. Above all, however, *both* companies became locked into a pattern of competitive rivalry with dynamics that significantly distinguished the post-war American industry from its pre-war predecessor.

The end of the war brought the culmination of long years of organizational degeneration at Ford. Intrigue and misrule at the top took the company close to collapse as Henry Ford clung on to power almost until his death in 1945. As Nevins and Hill demonstrated in their classic account, in his later years Henry had systematically undermined hierarchies and organization, and peremptorily driven out many of his best managers. Accounting and reporting structures had been thrown into disarray, and much of the plant was seriously run down. By the time that Henry Ford II was recalled from the Navy and managed to oust Harry Bennett and the quasi-criminal circles that had surrounded Henry in his last years, the organization was tottering.[68]

The conventional story of Ford's recovery in the next decade focuses on the critical reorganization of top management.[69] First, Henry II reassembled an effective management team from veteran Ford managers returned from war service, young financial and statistical controllers recruited from the Air Force (the 'Whiz Kids'), and a powerful cadre of experienced GM executives around former GM vice-president, Ernest Breech. On this basis, hierarchies were re-established, financial controls put in, and the company was thoroughly reorganized in conscious imitation of the management principles of GM. This

---

[68] Nevins and Hill, *Decline and Rebirth*, 229–68.

[69] Ibid.; Robert Lacey, *Ford. The Men and the Machine* (Boston: Little Brown, 1987); David Halberstam, *The Reckoning* (New York: Avon Books, 1986); John A. Byrne, *The Whiz Kids: The Founding Fathers of American Business—and the Legacy they Left* (New York: Doubleday, 1993).

managerial restructuring paved the way for the successful 'rebirth' of Ford over the next decade.

But it is important to qualify this picture in certain respects. There is no doubt that the organization of the firm was in chaos in 1945–6. When the 'Whiz Kids' conducted a department by department audit of methods and management structures their verdict, using a popular contemporary Air Force term, was that most departments were FUBAR ('Fucked Up Beyond All Recognition').[70] But the company retained substantial residual strengths. As a result of wartime profits it was financially highly prosperous, it had its pre-war range of cars to fall back on for some years, and its powerful dealer and brand name presence gave it great strength. The often quoted view of one leading manager that 'the company was not only dying, it was already dead, and rigor mortis was setting in' was something of an overstatement.[71]

Moreover, the extent to which Ford was remodelled on the lines of GM is also questionable. It is true that the leading GM managers brought with them, and consciously espoused, GM's principles of decentralized management and divisional autonomy.[72] They widely disseminated the GM organizational template as codified in Peter Drucker's 1946 study of GM, *The Concept of the Corporation*.[73] And they extended the process of divisionalization within Ford (already started by Henry Ford II before the GM managers came in) between 1946 and 1949. But the extent to which the organizational culture was actually remodelled along GM lines is less clear. David Hounshell has shown that 'the old order' or the 'Fordist ideal' proved remarkably resilient and in some senses finally triumphed.[74] The project of divisionalization could not be pushed far in a company dominated by one giant division (the Ford Division) and one mighty manufacturing facility (the River Rouge plant). Ford hesitated to create the wide product range that would make divisionalization most appropriate, and by 1949 an older organizational culture reasserted itself. Between 1949 and 1951 Ford's production engineering and finance functions curtailed the scope of divisional-

---

[70] See handwritten notes for C. Thornton, 'Organizational Problems of the Ford Motor Co.', 12 July 1946 (HFM, Acc. 881, Box 5); and Jim Wright, 'Reminiscences' (unpublished manuscript, HFM).

[71] John R. Davis quoted by Nevins and Hill, *Decline and Rebirth*, 294, and by Lacey, *Ford*, 441; David Hounshell, 'Planning and Executing "Automation" at Ford Motor Company, 1945–65: The Cleveland Engine Plant and its Consequences', in Haruhito Shiomi and Kazuo Wada (eds.), *Fordism Transformed. The Development of Production Methods in the Automobile Industry* (Oxford: Oxford University Press, 1995).

[72] See for example, L. D. Crusoe, 'Report on Progress in Organizational Development and Recommendations for Further Action', 1 May 1947 (HFM, Acc. 881, Box 5). See also Nevins and Hill, *Decline and Rebirth*, 313–31; Lacey, *Ford*, 430; Byrne, *Whiz Kids*, 126, notes that under Breech Ford was sometimes referred to as 'the Ford Division of General Motors'.

[73] Peter Drucker, *The Concept of the Corporation* (New York: John Day Co., 1946).

[74] David A. Hounshell, 'Assets, Organizations, Strategies, and Traditions: Organizational Capabilities and Constraints in the Remaking of Ford Motor Company, 1946–1962', in Naomi R. Lamoreaux, Daniel M. G. Raff, and Peter Temin (eds.), *Learning by Doing in Markets, Firms, and Countries* (Chicago: University of Chicago Press, 1999); id., 'Ford Automates: Technology and Organization in Theory and Practice', *Business and Economic History* 24, 1 (1995).

ization. Instead, the development of the company was focused on highly centralized manufacturing, specialization, and the pursuit of maximum production efficiency linked to the new highly automated plants at Cleveland and Buffalo.[75] These projects were 'pure Fordism' and symbolized the continuing power of the plants and the manufacturing function within the organization. Although elements of divisionalization persisted, this was a form of 'ersatz divisionalization' (Hounshell), using financial controls and profit centres, but eschewing real decentralization and autonomy.[76] Divisional decentralization came back on the agenda in the mid-1950s with the strengthening of the Mercury Division and the creation of the Edsel Division, but the disastrous failure of the key projects of these two divisions led to a further retreat from divisionalization.

Thus Ford retained an organizational culture and model quite distinct from that of GM. It was more centralized and more oriented to production efficiency and, especially, automation, than GM. It was also plagued by a continuing battle between financial controllers and plant managers.[77] In contrast, in GM, the Styling Department came to exercise a remarkable influence, and, despite serious compromises to divisional autonomy under Harlow Curtice's regime in the 1950s, decentralization remained a serious force.[78] Alfred Sloan had never given much priority to overseas operations, and he continued to leave them rather at arm's length in his later years. However, under Harlow Curtice, a new approach began to emerge, with overseas operations given a new prominence from 1955 onwards.

Despite the managerial and organizational differences between the two companies, however, their product and production policies had enough in common to make it possible to outline a broad 'American model' in the 1950s. After a brief period in the 1940s when they contemplated a possible need to develop small functional cars for a post-war age of austerity, both companies turned their back on the small car for a decade. They were more than happy to do so because it was evident that profit margins would have to be much thinner on small cars than on their preferred large vehicles. Beyond this, the industry also turned its back on the pursuit of low prices or fundamental product innovation. The car became a more expensive commodity relative to other consumer goods, and, in particular, the traditional low-priced car was pushed up-market. The size, power, and price of American cars all increased, giving rise to a bloated vehicle, loaded with ornamentation and accessories. The crucial competitive weapons became model turnover and style changes, whose punitive costs helped

[75] Hounshell, 'Planning and Executing Automation'.

[76] These practices are described in Kuhn, GM Passes Ford, 277, 290–1. Cf. also E. B. Rickard (Controller of Ford Division), 'A Study in Decentralization: Controllership in a Divisional Organization', NACA Bulletin 31 (1950), 567–78. For 'ersatz divisionalization' see Hounshell, 'Assets, Organizations, Strategies, and Traditions', 198–207.

[77] Vividly described in Halberstam, Reckoning, 212–14, 234–7.

[78] Stephen Bayley, Harley Earl and the Dream Machine (New York: Knopf, 1983); Ed Cray, Chrome Colossus: GM and Its Times (New York: McGraw-Hill, 1980); Kuhn, GM Passes Ford, 320–1.

to drive the Independents out of the market (though beneath the expensively differentiated skins the vehicles became more homogeneous than ever). The engineering emphasis was on more and more powerful engines, and big V8s more or
less drove out the six-cylinder engine in the horse-power wars of the 1950s,
while the functional performance of steering, brakes, or fuel economy was
neglected and the quality of manufacture, fit, and finish was often dubious.[79]
The excesses of the Detroit cruisers reached their apex with the over-inflated
Edsel in 1957. By this time, style and power excesses had produced a partial consumer revolt, and also opened up a gap for smaller imported cars to begin to
penetrate the market. One result, although it was relatively short-lived, was a
reawakened interest in smaller cars and the alternative product configurations
that had developed in post-war Europe.

In the companies' operations in Europe, the early post-war years saw substantial continuation of pre-war patterns in the relationship between the parents
and their subsidiaries. At Ford, centralized control and dogmatism in product
and production policy was mitigated only by occasional aberrations and *ad hoc*
departures. At GM a fairly relaxed arm's length control at times veered towards
abstention or indifference. But by the mid-1950s, the situation was changing.
Ford surrendered and withdrew from France. But in Britain, a well-embedded
local management began to exert more sustained influence, encouraging a more
creative 'hybridization' of Ford's methods and policies, but also a significant
degree of loss of central control. In Ford-Germany, a crisis of survival precipitated a wholesale reconstruction, combining a more considered attempt to
extend American influence with a new attention to building on European
resources or even learning from Europe. At GM, a period of benign neglect in
the 1940s and early 1950s left space for the resourceful Opel organization to consolidate its strong position in Germany, drawing selectively on American
resources, although the less self-sufficient Vauxhall suffered from its relative
isolation. By the late 1950s, however, GM was making concerted efforts both to
expand its European companies and to integrate them more effectively with its
US activities.

FORD IN FRANCE: DISSIMULATION AND RETREAT, 1948–1954

GM had stayed out of France before the war, and it showed no interest in enter-
ing it afterwards. But Ford came out of the war apparently well placed in
France. The shackles of the Matford joint venture had been shaken off, and its
Poissy plant (newly completed on the eve of war) was a fine facility that had suf-

---

[79] Charles E. Edwards, *Dynamics of the United States Automobile Industry* (Columbia, SC:
University of South Carolina Press, 1965); Abernathy, *Productivity Dilemma*; Brock Yates, *The
Decline and Fall of the American Automobile Industry* (New York: Random House, 1984);
Halberstam, *Reckoning*; Gartman, *Auto Opium*; Avner Offer, 'The American Automobile Frenzy
of the 1950s', in Kristine Bruland and Patrick O'Brien (eds.), *From Family Firms to Corporate
Capitalism: Essays in Business and Industrial History in Honour of Peter Mathias* (Oxford:
Clarendon Press, 1998).

fered little damage, in sharp contrast, for example, to the damage and disruption suffered by Peugeot.[80] Yet, as before the war, Ford struggled to find appropriate policies for its subsidiary, and oscillated between dogmatism and despair. Within a decade, Ford-France had drifted into withdrawal.

The new car for Poissy which had been designed by Dearborn before the war, based on the German Taunus, was never launched in France after the war. The tooling had been fully prepared but not shipped from the United States when war broke out. After the war, Dollfus decided to scrap it. He went to Detroit in 1946 and saw the plans for Ford's 'New Light Car' for the American market, and he decided to adopt this car for the French market (where it was named the Vedette). Contrary to Hounshell's description of the New Light Car as a 'European-style' car, it was not a small car in European terms and was not inspired by European cars. Instead, it was a quite substantial car (a 'French Buick', as Dollfus later described it) designed entirely with the American market in mind as a hedge against possible austerity conditions. Dollfus did not believe that Ford could make profits on small cars in France, and he preferred to pursue the benefits of sharing a common model programme with Detroit, especially since the car was already designed and engineered and involved no development expenses.[81]

The assessment of the French market on which this was based was entirely *ad hoc*. The project was 'conceived by ear' and Dollfus carried out no market research.[82] In contrast, pioneering market research by the French producers, in particular by Georges Toublan at Renault, showed that the buoyancy of demand for large 'American-style' cars just after the war rested heavily on short-term consumer liquidity, the black market, and the role of government 'licences to buy', but that the longer-term demand would be concentrated on much smaller and cheaper vehicles.[83] Accordingly, Renault and Citroën prioritized the development of frugal small cars.

Given its assessment of market prospects, however, Ford was very happy with the role designated for it in the government's key post-war planning document, the Pons Plan. It was, it believed, 'a very good thing'.[84] The Pons Plan gave Ford a major role in the large car segment and in trucks. Ford would make 35 per cent of *all* French trucks, and with half the pre-war truck stock destroyed

[80] Guaranty Trust Co. of New York (Paris Office), 'Report for the Ford Motor Co., responding to the former's letter of 6 December 1945', June 1946 (HFM, Acc. 713, Box 7).

[81] The V8 had a 112-inch wheelbase, the small V8 108-inch, and the Taunus 98-inch: the Light Car was 106-inch.

[82] Graeme K. Howard, 'General Summary of Position, Ford SAF, with Recommendations', 22 Nov. 1948 (FIA, AR 65-71, Box 30).

[83] Georges Toublan, 'Etude du marché de la 4CV', in *De Renault Frères à Renault Regie Nationale*, 17 Dec. 1978; extracts reprinted in Patrick Fridenson, 'La Bataille de la 4CV Renault', *L'histoire* 9 (Febr. 1979); Jean-Louis Loubet, *Citroën, Peugeot, Renault et les autres. Soixante ans de stratégies* (Paris: Le Monde Editions, 1995), 309; id., 'Les Grands Constructeurs privés et la reconstruction: Citroën et Peugeot, 1944–51', *Histoire, Economie et Societé* (July–Sept. 1990), 453–4.

[84] Elvinger, 'Etude de marché faite pour la Guaranty Trust Company of New York', March 1946 (HFM, Acc. 713, Box 7, p. 146).

and the government committed to support the purchase of trucks, the post-war prospects for Ford's 3.5-ton truck seemed excellent. Along with Citroën it would share the major large car segment.

But Ford's product strategy miscarried. Just as Renault and Citroën had forecast, the demand for large cars soon sagged. Even worse, without warning, Dearborn dropped the Light Car from its US programme in 1947, leaving Ford Société Anonyme Française (SAF) with an 'orphan' vehicle not linked to any ongoing developments in Dearborn. To compound this, its truck strategy ran into problems. French farmers wanted something much more basic than Ford's well-built trucks. Moreover, Ford also invested heavily in a diesel engine joint venture in the belief (which the French government encouraged) that diesel would dominate truck sales in the coming years. The project was a costly failure. Demand for diesel was limited and diesel trucks simply cannibalized sales of Ford's conventional V8-engined truck.[85]

By late 1948, a Dearborn review of the French market concluded that, 'were the Ford Motor Company examining the French market afresh and without commitments, it would not today embark on manufacture in France or make any investment in that country'.[86] Henry Ford II was convinced that it would be better to get out of France even at the expense of a major loss than to put any more resources in.[87]

Moreover, Dearborn was tormented not just by the economic prospects of its Ford subsidiary, but also by the political threat of France's 'going Communistic'. Paul Hoffman and other Marshall Plan administrators attempted to prevail on Ford to stay in France by stressing the important role of American investment in creating stability. But this simply made Ford even more anxious.[88] Nevertheless, such pressures, combined with legal difficulties in extricating the Ford name and rights, and fears of possible nationalization by the French government if it tried to pull out, made Ford accept the necessity of staying on for the time being.

From this time on, Dearborn's will to stay and build its presence had gone, yet it saw no ready way out of its French embroilment. Public acknowledgement that it wished to withdraw would, however, damage sales, status, and the viability of the French operation. The result was a prolonged period of sham and dissimulation. Enough had to be done to give the appearance of a long-term commitment to France, while in fact all corporate energies were devoted to limiting new commitments and preparing the ground for an orderly withdrawal, probably through the sale of the concern to another company.

[85] Jean Charpentier, 'Reasons Why the Diesel Engine Should Not be Built by Ford SAF', n.d. (1948) (FIA, AR 65-71, Box 14).
[86] Howard, 'General Summary of Position, Ford SAF, with Recommendations', 22 November 1948 (FIA, AR 65-71, Box 30).
[87] Henry Ford II to Breech, 1 Nov. 1948 (FIA, AR 65-71, Box 14).
[88] Henry Ford II to Breech, 3 Nov. 1948 (FIA, AR 65-71, Box 14).

In the meantime, a successor had to be found for the ageing Dollfus. In May 1949 he was replaced by François Lehideux, a consummate wartime collaborator, who had led the Vichy automobile manufacturers' association during the war. Lehideux demanded, and thought he had got, cast iron assurances that Ford intended to make SAF an important factor in the French auto industry. Lehideux wanted to extend the product range and strengthen the organization, but he got little support from Dearborn and soon became 'disgruntled and sour' and had to resort to frequent resignation threats to win quite small points.[89]

The key to any future development of the company was the question of a new model. The Vedette had good sales in its class, but a company based simply on Vedette and truck sales was not a viable unit. Viability would probably require a good 6–8CV car which would have to compete directly with Renault, Citroën, and Peugeot. But this would require an investment of $15m. to $20m. By the summer of 1951, Dearborn had concluded that: 'To make this magnitude of investment in France . . . is outside the realms of realism, or common sense, from a commercial standpoint. To make a capital investment of this magnitude in order to achieve a doubtful position in France is commercially unsound'. Instead, the aim had to be to get out 'lock, stock and barrel'. This could not be accomplished at once without 'disorderly collapse' but it should be possible to move towards 'graceful liquidation' in a few years' time through a programme of strict financial controls and cost management.[90]

Lehideux was not told of the withdrawal planning and was only told definitively that he would not get a new car in late 1951. The news came shortly before Lehideux travelled to the USA to represent Ford at the ECA's Operation Impact, a spectacular conference aimed, as Matthias Kipping has put it, at 'converting European employers to the American creed'.[91] It is a telling irony that Lehideux should be representing Ford at this epitome of the propaganda of 'Americanization' while squabbling with Ford for resources to survive, and with his parent company plotting to dispose of the company behind his back.

The path to 'graceful liquidation' was not smooth. In 1952, a further financial crisis forced Dearborn to reveal to Lehideux that they were considering the liquidation of SAF to avoid the emergence of an embarrassing position in discussion with its bankers. Lehideux 'erupted like a volcano' and 'screamed himself hoarse'—but then turned, like a good collaborator, to trying to negotiate good terms for himself in such a situation.[92] Lehideux was partly pacified, but he remained a dangerous loose cannon in the eyes of Dearborn. Even more awkward was the response of the US government to the news of a possible withdrawal. They let it be known in no uncertain terms that this would be highly

[89] Howard, 'Memorandum on François Lehideux', 10 May 1949; Arthur J. Wieland, 'Report on conversation with Lehideux', 27 Nov. 1950 (FIA, AR 65-71, Box 30).

[90] Wieland, 'Memo on Ford SAF', 23 May 1951; Lazard Frères to Breech, 15 Jan. 1951 (FIA, AR 65-71, Box 30).

[91] Matthias Kipping, ' "Operation Impact": Converting European Employers to the American Creed', in Kipping and Bjarnar, *Americanisation of European Business*.

[92] Bogdan to Wieland, 20 Aug. 1952 (FIA, AR 65-71, Box 30).

unwelcome in Franco-American relations, and that Washington might be 'on the other side of the fence' in such a situation.[93] The French banks and government had generally been strongly supportive of Ford's presence in the past, and faced with the depth of Ford's current financial problems, they now urged Ford to take a bigger role and 'assume almost dictatorial powers'.[94]

From 1945 to 1952, Ford had deliberately resisted putting in American managers or increasing its control and involvement. But, faced with the combined problems of Lehideux, Washington, and the banks, Ford once again backed away from liquidation. This time they decided that they would have to stabilize the situation more fully before they could extricate themselves. Ford now recognized that this would require a 'reorientation to the USA' and closer liaison with the parent company. It would require more 'bluff' and 'serenity' to successfully disengage.[95] Accordingly, Dearborn sent in a senior Detroit manager, Jack Reith, to smarten the company up for sale (ideally to Citroën) over a two-to-three-year period, and to 'steer Lehideux to the point where he would submit his resignation',[96] something Reith successfully accomplished within a year.

In an expanding market, Reith oversaw a sales revival of a facelifted Vedette, and even generated a cash flow substantial enough to start planning the long-resisted small car. However, Reith did not lose sight of his main objective, and on the back of improved results he seized the moment to sell the company to Simca in 1954, surprising observers who believed that Ford was just turning the corner in France, and winning a hero's reception on his return to Dearborn for extricating Ford from its French morass.[97]

The post-war performance of Simca, the company that acquired Ford's operations, suggests certain opportunities missed by Ford. In the late 1940s, Simca drew strongly and directly on American production technology and expertise at its Nanterre plant. In the early 1950s, as Citroën and Renault delayed bringing out new cars, Simca moved boldly to launch its own modern small car, the Aronde, which briefly became the number-one seller in France, indicating that an aggressive competitor had the scope to capture market share. The purchase of Poissy enabled it to escape from its cramped site at Nanterre and expand its output and model range. In 1958 Simca sold more cars than any French producer except Renault.[98]

Contrary to some views, central government planning did not freeze the structure of the French auto industry and confine Ford or other firms to strictly

---

[93] Bogdan to Wieland, 20 Aug. 1952 (FIA, AR 65–71, Box 30).

[94] Wieland to Bogdan, 8 Sept. 1952; 'Notes on Today's Conversation with Mr. Chevrier' (head of the syndicate of Ford's bankers), 16 Sept. 1952 (FIA, AR 65–71, Box 30).

[95] Wieland to Bogdan, 8 Sept. 1952 (FIA, AR 65–71, Box 30).

[96] Bogdan, 'Notes', 16 Sept. 1952 (FIA, AR 65–71, Box 30); Byrne, *Whiz Kids*, 199.

[97] J. R. Davis, 'Memo on the Poissy Situation', 23 Apr. 1953 (FIA, AR 65–71, Box 30); Wilkins and Hill, *American Business Abroad*, 394–5; Byrne, *Whiz Kids*, 210.

[98] Patrick Fridenson, 'L'Industrie automobile: la primauté du marché', *Historiens et Geographes* 361 (1998), 236; id., 'L'Industrie automobile française et le Plan Marshall', in Rene Girault and Maurice Lévy-Leboyer (eds.), *Le Plan Marshall et le relèvement économique de l'Europe* (Paris: Comité pour l'histoire economique et financière de la France, 1993), 285.

predetermined segments. Renault, Citroën, and Simca, all departed sharply from their initially assigned roles. Rather, Ford in France lacked both will and confidence in its own model. It no longer understood how to create mass motorization. It lacked a commitment to the French market. It paid lip-service to the 'American creed', but refused to put in resources to apply that creed within its own organization. Having stayed in France for twenty years with almost no profits, Ford left just as France was about to become one of the fastest growing markets in Europe, and, a few years later, a crucial gateway to the Common Market.

In contrast, it was Renault and Citroën that seized the real opportunities to 'Americanize' the French car market. Fridenson and Loubet have shown how these companies drew inspiration from older pre-war American models of mass production and economies of scale as a path to mass motorization. But they also creatively reworked these ideas in conjunction with new and distinctively French elements, related to a new paradigm of small functionally efficient cars at low cost, and then diversified production as incomes rose and the market evolved. Fridenson has described elements of both 'Fordism' and 'Sloanism' at Renault, and as Loubet put it: 'If Fordism is mass production of a single model for the largest possible market using rational organization of labour, Pierre Lefaucheux is perhaps the French Ford that André Citroën had hoped to become',[99] and, one could add, which Ford of France did not know how to be.

Germany: Reluctant Americanization, 1945–1962

Both Ford and GM were slow to recommit themselves to operations in Germany after the war.[100] Opel lost its Brandenburg truck plant and the blueprints and tooling for the Kadett as reparations to the Russians, and GM had serious doubts about the future of the German market and Opel's financial position. As Sloan put it, 'we were somewhat in the air about resuming control at Opel'[101] and top management disputed and vacillated over the issue for almost four years before finally resuming full control and responsibility.

Ford's German operations suffered much less damage, and they enjoyed privileged assistance from the British occupation authorities in restarting operations. But, unlike the fully integrated Opel factories, Ford's Cologne plant was no more than an assembly operation, depending heavily on weak outside suppliers. Ford's own internal problems in the United States absorbed nearly all its will and attention, and indecision and hesitation were largely responsible for allowing the company to incur huge financial losses during the currency conversion of 1948. Dearborn's main goal was not reconstruction, but the pursuit of a possible purchase of Volkswagen as a way of transforming Ford's position.

[99] Loubet, *Citroën, Peugeot, Renault*, 49.
[100] Steven Tolliday, 'Enterprise and State in the West German *Wirtschaftswunder*: Volkswagen and the Automobile Industry, 1939–1962', *Business History Review* 69, 3 (1995), 301–8.
[101] Sloan, *My Years*, 331.

Only when the proposed merger fell through did Ford rather reluctantly turn its attention to the other needs of its German subsidiary.

This indecision and delay by the two American multinationals, as I have shown elsewhere, created a vacuum which Volkswagen exploited to quickly consolidate a position as the market leader.[102] Ford at first took a very negative view of German prospects and continued to actively consider withdrawal until 1951. It limited its ambitions simply to 'produce an acceptable unit in Germany without the expenditure of a great amount of money'.[103] Dearborn quickly decided *not* to make a low-price car to challenge VW since it had neither a suitable product nor adequate manufacturing facilities (84 per cent of the cost of Ford's German vehicles was bought in). Instead, it relied on facelifting the prewar Taunus which it hoped would suffice against other competitors, since, except for VW, 'we are all old-fashioned together'. Meanwhile, Dearborn remained absolutely opposed to putting any new money into the subsidiary and relied on 'buy-as-you-go' investments in incremental plant and product improvements.[104]

Local management argued the case for expansion, new models, and new capacity for a rising market, but Dearborn took a resolutely pessimistic view. In 1953, a study concluded: 'auto production in Germany is not likely to increase drastically from now on . . . Our own opinion is that the German economy will have to slow down and that the levelling off may come sooner and be at a lower level than predicted in the [local management] study'.[105] Accordingly, Dearborn focused on 'a program of survival' with the aim of 'Stay No. 4 in Germany'.[106]

By the mid-1950s, German demand was soaring and cars of any sort were eagerly snapped up by buyers. Ford's lack of investment resulted in a serious lack of capacity, and, despite stretching its plant to the limit through long hours of overtime and Sunday working, Ford did not have enough cars to sell, and its market share tumbled from 10 per cent in 1952 to 7 per cent in 1956. Ford made big losses in Europe's fastest growing market. It lost fourth place to the Bremen Group and barely kept its nose in front of Auto-Union/DKW. By 1956 it was evident that even 'staying fourth' would require a doubling of capacity and an all-new model. But Dearborn felt it did not have the managerial or technical capacity to handle such developments in Germany. Instead, its conclusion was, 'we are too far behind the parade to go it alone', and a further fruitless search for a merger partner followed, vainly reinspecting the inadequate facilities of Borgward, Auto-Union, and Goggomobil.[107]

---

[102] Tolliday, 'Enterprise and State', 308–50.

[103] Wieland to Howard, 19 Aug. 1949 (FIA, AR 65-71, Box 30).

[104] 'A Program for a New Passenger Car to be Produced at Ford Werke AG, Cologne', 24 Feb. 1950 (FIA, AR 65-71, Box 30); A. J. Wieland, 'Report on Germany', 24 Jan. 1950 (FIA, AR 67-13, Box 1); Wieland, 'Notes on Germany', Dec. 1949 (FIA, AR 67-13, Box 2).

[105] J. W. Sundelson, 'Comments on Cologne's Integration Program,' 29 May 1953 (FIA, AR 75-62-616, Box 78).

[106] Bogdan to Wieland, 13 May 1953 (FIA, AR 75-62-616, Box 78).

[107] 'Ford Cologne', May 1956 (FIA, AR 67-13, Box 1).

Given Ford's debilitating dependence on outside suppliers, increased vertical integration was a precondition for more effective manufacturing. In the mid-1950s, local management had won some investment in press shops and stamping, but this did little more than prop up a deteriorating position. Local management also sought a more flexible approach to new models designed specifically for German needs, or greater autonomy for the German subsidiary. They complained that: 'too little consideration is given by the Americans that business can be done differently and still with the same success, as it is done in the States'. The examples of Renault and Fiat showed that independent approaches without support from the USA could be highly successful. 'So why not do in Rome as the Romans do?'[108]

Dearborn had no time for this. It was frankly contemptuous and mistrustful of local German managers, a feuding duumvirate comprising a wartime anti-Nazi, coupled, on Ford's insistence, with a leading former Nazi with an unpleasant war record.[109] But Dearborn was unwilling to Americanize management. While disparaging Cologne's engineering practice and financial controls, it confined American aid to occasional engineering visitations or financial inquisitions, and refused to put in substantial American management resources, even though it recognized that, 'unless we assume the burden of making our own investigation and analysis on the ground, I see no choice, for better or worse, but to accept as valid a large area of opinion developed by the Cologne management'.[110]

By 1958, the situation had become critical. By its own analysis, unless Ford could grow, it would be squeezed out and suffer 'the same fate as the Independents' in the USA. This took on a broader significance for Ford since, as it had withdrawn from France, the loss of Germany would leave Ford with no effective manufacturing capacity within the Common Market, an area which it predicted would be a key growth area in the coming years.[111] In face of this crisis, Ford finally decided to take full control of the situation and 'Americanize' top management, putting in a new president, John S. Andrews, along with several American directors and staff men. Andrews at once took steps to overhaul management methods, including cost controls, labour standards, design-for-manufacture, and vendor-assistance to stem the losses. But the fundamental problems of product and plant were more intractable.

Ford-Germany's key car was the Taunus 12M (and its 15M derivative) which were basically heavily facelifted versions of its pre-war car and hopelessly out of date. It quickly became clear to Andrews, as it had been clear to local management for a long time, that only a fully competitive successor to the 12M (a

[108] R. H. Schmidt to John S. Andrews, 16 Apr. 1958 (FIA, AR 65-71, Box 31).
[109] e.g. Bogdan to Wieland, 23 Apr. 1953 (FIA, AR 75-62-616); on former Nazi Robert Schmidt see Peter Lessman, 'Ford Paris Under the Sway of Ford Cologne in 1943', *German Yearbook of Business History* (1994).
[110] Bogdan to Wieland, 23 Apr. 1953 (FIA, AR 75-62-616).
[111] 'Cologne Facilities Expansion Program', 29 Apr. 1958 (FIA, AR 65-71, Box 31).

C-class car) could hope to secure a future for Ford in Germany. But how could Cologne develop a 12M replacement? It would have to compete head-to-head with the most popular cars in Europe like the VW and the Renault Dauphine. But developing such a car would be expensive and difficult, and would have no guarantee of big sales, even if the company were prepared to 'buy market share' by accepting large losses for several years. The engine and design of the new car would be wholly conventional and without new features, and there were serious doubts about whether Cologne could handle the development and engineering of even a conventional car, or bring it to market within a reasonable time frame. The creation of a car for the German market derived from the British Anglia/Prefect line was ruled out because of different styling, size, and power-to-weight ratios that made it incompatible with the way that the German market was segmented.[112] The result was an impasse and a full-scale survival crisis for Ford-Germany by 1958–9.

Ford-Germany escaped from this situation mainly through unparalleled injections of resources from the parent company in the following three years. But this striking change of strategy was not primarily driven by the needs of the ailing subsidiary. Rather, Ford's American priorities, partly fortuitously, created a temporary congruence of interest between developments in Europe and in America which helped the German company to escape from a potentially terminal crisis.

In the late 1950s, Ford had taken a battering in the US market from the unexpected success of the VW Beetle. Ford had to grapple with the idea, which it had long resisted, of producing a small car for its home market, despite serious scepticism about the potential profitability of such a car.[113] Dearborn considered building such a car in the USA, and also investigated importing a car from Britain or Germany. A European-built car would have the advantage that, even if it did not sell well in the USA, its base of European sales would probably ensure adequate overall volumes. Unfortunately, no suitable European car existed. The British Anglia was old-fashioned and, though it had sold some respectable numbers in the USA, was not a potential 'Beetle-fighter', and the Taunus was outdated. Moreover, Ford was wary of building European capacity that would depend heavily on the US market. Until 1958, Dearborn believed that *if* a small car was required for the US market, it should be specifically designed for the US market and built in the USA, possibly sourcing certain low-cost components from Europe.[114]

---

[112] 'Forward Product and Facility Proposal. Executive Committee FMC', 12 June 1959 (FIA, AR 65-71, Box 31); Carl F. Levy (Ford-International), 'Company-Wide Inter-relationships of Forward Car Programs', 3 Sept. 1959 (FIA, AR 67-14, Box 2).

[113] Product Planning Office (Ford-Division), 'Economy Car Report', 13 Nov. 1957 (FIA, AR 67-7, Box 1).

[114] Fred G. Secrest, 'Alternative Plans for marketing a Small Car in the United States', 17 Sept. 1957; J. W. Sundelson and C. Barion, 'Why FMC Should Consider Manufacturing Any Small Car for the US Market in Europe', 11 Oct. 1957 (FIA, AR 67-7, Box 2).

But, during 1958–9, detailed cost studies by Robert McNamara's product planners in Ford-Division challenged this orthodoxy. They showed that Europe, and Germany in particular, enjoyed definitive cost advantages in engine production. Indeed, McNamara concluded, a small car for the US market could *only* be made profitably if it were to be primarily sourced from Germany.[115]

By 1959, the cost comparisons for the US small car and the imperatives of the German crisis fused together in the 'Cardinal' project, for a small car to be launched simultaneously in the USA and Germany in 1962. Dearborn had rejected proposals for a C-class car for the German and European market alone. But if the German car piggy-backed on the new American small car, the economics of the project would be transformed. The 'Cardinal' would be designed and developed in Detroit. Its engine and powertrain would be primarily sourced from Germany, and it would be assembled at Cologne and at Louisville in the USA. It would be based on up-to-date front-wheel drive packaging (the PonyPac), which was familiar territory for European leaders such as VW, British Motor Corporation (BMC), and Citroën, but which was still largely unknown territory for Detroit. This was a risky technical exercise for the Dearborn engineers. Many in Ford feared that they could not design and build such a car at costs comparable to those of leading European firms, and that their imitative catch-up car could not justify being sold at a higher price than the established European competitors. However, American sales would offer good economies of scale which could offset high design and building costs. Detroit would get low-cost production and Cologne would get a new car and a new engine.[116]

The project was agreed in April 1960, with the car to be launched in summer 1962, based on an investment of $43m., including $14m. in Cologne. In the event, the American launch of the Cardinal was finally aborted at the last moment. But by then, nearly all the development and tooling had been done, and the car was launched as planned in Germany as the new 12M. The Cardinal project provided the backbone of recovery for Ford-Germany. US Ford threw immense effort into the development of German technical and production capabilities, since the resulting car was to be sold in US markets. Short-term needs for intermediate product facelifts and developments were underwritten from Detroit, and substantial investments in modern facilities and American tooling were made. This injection of support enabled Ford Germany to take advantage of remarkable growth in the German market, which doubled between 1958 and 1962. On the back of the 12M Ford increased its share of the German market from a meagre 8 per cent in 1958 to nearly 20 per cent in 1965. Ford International

[115] John S. Andrews to Tom Lilley, '1400lb Car Progress Report', 18 March 1959 (FIA, AR 67-7, Box 2); Ford-Division Purchasing Office, 'Progress Report on Procurement of Production Parts from Foreign Sources', 21 Sept. 1959 (FIA, AR 67-7, Box 1); T. O. Yntema, 'Renault Engine Study', 6 Nov. 1959 (FIA, AR 67-7, Box 2).

[116] Tom Lilley, 'Report on Market Potential of Class C Car and Evaluation of Proposed Product Objectives', 24 Apr. 1959 (FIA, AR 67-7, Box 2); 'Worldwide Passenger Car Specification Study (= "Paper Study of 1400lb Car")', Mar. 1959 (FIA, AR 67-7, Box 2).

also put substantial resources into building up the marketing of Ford-Germany's exports within the Common Market.[117]

Thus, for more than a decade, Ford developed no clear strategy for Germany. Cologne was neglected and lacked physical, technical, and managerial investment. Ford-Germany itself did not have the strength to pull itself up by its bootstraps, and its weak local management did not have the autonomy to push ahead of Dearborn. The operation might well have expired after 1959. It did not do so, primarily because, almost by chance, it fitted with a major strategic need for the US operations. Even though the US side of the project was eventually aborted, the resources and capabilities that Dearborn injected into the company during this episode facilitated a belated revival.

The trajectory of GM and Opel was quite different. Despite some difficulties, Opel prospered and was a strong second to VW in Germany in the 1950s. After the loss of the Kadett to the Russians during the war, Opel did not make a costly effort to re-enter the low-price market against VW. Instead it accepted the situation and concentrated on higher-priced cars in the segment above the Beetle, notably the Olympia, restyled as the Rekord from 1953. Opel dominated this segment almost as much as VW dominated the segment below with the Beetle, and with VW producing no other model, Opel enjoyed a degree of comfort and good profits in this segment.[118] Despite wartime losses, Opel retained extensive, integrated manufacturing facilities, making more than 50 per cent of its vehicles in house, compared to only 18 per cent at Ford. This manufacturing integration kept costs down and made Opel a keen price competitor. Meanwhile, retained profits sustained comparatively high levels of investment. Between 1948 and 1961, Opel accounted for 23 per cent of new investment in the German automobile industry, compared to 34 per cent at VW and 9 per cent at Ford.[119]

Opel concentrated its investment on new stamping and body facilities at Russelsheim in the early 1950s, introducing some of the most advanced automation and transfer machines in Europe, and benefiting from excellent design for assembly. It did not pursue assembly automation, however, preferring conventional lines which could assemble a variety of models. Although it drew freely on American ideas and GM technical resources in these investments, Opel retained strong autonomous capabilities. It was headed by an American (Edward W. Zdunek) between 1948 and 1961, but Detroit did not seek tight control and Opel made few demands on the parent for investment capital. Opel was allowed substantial autonomy and developed a fairly independent role in collaboration with its parent. Model development drew heavily on American

[117] Steven Tolliday, 'American Multinationals and the Impact of the Common Market: Cars and Integrated Markets, 1954–67', in Franco Amatori, Andrea Colli, and Nicola Crepas (eds.), *Industrialization and Deindustrialization in Europe* (Milan: Franco Angeli, 1999).

[118] Klaus Brandhuber, *Die Insolvenz eines Familien-konzernes. Der wirtschaftliche Niedergang der Borgward-Gruppe* (Cologne: Muller Botermann Verlag, 1988), 90.

[119] Yntema to Breech, 16 Apr. 1952 (FIA, AR 65-71, Box 31); K. W. Busch, *Strukturwandlungen der westdeutschen Automobilindustrie* (Berlin: Duncker & Humblot, 1966), 93 and Table 38; Brandhuber, *Die Insolvenz*, 185.

styling and the inspiration of 1950s Chevrolets with their low-slung shapes and wraparound windscreens, and there was substantial technical and styling assistance from GM's Technical Center in Warren, Michigan. But German engineers, led by the experienced technical director, Karl Stief, dominated the technical side, and Opel's independent capabilities grew in the 1950s to such an extent that, when Opel decided to enter the C-class to compete directly against VW in the early 1960s, Opel staff under Hans Mersheimer played the leading role in styling and engineering, creating the new Kadett as a distinctively European car.[120] Although more research is needed, Opel seems to have operated as a resourceful German company, drawing selectively on support and resources from its US parent.

BRITAIN: RELUCTANT HYBRIDIZATION, 1945–1962

GM's relationship with its British subsidiary had certain features in common with the Opel story. But in the British case, Vauxhall lacked the local strengths and capabilities for independent initiative that Opel exhibited. The Detroit-inspired introduction of unitary construction in the 1930s had given Vauxhall an important technical advantage. However, GM committed no resources to the long-term development of the new technology, in effect isolating its European operations from its technical mainstream. While Opel had sufficient local engineering capabilities to at least partially carry forward its development, Vauxhall became something of a technical backwater.

Vauxhall's pre-war cars were still technically advanced in the late 1940s. But until the early 1960s Vauxhall produced nothing creative or new. In 1947 the small Vauxhall 10 was dropped, and Vauxhall did not re-enter small cars until the Viva in 1963, despite the fact that its unit-construction technology potentially offered its greatest advantages in the manufacture of smaller cars. Instead, Vauxhall produced a restricted range of medium to large cars using rigorous production rationalization. Between 1948 and 1957 it pursued a 'one-body policy', making just two cars (the four-cylinder Wyvern and the six-cylinder Velox) that shared a common body, gearbox, and brakes. This policy permitted extremely rationalized production, and, despite low annual volumes, Vauxhall achieved economies of scale by running its unchanged basic model over an extended period of time (often seen as the classic 'European' strategy of scale economies in this period). It retained only a small share of the British market during the 1950s (10 per cent in 1947 and 11 per cent in 1960), but it generated the highest profits per unit in the industry.[121]

[120] D. S. Harder and D. J. Davis, 'Notes on European Trip', 6 Sept. 1957 (FIA, AR 68-5, Box 3); Busch, *Strukturwandlungen*, 123; Ludvigsen, *Opel. Wheels to the World*, 66–70; Sedgwick, *Motor Car*, 126–7.

[121] £80 per vehicle in 1956 versus Ford, £45, BMC, £35 and Standard, £30 (D. G. Rhys, *The Motor Industry. An Economic Survey* (London: Butterworth, 1972), 363.

In part, Vauxhall adopted this policy because its adoption of unit construction cut it off from the mainstream of GM's engineering development. There was no commonality between Vauxhall and GM bodyshells, and Vauxhall could not benefit from spin-offs from major GM product developments. Moreover, the adoption of unit construction had locked Vauxhall into in-house production of its own bodies and constrained output.[122] In other respects, however, Vauxhall remained rather slavishly dependent on US engineering. It faithfully duplicated the Detroit basics (three-speed gearbox, mechanical pump-feed, spongy steering, and ultrasoft front suspension), producing cars that were staid and technically stagnant by European standards. Vauxhall sought to offset this by heavy reliance on styling changes to attract customers. Mostly these were superficial. One result was a rather odd conjunction of unadventurous engineering with garishly 'modern' American styling. Vauxhall's periodic restyles have been described as 'Atlantic drift', with their growing emphasis on chrome, wraparound windscreens (personally mandated by GM President Harlow Curtice in the mid-1950s), and tailfins.[123]

GM was content with Vauxhall's sluggish sales performance in a comfortable and profitable market niche. Detroit tightly controlled the basic lines of policy. In the 1940s, Vauxhall's British managing director, Charles Bartlett, disputed the 'one-body' strategy and wanted more model diversity. In response, the directors of GMOO (GM Overseas Operations), Ed Riley and Walter Hill, took a more direct grip on the company and increasingly marginalized Bartlett until his retirement in 1954. He was replaced by American managers, Walter Hill (1953–5) and Philip Copelin (1955–61).[124]

From 1955, GM President Harlow Curtice began to give greater prominence to overseas operations, and to press Copelin for expansion at Vauxhall, which would necessarily involve more models and more frequent model changes. This created great problems for Vauxhall, which had lost its ability to deal with change. The rather modest attempt to widen the model range with the introduction of the Victor in the 1954–7 expansion programme strained Vauxhall's engineering capabilities. Serious leak and corrosion problems on the new Victor badly damaged Vauxhall's reputation, and much time and money were spent in the late 1950s not only fixing the problems but also restoring Vauxhall's reputation.[125]

The regeneration of Vauxhall's engineering capabilities from the late 1950s was achieved not so much by 'Americanization' as by putting Vauxhall more

[122] GM did not agree to expansion of body capacity at Luton until 1954–7 when Vauxhall enlarged its range to two bodyshells, with the introduction of the Victor. Platt, *Addiction to Automobiles*, 165–6; Philip W. Copelin, 'Development and Organisation of Vauxhall Motors Limited', in R. S. Edwards and Harry Townsend (eds.), *Studies in Business Organisation* (London, Macmillan, 1961), 84.

[123] Georgano *et al.*, *Britain's Motor Industry*, 121; Copelin, 'Development and Organisation', 82. Trevor Alder, *Vauxhall, The Postwar Years* (Sparkford, Somerset: Foulis, 1991) calls them 'scaled down caricatures' of big American cars.

[124] Platt, *Addiction to Automobiles*, 148, 156–9.   [125] Platt, *Addiction to Automobiles*, 178.

under Opel's wing. The crucial move in Vauxhall's expansion was the creation of the 1963 Viva, Vauxhall's first all-new post-war car, which took it back into the small car segment for the first time since 1947. There was little US input into the Viva. Instead, it was derived from the Opel Kadett, launched the previous year, and as such was a first step towards Opel/Vauxhall rationalization, with Opel as the senior partner. The Viva was a highly conventional boxy front-engine rear-drive car, at first with a 'European' look and later with more American 'Coke-bottle' styling, but it achieved respectable sales against the Austin A40 and the Ford Anglia. The Opel leadership and derivation of Vauxhall models became increasingly the norm in the late 1960s and early 1970s.[126]

In contrast, Ford's pre-war pattern of tight control of British product and production strategy from Dearborn continued after the war. In 1947–8 Dagenham lobbied desperately for Dearborn to provide a new small car for what it predicted would be the largest and fastest-growing segment of the market. It wanted a low-weight, high-performance, four-cylinder car with a 94-inch wheelbase and 1200cc engine, rather than a continuation of the existing larger 'compromise' cars, which covered both 1200cc and 1500cc segments. But Dearborn refused, and, despite Dagenham's protests, discontinued all development work on a small car for England.[127] Instead, Dearborn opted to avoid intense competition in the small-car sector and to focus on larger higher margin products which it understood better.

The result was the highly successful Consul/Zephyr line of cars, launched in 1950. These cars were almost entirely developed and engineered in Dearborn. Dagenham had no role before the layout drawing stage, and even at the detailing stage was confined to work on the electrical systems, soft trim, and wooden body parts. Such extreme centralization was often counterproductive. Dearborn specified many items not easily available in Britain, and also isolated Ford from the government drive to increase component standardization in the car industry. Major items, like front sheet metal, had to be redesigned at a late stage to accommodate the only available British lamps. As a result, Dearborn concluded that, in future model developments, it would be essential to give Dagenham a larger role in body and chassis engineering.[128]

The new cars were 'well-balanced scale-downs of traditional Dearborn 3-box designs', but they involved important product innovations in the UK context, and were linked to the development of new production and assembly techniques. Extensive transfer machining was introduced for cylinder blocks and crankshafts for the new engines, and the assembly lines eliminated 'pit work' by

[126] Michael Allen, *British Family Cars of the Early Sixties* (London: Guild Publishing, 1989), 169; Michael Sedgwick, *Cars of the Fifties and Sixties* (Göteberg: Nordbok, 1983), 212; Donner, *Worldwide Industrial Enterprise*, 58; Alder, *Vauxhall*, 67, 92.

[127] V. Y. Tallberg, 'Memo on Telephone Conversation re: New English Car Models', 20 May 1948; Hennessy to Tallberg, 14 May 1948; '1200/1500/2250cc Engines: Report on Conference', 29 June 1948 (HFM, Acc. 480, Box 11).

[128] Tallberg, 'Design of English Car', 16 July 1948 (HFM, Acc. 480, Box 11).

moving cars along at different levels for ease of assembly. These product and production changes generated major price advantages over competitors in this class of car, and Ford quickly dominated the medium-priced sector, particularly because the product responses from Morris and Austin were laboured and old-fashioned. The Consul/Zephyr line became the backbone of Ford's profits in 1950s.[129]

Meanwhile, in its smaller cars, Ford stuck with barely facelifted versions of its pre-war Anglia and Prefect until 1953. These models lagged behind the technical development of rivals like the Morris Minor (launched in 1948), but continued to sell well because cars were in short supply and because Ford could offer low prices since all the development and tooling costs had long since been amortized. However, in the early 1950s, their market share began to tumble in face of more modern rivals. The Dearborn development studios under Walter Appel responded with a 'startling and complete' transformation and relaunch of the Anglia and Prefect in 1953.[130] Like the Consul/Zephyr line, the new cars were distinctively modern in style and based on unitary construction and MacPherson strut suspension, with dramatically improved mechanical features. According to one commentator, 'Apart from the model names, nothing was carried over from the old cars'.[131]

Yet, although the Anglia and Prefect were radical in some respects, this was not the whole story. Dearborn was very concerned with the profit implications of killing off the company's 'cash-cow' models, and undertaking a high level of investment in essentially low-margin cars. Logic pointed to a new overhead valve engine for these cars, in order to match the new standards in fuel economy, acceleration, and gearing achieved by Austin, Morris, and Standard. Ford certainly had the engineering capacity to do this. But it chose not to. Instead, it developed a new engine 'on the cheap'. Dearborn engineers ingeniously reworked the pre-war 1172cc sidevalve, which dated back to the Model C, achieving dramatic improvements in performance by careful redesign of every part of the engine. This enabled them to carry over a remarkable 70 per cent of the tooling from the old engine to the new, and achieve dramatic capital cost savings because the old tools 'have unquestionably been amortised several times over'.[132]

Even so, the investment costs were still high, profitability fell, and Ford could not match the low prices of the new cars from Austin and Morris. But Ford sus-

[129] Michael Allen, *Consul, Zephyr, Zodiac Executive. The Big Fifties Fords* (Croydon: Motor Racing Publications Ltd., 1983; 2nd edn., 1990), 11–18, 60–1, 104–12.

[130] W. D. Appel to Wieland, 13 June 1951; 'A Program for a Light Passenger Car for FMC Ltd., Dagenham', June 1951 (FIA, AR 75-62-616, Box 71); T. O. Yntema, 'Dagenham's Project Request for Development and Production of a New Light Car and Van', 12 Feb. 1953 (FIA, AR 65-71, Box 29).

[131] Turner, *Ford Popular*, 42, 53–63, 87, 104, 107.

[132] Appel to Olle Schjolin (Ford-Dagenham chief engineer), 7 Aug. 1951 (FIA, AR 75-62-616, Box 71); E. L. G. Robbins, 'Estimated Capital Expenditure for the New Small Car', 9 Sept. 1951 (FIA, AR 65-71, Box 29); Turner, *Ford Popular*, 153.

tained its overall profits in the following years by ingenious opportunism. Rather than scrapping the old Anglia, Ford completely stripped it down, reduced its price far below that of all other cars of its type, and offered it as the 'Popular',[133] which sold profitably in substantial volumes for three more years.

Thus Ford developed a distinctive strategy for the UK. High-margin larger vehicles and stripped-down small ones maintained profitability. Automation and mechanization were pursued, but on a more flexible basis than in the USA, with common machines being used for several variants of engines and major mechanical parts to offset lower volumes.[134] Meanwhile, in contrast to its pre-war approach, Ford used engineering capabilities rigorously to minimize costs through design and frugal adaptation to the demands of the local context—although it still deliberately steered clear of pursuit of a European mass-produced small car. Perhaps above all, however, Ford benefited from the plentiful space, integration, and layout that its Dagenham plant brought with it from the 1930s. In those years, the giant plant had suffered from costly overcapacity and had run at a substantial loss, but in the first post-war decade it provided a productive potential which none of its rivals enjoyed, and imposed few constraints on production. For example, it allowed Ford to keep a model like the Popular running almost as a 'free good' alongside its own replacement. Even so, however, Ford's pattern of relatively large capital investment in modern factory equipment for its new models was not necessarily wholly effective: in 1956, BMC held a 39 per cent market share with net assets of £62m., while Ford used more assets (£65m.) to take only 27 per cent of the market.[135]

Although Dearborn retained tight control over product and investment policy, it did devolve much wider responsibility to local managers than before the war. Dearborn was very impressed by the quality of Dagenham's top managers, the 'three knights' (Perry, Cooper, and Hennessy). But increased decentralization also stimulated more demands for a wider local role. Hennessy, in particular, objected to what he regarded as overfussy and onerous reporting procedures. He believed that many of the cost comparisons which Dearborn required were meaningless, and that excessively strict 'harmonization' of financial procedures (notably the strictures of the notorious Section 6, Chapter 30 of the FMC Controllers' Manual) were primarily designed to shackle him. He also chafed at Ford's rigid planning procedures based on formal forecasting tied to capital budgeting. In relation to export markets, he argued that: 'There is little point in constant market research, and we can only assume perhaps that what we lose on the swings will be balanced by what we gain on the roundabouts'. He

[133] All the items deleted from the obsolete Anglia were available as optional extras to bring the Popular back up to the old Anglia standard of equipment. For example, no traffic indicators (£2.75 extra); ashtrays (38p); interior light (42p) etc. Turner, *Ford Popular*, 68.

[134] On the relationship between US and British technology strategies, see also Jonathan Zeitlin, 'Reconciling Automation and Flexibility? Technology and Production in the Postwar British Motor Vehicle Industry', *Enterprise and Society* 1 (2000).

[135] George Maxcy and Aubrey Silberston, *The Motor Industry* (London: George Allen & Unwin, 1959), 117, 178.

was wary of overinvestment which could result from formal forecasting techniques. His preference, which may have reflected the resources of the Dagenham site, was to rely on conservative estimates, and then: 'If we can sell more, we have the capacity to meet demand, and existing facilities can be easily stretched with small additional investments'.[136]

Local management was able to increase its scope because Ford did little to develop its apparatus of control over its overseas operations before the mid-1950s. After a brief and unsuccessful flirtation with the creation of a much stronger International Division in the late 1940s, Ford-International, though still technically a division, was demoted to subordinate status, and was often bypassed by Dearborn in dealings with the major European subsidiaries. Key product and investment decisions were made by top management in Dearborn, while Ford-International played a liaison role. Ford-International lacked effective manufacturing or engineering staff of its own, and between 1950 and 1956, its chief manufacturing officer, J. J. Welker, dealt with European manufacturing matters on a part-time basis, travelling to Europe when necessary, while permanent American manufacturing staff were not stationed in Europe.[137]

Until 1955 it was possible to 'get by' in this way because little was being done abroad in terms of expansion, but by this time, Dearborn was becoming aware of the need for much bigger forward programmes in Europe. In this context, Welker argued that divisionalization now needed to be taken seriously in international matters: 'If we agree that these plants cannot be properly operated from the US, then the only possible solution is organization at the local level following the well-known principle of centralized policy and decentralized operation.' This would involve both expanded full-time staff at Ford-International and permanent US executives with more status and authority, including manufacturing personnel, in Europe.[138]

The forward product programmes of the mid-1950s had to deal with the continued rapid expansion of European demand, which was now making even Dagenham's capacity inadequate. Ford would require major new plant and new models in Britain and Germany within the next five years, and the related planning opened up a succession of important issues about products, international structures of control, European strategy, and the relations between national subsidiaries and their product programmes.

The key issue which crystallized discussions was the replacement programme for the Anglia/Prefect which, following a normal five-year product cycle, would be due in 1958–9, and which would have to be linked to major capacity expansion. Dagenham, it will be recalled, had not initially favoured the product configuration that Dearborn decided on in the late 1940s, but had by now become wedded to it. The Anglia/Prefect price-model configuration fitted well with the

[136] Hennessy to Wieland, 30 Apr. 1953 (FIA, AR 75-62-616, Box 71); G. E. Altmansberger (director, budget and costs) to Breech, 18 Mar. 1949; 'Comparative Cost Exhibits' (FIA, AR 65-71, Box 28).
[137] J. J. Welker to Wieland, 8 May 1956 (FIA, AR 78-13, Box 2).   [138] Ibid.

distinctive pattern of market segmentation that had developed in post-war Britain. It was a product package that precisely targeted its British competitor models from BMC and Hillman, and Dagenham believed that the replacement model (105E) needed to continue to fulfil this function. A secondary aim would be to compete with VW or the Opel Rekord in export markets. But this, Dagenham argued, should not primarily shape the design of the car. Dagenham was prepared to accept relatively low rates of export (about 40 per cent of its output) and to live with heavy dependence on the home market.

The Anglia/Prefect package was not ideally suited to continental markets. It did not directly target European low-price cars (Renault 4CV, Fiat 600, Lloyd, etc.), which were much smaller and strictly frugal, nor did it meet the more powerful and better equipped 1200–1500cc range at higher prices, such as the VW and the Rekord. Instead, it sold at a price close to that of the small European cars, but was larger and more spacious. This took it close to the VW and Rekord in size, but left it underequipped and underpowered in comparison. Its size/power combination was effective in Britain and a number of Commonwealth export markets, but in Europe, where **either** smaller **or** more powerful cars were required, it had underperformed. Moreover, because it was a larger but underpowered car selling in a small-car price segment, its profit margins had been quite narrow.[139]

In 1956–7, Dearborn raised the question of whether this sort of product package could prosper in changing European conditions. Important figures in Ford-International believed that, in the future, the necessary economies of scale for new models would require a sales base much larger than any national market could offer, and they also doubted if, in any case, the relative isolation of the British market would last. J. W. Sundelson of Ford-International argued that free trade developments in Europe, associated with the emerging Common Market, were likely to become of great importance over the next five-year product cycle.[140] A rapid opening up of European trade would leave Dagenham exposed: 'It would not appear wise to assume that a product that might not stand up to the VW or the Dauphine should believe that these and other hot cars would not also come into the hitherto protected UK domestic market as well. The trend is inevitable, the timing is vague.'[141] Dagenham, in contrast, maintained that free trade developments in Europe in 1956–7 were of marginal importance, and should not distract from the primary importance of having the right product for the British market.

Dearborn, however, pressed the issue of whether Dagenham should break from its UK market focus and instead develop a pan-European car based on mass production and economies of scale. In December 1956, it proposed that Dagenham should postpone its proposed replacement of the Anglia/Prefect (the

---

[139] J. Wilner Sundelson to Tom Lilley, 26 Nov. 1956 (FIA, AR 75-63-430, Box 67).

[140] Sundelson, 'A Review of Dagenham Presentation Materials', 12 June 1957 (FIA, AR 75-63-430, Box 67).

[141] Sundelson to Andrews, 28 Mar. 1957 (FIA, AR 75-63-430, Box 67).

105E) in 1958/9 and redesign it as a 1500cc VW-class car for launch in 1962. This would involve a leap to a wholly new 'dream car' based on front-wheel drive and which would be fully competitive with its leading European rivals in specifications and price. In conjunction, Dagenham might either drop out of the low-price segment altogether, or develop a genuinely small 1000cc New Light Car for October 1958, which could challenge the Fiat 600, Renault 4CV, or the Lloyd. In December 1956, Dearborn released funding for long lead-time tooling for such a car.[142]

Dagenham was ready to accept development of a 'new light car' as an additional model, though it warned that this would require high levels of transatlantic co-operation and might be hard to achieve. But it strongly resisted cancelling the 105E. It argued that delaying an Anglia/Prefect replacement pending the preparation of a 'dream car in 1962' would seriously damage its position in the British market. For key figures in Ford-International, however, this was a hopelessly compromised position. Dagenham's approach, they concluded, was 'the dental school of product planning', the main criterion of which was to fill every gap or cavity in the market. Instead, 'the merits of a real economy of scale versus the dental approach' should now be the focus.[143]

But Ford-International lacked power and determination on these issues. These discussions came two years before McNamara's cost studies in the US Ford-Division opened up the prospect of developing a European small car with the American market as a primary target. In the absence of weighty backing or a product champion for the 'dream car' or the 'new light car' in the top echelons of the company, Dagenham was able to fight off these proposals and reinstate the already well-advanced 105E programme.

Dagenham was able to effectively fight its own corner for the 105E project partly because it had substantially advanced its own engineering capabilities and could now handle such a project with quite limited American input. In particular, since 1953, its R&D and product planning group under Terence Beckett had extensively recruited engineering graduates, and become a very capable unit.[144] This not only increased Dagenham's autonomy, but also deepened its commitment to a distinctive technological trajectory. In engines, resources were focused on the development of a linked series of in-line overhead valve engines ('Kent engines') with an innovative approach to commonization which enabled a wide variety of engines to be made on the same transfer machines. In drive configurations, the focus was on front-engine rear-wheel drive layouts. Such a line of development was radically counterposed in engineering terms to the front-wheel drive and V4 engine configuration (the 'PonyPac') that Ford settled on for the Cardinal.[145]

---

[142] Sundelson to John Smith (assistant managing director, Dagenham), 4 Dec. 1956 (FIA, AR 78-13, Box 2); Sundelson to Andrews, 28 Mar. 1957 (FIA, AR 75-63-430, Box 67).

[143] Sundelson to Andrews, 19 Mar. 1957 (FIA, AR 75-63-430, Box 65).

[144] Jonathan Wood, The Ford Cortina (London: Osprey, 1984), 22.

[145] Cardinal Program Project Officers, 'Ford-Germany C-car report', 25 Apr. 1960 (FIA, AR 67-14, Box 2).

Within two years, however, Dearborn's plans for the Cardinal highlighted the potential of linking European and American markets around a front-wheel drive package. It also underlined the isolation of the British 105E development and its inability to provide a platform for the next generation of small European cars. Once the Cardinal project was under way in 1960, the thoughts of Ford International turned again to drawing Dagenham into this stream of development.

Hennessy's response was not only to resist fiercely such collaboration, but also to initiate a British *alternative* to the Cardinal, based on Dagenham's conventional layout, which would be developed very rapidly to beat or meet the launch date of the Cardinal. Dagenham pointedly adopted the code-name 'Archbishop' for its response to the Cardinal.[146] Dagenham's initial idea was to replace the Anglia with a conventional C-class car, with the old Anglia being stripped down to sell as a Class B car once the 'Archbishop' was launched.[147] But, during the process of development, the concept took an innovative turn. The Cortina (as the car was finally known) was positioned as a car that spanned the C-class and larger D-class segments. It offered the performance and space of a medium-sized family saloon at prices previously associated with much smaller cars, and it combined the higher margins of the larger car with the higher volumes and production economies of the medium-sized car. In conjunction with the panache of its marketing and image, the Cortina created a new segment in the British car market and became the fastest-selling car in British auto history.[148]

But the history of the product planning and engineering behind the Cortina also reflected the changing dynamics and balance of forces within Ford's international organization. First, the development of the Cortina represented the creative application of Dearborn methods to an English problem under English leadership, and the evasion of Ford International's belated turn to more 'Americanizing' ideas, such as the quest for common mass-production platforms for European sales.

The Cortina was *not* a mechanically innovative car. As Terence Beckett put it, the Cortina was 'in overall specifications a rather ordinary motor car'. Less charitably, according to one critic, the Cortina was 'without an original thought in its conception' and 'pioneered nothing yet sold to everyone'.[149] It had none of the engineering innovations that characterized the rival BMC 1100, a technically advanced car of engineering excellence, which was launched at the same time, with front-wheel drive, transverse-mounted engine, disc brakes, hydrolastic suspension, Pininfarina styling, and advanced internal packaging.[150] Rather, the

---

[146] Dagenham Product Staff, ' "Archbishop" Car Study', July 1960 (FIA, AR 67-14, Box 2); Wood, *Cortina*, 32.

[147] See the diagrams reproduced in Wood, *Cortina*, 35, from the Product Planning Committee Document, Jan. 1961.

[148] Wood, *Cortina*, 39, 94, 45, 22; Allen, *British Family Cars*, 72–5.

[149] Beckett quoted in Wood, *Cortina*, 60–2; Sedgwick, *Cars of the Fifties and Sixties*, 50, 205.

[150] The 1100 outsold the Cortina in the UK, but the Cortina did better in exports. Kenneth Ullyett, *The 1100 Companion* (London: Stanley Paul & Co., 1967).

Cortina's achievement was to create an innovative package out of conventional no-nonsense mechanics and cost-reducing engineering.

Its product planning team, headed by old Etonian Hamish Orr-Ewing, vigorously applied Dearborn methods in competitive rivalry with the American/German Cardinal project. The key to the Cortina's success was dramatic weight reduction through an innovative bodyshell. This was achieved through strict implementation of Dearborn's 'Red Book' planning procedures in product development controls, and extensive use of (then unfamiliar) concurrent engineering practices.[151] Using aircraft stressing techniques, Dagenham's designers took out 150 lb. of weight and 20 per cent of parts from the Consul Classic body that they used as their base.[152] Meanwhile, Dearborn's input was minimized. Although Dagenham's design team was now supervised by Roy Brown, formerly Ford's chief stylist for the Edsel, Detroit's direct influence was largely confined to sheet metal styling. But above all, the Cortina was designed and developed with remarkable speed. The clay model was approved in November 1960 only nine months after design started and the car duly caught up with the Cardinal and was launched in September 1962.

As a car, the British Cortina had a much greater impact than the Cardinal. It created a new market segment, it linked product innovation to low-cost production through effective design for manufacture, and it consolidated Dagenham's mastery of its rear-wheel drive in-line engine configurations that were enormously successful in the 1960s. In contrast, the Cardinal was a guinea-pig for front-wheel drive for Ford. But Ford did not go on to develop and build on the Cardinal technology. The system was never used in any other model, and it was allowed to die in 1969. Nor was the Cardinal a great sales success. It sold well and contributed to the dramatic revival of Ford-Germany, but the old-fashioned D-car which was 'refreshed' as a stopgap alongside it actually matched its sales in a market clamouring desperately for vehicles of any sort. Instead, the real significance of the Cardinal project was the injection of technical and engineering capabilities into an organization that previously had had little prospect of developing them.[153]

Thus, by the early 1960s, Ford of Britain had developed significant autonomy and a will of its own, symbolized by the Cortina. Its long encounter with American engineering and Ford methods had permitted the creation of a powerful 'hybrid': a British organization that could use its capabilities at times against the wishes of the US parent. By the time that Ford-International began to look beyond planning on the basis of the confines of national markets, Ford of Britain had developed its own conception of its interests and an ability to

[151] H. E. Lewis (product timing manager Ford-Division), 'Review of Ford-England Timing Plans', 27 Dec. 1960 (FIA, AR 67-14, Box 2).
[152] Dagenham Product Staff, '"Archbishop" Car Study', July 1960 (FIA, AR 67-14, Box 2); Wood, *Cortina*, 39 shows diagrams comparing Cortina and Classic bodies.
[153] Cologne engineers went on to effectively develop their own V-engine series. Sedgwick, *Cars of the Fifties and Sixties*, 212; for sales data, see Hans-Peter Rosellen, *Ford-Schritte: der Wiederaufstieg der Ford-Werke Koln von 1945 bis 1970* (Frankfurt: Zyklam Verlag, 1988).

defend them. The Cardinal project revived another European subsidiary that had appeared dead in the water. But Ford lacked the intense and powerful commitment to the Cardinal and its technology to carry it over into being the core of a broader European strategy. Instead, Germany emerged as a rival national centre to Britain within Ford, while broader notions about pan-European mass production of common models, which Dearborn had started to put on the agenda in the mid-1950s, remained for the time being on the sidelines.

3.4 Conclusions

Both before and after the Second World War, Ford and GM possessed exceptionally powerful resources and capabilities. On those occasions when they brought those capabilities to bear in Europe in an effectively targeted fashion (such as the Model Y, unitary construction, the Zephyr/Zodiac line, or the Cardinal project) they could achieve impressive results. But often there was a large gap between their potential and their ability to apply it effectively. In Britain, France, and Germany, it was arguably the domestic companies, such as Citroën, Renault, VW, Austin, and Morris, that were better able to draw selectively upon American experience and apply it in the European context. Moreover, during the 1950s, European companies began to develop an alternative paradigm of mass motorization based on smaller and lighter vehicles, fuel economy, and the pursuit of low running costs through engineering innovation, to which the American companies were slow to respond.

Instead, for long periods, the US producers seemed unable to apply their strengths creatively to the full. Ford, as we have seen, was at times a slave to its own preconceptions of product and production in the car industry. Under Henry Ford before the war, it frequently took evangelistic or even impossibilist stances on model and production policy, at times only being saved from the consequences of its own folly by the pressure of local managers or the application of its own deep reserves to produce improvized solutions *in extremis*. After the war, it retreated from these extreme positions, but struggled to devise a more appropriate stance for its European operations. In France, it was left with little more than a rhetoric of Americanization used to cover retreat. In Germany, it eventually pulled out of thirty years of drift only because certain of its needs in America rather fortuitously coincided with those of its struggling German subsidiary. In England, local management and engineering gradually emerged from under the control and tutelage of Dearborn to become a learning partner with significant powers of initiative of its own. In a sense this created a robust hybrid organization, combining elements of good American and British practice (albeit in somewhat unplanned and conflictual ways). But it also created an organization with a mind of its own, which posed problems for Ford when, in the late 1950s and early 1960s, it began to try to puzzle out new ways to integrate and develop its international organization.

GM's approach to product and production strategy, and to organizational structures, was very different. Before the war, as a result of internal indecision and its own traditions of decentralization and growth through acquisition, GM bought and ran Opel and Vauxhall as fairly autonomous European companies, yet also found pragmatic and effective ways to inject American capabilities and resources into them. After the war, however, the results were more mixed. Vauxhall was allowed to stray into a backwater with little purposive development for a considerable time. However, at Opel, a strong German-based organization flourished, drawing on inputs of American technology and organization, but not depending on them.

The divergence between the Ford and GM histories poses serious problems for over-simple notions of an 'American model'. Clearly, the two companies were pursuing radically different strategies and different models of product, production, and organization. Were they, nevertheless, sufficiently similar to be described under a common rubric of an 'American model'? In certain respects, both companies exhibited elements of pervasive 'American' ideal types that sprang to the mind of contemporaries, involving notions of mass production, standardization, automation, deskilling, and systematic management. On the other hand, the application of these general notions resulted in contrasting practices, and in both companies the 'core model' of product and strategy changed substantially over time.

What constituted 'Americanization' in this context? Was it a single-model policy, or a full-range policy of diversification (Fordism or Sloanism)? Did it involve cheap cars for the masses, or high value-added vehicles for the affluent? (Both companies rose by making cheap cars to diffuse mass motorization but then moved on to other priorities.) Did it imply the priority of cost reduction, standardization, and scale economies: or the pursuit of power, comfort, and stylistic diversity? Should the technological trajectory focus on a pursuit of innovative capabilities, or on the spread of a dominant US technical configuration or paradigm? Should the organizational trajectory imply increasing control and conformity with US headquarters, or should it involve the creation of high-level technical and managerial competences in subsidiaries—which could then lead to autonomy or a departure from US practices? At a more detailed level, the relationship between specific practices and the notional models could be highly ambiguous. For example, was Vauxhall in the 1950s conforming to American-style standardization and a single-model policy, or was it engaged in niche retreat? And Ford continued to be perceived in Europe as a model of effective hierarchical and bureaucratic organization, even while Henry Ford was internally providing advanced lessons in the art of dis-organization.

All of this should make the historian wary of over-hasty assumptions about American 'ownership advantages' and their transferability. The advantages, and the costs and benefits of their transfer, need to be carefully examined and put in context. This study has opened up some of these points for the automobile industry. But their wider significance requires more extended comparative treatment, and, in particular, similar studies of other large US multinationals in Europe in this period.[154] Such studies will need to put into perspective both the achievements, and the absences and failures, of US multinationals. As we have seen, the automobile industry, one of the largest sectors of US foreign investment in Europe, provides a difficult and ambiguous picture at close quarters. Other chapters in this volume also shed new light on these issues. Paul Erker's study of the tyre industry offers important parallels. Among other things, he highlights the persistently problematic nature of 'direct' transfers from the USA, and the necessary role of receptive and capable local partners. As in autos, the American tyre companies often construed their 'ownership advantages' in myopic and misleading ways. But in tyres, the consequences were even more serious for Goodyear and Firestone. Thoroughgoing Americanization, which seemed the path of greatest advantage in the 1950s, became a roadblock to innovation with the arrival of the European-driven 'radial revolution' in the 1960s and 1970s, resulting in the defeat and withdrawal of the American companies. It also appears that other US companies which dominated their markets and technologies within the USA had serious difficulties in directly transferring them to Europe.[155] Perhaps more wisely, as Matthias Kipping and Ruggero Ranieri show, the US steel companies, despite the dominance of their best practice in European reconstruction, avoided direct involvement in Europe, with the notable exception of Armco, which specifically built its business around the selective transmission of technology and organization.

The broad-brush images of the ownership advantages of US multinationals and their effectiveness as 'transmission belts' need careful examination. In this case, as in the case of other putative transfer mechanisms of Americanization examined in this book (Marshall Aid, technical assistance, or productivity missions), direct transplantation had a limited impact and the more critical factors may well be much broader processes involving European manufacturers, selectively drawing from an ambiguous and multiple palette of American ideas, and reworking them in conjunction with indigenous technologies and organizational forms.

[154] Between 1945 and 1965, the USA accounted for 85 per cent of all new direct foreign investment flows in the world, and this investment was massively concentrated in Europe. By 1957 the largest 50 US foreign investors controlled 50 per cent of US foreign assets and sales: Carl Dassbach, *Global Enterprises and the World Economy* (New York: Garland, 1989), 1–3.

[155] For example, GE had an unhappy history in Europe, and the development of European operations by the big US chemical companies (Du Pont, Dow, Monsanto) was belated and awkward. On GE, see: Chandler, *Scale and Scope*, 351–3; Wilkins, *Maturing of Multinational Enterprise*, 67–8. On the chemical companies, see, Graham D. Taylor and Patricia E. Sudnik, *Du Pont and the International Chemical Industry* (Boston: Twayne, 1984), 187–94; Wilkins, *Maturing of Multinational Enterprise*, 376.

PART TWO

Reworking US Technology and Management: National, Sectoral, and Firm-Level Variations

Chapter 4

Americanizing British Engineering? Strategic Debate, Selective Adaptation, and Hybrid Innovation in Post-War Reconstruction, 1945–1960

JONATHAN ZEITLIN

4.1 Introduction

Among the most well-worn tropes in the long-running debate on British economic decline is the failure of domestic manufacturers to embrace American methods of mass production and systematic management with greater vigour and enthusiasm. Since the emergence of a 'declinist' discourse in the late 1950s and early 1960s against the backdrop of contemporary anxieties about slow relative growth and falling international trade shares, insufficient Americanization has formed a major count in indictments of British economic performance. Much of the historiographical debate initially focused on the pre-1914 and inter-war periods, pitting against one another proponents of rival interpretative approaches based on entrepreneurial failure, rational decision making under given constraints, and in a subsequent phase institutional rigidities, in assessing and explaining the UK's relative economic performance.[1]

---

[1] For representative statements of the entrepreneurial failure, rational decision-making, and institutional rigidities approaches to the history of British economic performance since 1870, see, respectively: D. H. Aldcroft and H. W. Richardson, *The British Economy 1870–1939* (London: Macmillan, 1969); Donald N. McCloskey, *Enterprise and Trade in Victorian Britain* (London: Allen & Unwin, 1981); Bernard Elbaum and William Lazonick (eds.), *The Decline of the British Economy* (Oxford: Oxford University Press, 1986). For a brief overview of the debate, see Michael Dintenfass, *The Decline of Industrial Britain, 1870–1980* (London: Routledge, 1992); for an authoritative survey of current views among professional economic historians, see Roderick Floud and Donald McCloskey (eds.), *The Economic History of Britain from 1700* (2nd edn., Cambridge: Cambridge University Press, 1994), vols. 2–3. For critical perspectives on 'declinism' in British economic historiography, see David Edgerton, *Science, Technology, and the British Industrial 'Decline', 1870–1970* (Cambridge: Cambridge University Press, 1996); Barry Supple, 'Fear of Failing: Economic History and the Decline of Britain', *Economic History Review*, 2nd ser., 47 (1994), reprinted in Peter Clarke and Clive Trebilcock (eds.), *Understanding Decline: Perceptions and Realities of British Economic Performance* (Cambridge: Cambridge University Press, 1997), 9–31. For a pioneering discussion of the emergence of the idea of relative economic decline in Britain during the 1950s and 1960s, see Jim

More recently, the debate has focused on the period since 1945, in the context of competing efforts to account for Britain's lower rates of economic growth and productivity catch-up during the post-war 'golden age' compared to those of other Western European countries and Japan. Historians and economists from very different ideological and methodological standpoints such as Corelli Barnett, Nicholas Crafts, Steven Broadberry, Nick Tiratsoo, and Jim Tomlinson all appear to concur that British manufacturing industry was relatively resistant to the adoption of American production and management methods during the immediate post-war era, and that inadequate take-up and implementation of US techniques plays a major part in explaining the relative deficiencies of the UK's growth and productivity record in these years. At the same time, however, these authors also disagree sharply with one another in allocating responsibility for the limits of Americanization in post-war Britain among various groups of contemporary actors, notably government, business, and labour.[2]

Much of the recent debate on British manufacturing and post-war Americanization has been conducted at a high level of aggregation, with little attempt to examine the relationship between theory and practice in concrete industrial settings. Much of this literature also suffers from the absence of a well-developed comparative framework, failing to apply the same yardsticks and assessment criteria to Britain and to other countries, including the United States, which is transformed in some recent accounts into an industrial variant of Garrison Keillor's fictional Lake Wobegon where all the firms are 'above

Tomlinson, 'Inventing "Decline": The Falling Behind of the British Economy in the Postwar Years', *Economic History Review*, 2nd ser., 49 (1996), 731–57.

  [2] Corelli Barnett, *The Audit of War* (London: Macmillan, 1986); id., *The Lost Victory* (London: Macmillan, 1995). N. F. R. Crafts, ' "You've Never Had It So Good?": British Economic Policy and Performance, 1945–60', in Barry Eichengreen (ed.), *Europe's Post-War Recovery* (Cambridge: Cambridge University Press, 1995), 246–70; S. N. Broadberry and N. F. R. Crafts, 'British Economic Policy and Industrial Performance in the Early Post-War Period', *Business History* 38, 4 (1996), 65–91; Stephen N. Broadberry, *The Productivity Race: British Manufacturing in International Perspective, 1850–1990* (Cambridge: Cambridge University Press, 1997). Nick Tiratsoo and Jim Tomlinson, *Industrial Efficiency and State Intervention: Labour 1939–51* (London, 1993); id., *The Conservatives and Industrial Efficiency, 1951–64: Thirteen Wasted Years?* (London: Routledge, 1998); id., 'Exporting the "Gospel of Productivity": United States Technical Assistance and British Industry, 1945–1960', *Business History Review* 71 (1997), 41–81; Jim Tomlinson and Nick Tiratsoo, 'Americanisation Beyond the Mass Production Paradigm: The Case of British Industry', in M. Kipping and O. Bjarnar (eds.), *The Americanisation of European Business, 1948–1960: The Marshall Plan and the Transfer of US Management Models* (London: Routledge, 1998), 115–32; Nick Tiratsoo and Terry Gourvish, ' "Making It Like in Detroit": British Managers and American Productivity Methods, 1945–c.1965', *Business and Economic History* 25, 1 (1996), 206–16. Cf. also Paddy Maguire, 'Designs on Reconstruction: British Business, Market Structures and the Role of Design in Post-War Recovery', *Journal of Design History* 4 (1991), 15–30. For disagreement about the respective roles of government, business, and labour in limiting the take-up of American methods in the post-war era, see Jim Tomlinson and Nick Tiratsoo, ' "An Old Story, Freshly Told"? A Comment on Broadberry and Crafts' Approach to Britain's Early Post-War Economic Performance', and S. N. Broadberry and N. F. R. Crafts, 'The Post-War Settlement: Not Such a Good Bargain After All', both in *Business History* 40 (1998), 62–79.

average'.[3] Much of the debate, finally, is marred by teleological and determinist assumptions about the intrinsic superiority of American-style mass production and professionalized management during the post-war era, in which the persistence of 'pre-Fordist' methods is taken to be a self-evident indication of backwardness and complacency, an avatar of and contributory factor in the subsequent decline of domestic manufacturing.

Chapter 1 of this volume attempts to place British responses to post-war Americanization in a broader comparative context, drawing out their similarities and differences in relation to those of other European countries and Japan. This chapter instead seeks to sketch out the contours of British debates about Americanization and reconstruction in a key sector of manufacturing—the engineering or metalworking industries. Its central concerns are threefold. First, contrary to the claims of some recent historians such as Corelli Barnett, it highlights the determined efforts during the immediate post-war years—above all by the Attlee Labour governments—to push British industry towards the adoption of American-style mass-production and management methods. Second, it re-examines contemporary objections to these proposals—some more prescient than others—and reassesses the practical impact of both on the reconstruction of British engineering. Often, as we shall see, there were significant practical obstacles in both the short and long term to the wholesale adoption of the American model. Often, too, however, British manufacturers selectively adapted elements of US techniques to fit with their existing product and production strategies aimed at smaller and more diverse markets; in some cases, moreover, their creative modifications of transatlantic methods generated innovative hybrid forms of flexible manufacturing which anticipated in important respects those later made famous by the Japanese. As a result, finally, this chapter calls into question the causal link between the limits of post-war Americanization and the subsequent decline of British manufacturing. For as Chapter 1 points out, in an era when American manufacturers have themselves struggled to respond to the challenges of new competitive strategies based on greater product diversity and productive flexibility, there can be little justification for considering mass production and systematic management as they were

---

[3] For particularly egregious examples, see Tiratsoo and Tomlinson, *Conservatives and Industrial Efficiency*; id., 'Exporting the Gospel of Productivity'; Tomlinson and Tiratsoo, 'Americanisation Beyond the Mass Production Paradigm'; and Tiratsoo and Gourvish, ' "Making It Like in Detroit" ', which use surveys and reports from the 1960s and 1970s on the uneven application in mainly small and medium-sized firms of practices such as work study, standardization, mechanical handling, and statistical quality control to argue that the limited post-war adoption of 'well-proven American techniques' significantly contributed to the subsequent decline of British manufacturing, without any apparent recognition of the wide disparity in production methods and productivity levels by firm size and sector within the United States itself during the same period. For introductions to the voluminous literature on industrial dualism in the post-war American economy, see Robert Averitt, *The Dual Economy* (New York: Norton, 1968); Peter B. Doeringer and Michael J. Piore, *Internal Labor Markets and Manpower Analysis* (Lexington, MA: Heath, 1971); Susanne Berger and Michael J. Piore, *Dualism and Discontinuity in Industrial Societies* (Cambridge: Cambridge University Press, 1980).

practised in the United States during the 1940s and 1950s as a universal model of industrial efficiency which other nations failed to embrace at their peril.

4.2 Promoting Productivity, Emulating America

In Britain as in other European countries, domestic debates about Americanization of industry had a long history prior to 1945. From the 1890s onwards, British observers closely monitored the evolution of mass-production techniques in American industry and carefully considered their applicability to domestic economic conditions. While recognizing the cost reductions and economies of scale theoretically obtainable by following the American model, participants in these discussions drew attention to the practical advantages of more flexible methods based on the use of skilled workers and general-purpose machinery in catering for the varied and fluctuating markets characteristic of British engineering. They were also concerned about the inflexibility of special-purpose equipment, the high overhead costs of bureaucratic management, and the restrictive impact of standardization on product innovation. At the same time, however, British metalworking firms selectively appropriated certain features of American practice which could enhance product performance or productive efficiency without excessive loss of flexibility—such as new machine tools, payment systems, and interchangeability of key parts—and used these piecemeal innovations to maintain or expand their position in domestic and international markets.[4]

The vast output and productivity of American industry during and immediately after the Second World War reopened this debate and gave renewed impetus to British advocates of mass production. During the later years of the war, government planners, trade unionists, and progressive industrialists put forward a variety of more or less far-reaching proposals for the reconstruction on mass-production lines of important sections of British engineering such as motor vehicles and agricultural machinery. But it was during the post-war productivity drive that official efforts to promote the adoption of American methods in British manufacturing reached their apogee through government standardization campaigns, the transatlantic missions organized by the Anglo-American Council on Productivity (AACP), and the procurement policies of the nationalized industries.

During the initial phase of post-war reconstruction, the Labour government's central priorities were to maintain full employment and boost output by 'manning

---

[4] This chapter draws on material from the author's forthcoming book, *Between Flexibility and Mass Production: Strategic Debate and Industrial Reorganization in British Engineering, 1830–1990* (Oxford: Oxford University Press), which will contain fuller discussion and references. On the British reception of and response to American manufacturing methods before the First World War, see also the author's article, 'Between Flexibility and Mass Production: Strategic Ambiguity and Selective Adaptation in the British Engineering Industry, 1830–1914', in Charles F. Sabel and Jonathan Zeitlin (eds.), *World of Possibilities: Flexibility and Mass Production in Western Industrialization* (Cambridge: Cambridge University Press, 1997), 241–72.

up' key industries. By 1947, however, it had become apparent that labour reserves were drying up, while capital investment was constrained by material shortages and export priorities and could not be expected to yield rapid results in any case. In the short term, therefore, increased output could only come from more efficient use of existing resources, and raising productivity accordingly moved to the centre of Labour's economic objectives. The resulting productivity drive was spearheaded by Sir Stafford Cripps, Chancellor of the Exchequer and long-standing enthusiast of scientific management, who believed that factory reorganization and the diffusion of best practice could yield 'twenty, thirty or even fifty per cent increases in output with lowered costs and higher pay for the operatives'.[5]

The productivity drive encompassed a wide range of policies aimed at improving the efficiency of British industry in both the short and long term, from propaganda and exhortation through the diffusion of work-study, operations research, and human relations techniques to public support for research and development. Among the most important initiatives specifically directed at engineering was the promotion of product standardization. In November 1947, Herbert Morrison, chairman of the Cabinet Committee on Production, proposed that the Board of Trade should consider 'measures for a speed up in standardization and variety reduction in British industry', including compulsory enforcement of dimensional standards for certain products. The Board of Trade responded by commissioning a detailed inquiry into the scope of efficiency gains achievable through standardization. Despite its authors' evident enthusiasm for all forms of rationalization, the inquiry concluded that reduction of product variety by itself yielded cost savings and output gains of only 5–10 per cent; larger savings of 25–50 per cent required much longer runs and heavy capital investment in new plant. 'Standardization by the reduction of variety', concluded the Board, 'is not a panacea for the lessening of costs'; and its report also underlined the attendant dangers of 'sterilization of design', reduced consumer choice, and loss of ability to cater for diverse export markets. True to its voluntarist traditions, therefore, the Board of Trade firmly rejected compulsory enforcement of standards except in cases of government procurement or public health and safety.[6]

The attitude adopted by the Ministry of Supply (MoS), the department directly responsible for most of engineering, was altogether more positive. While well aware of the dangers cited by the Board of Trade, the Ministry was convinced that 'the degree of standardisation so far achieved in . . . engineering . . . falls short of what can be done with advantage to productivity'. In November 1948, therefore, the Ministry set up a Committee for Standardization

---

[5] Tiratsoo and Tomlinson, *Industrial Efficiency and State Intervention*, ch. 4; UK Public Record Office (PRO), BT 195/19: 'Productivity Drive—Autumn 1948'.

[6] PRO BT 64/2399: 'Standardisation', memo. by the president of the Board of Trade to the Cabinet Production Committee (Sept. 1948). Tomlinson and Tiratsoo cite the conclusions of this document, but fail to note its critique of Cripps' original claims about the productivity gains achievable through standardization and reduction of variety: see 'Americanisation Beyond the Mass Production Paradigm', 118.

of Engineering Products under Sir Ernest Lemon, a pioneering architect of railway standardization, to investigate the scope for the reduction of variety in the industry and the role of government action in effecting it. After an extensive survey of domestic practice, which revealed a broad spectrum of sectoral variation, the Lemon Committee came down unequivocally in favour of greater standardization in British engineering. 'There can be no doubt', argued its 1949 report, 'that the relatively high degree of specialization and simplification in US industry (including the smaller firms), is a major reason for their higher industrial productive efficiency.' Correctly applied, the committee contended, specialization, simplification, and standardization should neither inhibit technical progress in design, impose uniformity on consumers, nor result in a loss of markets at home or abroad. Thus designs could be periodically altered to take account of technical advance; a reasonable variety of final products could be built up from standardized materials, parts, and components; and customers could be induced to accept standard articles through the vigorous use of price incentives. Despite these robust conclusions, however, the policy recommendations advanced by the Lemon Report remained decidedly modest. Like the Board of Trade, the Lemon Committee rejected compulsory enforcement of industrial standards as both impracticable and likely to produce 'serious rigidities and inefficiencies'. The British Standards Institution (BSI) should expand its staff, streamline its procedures, and play a more active role in initiating new standards, receiving in return an increased public grant, but industrial standardization must ultimately depend on the voluntary consent of private manufacturers. Government, the committee concluded, should encourage standardization through a combination of public exhortation, tax allowances for capital investment, and above all the purchasing policies of the nationalized industries, which consumed some 30 per cent of non-exported engineering.[7]

Closely associated with the standardization campaign was the work of the Anglo-American Council on Productivity. This was established in 1948 as a response to US criticisms of British industry in the context of Marshall Aid. The Attlee government also saw the AACP as a means of devolving greater responsibility for the success of the productivity drive on to business and labour organizations. The AACP's central activity consisted of organizing joint missions by British businessmen and workers to visit US plants to investigate the sources of superior American productivity and disseminate their findings. Most of the missions focused on individual industries, including a wide range of engineering sectors, but a number of specialist teams were also set up to study key aspects of American practice such as product simplification, design for production, materials handling, production control, and management accounting.[8]

---

[7] PRO CAB 134/639: 'Standardising and Simplifying Engineering Products', memo. by the Minister of Supply to the Cabinet Production Committee, 8 September 1948; MoS, *Report of the Committee for the Standardization of Engineering Products* (London, 1949).

[8] Anthony Carew, *Labour under the Marshall Plan: The Politics of Productivity and the Marketing of Management Science* (Manchester: Manchester University Press, 1987), ch. 9;

Broadly speaking, the AACP missions concerned with engineering confirmed prior expectations of a substantial gap between American and British productivity and practice. At the same time, however, the size of the gap identified was far from uniform across engineering as a whole: thus the standard contrast between US mass production and British craft practice was sharpest in diesel locomotives and internal combustion engines, but scarcely evident in metalworking, machine tools, and woodworking machinery, with other sectors such as valves, pressed metal, and electrical control gear somewhere in between. While many of the teams paid tribute at least in passing to the virtues of the 'American way of life' and the 'efficiency-mindedness' of US workers, the main factors adduced to explain Anglo-American productivity differences were altogether more concrete. Contrary to the recent claims of historical economists such as Broadberry and Crafts, however, the AACP engineering reports placed little emphasis on restrictive labour practices as a source of transatlantic variations in productivity. Far more crucial, in their view, were the larger size and greater homogeneity of the American market, which made possible the longer runs that justified the fuller use of mass-production techniques such as special machinery and tooling, mechanical-handling devices, and subdivision of labour. But the US productivity lead was not attributed simply to the influence of favourable environmental conditions. Thus the reports emphasized the active commitment of American engineering companies to the 'three charmed "Ss" of high productivity': simplification, standardization, and specialization. Even smaller firms, they found, tended to specialize on a narrow range of products, and high-production tooling was common on much smaller batches than in Britain. US manufacturers, moreover, constantly sought to expand demand for their standard lines by 'making the market' through a combination of research, advertising, and price discrimination against special orders. The effectiveness of mass production in US industry, the reports further argued, also depended on the careful application of systematic management methods such as standard costing, work study, and production planning, as too did the success of labour policies such as wage incentives and job evaluation.[9]

But the practical message conveyed by the AACP productivity missions for the reorganization of British industry was decidedly contradictory. Like the productivity drive itself, the AACP reports held out the prospect of large and rapid efficiency gains through the adoption of American production methods; but the more radical the changes proposed in domestic practice, the more distant the potential benefits. On the one hand, therefore, the productivity teams

Tiratsoo and Tomlinson, *State Intervention*, ch. 7; id., 'Exporting the Gospel of Productivity'. For a complete list of the productivity team reports, see AACP, *Final Report of the Council* (London: AACP, 1952).

[9] For a synthetic analysis of the reports commissioned by the AACP itself, see Graham Hutton, *We Too Can Prosper* (London: George Allen & Unwin, 1953). For contrasting views on the role of restrictive labour practices in the AACP reports, see Broadberry and Crafts, 'British Economic Policy and Industrial Performance'; Tomlinson and Tiratsoo, ' "An Old Story, Freshly Told"?'; Broadberry and Crafts, 'Post-War Settlement'.

highlighted the importance of inherently long-term measures such as reduction of product variety, greater mechanization and automation, and the development of systematic management capabilities. On the other hand, however, many of their recommendations understandably tended to focus on short-term measures which could be introduced within the existing organization of production, such as better tooling, work study, or incentive payment schemes. Paradoxically, therefore, at the same time as the AACP missions apotheosized American practice as a model for British industry, they also helped to circumscribe the scope of domestic reform initiatives.

Bringing it All Back Home

How did the various sections of the British engineering community respond to the Labour government's productivity drive, with its overriding emphasis on the benefits of standardization and Americanization?[10] On the business side, the reactions of peak associations like the Federation of British Industries (FBI) and the employer representatives on the Engineering Advisory Council (EAC) were strongly coloured by their anxiety to limit the scope of state intervention in the private sector. The FBI, as Tomlinson and Tiratsoo have shown, regarded much of the AACP's work as a thinly veiled criticism of British management, while simultaneously seeking to use it as a means of deflecting government productivity initiatives into a less threatening bipartite forum. Both the FBI and the employers on the EAC were extremely hostile towards any move in the direction of compulsory standardization, seen as the thin end of the wedge of state control over their product policies, opposing for example the Lemon Committee's proposal to give the BSI greater initiatory powers in drawing up industrial standards. Employer representatives on the EAC also warned that 'it was particularly important to retain flexibility of production in order to combat American competition' and 'the standardisation of end products would be fatal from the selling point of view'. The employers' side thus agreed to recommend that engineering trade associations set up co-operative schemes for increasing production efficiency—including through standardization—where these did not already exist, while insisting that individual firms remain free to participate or not as they chose.[11]

Trade unions, by contrast, were generally more supportive. Like the FBI, the Trades Union Congress (TUC) sought to dispel what it saw as ill-informed foreign criticisms of domestic industry and emphasized the key role of higher

[10] I have borrowed the title of this section not only from Bob Dylan but also from Bryn Jones, *Forcing the Factory of the Future: Cybernation and Societal Institutions* (Cambridge: Cambridge University Press, 1997), ch. 3, which includes a useful analysis of the AACP reports.

[11] Tiratsoo and Tomlinson, *Industrial Efficiency*; PRO BT 195/1, EAC Minutes 3 March 1949; BT 195/19: 'Production Efficiency: Memorandum by Employer Members', EAC(49)67, June 1949; SUPP 14/141: 'Report of the Lemon Committee for Standardisation of Engineering Products: Comments by Employer Members on the EAC', 2 January 1950; SUPP 14/333: 'Production Efficiency: Memorandum by Employer Members', 28 July 1950.

capital investment in raising productivity. But the TUC General Council also strongly pressed member unions to take an active part in AACP work, and following a US team visit of its own, the TUC established an in-house production department to train British trade unionists in productivity-enhancing techniques such as work study. The reactions of individual unions were more complex but still, on balance, strongly positive. Jack Tanner, President of the Amalgamated Engineering Union (AEU), chief union spokesman on the EAC and a member of the TUC General Council, was an ardent supporter of the productivity drive, often pressing government and employers alike for more vigorous measures. Thus, for example, Tanner criticized the employers' disappointing response to the Lemon Report, while also underlining the need to show the Americans on the AACP that the British were taking effective measures to increase standardization, an objective which had figured prominently in the AEU's earlier proposals for post-war reconstruction. Not all metalworking trade unionists were equally enthusiastic, however, particularly about the AACP: a notable case in point was the National Union of Foundry Workers, which fell out with employers and the TUC over the composition of one of the productivity teams. Even within Tanner's own AEU, the onset of the cold war in 1947–8 prompted Communist members of the Executive Council to propose unsuccessfully that participation in the AACP be postponed until full consideration had been given to the union's 'four-year plan' for the industry; but as we shall see shortly, this alternative 'plan' was scarcely less productivist than the official productivity drive itself.[12]

In 1945, the National Engineering Joint Trades Movement (NEJTM), an umbrella organization uniting the AEU and the Confederation of Shipbuilding and Engineering Unions (CSEU), had vigorously campaigned for the continuation of wartime planning methods to promote productive efficiency and secure full employment after the return to peace. At the heart of the unions' analysis lay a fervent belief in the superiority of mass production and the urgent need for technical modernization in British industry. Given the high levels of unemployment between the wars, the NEJTM argued, British engineering had understandably become 'cautious and restrictive in its outlook. In technique, equipment and output per head many sections were far behind, for example, American standards.' There had been a great improvement in productivity during the war, principally because the government became the main purchaser of engineering products, but also because 'the initiative of the organized workers played an important part in the drive for output and efficiency'. Without guaranteed bulk orders for engineering products such as machine tools, housing

---

[12] Tiratsoo and Tomlinson, *Industrial Efficiency*, 133–42, 164–5; Carew, *Labour under the Marshall Plan*, 147–56; Alan Booth, 'Corporate Politics and the Quest for Productivity: The British TUC and the Politics of Industrial Productivity, 1947–1960', in Joseph Melling and Alan McKinlay (eds.), *Management, Labour and Industrial Politics in Modern Europe* (Cheltenham: Edward Elgar, 1996), 44–65; PRO BT 195/1: EAC Minutes, 3 Mar. 1949; SUPP 14/141: EAC Minutes 1 Feb. 1950; BT 195/19: Leslie to Trend, 7 Jan. 1949.

components, and standardized consumer durables, the unions concluded, 'it is most difficult to see how full advantage can be derived from planned mass production'. The NEJTM therefore demanded that the new Labour government create a tripartite Engineering Board with strong executive powers to oversee the production and investment plans of private enterprises; and this national body was envisaged as the apex of an articulated system of participatory planning stretching down to similar boards in the regions and Joint Production Committees (JPCs) in individual factories. Nor, finally, was such planning to be confined to capital alone: the unions specifically requested that wartime manpower-budgeting techniques also be carried over into post-war economic management. The engineering unions' major response to the productivity drive was to relaunch their campaign for participatory planning. 'Engineering and the Crisis', issued by the CSEU in December 1947, attributed the government's economic difficulties to a 'lack of plan', which allowed scarce raw materials and skilled labour to be 'spread over too many non-standardised products often far from really essential', resulting in 'idle time coupled with low productivity' and 'ignorance of, or indifference to, the purpose of the production drive on the part of sections of the workers'. The CSEU accordingly reiterated its demands for the transformation of the EAC into an effective planning body, buttressed by a vertical pyramid of sectoral, regional, and district committees to provide detailed information and advise on the implementation of its programmes. Hand in hand with the renewed planning campaign went a demand for statutory JPCs, for as Tanner told the EAC in April 1948, 'there would not be a real response from the workers [to the productivity drive] unless they felt that they were not merely being consulted but were participating in the organisation of production'.[13]

The Labour government's reaction to the unions' proposals throughout this period was largely unsympathetic. Unlike iron and steel, contended the Ministry of Supply, engineering was not a basic industry, while sectors like locomotives and tin cans were far too diverse to be governed by a single planning agency. Whereas the government actively embraced 'strategical' or 'democratic' planning for engineering, seeking to guide the industry towards broadly defined objectives through administrative levers such as export quotas and raw material allocations, it firmly rejected any suggestion of 'tactical' or 'totalitarian' planning aimed at specifying the types of product, quantities, and delivery dates in particular factories. Britain, officials emphasized, was not a closed economy, but depended on being able to provide goods overseas customers were willing to buy in a competitive market in exchange for vital imports: 'Consequently, there

---

[13] PRO SUPP 14/137: NEJTM, memo. on 'Post-War Reconstruction in the Engineering Industry', June and Dec. 1945; SUPP 14/138, NEJTM to J. Wilmot (Minister of Supply), 28 Feb. 1946; TUC Archives, Modern Records Centre, University of Warwick (hereacter MRC), MSS 292/615.2/5: CSEU, 'Engineering and the Crisis', 8 Dec. 1947; PRO BT 195/6: EAC Minutes, 7 Apr. 1948; James Hinton, *Shop-Floor Citizens: Engineering Democracy in 1940s Britain* (Aldershot: Edward Elgar, 1994), chs. 8–9. The AEU formally rejoined the CSEU in Dec. 1946.

had to be a good deal of flexibility, both in the broad outlines of planning and even more in the detailed picture.' Nor could ministers contemplate devolving executive powers to a tripartite body outside their own direct control: such an action would represent at once 'an abrogation of government' under British doctrines of parliamentary sovereignty, and an unacceptable infringement of managerial prerogatives in privately owned firms. While ministers were unwilling to give unions more than an advisory role in industry-level planning, proposals to extend joint consultation in the factories enjoyed much wider support, and Cripps himself threatened employers with legislative sanctions if they did not voluntarily co-operate in the re-establishment of wartime JPCs. In the event, however, the government backed away from statutory compulsion as a result of opposition from the Ministry of Labour and, after some inter-union wrangling, from the TUC as well to such a permanent breach of voluntarist principles in peacetime. In hopes of averting more vigorous state intervention, the Engineering Employers Federation (EEF) prudently encouraged its members to take an active part in the reestablishment of JPCs, whose numbers rapidly returned to peak wartime levels. But in the absence of any organic connection to wider economic planning or rights to information about the export targets fixed for individual firms, this second wave of JPCs inspired little enthusiasm among most engineering workers and shop stewards.[14] Rejected by the Labour government, the CSEU drew up its own 'Plan for Engineering', which it sought to promote through the TUC, with sector-by-sector proposals for the extension of public control, including bulk purchasing of standardized utility products, the creation of statutory boards or development councils, and selective nationalizations.[15]

But the most intensive and wide-ranging debate about standardization, Americanization, and productivity during this period was that conducted by trade associations, engineering societies, and the trade press. As one might expect, there were pronounced sectoral variations in attitudes towards the 'three Ss': thus, for example, as a small survey of top industrial managers conducted by a group of Oxford economists found, manufacturers of intermediate goods or components for other businesses saw greater scope for standardization and reduction of variety than did those making complex capital equipment and finished products for consumer markets. Beneath these sectoral variations, however, ran some common threads. British businessmen and engineers were well aware of the potential cost and productivity advantages of mass production, and the overall tenor of the debate was far more favourable to standardization, simplification, and specialization than it had been in previous periods. At the

---

[14] Hinton, *Shop-Floor Citizens*, 170–6, 184–94.

[15] SUPP 14/141: memoranda and notes of meetings with unions and employers, 1946–8; TUC Archives, MRC MSS 292/615.2/5: 'The Engineering Worker and Economic Recovery', 5 Dec. 1949; TUC, Engineering and Shipbuilding Industries National Advisory Committee, 'Minutes of First Meeting', 1 Mar. 1950; CSEU, *Plan for Engineering* (1951; reprinted with further proposals by the Association of Engineering and Shipbuilding Draughtsmen, 1953).

same time, however, pre-war reservations about their commercial and technological limitations under domestic conditions had by no means been dispelled
even in the most self-consciously progressive sectors. Many respondents to the
AACP and Lemon Reports emphasized the difficulties of imposing standard
designs on varied export markets and the risks of excessive reliance on a narrow
range of products under rapidly changing competitive conditions. Other commentators argued that 'the conditions which created a demand before the . . .
war for a variety of products would return once the seller's market gave place to
a buyer's market', and that the ability to produce equipment like transformers
or diesel locomotives to almost any specification was 'a priceless asset for
overseas trade where individuality in requirements is even more marked than at
home'. Another recurrent theme in these discussions was the constraints
imposed by excessive standardization on technological innovation and design
changes. Such reservations were often reinforced by a closer acquaintance with
American practice: thus one AACP team found that US woodworking machinery manufacturers froze their designs for up to ten years in order to amortize the
costs of special tooling, while another reported that: 'A consequence of the large
scale of production and widespread use of special purpose machinery is
undoubtedly a tendency for design to become stereotyped and the combustion
systems used by certain engine builders would certainly be considered out of
date by British standards'.[16]

For some contemporary commentators like *The Economist*, 'the case for
reducing the variety of products of the average British factory [was] unchal-
lengable', constituting 'one of the cheapest and most rapid ways of raising
national productivity'. But even engineers strongly favourable to the 'three Ss'
like D. G. Daglish of English Electric's internal standards section cautioned
against excessive product simplification: 'Merely because 40% of the items
show only a 5% of sales, it does not follow that all can be arbitrarily cancelled.'
'For any current range of varieties of a complex product', likewise observed H.
E. Merritt, chief research officer at the British Transport Commission, 'it is
probable that although a few marginal variations in small and unprofitable
demand may existing, deletion of the remainder might leave a gap, embarrassing to the sales department', which could only be overcome through a long-term
redesign project 'in order to cover the market requirement with fewer varieties
of greater versatility'. Even more fundamentally, argued the *Engineer*, among
the low-selling items 'may be found some products recently introduced which
are destined to prove in the future really profitable for increased production. All
new products, after all, are "odd" products to begin with.' By the later 1950s, as

[16] P. W. S. Andrews and E. Brunner, 'Productivity and the Businessman', *Oxford Economic
Papers*, n.s. 2 (1950), 197–225; PRO SUPP 14/141: industry comments; British Productivity Council
(BPC), *The Wheels of Progress. A Review of Productivity in the Diesel Locomotive Industry*
(London: BPC, 1954), paras. 35–6; PRO SUPP 14/330: 'Notes on Meeting of the Political and
Economic Planning (PEP) Engineering Group, 31/5/48'; AACP, *Woodworking Machinery* (London:
AACP, 1952), 10; id., *Internal Combustion Engines* (London: AACP, 1950), 40.

Tiratsoo and Tomlinson acknowledge, the British Productivity Council (the AACP's successor body) had itself come to emphasize 'the need to cut out *wasteful* variety, not to eliminate variety altogether', since that would lead to 'the narrowing of consumer choice', and the deterioration of 'skill in design and inventiveness'; but distinguishing the former from the latter, as the preceding discussion indicates, was anything but a straightforward or consensual matter.[17]

Faced with these objections, advocates of standardization tended to shift their ground from end-products to components. But despite its attraction as a means of reconciling production economy with product diversity, standardization of components raised nearly as many problems as it resolved. Not only did component standardization increase production and inventory costs in the short run, but as contemporaries argued and subsequent experience confirms, it could also inhibit the development of new models because of the interdependence among their constituent parts. 'Even the standardisation of components for use in locomotives of different overall design presents great difficulty', contended British manufacturers, 'since the modern diesel locomotive is so compactly built that even minor variations to meet a special requirement can cause repercussions throughout the structure.' Similar reservations were also voiced even by some of the most ardent domestic proponents of rationalization and large-scale production. Thus Sir Patrick Hennessy of Ford UK argued that 'if . . . all manufacturers had the same wheels, someone with a bright idea for an improved model might come along and be prevented from succeeding because he needed different wheels'. A preferable solution, a Board of Trade study concluded, was 'to decide on a good design and to make it in very large numbers which would give economies of scale both in components and in assembly, rather than attempt to make it with standard parts in a number of different designs'. Such arguments highlighted the underlying tension between individual and collective standardization, which meant, as contemporaries were all too well aware, that any vigorous attempt to pursue the one was likely to undercut the other, at least in the medium term.[18]

Despite their unquestioned admiration for the achievements of American productive organization, moreover, British commentators remained critical not only of its inflexibility but also of the high levels of stocks and work in progress on which it depended, anticipating more recent critiques of US manufacturing

[17] 'The Cost of Variety', *The Economist*, 29 Dec. 1951, 1595; D. G. Daglish (English Electric), 'An Organization for Internal Standardization in a Large Manufacturing Company', *Journal of the Institution of Electrical Engineers* 98, 1 (1951), 198; H. E. Merritt (British Transport Commission), 'Simplification Creates New Problems for Top Management', *Proceedings of the Institution of Mechanical Engineers* 162 (1950), 188; id., 'Simplification and Management', *Engineer*, 24 Feb. 1950, 242; BPC, *Better Ways. Nineteen Paths to Higher Productivity* (London: BPC, 1957), 34, quoted in Tiratsoo and Tomlinson, 'Exporting the Gospel of Productivity', 70–1; cf. also Tomlinson and Tiratsoo, 'Americanisation Beyond the Mass Production Paradigm', 118–19.

[18] BPC, *Wheels of Progress*, 9; PRO BT 64/2379: C. Gordon and T. E. Easterfield, 'Standardisation, Specification and Production of Variety' (1948) ; SUPP 14/330: Hennessy to PEP Engineering Group.

practice in comparison to that of the Japanese. Thus B. E. Stokes, a British production engineer recently returned from an extensive study tour of US factories, in 1952 found 'embarrassingly large stocks of material and work in process' in many cases, an observation which echoed a number of the AACP teams. Behind this phenomenon lay what he called 'the problem of the so-called "economic" batch': an 'erroneous idea of machine utilisation in which the machine requirements dominate the company requirements', leading production managers to insist on manufacturing components in long runs without reference to changing patterns of customer demand, the relative value of special and standard products, or the costs of immobilized capital. Stokes's proposed solution, as in the 'just-in-time' system being developed by Toyota's Taiichi Ohno at that very moment, was to tailor production more closely to demand and reduce the costs of small-lot manufacture by increasing the speed and efficiency of machine changeovers. Such a strategy, however, required a system of production control capable of dealing with greater complexity and variety. 'American companies', Stokes observed,

have avoided many of the administrative problems arising from complex manufacturing activity by not becoming involved in many factors which create complexity—that is, the handling of wide diversity of product within the manufacturing unit with its consequent wide range of material requirement, large variety of component usage with attendant service problems, inability to take advantage of specialised and more efficient machinery and the flow layout of process which is largely self-progressing.

But the general tendency in the USA, Stokes contended, was 'to increase diversification of product and complexity of manufacture, a trend which managements generally explained as necessary in order to attract more customers . . . increase resistance to economic fluctuations', and respond to 'changing customer tastes'. As the AACP reports confirmed, however, few American firms had developed planning systems suitable for handling a greater variety of products manufactured in smaller quantities, while 'their immediate reaction, "increase the progress staff" [was] not unknown in Britain!'[19]

These criticisms of the post-war Americanization drive received their fullest and most cogent formulation in the influential trade journal, the *Engineer*. While welcoming the search for domestic productivity improvement through the study of foreign industry, the journal was sceptical whether 'knowledge gained of American methods of production is necessarily applicable to British

---

[19] B. E. Stokes, 'The Organisation of Production Administration for Higher Productivity', *Journal of the Institution of Production Engineers* 31 (1952), 201–4, 211–14; BPC, *Production Control* (London: BPC, 1953). On Ohno and the origins of 'just-in-time' production at Toyota in the 1950s, see Michael Cusumano, The *Japanese Automobile Industry* (Cambridge, MA.: Harvard University Press, 1985); Koichi Shimokawa, 'From the Ford System to the Just-in-Time Production System', *Japanese Yearbook on Business History* 10 (1993), 84–105; Steven Tolliday, 'The Diffusion and Transformation of Fordism: Britain and Japan Compared', in Robert Boyer, Elsie Cherron, Ulrich Jurgens, and Steven Tolliday (eds.), *Between Imitation and Innovation: The Transfer and Hybridization of Productive Models in the International Automobile Industry* (Oxford: Oxford University Press, 1998), 57–95.

practice' while also doubting whether 'all methods that prove profitable across the Atlantic will also prove possible here'. 'In the US', it went on to observe:

there has never existed that great pool of skilled labour that exists in Europe. The American manufacturer has thus often been forced by circumstance to adopt mechanized methods of production by unskilled or partly skilled labour. High capital costs have been involved and the price of the articles produced has not always been low.[20]

Even more important than these differences in British and American patterns of labour supply, contended the *Engineer*, were differences in the structure of their product markets:

The American manufacturer has available to absorb his products a very large and homo-geneous home market. He can therefore afford to rationalise, to invest large capital in specialist plant and to increase output per head thereby. But the British manufacturer serves a relatively small home market and has also to depend to a far higher degree than the American on satisfying variegated export markets which look to European rather than American manufacturers to supply them with the 'tailor-made' goods they require.

In many sectors, the journal argued, Britain's real competitors were other Europeans rather than the United States, and the system of sending out teams from British industries to learn the methods of other countries needed to be extended geographically. 'Much of real value', it concluded, could

be learnt from a study of other plants in Western Europe, which are confronting prob-lems which are much more similar to our own than those facing American producers. After all, America is not the only prosperous country in the world. The Swiss, for ex-ample, have built up a rich and varied industrial culture at least as attractive as the American.[21]

Among the central paradoxes of the productivity debate was its comparative neglect of Germany, Britain's major pre-war competitor in most of the engin-eering industries. Post-war intelligence missions which inspected German facto-ries were impressed by the skills and training of the labour force at all levels, from engineers and technicians to foremen and manual workers, but not by pro-duction methods and organization, which they regarded as inferior to those of not only American firms but also well-run British companies. 'Thus as regards the internal organisation of German industrial concerns we have very little to learn', concluded one specialist investigation of production control; and this finding was echoed in the reports of other intelligence teams dealing with

[20] 'The Value of Productivity Teams', *Engineer*, 5 May 1950; 'Diesel Locomotive Production', *Engineer*, 22 Dec. 1950.
[21] 'Value of Productivity Teams'; 'Anglo-American Council on Productivity', *Engineer*, 6 May 1949. Cf. also 'American and European Productive Methods', 4 June 1948; 'Productivity', 12 Nov. 1946; 'Mission to Canada', 18 Feb. 1949; 'American and British Productive Methods', 19 Aug. 1949; 'Simplification and Management', 24 Feb. 1950; 'Simplification in Industry', 22 Sept. 1950; 'Anglo-American Council', 16 May 1952. For a parallel invocation of Switzerland's 'organized craftsman-ship'—the manufacture of specialized, high-quality products in small and medium-sized firms—as an alternative to the American mass-production model by Italian engineers and managers during and immediately after the Second World War, see Bigazzi's chapter in this volume.

individual metalworking sectors such as aircraft instruments, machine tools, and power presses. Yet German engineering exports grew extremely rapidly during the 1950s, particularly to Western Europe, and by the middle of the decade at the latest, German machinery manufacturers had recaptured their pre-war share of world markets, largely at British expense.[22]

4.3   From Theory to Practice

Reconstruction planning and the post-war productivity drive touched off a far-reaching debate about the efficiency and organization of British engineering, in which participants gauged the distance separating their indigenous manufacturing methods, for good and ill, from American mass production on the one hand and German craft production on the other. But what was the impact of this debate, and the official campaigns which inspired it, on the evolution of product and production strategies in the British engineering industries during the immediate post-war years? How did metalworking manufacturers decide in practice which American experiences were worth while and in what manner they could be adapted to domestic conditions? To what extent did British engineering firms succeed in reconciling the countervailing demands of product diversity and higher productivity, complexity of manufacture and cost consciousness, flexibility and mass production?

Many of the most ambitious experiments with standardization and mass production, predictably enough, were undertaken by the public sector. Both the Ministry of Supply and the Central Electricity Board, for example, saw design standardization as a key to overcoming backlogs in deliveries of power-station equipment, and in November 1947 the MoS issued a statutory order limiting turbo-alternators for the domestic market to two standard sizes, 30 and 60 megawatts (MW). Similarly, among the first acts of the newly nationalized British Railways in 1948 was to develop a standard family of steam locomotives and coaching stock for use on all its constituent lines, followed by flow-line production of standard electric carriages for its southern commuter services during the early 1950s.[23]

[22] British Intelligence Objectives Sub-Committee, *Investigation of Production Control and Organisation in German Factories*, Final Report 537 (London: HMSO, 1947), 8; Alan Kramer, *The West German Economy, 1945–1955* (Oxford: Berg, 1991), ch. 6; Alan S. Milward, *The European Rescue of the Nation-State* (London: Routledge, 1992), 396–424. A fuller discussion of post-war British perceptions of German manufacturing practice will appear in this author's *Between Flexibility and Mass Production*, ch. 7.

[23] Martin Chick, *Industrial Policy in Britain, 1945–1951: Economic Planning, Nationalisation and the Labour Governments* (Cambridge: Cambridge University Press, 1998), ch. 6; Leslie Hannah, *Electricity Before Nationalization* (London: Macmillan, 1979), 322; id., *Engineers, Managers and Politicians: The First Fifteen Years of Nationalised Electricity Supply in Britain* (London: Macmillan, 1982), 24–5; I. V. Robinson, 'Standardization of Steam Turbo-generating Plant', *Proceedings of the Institution of Mechanical Engineers* 164 (1951); T. R. Gourvish, *British Railways, 1948–1973* (Cambridge: Cambridge University Press, 1986), 87–90; J. Johnson and R. A. Long, *British Railways Engineering, 1948–80* (London: Mechanical Engineering Press, 1981); *Engineering*, 1 Dec. 1950; Frank Woollard, *Principles of Mass and Flow Production* (London: Iliffe, 1954), 35–42.

Yet the results of these initiatives often proved disappointing even to their architects, vindicating many of the economic and technological objections raised by contemporary critics. Thus as a number of commentators forewarned, the sizes and steam conditions selected for standard turbo-alternators turned out to be too small and too conservative to give maximum operating benefits, handicapping the thermal efficiency of British electricity generation during the 1950s and 1960s, in contrast to the French, who based more of their post-war re-equipment programme on sets of 100 MW and above before standardizing on 125 MW during the early 1950s. Real capital costs per unit of generating capacity declined substantially between 1948 and 1953, but mainly because of the nationalized electricity authority's growing emphasis on more advanced 60 MW and (after the withdrawal of the standardization order in 1950) 100 MW sets. Little of this gain, conversely, appears to have been due to cost reductions associated with the replication of standard designs, since production economies of scale were undercut by the division of orders among a multiplicity of separate manufacturers, who remained reluctant in any case to offer quantity discounts on collectively agreed prices. In 1950, the electricity authority therefore persuaded one major manufacturer of power-generating equipment, C. A. Parsons Ltd., to withdraw from the group price-fixing agreement in exchange for a guarantee of 'solid block' orders for standard designs. But the savings obtained thereby satisfied neither party, and turbo-alternator manufacture at Parsons during the mid-1950s continued to be organized on jobbing lines, with a predominance of flexible, general-purpose machinery the output of which was 'determined, to a large extent, by the skill, dexterity and, indeed, willingness of the operator to use [it] to the best advantage'.[24]

On the railways, too, investments in a new generation of standard steam locomotives helped to slow down the introduction of more efficient diesel models, raising rather than reducing operating costs in the longer term. The key problem here, as contemporaries pointed out, was uncertainty about the future balance of advantage among competing traction technologies. During the 1940s, diesel locomotives had rapidly replaced steam on American railways, where savings on maintenance and fuel charges more than offset their higher first cost.

---

[24] Chick, *Industrial Policy in Britain*, 138–57; Hannah, *Engineers, Managers and Politicians*, 104–10; *Engineer*, 21 Nov. 1947, 17 Mar. 1950; Robinson, 'Standardization of Steam Turbo-generating Plant'; J. Henderson (Parsons), 'Some Problems Associated with the Selection of Machine Tools for Heavy Engineering Production', *Transactions of the Manchester Association of Engineers* (1954–5), 323; Monopolies and Restrictive Practices Commission, *Report on the Supply and Exports of Electrical and Allied Machinery and Plant* (London: HMSO, 1957), 49–53, 82–6. For the contrast with France, see also F. P. R. Brechling and A. J. Surrey, 'An International Comparison of Production Techniques: The Coal-Fired Electricity Generating Industry', *National Institute Economic Review* 36 (1966), 30–42; there, too, however, standardization of generating equipment posed a similar trade-off for the nationalized electricity industry between suppliers' manufacturing costs on the one hand and its own operating efficiency on the other, and *Électricité de France* privileged the latter by introducing design innovations with each successive unit of the same capacity until the mid-1960s: see Robert L. Frost, *Alternating Currents: Nationalized Power in France, 1946–1970* (Ithaca, NY: Cornell University Press, 1991), 176–7.

But dieselization seemed less suitable for British conditions, with their shorter distances, denser and more diverse traffic, and scarcer oil supplies; only in the shunting yards, the near-continuous operations of which ensured cost effectiveness, had diesel-electric engines been widely adopted. R. A. Riddles, the Railway Executive's Chief Mechanical Engineer, believed that steam would remain the power plant of choice for British locomotives until the relaxation of capital constraints made possible its gradual displacement by electricity, as in continental Europe. Hence Riddles refused to compromise the proven benefits of standardization, which he had personally experienced as a designer of wartime austerity locomotives for the MoS, by authorizing large-scale trials of diesel engines on main-line services. Within a few years, however, changing patterns of fuel availability, growing power requirements, and other unforeseen developments drove British Railways to accept the need for speedy conversion to diesel traction, retrospectively confirming the earlier judgement of critics like the *Engineer* that a less ambitious standardization scheme focused on fittings and components rather than complete steam locomotives might have been more prudent.[25]

The impact of Americanization and the productivity drive on the private sector can be examined at both a collective and an individual level. At the level of collective organization, trade associations in a variety of sectors from motor vehicles, internal combustion engines, and radio equipment to machine tools, textile machinery, and railway rolling stock established special committees to promote standardization and productive efficiency through exchange of technical information and agreement on joint component specifications. In motor cars, for example, a committee of chief engineers and purchasing managers from the 'Big Six' manufacturers began to publish common standards books for components such as brake shoes, the designs of which were reduced from 136 to 10 basic sizes; while rolling-stock manufacturers worked together to simplify components and foster the acceptance of standardized designs by colonial railway authorities. Trade associations in some sectors like aircraft, locomotives, and cable manufacturing carried interfirm collaboration considerably further through the provision of common services such as testing, procurement of raw materials, sharing of equipment and tooling, and collection of information on foreign markets. Perhaps the most widespread such service was co-operative R&D, with the formation during or immediately after the war of new research associations for sectors such as motor vehicles, machine tools, radio equipment, internal combustion engines, and marine engineering. Many of these voluntary collective initiatives, to be sure, were aimed at forestalling the possibility of state compulsion, and they proved most effective when the authority of trade associations could be reinforced by that of government, as in the case of aircraft where the MoS required the use of standard components on all military contracts. In

---

[25] Chick, *Industrial Policy in Britain*, 157–64; Gourvish, *British Railways*, 88–9; Johnson and Long, *British Railways Engineering*; PEP, *Locomotives* (London: PEP, 1951), 10–11; *Engineer*, 21 Jan. 1949, 21 Sept. 1951.

motor vehicles, by contrast, designers remained free to deviate from common component specifications by invoking a 'failure to agree' provision; while Nuffield and Austin broke off arrangements for the exchange of confidential information in 1949, following the revival of long-standing hostilities between their chief executives and the Labour Party's decision not to include the industry among its candidates for future nationalization.[26]

At the level of the individual firm, as one might expect, the practical impact of Americanization varied widely, depending, among other things, on sectoral patterns of demand and the strategic choices of top management. Thus, for instance, AACP follow-up reports found much greater willingness to standardize and invest in special-purpose machinery and tooling among manufacturers of valves, an intermediate good for which demand was large and rapidly expanding, than among makers of diesel locomotives, a complex finished product for which the domestic market, unlike that in the USA, remained extremely narrow.[27] Within each sector, too, some firms were vastly more enthusiastic than others about the commercial prospects of mass-produced goods. One noteworthy case was Standard Motors, which reconverted its wartime shadow aircraft factory for high-volume production of standardized cars and tractors using a single interchangeable engine for both types of vehicle and large numbers of automatic machine tools imported from the United States. To ensure adequate labour supplies and workforce commitment for the anticipated quantum leap in productive scale, moreover, the firm broke with the local engineering employers' association by agreeing a new wage structure with the unions based on a drastic reduction in job grades, shorter hours, a guaranteed weekly rate, and high output bonuses for large, self-organizing gangs.[28]

But even where top management was less sanguine than Standard's Sir John Black about betting the company on the success of a single design, the reconstruction period saw a widespread rationalization and narrowing of product ranges, often involving mergers and amalgamations as well as the reorganization of existing firms. These tendencies could be seen most clearly in fast-growing

[26] PRO SUPP 14/141: industry comments; SUPP 14/333: 'Production Efficiency', memos. by employer members of the EAC, July 1949 and June 1950; minute by W. C. Wallis, 22 Dec. 1950; 'Committee for Standardisation of Engineering Products: Second Review of Action Taken on the Committee's Report, Nov. 1949'; AACP, *Simplification in British Industry* (London: AACP, 1950), 9–10; PEP, *Motor Vehicles* (London: PEP, 1950), 136; PEP, *Locomotives*, 20; *Engineer*, 28 Jan. 1949, 31 Mar. 1950; *Automobile Engineer* (AE), Jan. 1952; BPC, *Wheels of Progress*, 11; Timothy R. Whisler, *At the End of the Road: The Rise and Fall of Austin-Healey, MG and Triumph Sports Cars* (Greenwich, CT: JAI Press, 1995), 79–80.

[27] BPC, *Top of the Shop: A Productivity Review in the Valves Industry* (London: BPC, 1954), esp. 4–5; BPC, *Wheels of Progress*.

[28] Nick Tiratsoo, 'The Motor Car Industry', in Helen Mercer, Neil Rollings, and Jim Tomlinson (eds.), *Labour Governments and Private Industry: The Experience of 1945–51* (Edinburgh: Edinburgh University Press, 1992), 170–3; Steven Tolliday, 'High Tide and After: Coventry Engineering Workers and Shopfloor Bargaining, 1945–80', in Bill Lancaster and Tony Mason (eds.), *Life and Labour in a Twentieth-Century City: The Experience of Coventry* (Coventry: Cryfield Press, 1985), 209–10; Seymour Melman, *Decision-Making and Productivity* (Oxford: Blackwell, 1958), 32–40, 154–5.

sectors like motor vehicles, internal combustion engines, and related components. Thus the National Advisory Council for the Motor Manufacturing
Industry (NACMMI) claimed that the number of basic models in 1948 would be
reduced to 42 from 136 in 1939, with an even greater drop in the number of body
variations from 299 to 40. During the same years, Austin replaced eight basic
models and seven engines with three of each; while Nuffield pruned its more diffuse offerings from 17 to 9 basic models, though the number of engines fell only
from 10 to 9. Despite Nuffield's open hostility to the 'one factory, one model'
campaign of the MoS, moreover, its Board privately conceded that there was 'no
doubt' that rationalization was the 'right course to pursue, both economically
and politically'. In 1949, therefore, the group appointed a corporate standards
engineer and undertook a major reorganization of its production facilities, transferring the manufacture of Wolseley cars to Cowley and concentrating output of
rear axles and front suspensions for the full range of Nuffield vehicles at the
newly vacated Washwood Heath plant near Birmingham. After the two rivals
merged in 1952 to form the British Motor Corporation (BMC), its new chief
executive Leonard Lord swiftly reorganized production of mechanical parts,
reducing the number of engines from nine to three basic sizes, while retaining a
larger number of body shells for the separate Austin and Morris marques.[29]

Rather more ambitious were the new model strategies pursued by the
American-owned motor vehicle firms. Before the war, Vauxhall, General
Motor's British subsidiary, had followed a diversified product policy comparable to that of its indigeneous competitors, manufacturing five distinct engine–
chassis combinations and ten body shells for an output of some 35,000 cars and
25,000 Bedford commercial vehicles in 1938. From 1947 onwards, however,
Vauxhall concentrated production on just two new car models, the Wyvern and
the Velox, with a single four-seater body and maximum use of common components and dimensions (including bores and strokes) across their four and six-
cylinder engines. Ford, the UK product range of which was already distinctively
concentrated before 1939, adopted a similar design approach not only to the
engines for its new Consul and Zephyr models, introduced in 1950, but also to
those for its Fordson tractors, whose petrol, diesel, and vaporizing oil versions
shared a large number of common parts, including cylinder blocks, crankshafts,
and connecting rods.[30]

A major impetus towards standardization came from component manufacturers themselves. Already before the war, top managers at Lucas, the dominant
supplier of electrical and fuel-injection equipment for British motor vehicles,
were deeply impressed by the efficiency of American component firms like

[29] MoS, *National Advisory Council for the Motor Manufacturing Industry: Report on
Proceedings* (London: HMSO, 1947), 3, 10; PEP, *Motor Vehicles*; R. J. Wyatt, *The Austin,
1905–1952* (Newton Abbott: David & Charles, 1981), 247, 287; Whisler, *At the End of the Road*, 22,
79–80; *Engineering*, 10 Nov. 1961.

[30] PRO CAB 124/626: John Jewkes, 'Post-War Resettlement of the Motor Industry', 4 Dec. 1944,
para. 9; BT 64/2314, tabular analysis of motor vehicle manufacturers' model policies, Dec. 1938;
PEP, *Motor Vehicles*, 26, 30; *AE*, July 1951, Feb. 1952, June–July 1952.

Delco-Remy, the manufacturing methods of which they sought to emulate within their own plants as far as possible. After the war, the company, whose staff made 85 visits to the USA between 1945 and 1949, formulated ambitious plans to double direct operator productivity through further standardization of design. Working through the NACMMI and the 'Big Six' standardization committee, Lucas vigorously pressed car manufacturers to adopt common specifications for components such as dynamos, starters, batteries, coils, and headlamps. By such means, managers envisaged, 168 current models could be replaced by just 41 standard designs, yielding longer runs and lower costs, a point driven home by the imposition of substantial price penalties on non-standard products. Lesser component firms like Glacier Metal, suppliers of plain engine bearings for a variety of applications, likewise aimed to reduce costs, improve deliveries, and justify investment in high-performance tooling by replacing some 20,000 customers' drawings with a standard list of designs and sizes manufactured for stock.[31]

Even in more specialized sectors where opportunities for full-scale rationalization appeared more limited, manufacturers none the less sought to reduce the complexity of their product pallettes through partial standardization. Thus machine-tool firms like H. W. Kearns, Joseph Stubbs, and William Asquith introduced unit-construction systems whereby families of equipment such as horizontal borers or vertical millers in different sizes and specifications could be derived from standard, interchangeable components like spindle slides, tables, uprights and beds; in one of these plants, the beds themselves were broken down into modular units to simplify the production of castings. More innovatively still, British manufacturers like Asquith, Archdale, and Austin developed automatic multifunction and transfer equipment constructed from standard, recombinable mechanical units and special machining heads. In textile engineering, similarly, the TMM combine (Textile Machinery Manufacturers Ltd.) designed its new models so that 'machines of different sizes' could be 'built up from suitable combinations of basic units', as well as concentrating production of each major type of equipment in a separate operating company.[32]

Such applications of the 'three Ss' often paved the way, as apostles of the American productivity gospel hoped, for a more or less far-reaching transformation of manufacturing methods. Thus in a number of sectors, longer production runs and rapidly rising demand encouraged firms to install high-throughput special machinery and tooling, especially after constraints on new

[31] Harold Nockolds, *Lucas: The First Hundred Years, vol. II: The Successors* (Newton Abbott: David & Charles, 1978); AACP, *Simplification in British Industry*, 10; PRO BT 64/2314: T. E. Easterfield, 'Standardisation as an Aid to Productivity' (1949); *Engineer*, 5 Aug. 1949.

[32] C. A. Sparkes, 'Design in Relation to Manufacture', *Transactions of the Manchester Association of Engineers* (1950–1); T. P. N. Burness (Asquith), 'Modern Machine Tool Practice', *Transactions of the Institution of Engineers and Shipbuilders in Scotland* (1945–6); *AE*, Feb. and June–July 1952, Mar.–Apr. 1953; AACP, *Simplification in British Industry*, 11; MoS, *Interim Report of the [Evershed] Committee of Investigation into the Cotton Textile Machinery Industry* (1947), paras. 31, 44.

capital investment began to ease in the late 1940s and early 1950s. The most spectacular example was the automatic transfer equipment introduced by car manufacturers like Ford, Vauxhall, Austin, and Nuffield, in which mechanical handling devices moved major components such as engine blocks, cylinder heads, and gearboxes through a series of multifunction machining stations without any direct intervention by the operator. Austin was especially active in extending the use of both in-line and rotary-transfer machinery; by 1953, it was claimed, the company had applied such equipment to a much wider range of parts than had been done in any other British automobile factory, stealing a march on the USA itself in the case of crankshafts. Where production volumes were insufficient to justify this level of mechanical integration, automatic multifunction machinery could none the less be employed on a stand-alone basis, as in the manufacture of fuel-injection equipment at Leyland and CAV, Lucas's commercial vehicle subsidiary. In bodywork, similarly, these motor vehicle firms invested heavily in special welding machinery, elaborate press tools, and mechanical handling devices, as well as new automatic rustproofing plants and conveyorized paint shops. So did US-owned manufacturers of electrical appliances, like Frigidaire (a GM subsidiary), which retooled its Kingsbury factory with template dies, tangent benders, and automatic spot-welding machinery for high-volume production of refrigerator cabinets; or Hoover, whose new washing-machine factory at Merthyr Tydfil incorporated an automatic anodizing plant along with a host of special-purpose equipment for pressing, shaping, and welding the main shells. Even in textile engineering, where growth prospects appeared less rosy, TMM re-equipped a wartime shadow factory for fully jigged machining of standardized parts to fine limits by semi-skilled operators.[33]

Less dramatic but no less significant in many cases was the reorganization of production layouts along flow-line principles. Here again, Austin provides an outstanding example: between 1946 and 1951, the company comprehensively replanned and mechanized its stores and assembly shops at Longbridge, integrating a maximum number of operations on to the lines themselves, and controlling the flow of materials by means of punched-card readers in each department. But many large manufacturers of cars, commercial vehicles, internal combustion engines, and electrical appliances such as Rover, Leyland, Perkins, Hoover, and Frigidaire also introduced power conveyors and mechanical handling systems to ensure progressive movement of work through assembly, machining, and foundry processes. Other firms, however, often sought to obtain the benefits of flow-line layout through the use of less costly devices like roller conveyors and gravity slides, as in the cases of diesel-engine erection at

---

[33] Woollard, *Principles*, 30–3, 147–50; *AE*, Nov. 1947, July–Oct. 1948, May 1950, July and Oct. 1951, June–July 1952, Mar.–Apr. and Aug. 1953; BPC, *Applying the Pressure: A Review of Productivity in the Pressed Metal Industry* (London: BPC, 1953); H. J. Graves, 'An Outline of BMC Developments in the Field of Automation', *Journal of the Institution of Production Engineers* 36 (1957); *Engineering*, 10 Mar. 1950; MoS, *Interim Report of the [Evershed] Committee*, para. 30.

Brush's McLaren subsidiary, twist-drill machining at the English Steel Co., electric cooker manufacture at GEC's Swinton plant, or even cylinder-head machining for Vauxhall's Bedford trucks.[34]

Probably the most widespread transatlantic import of all was work study, which diffused rapidly through British manufacturing industry during the 1940s and 1950s. While time study for rate-fixing and production control had long been familiar in UK engineering circles, and motion study became more so as a result of wartime government initiatives like the Production Efficiency Board and the USA-originated Training Within Industry programme for supervisors, the AACP missions and the post-war productivity drive helped to popularize related American techniques such as work simplification and predetermined task allowances. Some domestic companies introduced work-study methods as a basis for production planning as well as for incentive payment systems, as in the case of valve manufacturers such as Glenfield & Kennedy and Walker, Crossweller & Co. or the motor vehicle firms discussed above. Occasionally, as at Perkins' Peterborough diesel-engine factory, rationalization of production, job analysis, and predetermined output targets could enable management to eliminate incentive bonuses altogether in favour of a flat hourly rate. More often, however, as with inter-war task measurement and incentive systems like Bedaux, British engineering firms looked to work study instead as a means of raising output and effort levels which could easily be 'added on' to existing production arrangements without the need for far-reaching reorganization.[35]

4.4 Obstacles to Americanization: Supply and Demand

Thus even in the private sector, a significant stratum of British engineering manufacturers responded positively to official initiatives and the allure of Americanization in the hope of raising productivity and tapping into vast potential markets for mass-produced goods. Some of these, to be sure, were local subsidiaries of American-owned multinationals, presaging a growing trend towards US direct investment during the 1950s and 1960s; but others like Austin, Lucas, Perkins, or GEC ranked among the leading indigenous companies in their sectors. At the same time, however, Britain's dire post-war economic position and the Labour government's broader policy choices perversely combined to limit the scope for Americanization in the short and medium term. Foremost

[34] Woollard, *Principles*, 23; *AE*, Sept. 1946, Mar. 1947, Sept. 1951, Feb. 1952, and Aug. 1953; *Engineering*, 18 Nov. 1949, 12 Jan. and 13 Apr. 1951, 21 Mar. 1952; AACP, *Final Report*, 33.

[35] For work-study and payment systems at Glenfield & Kennedy, Walker, Crossweller & Co., and Perkins, see BPC, *Top of the Shop*, 20–1, 27; 'Re-organization of a Diesel Engine Works', *Engineering*, 3 July 1953. For extended discussions of the diffusion of work-study techniques in British industry during the 1940s and 1950s, see S. J. Dalziel, 'Work Study in Industry', *Political Quarterly* 27 (1956), 270–83; Carew, *Labour under the Marshall Plan*, 168–9, 199, 207–10; Tiratsoo and Tomlinson, *Industrial Efficiency and State Intervention*, 148–50; id., *Conservatives and Industrial Efficiency*, 53–4; Howard F. Gospel, *Markets, Firms and the Management of Labour in Modern Britain* (Cambridge: Cambridge University Press, 1992), 118–20.

among the obstacles facing prospective mass producers were widespread but unpredictable shortages of vital inputs such as fuel, power, labour, and raw materials—above all steel—which inhibited expansion plans and disrupted manufacturing programmes. Closely related were official restrictions on capital investment, based on an explicit policy of 'make-do-and-mend' favouring expenditure 'directed to getting more output from existing plant rather than the creation of fresh capacity'. These restrictions and shortages, which had helped to inspire the productivity drive in the first place, were further reinforced by the scarcity of foreign exchange for imported equipment and an export quota of 50 per cent for British machine-tool output. Understandable though they were in light of Britain's precarious balance of payments, export quotas of 50–75 per cent for many engineering goods, backed up by controls over steel allocations and adjusted at short notice, not only prevented manufacturers from building up stable home demand for standardized products, but also forced them to disperse their output across a wide range of foreign markets, often based on short-run currency considerations rather than long-run commercial strategy. Although supply-side bottlenecks began to ease after 1948, permitting a relaxation of physical controls, they returned with a vengeance in 1950 as a result of the crash rearmament programme triggered by the outbreak of war in Korea. Defence expenditure on metal and engineering goods more than doubled over the next two years to absorb nearly 16 per cent of British output and 18 per cent of employment by 1952–3, leading to acute shortages of labour and steel, as well as still higher export quotas and a further squeeze on the home market for consumer products like motor cars and electrical appliances.[36]

Such pervasive macroeconomic uncertainty magnified the risks and exacerbated the difficulties inherent in the pursuit of mass-production strategies, as a number of their most ardent British protagonists painfully discovered. Among the most disappointing experiences was that of Standard Motors. After initial successes in export markets during the late 1940s, the firm's medium-sized Vanguard car encountered growing consumer resistance abroad during the early 1950s as competing models became more widely available, not least because its design had been shaped, as The Economist caustically noted, 'less by market research than by the desirability of using the same engine in it as in the Ferguson tractor'. By 1952, Standard had the worst export record of any of the 'Big Six' manufacturers, only managing to avert a threatened cut in its steel allocations by the introduction of a new sports car model, which unlike the firm's saloons rapidly carved out a lucrative niche in the North American market. In the case of the Ferguson tractor, development costs proved considerably higher than anticipated, while production also increased more slowly and had to be

---

[36] For fuller analysis of the impact of supply constraints, official controls, and rearmament, see Alec Cairncross, The Years of Recovery: British Economic Policy, 1945–51 (London: Methuen, 1985); Mercer et al., Labour Governments and Private Industry, esp. chs. 2 and 5 (quotation, 80); PRO BT 258/922: MoS, 'The Impact of the Defence Programme on Industry', 18 Sept. 1956 (figures, table 3).

sharply curtailed in 1949 due to a collapse in sales. Since more diversified firms like David Brown and others managed to increase tractor output during the same year, whereas that of Ford and Standard fell by a similar proportion, this reverse led some contemporaries to question 'whether the production programmes of the two largest manufacturers are too rigid', while highlighting for others the dependence of successful product specialization on 'an accurate assessment of market requirements'. Even when sales did take off during the early 1950s, moreover, the tough licensing terms negotiated by Harry Ferguson ensured that Standard's profit margins remained thin. Although the company's innovative labour agreement delivered tangible benefits in terms of industrial peace, workforce stability, reduced overheads, and increased productivity, Standard never made the expected jump in volume or market share, and its wage levels gradually fell back towards those of other Coventry motor firms.[37]

Other would-be mass producers encountered analogous if less damaging setbacks. At Glacier Metal, for example, 28 per cent of production time in 1948 was spent on runs of ten days or less, for which elaborate press tools were uneconomic, as customers insisted on immediate delivery and were not prepared to wait for special machines to be built and set up, or demanded deliveries of small quantities over a period of time to avoid carrying large stocks. Nor could the firm's efforts to promote its standard bearings through price differentials have much incentive effect, given the small share of this component in the total cost of a motor. Even firms with greater market clout like Lucas found their standardization programmes impeded by the necessity of manufacturing components and spares for the wide range of countries, often with different technical specifications, to which their customers' products were exported in hopes of meeting official targets. At Nuffield's Cowley plant, similarly, a combination of steel shortages, restrictions on domestic sales, and difficulties in export markets reduced capacity utilization on special-purpose equipment installed for the production of its Morris Minor and Oxford models to just 42 per cent in 1951 and 52 per cent the following year. Manufacturers of electrical appliances such as cookers, refrigerators, washing machines, heaters, and vacuum cleaners likewise found their anticipated ability to achieve long runs and production continuity impeded by materials shortages, erratic power supplies, export quotas, and fluctuations in purchase tax on domestic sales during this period.[38]

4.5 Selective Adaptation and Hybrid Innovation

In this volatile environment, it is scarcely surprising that many British metal-working industrialists preferred a more selective and incremental approach to

[37] Tiratsoo, 'Motor Car Industry', esp. 178; *Engineering*, 27 Jan. 1950, 89.

[38] Easterfield, 'Standardisation as an Aid to Productivity', app. A; Nockolds, *Lucas*, vol. II, 92; Whisler, *At the End of the Road*, 110–11; T. A. B. Corley, *Domestic Electrical Appliances* (London: Jonathan Cape, 1966), 41–8.

the adoption of American-style manufacturing methods. Hence even those British metalworking firms which embraced mass production most enthusiastically often creatively modified US manufacturing methods to handle a wider range of products in smaller quantities through hybrid innovations similar to those later made famous by the Japanese. Thus Austin built up its transfer machinery from various combinations of special unit heads manufactured in-house with interchangeable Archdale bases and mechanical handling devices; not only could these standard elements be reconfigured for work on new models with a minimum of scrapping, but design changes could be accommodated on existing equipment by adding new sections. In body manufacture, similarly, Nuffield Metal Products developed 'special-purpose' welding machines composed of as many standard components as possible, some 80 per cent of which could be reused after each run. Even firms with relatively short runs of repetition work had begun to utilize flow-production techniques, observed Frank Woollard, a leading automotive engineer, by grouping similar components such as shafts, pulleys, gears, and cams 'to permit a layout of the machines in operation sequence', and then adapting fixtures, jigs, tools, and gauges 'to reduce the changeover time to a minimum'.[39]

Nor were such modifications of American mass-production techniques to deal with smaller and more diverse markets confined to British-owned enterprises. While its US parent company pursued the dedicated automation strategy which culminated in the highly productive but rigid Cleveland engine plant of 1951, Ford UK built a single transfer line to machine the four- and six-cylinder blocks for its Consul and Zephyr models, thereby spreading the costs over a larger number of units, with no more than 20 minutes needed for a complete change-over between the two. For the Fordson tractor, with its lower volumes and three different varieties of engine, Dagenham preferred more adaptable multifunction machines with special fixtures designed 'to allow change-over from one type to another . . . with a minimum of lost time'. Vauxhall, with a motor vehicle output considerably lower than that of Ford UK, rejected transfer lines altogether in favour of multi-head automatics for its four- and six-cylinder engines, with separately-tooled equipment on parallel tracks to cope with unavoidable variations in machining operations between the two. Although this policy entailed higher initial capital expenditure and more complex work routing, countervailing advantages included 'the much greater productive flexibility that comes from being able to machine either size in any desired order', reduction in idle time 'since changes are made only when tool deterioration makes them necessary and not at the end of every batch', and

---

[39] *AE*, Mar.–Apr. and Aug. 1953; Graves, 'BMC Developments in the Field of Automation'; BPC, *Applying the Pressure*, 19–20; *Engineering*, 3 July 1953; Woollard, *Principles*, 42–3, 83, 150. During the mid-1950s, Nissan adapted the design of transfer machinery from Austin with which it had a production tie-up and technical assistance agreement; while Toyota's Ono first encountered stamping presses equipped for rapid die-changes in the United States, where this feature was rarely used: see the chapter by Wada and Shiba in this volume.

lower levels of work in progress by removing 'the need for contingency stocks to meet sudden changes in demand'. Like Ford and Austin, moreover, Vauxhall developed these innovative engine-manufacturing lines in close collaboration with British machine-tool firms such as Asquith and Archdale, as well as the local subidiary of Cincinnati Milling; like Austin, too, its automatic machines were fabricated from standard drive units and special multi-spindle heads, thereby simplifying and cheapening 'modification to suit design changes'.[40]

In assembly as in machining processes, British engineering manufacturers adapted standard mass-production methods to accommodate greater product variety and shorter runs through practices similar to those developed by their Japanese counterparts. At Austin's new mechanically controlled assembly plant, opened in 1951, three different body and chassis types, as well as right- and left-hand steering, were put together on a single track; normal variations in assembly sequence could be introduced in 90 to 120 minutes, while colour changes involving operations remote from the line took no more than eight hours. At Nuffield Metal Products, too, four different body types could be assembled on multipurpose jigs without any resetting, allowing transfers of urgent work from one line to another in case of breakdowns. Rover carried this approach a step further by assembling several distinct models simultaneously on the same tracks, as did Leyland, which built built all types of four-wheel chassis on a single three-section line, with extra stations for more complex models. Even in the case of more specialized products like machine tools, BSA Tools managed to assemble many different sizes and types of equipment on a single manual track, with machine bases pushed on trolleys from station to station.[41]

More significantly still, a number of British engineering firms introduced production control systems aimed at minimizing internal buffer stocks and work in progress, in line with contemporary critiques of American practice discussed earlier. Thus Austin, for example, reduced floating stocks on the assembly lines, cut contingency stores to two days' requirements, and achieved stock turnover ratios in some departments of more than 50 times a year by issuing parts for each vehicle in individual sets from a common 'marshalling yard', which served as 'an automatic progress chaser, a balancing mechanism and a reservoir with a small but effective "head"'. Similar systems were introduced at both Rover and Leyland, where sub-assemblies were produced only as required for immediate use on the chassis assembly lines, while Perkins, too, deliberately maintained 'only small buffer stocks of components and sub-assemblies' in order to maximize the floor area devoted to actual production processes. Despite materials shortages and other disruptions associated with post-war reconstruction,

---

[40] AE, Sept. and Oct. 1951, Feb. 1952, 66, and June–July 1952, 211, 261. On the genesis and inflexibility of Ford's Cleveland engine plant, see David Hounshell, 'Planning and Executing "Automation" at Ford Motor Company, 1945–65: The Cleveland Engine Plant and its Consequences', in Haruhito Shiomi and Kazuo Wada (eds.), Fordism Transformed: The Development of Production Methods in the Automobile Industry (Oxford: Oxford University Press, 1995), 49–86.

[41] Woollard, Principles, 45–7, 84, 173–5; BPC, Applying the Pressure, 20; AE, Sept. 1946, Feb. 1951.

moreover, both Austin and Rover sought to extend these systems of low-inventory manufacture beyond their own plants through 'frequent deliveries of relatively small quantities from outside suppliers'. While Austin did not always find it easy 'in the circumstances existing to-day . . . to obtain the necessary degree of co-operation', *Automobile Engineer* reported, 'several companies have already realized that the system is good for both the supplier and user of the materials', and the car firm had designed special standardized trailers for transporting parts from outside factories.[42]

Although these British manufacturers were still attempting to push rather than pull components through the production process, their rationale for cutting inventories and stripping out buffer stocks closely resembled that of the 'just-in-time' system pioneered by Toyota and later adopted by other Japanese automobile companies. First and foremost, such British firms, like their Japanese counterparts, sought to hold down capital investments in stocks and work in progress, particularly since materials costs had risen much faster than those of other inputs in relation to pre-war prices. Like Toyota, too, however, a number of British engineering companies also saw the reduction of inventory as a mechanism for inducing continuous improvements in product quality and productive efficiency. At Austin, for example, managers expected that the new assembly lines would exert an indirect pressure on upstream production processes 'since there is no means of switching a faulty component or sub-assembly and replacing it by another one', so that 'the quality of the product . . . will be obtained with much less rectification than hitherto'. At Rover, similarly, the firm believed that the elimination of buffer stocks would 'inevitably engender in the personnel of the factory a sense of care that will go far to prevent interruption of the work flow'; while at BSA Tools, 'the major advantage' of flow production was held to be that 'it throws into high relief any shortages which might jeopardize the time schedule for the production of the machine'. Among the benefits of manufacturing components on the 'group system', Woollard likewise noted, was that 'faulty work will be found almost immediately and the necessary adjustments made . . . since inspection will naturally be in line with the machines'.[43]

Such production systems typically required a considerable measure of functional flexibility on the part of even semi-skilled operatives to cope effectively with the variations in tasks and times involved in turning out different products on the same lines. '[I]n manning a continuously operated plant', advised Woollard, it was 'of very considerable help' to train 'operators to do several jobs . . . Operators who have had the opportunity of working on many and various

[42] *AE*, Sept. 1946, Mar. 1947, Sept. 1951, 329, Aug. 1953, 344; Woollard, *Principles*, 21, 172.
[43] *AE*, Sept. 1946, 395, Mar. 1947, Sept. 1951, 329; Woollard, *Principles*, 42–3, 47, 172. According to Cusumano, *Japanese Automobile Industry*, 315, Nissan adopted 'an American technique called "marshalling"' in 1955 to improve the coordination of component supply in its Yokohama factory, but given the firm's tie-up with Austin, the source of this borrowing is more likely to have been British. Cf. also the chapter by Wada and Shiba in this volume, n. 74.

machines and assembly lines make ideal flying-squad men', who could be frequently redeployed to overcome bottlenecks and imbalances in production.
Often, however, demands for versatility extended beyond small squads of multi-
skilled operatives to embrace whole sections of the work-force; and many
British manufacturers of motor vehicles, internal combustion engines, electrical
appliances, and other engineering products found the solution to this problem
in group bonus systems which tied the earnings of each worker to the output of
an entire production line. At Leyland, for example, not only assembly-line
workers but also machine operators were remunerated according to 'a pooled
bonus scheme based on the group output of completed parts. When necessary
the operators move freely from one machine to another in the group. This is an
important factor in maintaining an even flow of parts and in preventing an accumulation at any stage along the machining line.'[44]

4.6   Conclusions

Despite the Labour government's staunch commitment to planning and stability as the basis for full employment and industrial modernization, few periods
have seen greater turbulence in product markets, factor supplies, and technologies than that of post-war reconstruction. Under these volatile conditions, a
number of the most single-minded experiments with Americanization and mass
production came wholly or partially unstuck in both the public and private sectors, substantiating the advance reservations expressed by contemporary critics.
Often, too, collective schemes for co-ordinated specialization or industry-wide
standardization broke down in the absence of effective governance mechanisms
capable of ensuring the compliance of individual enterprises, as for example in
motor vehicles. Hence most British metalworking firms, the available evidence
suggests, tended to adopt incremental strategies of productivity improvement in
which transatlantic techniques were selectively grafted on to indigenous manufacturing practices, as in the case of work study. But as we have seen, a significant minority of domestic engineering manufacturers, including the local
subsidiaries of US-owned multinationals, creatively modified American-style
mass-production methods to accommodate greater variety and shorter runs,
thereby adjusting to differences in the structure and stability of British markets.
In some instances, these hybrid innovations built on practices known to
American production engineers but rarely applied in US industry; in many
respects, too, they closely resembled and may even have helped to inspire the
techniques of reconciling product variety with productive efficiency later
deployed so successfully by Japanese manufacturers like Toyota and Nissan

[44] Woollard, *Principles*, 82; *AE*, Jan. 1951, Feb. 1951, 65, 67–8; *Engineering*, 2 Jan. 1951; AACP, *Final Report*, 34; BPC, *Applying the Pressure*. For a comparative analysis of group piecework sys-tems in Coventry engineering plants during the post-war period, see Tolliday, 'High Tide and After'.

such as quick tooling changes, mixed-model assembly, and just-in-time component supply.[45]

Despite its enormous ideological influence, therefore, the practical impact of the Americanization drive on British engineering remained surprisingly limited during the first post-war decade. It was only between the mid-1950s and the mid-1970s, as a massive wave of government-promoted mergers and takeovers transformed the structure of the industry, that British engineering companies decisively abandoned their indigenous model of productive organization in favour of imported management techniques such as multidivisionalization and measured day work. Far from reviving its competitive fortunes, however, this putative Americanization of British engineering was associated instead with a rapid loss of market share both abroad and at home, resulting in a steep decline of domestic production and employment. By the 1980s, ironically, the competitive difficulties of British engineering firms, like those of the Americans themselves, were frequently attributed to their inability to match the standards of product innovation and productive flexibility set by the Germans and the Japanese in meeting the demands of increasingly diverse and volatile international markets.[46]

[45] For a fuller discussion, see the present author's 'Reconciling Automation and Flexibility? Technology and Production in the Postwar British Motor Vehicle Industry', forthcoming in *Enterprise and Society* 1 (2000).

[46] These arguments will be developed and documented more fully in this author's *Between Flexibility and Mass Production*, pts. III–IV. For a preliminary overview, see Paul Hirst and Jonathan Zeitlin, 'Flexible Specialization and the Competitive Failure of UK Manufacturing', *Political Quarterly* 60 (1989), 164–78. For recent discussion of the negative impact of the adoption of the American multidivisional model on the post-war performance of large British manufacturing companies, see Geoffrey Jones, 'Great Britain: Big Business, Management, and Competitiveness in Twentieth-Century Britain', in Alfred D. Chandler Jr., Franco Amatori, and Takashi Hikino (eds.), *Big Business and the Wealth of Nations* (Oxford: Oxford University Press, 1997), esp. 122–36; id., 'Global Perspectives and British Paradoxes', *Business History Review* 71 (1997), 290–8.

Chapter 5

Failure to Communicate: British Telecommunications and the American Model

KENNETH LIPARTITO

5.1 Introduction

Throughout the twentieth century, British telecommunications has remained substantially behind the United States in measures such as productivity and telephones per capita. Britain also failed to adopt advanced electronic switching technology for nearly two decades after the United States. It lagged behind other nations as well, such as Sweden and Japan. Nor did telecommunications manufacturing in Britain keep pace with global leaders. In attempting to redress these failures, however, British policy makers made two mistakes. They ignored powerful international tendencies embodied in 'best practice' nations such as the United States. Yet they also responded to backwardness by adhering too closely to policies devised for other places and other times. So both those who affirm and those who deny that America represented a single, international best way will find lessons here.

Telecommunications combines a giant system of distribution with large-scale equipment production. Crucial to the growth of this industry are the links between the network and manufacturing. Items of equipment, including telephones, are in essence capital goods. They are inputs into the provision of communications services. But demand for communications services cannot be conveyed directly between users and producers. Communications demand is interdependent, meaning that it is closely controlled by the number of other users. Networks are the result of this interdependency, and they are the 'market' to which innovators, manufacturers, and consumers alike look when making decisions.

The network largely controls changes in technology, service, and equipment. Though breakthroughs in technology may come from unexpected sources, advances in design and manufacturing still require strong demand-side links and motivations. The size and rate of growth of the network tells manufacturers what to make and how much to invest. The need to solve operational and technical problems stemming from network growth provides the key 'reverse salients' for innovators. No manufacturing firm in telecommunications can succeed without the right links to a network.

These relationships had been worked out clearly in the United States by the 1920s. There, private ordering through vertical and horizontal integration plus government regulation assured rapid and co-ordinated advances in network demand and new products and services. This model, however, was not the only possibility. Germany and Scandinavian nations, for example, had publicly owned networks supplied by private firms. They devised means of co-ordination different from those in the United States. Japan developed still another set of successful institutional arrangements. In Britain, structure and policy were most problematic.

Britain's fundamental mistake came from a policy of underinvestment in the domestic telecommunications network. To some extent, this policy reflected constraints from outside the nation, though much of the blame must lie with the British Treasury and Post Office. Through the 1960s, these agencies seriously underestimated demand.[1] They did so for what can be called cultural reasons. As long ago as the late nineteenth century, British policy makers had placed telecommunications in a far more restricted framework than did American and Scandinavian telecommunications advocates.

Concerned about falling behind, Britain sought to shake up the bureaucracy, catch up to America, and inject competition in the supplier end of the business. These efforts led to a second mistake. It tried to imitate American practice, just at the moment when the American model had passed its prime. Successive British governments, even before the Second World War, had consistently misconstrued the very sources of American strength. They tended to be captivated by the advantages of size, scale, monopoly, and vertical integration. But they ignored alternative lessons that stressed the role of competition, users, and informal governance structures, even in the supposedly centralized Bell System. Comparisons show that other nations were more successful in telecommunications when they adapted practices from elsewhere to local circumstances. Success required a healthy balance between taking up global best practice and creating indigenous capabilities.

5.2   The Demand Side: A Failure of Market Imagination

In 1946, British telecommunications, like the British economy generally, stood at a crossroads. War had diverted substantial *matériel* and equipment from the normal needs of civilian communications, making for a deep deficit in investment. Yet war had also improved selected areas of telecommunications technology. Long-distance voice channels, for example, had increased markedly, as

---

[1] This was admitted by the Post Office in 1967. London School of Economics Library, Misc. 332/71, *Seminar on Problems of Industrial Administration*, 'The General Post Office: Its Tasks and Problems', 1967.

circuits were constructed for military needs. Breakthroughs such as microwave offered new alternatives to land-lines, while improvements in coaxial and carrier transmission systems provided greater bandwidth on existing circuits. Radio technology also benefited from military needs, and during the war, Britain captured or diverted enemy cables, adding to its impressive international telegraph network.

Post Office authorities faced the initial post-war years confidently. They foresaw taking advantage of these improvements in technology. They expected investment to recover, allowing them to rebuild parts of the network that had been destroyed and make the additions necessary to serve a reviving civilian economy. Transition to a post-war economy, however, depended on making new investments or different investments in technology and equipment. Waiting lists for residential service had grown during the Depression and war years, when fewer customer installations could be made. Advances in long-distance communications for military purposes did not necessarily serve the interests of civilian customers. Additions to transmission capacity had to be balanced by improvements in switching, and additions to local loops.

Two decades after war ended, few of these goals had been met (as is indicated by Figure 5.1). The fate of telecommunications was foreshadowed in the very first sterling crisis of 1947. Recommendations by the Plowden Economic Planning Council reduced investment sharply, resulting in a big increase in waiting lists for service, a constant feature of the British scene in the 1950s and early 1960s. Expansion of rural service was curtailed, and even investment in business lines, the top civilian priority, was scaled back.[2] In the following years, the Conservative government also starved telecommunications of capital, largely for reasons of fiscal restraint. These were years of 'stop–go' economic policy, with real constraints from trade deficits and pressure on the pound. The government was determined to revive the British economy by cutting public expenditure. Public industries such as the Post Office became tools of demand management, their budget priorities set in accordance with macroeconomic needs. This practice only stopped after the 1961 Plowden Report showed it to be ineffective. Telecommunications faced a second set of constraints, however, from the cold war, which diverted investment to military communications.[3]

Investment restrictions were especially harmful in a capital-intensive industry. Capital equipment was exactly what was needed, in the form of upgraded switches, more lines, and more channels on existing lines. The post-war years in

[2] British Telecom Archives, London, UK (hereafter BT), Post 70, Standing Telephone Advisory Committee (1947–1950). Waiting lists rose and fell, but even as late as 1960, some half a million people were wanting service. At the worst, the waiting time became so long that no more requests for service were taken.

[3] BT Post 70, Standing Telephone Advisory Committee (1953), on military spending. BT Post 70, Standing Telephone Advisory Committee (1960) on macroeconomic policy and the Post Office. Jim Tomlinson, *Public Policy and the Economy since 1900* (Oxford: Clarendon Press, 1990), 172–276. Peter Hall, *Governing the Economy: The Politics of State Intervention in Britain and France* (Oxford: Oxford University Press, 1986).

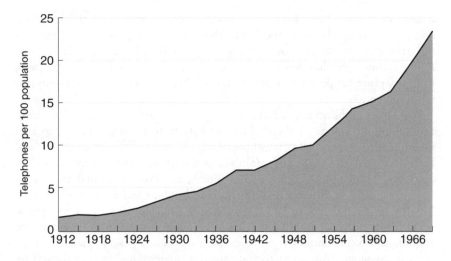

Figure 5.1. Telephone Density in Britain, 1912–1969

Source: B. R. Mitchell, *British Historical Statistics* (Cambridge: Cambridge University Press, 1988).

fact saw telecommunications become even more capital intensive, with greater potential economies of scale.[4] Improvements in capacity, speed, and network capabilities were largely driven by technology, while the role of labour, such as human operators, shrank.[5] Yet capital for public sector investment was scarce during the years of international crises that rocked Britain in the 1950s.

Budget reductions came from areas that directly affected the consumer. While contracts with suppliers and engineering work were partly shielded, deep cuts came in 'non-contract' engineering, which meant fewer telephone installations and delays in upgrading plant and opening new central offices. With Treasury support, the Post Office preserved longer-term investments against budget cuts. So research was generally protected, if not lavishly supported, and certain crucial long-term capital expenditures were allowed to proceed. But these choices placed customer service under even more pressure, since these long-term projects offered little immediate relief to overburdened facilities.

---

[4] As early as 1951, the small size and slow rate of growth of the network was causing under-utilization of capacity. Public Records Office, Kew, UK (hereafter PRO), T219/685, 'Post Office Charges, General Policy' (1951).

[5] Initially, labour was also in short supply, which frustrated the policy of making existing equipment last. Government policies discouraged the firing of workers, especially older ones, for fear that they would not find work again. Coming in conjunction with limits on capital, this policy hurt productivity in telephone service. Older workers, lacking training or knowledge of newer equipment, had to be kept on. New workers could not be hired. Engineers were demanded by higher-priority manufacturing firms. Shortages of vehicles meant that even when workers and materials were available, it was impossible to send them out to job sites. BT Post 70, Standing Telephone Advisory Committee (1952).

In order to deal with shortages of capital, as well as limits on labour, managers adopted a triage strategy. They concentrated scarce resources on the most vulnerable parts of the network, reasoning that if key switching points were unable to function, the entire network would suffer. As a result far more effort went into reconstructing London, which had the greatest concentration of users and most important business clientele, than went into rural areas or other cities. It was a policy that helped to increase London's attractiveness as a commercial, financial, and population centre, at the expense of other places.[6]

Though British capital investment grew during the 1950s and 1960s, it did not go into industries like telecommunications. Between 1950 and 1964, in fact, investment in telecommunications fell by nearly 7 per cent.[7] Existing technology and capacity were squeezed and stretched. Customers were kept in queues for long-distance circuits. Waiting lists for new service lengthened. Old equipment was kept functioning by spending more on maintenance.[8] Rates were raised and discounts eliminated, while service quality fell.[9] Net revenues actually increased, however, with telecommunications making a 5–7 per cent profit.[10] Those revenues proved extremely tempting to both Conservative and Labour governments. Post Office surpluses were generally expropriated by Parliament and placed in the general revenue account.[11]

The siphoning off of revenues and limits on investment led to new problems when economic growth picked up in the 1960s. Rising incomes and increasing demand (see Figure 5.2) caught the Post Office unprepared, behind schedule on modernization. Service quality deteriorated, outdated equipment proved inadequate, and personnel were stretched to the limit. Waiting lists again lengthened, with little prospect for relief in sight.[12] Only long-distance service was in reasonably good shape. It had benefited from the wartime and cold war spending on research. Rapid advances in transmission technology and the introduction of customer dialling for long-distance calls meant that by the late 1950s, the cost

[6] London in fact became one of the most telecommunications-intensive locations in the world by the early 1970s, but this demand alone did not assure the modernization and improvement of the British network. Military spending also tended to favour long-distance over local service, and centralized urban nodes over the suburbs or countryside.

[7] Jill Hills, *Information Technology and Industrial Policy* (London: Croom Helm, 1984), 112.

[8] BT Post 70, Standing Telephone Advisory Committee (1950). Piecemeal additions to the London trunk exchanges, for example, made it harder to reconstruct these exchanges later for direct long-distance dialling.

[9] Quality of service is hard to capture. But during the years of fiscal restraint, connection time either increased or showed no improvement. For most of the 1950s, it was about one second (15 per cent) higher than it had been in the 1920s. In 1951, answering time for London trunk calls shot up to more than twice the national average. BT Post 70, Standing Telephone Advisory Committee (1960), 'Quality of Telephone Service'.

[10] James Foreman-Peck and Robert Millward, *Public and Private Ownership of British Industry* (Oxford: Oxford University Press, 1994), 305.

[11] PRO T219/869, Reorganization of Post Office (1960). This practice was stopped for a few years after the 1932 Bridgeman Report, but it returned with the outbreak of war. The Bridgeman policy was reinstated after 1961, but only on a limited basis.

[12] BT Post 70, Standing Telephone Advisory Committee (1961).

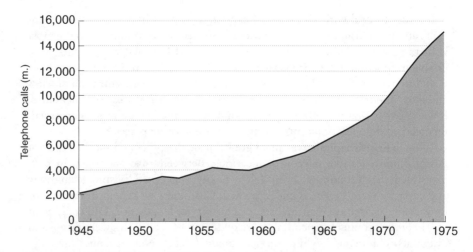

Figure 5.2. Total Telephone Calls in Britain (millions), 1945–1975
Source: B. R. Mitchell, *British Historical Statistics* (Cambridge: Cambridge University Press, 1988).

of long-distance circuits was dropping.[13] But because profits were high on long-distance lines, it was tempting to raise rates and increase revenues here to compensate for losses elsewhere. The profits on residential and especially rural service were estimated to be much lower than those on long-distance and business lines.[14] Ironically, the public Post Office in Britain was behaving as, in essence, a discriminating monopoly, taking profits in growing markets, underinvesting in lower-profit ones.

By 1965, the long-term costs of British telecommunications policy had become clear. Though in other utilities, Britain closed the productivity gap with the United States between 1950 and 1980, in telecommunications that gap actually widened. The American advantage may have been due in part to larger-scale operations and a larger network.[15] New technology was reducing the marginal cost of network expansion. The cost of telephone service was also falling much faster than that of other forms of communications on which the British public had been forced to rely—telegrams, teletype, mail.[16] The next generation of transmission technologies, PCM (pulse-code modulation, a type

[13] PRO T219/692, 'Relationship with Treasury and Financial Control, Report on Post Office Development and Finance' (1955).

[14] The attitude is succinctly expressed in PRO T219/692, 'Relationship with Treasury and Financial Control, Report on Post Office Development and Finance' (1955).

[15] Foreman-Peck and Millward, *Public and Private Ownership*, 308, 313, and 343 on importance of scale.

[16] BT Post 70, Standing Telephone Advisory Committee (1965), 'Investment Review and Industrial Inquiry'; Richard Pryke, *The Nationalised Industries: Policies and Performance since 1968* (Oxford: Martin Robertson, 1981), 251.

of digital transmission), waveguides, and advanced coaxial systems, portended an explosion of bandwidth. This new equipment required a far larger subscriber universe, but the stunted British network prevented full exploitation of scale economies.[17]

Matching demand (network size) to scale might have seemed simply a matter of more investment in user facilities. It was not, however; investment decisions on telecommunications were taken in a powerful cultural context. That context had grown up over decades of prior experience. It formed a series of unspoken assumptions about what sort of network the British public needed. Officials had long either assumed that customer demand was more limited than it was, or else deliberately stifled residential and, to a lesser extent business demand. History and culture, in short, restrained improvement of this British industry.

At the start of the twentieth century, Britain had taken a very different path in telecommunications than had the United States. In America, an outbreak of competition, combined with government regulation, prompted AT&T to adopt a bold new policy known as universal service. This phrase was loaded with implications, but one thing it came to mean was continual improvements in capital equipment in order to accommodate growing numbers of users at lower and lower rates. In a highly competitive telecommunications market, the base of telephone users expanded dramatically between 1894 and 1914. The American telecommunications network grew much faster than those in the rest of the world (see Figure 5.3). Although competition soon came to an end, the die was cast. There were simply too many people with telephones to go back to the old monopoly days of slow growth in selected, high-profit markets.

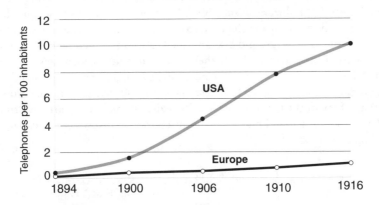

Figure 5.3. Telephone Growth, USA and Europe, 1894–1916
Source: Bell Telephone Quarterly (New York: AT&T, 1922) 128, 226.

---

[17] BT Post 70, Standing Telephone Advisory Committee (1965), 'Investment Review and Industrial Inquiry'.

In essence, competition had created a huge demand-side problem for AT&T. It forced the corporation to meet growing customer demands for more service, better service, and lower-cost service. The company responded by investing in research and emphasizing large-scale manufacturing. Universal service policy gave birth to a telecommunications analogue to the capital-intensive American system of manufacturing.

Nations like Britain, which never experienced competition, took a different road. They managed demand rather than expand supply. British authorities looked to America many times in the twentieth century. Each time they took away exactly the wrong lesson. Instead of seeing the increase in subscribership caused by competition, they lauded the Bell System monopoly. Not surprisingly, bureaucrats at the Post Office, and before them, the private, monopoly National Telephone Company, identified closely with the huge Bell corporation, far more than with small, independent operators which had flourished during competition.[18]

One revealing example of this selective borrowing from the United States is measured service. British officials insisted, with expert backing, that charging customers for each call, for the duration of calls, and for the distance called, made perfect economic sense. Similar arguments were put forth by AT&T. But in America, measured service was never popular with consumers. When given a choice, the public consistently opted for the carrier which provided flat rates, even though flat rates meant a higher basic service charge. Economists and telephone managers agreed that most people made too few calls to benefit from flat rates. Flat rates had to be set high enough to recoup the costs imposed by those who made a large number of free calls. But customers still preferred them.[19]

Eventually AT&T devised a schedule of charges that gave households flat-rate service, while building in safeguards to prevent abuse. The flat-fee structure generally encouraged usage, and also appealed to sceptical members of the public who worried about how much they would actually pay for service each month. All in all, flat rates worked to increase the number of network members. In response, AT&T was forced to invest in facilities sufficient to handle growth and a rising flow of traffic. By contrast, the regime of measured rates in Britain served as a tool of demand management, discouraging investment in capacity-expanding technology.[20]

---

[18] On the Bell System and British observations, see Kenneth Lipartito, '"Cutthroat Competition", Corporate Strategy, and the Growth of Network Industries', *Research in Technological Innovation, Management and Policy* 6 (1997), 1–53.

[19] These issues are discussed in Lipartito, '"Cutthroat Competition"', 38.

[20] New technologies often require creative pricing schemes to encourage users to accept them. Lowering the initial cost of service, by building connection and capital costs into monthly rates, reduces the fear and scepticism of new customers who might worry that they will end up paying more than they want for a new technology. Such practices were used effectively to market cellular telephones in the 1980s. Flat rates and line rentals had similar advantages: users knew exactly what they would be paying each month, and they did not have to bear high initial costs for equipment and connections.

British authorities often lamented that the nation was slow in picking up the 'telephone habit'. Engineers and policy makers could do little more than attribute the lower level of usage to social and cultural factors beyond their control. Part of the problem, however, may have been rigid pricing schemes, such as measured service. It had been a long-standing policy in Britain to charge for the actual cost of installation, which could be quite steep in remote rural areas. Reducing such charges, telephone managers reasoned, would add users without increasing net revenues, at least in the short term. Political pressures prevented the Post Office from making any moves that would lose money. Between 1955 and 1965, connection charges increased by 400 per cent. New customers, who were likely to be somewhat unsure of the value of a telephone, were dissuaded by the prospect of a large upfront payment.[21]

As a public bureaucracy rather than commercial enterprise, the Post Office was not market oriented. 'The telephone is not a public necessity like water supply, or even carriage of letters,' responded officials to critics. 'It is a business and private convenience, which under any circumstances only a minority of the public will be able to afford.'[22] Selling was limited to canvassing for subscribers, and these efforts were on a small scale. Nor was the Treasury particularly happy about expenditures for such activities, which it deemed a frivolous waste of money. The assumption was, either people wanted a telephone in the house, or they did not. Indeed, Treasury attitudes toward public telephone service bordered on contempt: 'We should all be forced to reconsider the value of the various amenity services . . . and should be compelled to realize that if we want these services, we must pay for them,' wrote one career civil servant. 'Perhaps as a result, we may want them rather less; in which case the burden on the Post Office will be reduced.'[23]

When funds were tight and waiting lists for new service lengthened, there seemed little logic in trying to entice more of the public to the network. Immediate constraints meant continuation of the old demand management strategy, with higher prices, metering, and measured service to discipline usage.[24] For a time the number of calls actually fell (see Figure 5.2). The net results of these demand-side policies were a smaller network and a greater reluctance among the British public to form a telephone habit. Government policy set up a vicious cycle of self-reinforcing expectations that kept telecommunications low on everyone's list of priorities.

[21] The origins of this attitude go back to the pre-war decades, but they were still deemed relevant after the Second World War: PRO BT56/10/B/377, 'Telephone Development in the UK and Reorganization of Industry in the Empire'. On connection charges, see Jill Hills, *Deregulating Telecoms: Competition and Control in the United States, Japan and Britain* (Westport, CT: Quorum Books, 1986), 85.

[22] PRO BT56/10/B/377, 'Telephone Development in the UK and Reorganization of Industry in the Empire'.

[23] PRO T219/685, 'Post Office Charges: General Policy' (1952). On Treasury resistance to expanding the network to more of the public, as late as 1960, see PRO T219/919, 'Post Office: Charges for Telephone Services to Residential Subscribers'.

[24] PRO T219/684, 'Post Office Charges: General Policy' (1950).

Even when policy makers turned their eyes favourably on telecommunications, they faced a deeply entrenched civil service with conservative ideas about
technology and the network. Direct Distance Dialling, which automated long-
distance service, began in the United States in 1951. British Post Office enthusiasm for this technology was tempered by Treasury officials, however, who
required new technology to be justified by narrow definitions of cost savings.
Improvements such as convenience or new service features were accorded at
best secondary consideration. Treasury kept a tight rein on Post Office purse
strings, to prevent 'over-investment'. Were the Post Office required to raise
money in private capital markets, they argued, it would better appreciate the
'merits of going slow'. In fact, however, if recent history is any indication, private capital markets would have been more than willing to finance new telecommunications technology that promised to reduce costs, enhance service, and
enlarge the network.[25]

Immediate post-war recommendations would have delayed subscriber long-
distance dialling by twenty years.[26] It soon became apparent, however, that
trunk switches required significant upgrading. As long-distance usage grew in the
late 1950s—as a result of falling transmission costs—pressure to automate trunk
exchanges increased. Even though Post Office projections showed that this
investment would pay off quite well, the Treasury insisted that any funds for this
purpose would have to come out of existing appropriations. There would be no
increase for the new technology. Favourable estimates by Post Office engineers
were discounted as mere 'enthusiasm for mechanization'. Initial efforts called for
the avoidance of 'unduly expensive and complicated equipment', and the building of a new system that could be grafted on to the old.[27]

With Labour back in power after 1964, modernization of the network went
forward more aggressively. Britain finally adopted the decades-old crossbar
exchange technology and was investigating newer electronic designs.
Productivity improved in these years, costs fell, and subscribership and usage
increased, though the gap between Britain and the United States did not close.[28]
Nor did all of these changes reflect a bold policy departure. Much of the momentum for Direct Distance Dialling, for example, had been built up in the preceding decade. As early as 1965, moreover, it was apparent that projections of
demand were still woefully inadequate. Even Labour's enhanced telecommunications investment programme was not going to clear the waiting list backlog.
During the early 1960s, the British public had developed a more favourable atti-

[25] PRO T219/506, 'Introduction of Subscriber Trunk Dialling and Group Charging' (1957).
[26] BT Post 70, box 4/156, 'Post War Planning' (1945).
[27] BT Post 70, Standing Telephone Advisory Committee (1951, 1954, 1960). On Treasury attitudes, see PRO T219/692, 'Relationship with Treasury and Financial Control, Report on Post Office
Development and Finance' (1955); PRO T219/506, 'Introduction of Subscriber Trunk Dialling and
Group Charging'; PRO T219/695, 'Review of the Financial Relations of the Post Office with the
Exchequer, the Treasury and Parliament'.
[28] Pryke, The Nationalised Industries, 164–80. Foreman-Peck and Millward, Public and Private
Ownership, 308–12.

tude toward the telephone, while rates had remained stable in nominal terms, falling in real terms. Among households, future growth was expected to be higher than past trends, as the post-war population boom reached adulthood. This generation looked upon the telephone as much more of a necessity than had the preceding one. With economic growth and the completion of network automatization, demand was expected to increase further. So the very success of any Labour government modernization scheme would require still higher levels of investment to feed the booming appetite for connections and handle increased traffic.[29]

Labour efforts to meet these demands ran headlong into another balance of payments crisis. Emergency measures in July 1966 forced the Post Office to revise its five-year investment projections downward by nearly 5 per cent.[30] It was unclear if Labour could dramatically change policy patterns that had been in effect since 1945, or earlier. Treasury staff were exceedingly resistant to losing control over the Post Office purse strings. Post Office workers were well organized, and enjoyed civil service status. Only in 1969 did the Post Office become an independent public corporation. This devolution of power, however, did not end Treasury oversight of the budget. The new public corporation also retained much of the old functional structure that placed the Director in charge of all postal services, with no single operational head of telecommunications. In any case, these changes may have come too late.[31]

By 1969, lack of prior investment had saddled Britain with much antiquated technology. Despite improvements, British productivity had not caught up to the leading nations of the United States, Sweden, or Japan during this time of great productivity increase.[32] The network needed to grow a great deal larger. It also had to accommodate the rapid increase in data transmission and new demands from business users for private networks, while making the shift to electronic equipment. Until well into the 1970s, nearly all British switches were either of the old Strowger design, or ageing reed relay and crossbar models. America had been using semi-electronic switches in quantity since the mid-1960s and installed the first fully digital exchange in 1976. But the British economy remained affected by the international crisis. Investment was expected to fall short of the original five-year projections until at least 1970, not taking into account inflation.

For all its shortcomings, British telecommunications policy can be explained less as a matter of 'mistakes' or ignorance and more as part of a consistent economic strategy dating back to the nineteenth century. In the United States, Sweden, and to a lesser extent Germany, telecommunications technology served largely national purposes. Like railways, the telephone and telegraph systems

[29] BT Post 70, Standing Telephone Advisory Committee (1965). On Labour's economic policies, see Tomlinson, *Public Policy and the Economy*, 238–76; B. W. E. Alford, *British Economic Performance, 1945–75* (London: Macmillan, 1988).

[30] BT Post 70, Standing Telephone Advisory Committee (1967).

[31] Hills, *Deregulating Telecoms*, 85–6.   [32] Pryke, *The Nationalised Industries*, 178.

were bonds of nationhood, designed to promote unification and to expand the domestic market. Such domestic concerns never carried as much weight in Britain. Instead, the more internationally oriented British economy used telecommunications to expand and serve its numerous foreign investments and large multinational firms.

From long experience with undersea telegraph cables, Britain established the world's most extensive international communications system in the decades before the Second World War. As Daniel Headrick has noted, this undersea cable network served as an 'invisible weapon', benefiting Great Britain in two world wars.[33] Supplemented by aggressive development of wireless technology, it also performed as a tool of empire, assisting management of far-flung British interests. In a like manner, international communications technology permitted British firms and financiers to manage the nation's complex portfolio of investments around the world.

From the middle of the nineteenth to the middle of the twentieth century, cable and telegraph were the media of international communications. Radiotelephony, a service promoted heavily by AT&T in the 1920s, was extremely expensive in the early decades. And undersea telephone cables for long distances were not possible before the transatlantic telephone (TAT) lines in the 1950s. Britain, Germany, France, and other nations sent telephone lines under smaller bodies of water, such as the English Channel and the Mediterranean. So while Europe was in telephonic communications with itself, it was linked to other parts of the world mainly by telegraph and telex.

This international communications system worked extremely well for British purposes. Telegraphs were excellent instruments for transmitting standardized bulk data quickly over long distances. In the nineteenth century, the heaviest users of telegraphs were the commodity and financial exchanges. It was relatively easy to reduce stock, bond, or commodity prices to code for transmission. Printing telegraphs, which produced tapes of dots and dashes, did not even require trained operators.

Telephones, by contrast, are the instruments for communicating detailed instructions. They are most useful for direct management of distant operations, rather than portfolio management using financial or other standardized data. It is a commonplace of economic history that the telegraph permitted the rise of large-scale, far-flung, bureaucratic enterprise. Telegraphs evolved in tandem with railways, in order to manage the complex movement of trains along track which stretched vast distances. But in fact, the telegraph was rather poorly suited to such detailed management. When telephones became available, American railway operators quickly switched to them, and found that they greatly increased their ability to manage the movement of freight.[34]

---

[33] Daniel Headrick, *The Invisible Weapon: Telecommunications and International Politics, 1851–1945* (New York: Oxford University Press, 1991).

[34] *Railway Gazette*, 3 Feb. 1905, 67; 14 Feb. 1908, 142; 28 Sept. 1928, 383. Philip Burtt, *Control of the Railways: A Study in Methods* (London: George Allen & Unwin, 1926).

American firms in general took advantage of America's faster developing, more extensive domestic telephone network to manage corporate operations across the nation from a central headquarters. Telephone technology was well matched to America's need for direct management of the departments of huge, administratively centralized, multi-unit firms. By contrast, a weak domestic British telephone network seems to have stifled this sort of use of the telephone. Britain's rail operator was a latecomer in making the switch from telegraph to telephone control. British wholesalers and retailers did not begin emphasizing telephone connections to customers until much later than their counterparts in America.

On the other hand, international cables and telegraphs were perfectly matched to British overseas needs. Britain earned substantial income from overseas investments in the early twentieth century. In order to finance long-term investment, and also to relieve pressure on the pound, it depended on sterling balances kept in London by foreign investors and governments.[35] In addition, the invisible items on Britain's foreign accounts helped to offset current account trade deficits. Dividends from long-term capital investments, combined with income from shipping, insurance, and underwriting, propped up Britain's shaky international position.

Undersea cables provided secure channels for transmitting data between London and countries that received British investment, reinforcing London's position as the centre of world finance. Later developments in telex technology supplemented and complemented telegraphic communications. Telex allowed more subtle and complex communications, and provided a good written record for bureaucratic administration. Britain's Eastern and Associated Companies dominated these services, and British Insulated and Callander's Cables (later BICC) were among the leaders in cable manufacture and research. Elsewhere shortwave wireless provided an inexpensive means of linking distant locations in the Third World with the metropole. The British Marconi Company was the world leader in shortwave radio.

In many ways, British policy towards international communications in the inter-war years matched American policy for domestic communications. In 1928, Cable & Wireless (C&W), a new British company, was formed with support from the British government. Much like the Bell System in America, this quasi-public entity served as the agent of network development. It combined the cable holdings of a variety of British firms with wireless properties of the Marconi Company and the British Post Office. Competition between older cable and newer wireless services motivated this combination. But it would be a mistake to dismiss the corporation as merely a conservative enterprise designed

[35] Michael Edelstein, 'Foreign Investment and Accumulation, 1860–1914', in Roderick Floud and Donald McCloskey (eds.), *The Economic History of Britain from 1700* (2nd edn., New York: Cambridge University Press, 1994). Ian Drummond, *British Economic Policy and the Empire, 1919–1939* (London: Allen & Unwin, 1972).

to protect existing investment from the creative-destructive potential of new technology.[36]

In preserving the cable network from wireless competition, C&W accomplished a number of goals. Some were strategic and military, resting on the belief that cable was a more secure medium than wireless for sensitive military communications. But others related to the construction of an extensive and efficient international medium for commercial transactions. Wireless technology was ideal for sending traffic between two central points—metropolises such as London and Hong Kong. But wireless towers alone could not collect traffic from outlying points. Cables were needed here.

From the point of view of constructing a broad-scope communications network, wireless and wired media were complements. But under competition, they could prove destructive to each other, and to the goal of network building. Wireless channels between metropolitan centres would take the highest valued traffic from cables, leaving them only with the less profitable traffic that they collected from outlying points. Since cables had to make landfall at various intermediate destinations, it was worth while to collect this traffic, so long as it was supplementary to the main traffic between urban centres. But without the high-end traffic, there would not be sufficient revenue to pay the cost of the cable. In the 1920s, wireless threatened to do to cable communications what MCI would in the 1960s do to the old Bell System—skim off the most profitable traffic without paying the cost of supporting the rest of the network.

The British response to this threat was to work out a non-competitive structure that supported the network. Like the Bell System, C&W substituted planned investment and co-ordination of technologies for competitively driven investment. Like the Bell System—with its strong commitment to building a larger subscriber base—C&W had incentives to behave as something other than a protective monopoly. It was guaranteed a 4 per cent return on investment. All profits above 4 per cent were split, with half paying for reductions in rates, and half retained by the corporation. Nor did the competitive threat fully disappear in the international arena. American wireless services kept up the pressure on C&W. Since wireless technology used high frequencies, it could take advantage of multiplexing—the creation of more than one channel on a single circuit. In response, undersea cables were rebuilt using coaxial technology, which replaced lower-capacity copper wires. Pressure from the British Post Office, which insisted on cheap letter-telegrams and discounted weekend rates, forced C&W to improve the speed and capacity of its service. Airmail, radiotelephony, and facsimile all remained long-term threats to cable service. In response, C&W used excess capacity in off-peak hours to provide competitive low-rate services. There were bulk discounts for commercial, government, and news organiza-

---

[36] University of Warwick, UK, Cable & Wireless Collection (hereafter C&WW), 86/04/01, 'Arbitration Tribunal'. This document contains much of the information for the following section.

tions. Rates of international communications continued to fall, and the speed of transmission increased.[37]

With its strategic and economic importance, international communications received much more generous treatment than did domestic telecommunications. In international communications, British service providers and manufacturers cultivated the capabilities to be dominant firms in the world market. In contrast to post-war assumptions, inter-war international experiences suggest that telecommunications was an investment that could well have contributed to the growth of the British economy.

5.3   The Supply Side: Misconstruing the Lessons of the Bell System

Though domestic telecommunications service remained under severe constraints in the post-war decades, manufacturing of telecommunications equipment stood much higher in the minds of British policy makers. It was one of those industries that the government believed could provide needed foreign exchange. During the inter-war period, British telecommunications manufacturers had done quite well, supplying equipment to colonies in India and Asia and penetrating markets in South America. Even in manufacturing, however, British policy took a wrong turn that owed itself in part to a misreading of the map of American experience.

Britain assumed, wrongly, that after the Second World War, exports would pick up in sterling areas, the prime markets for Britain in the pre-war period. In fact revival of the European economy made the correct target for export the First, not the Third World. Colonial markets needed equipment of older design, which would not help British manufacturers make the leap to the coming era of electronics in telecommunications.

The falling-off of traditional export markets should have turned British manufacturers to domestic and European ones. Here, however, underinvestment in the domestic network had devastating consequences. Because modernization at home was slow, domestic demand for equipment remained low. At home too, much of the demand was for older, outdated equipment. In the 1950s and 1960s, manufacturers kept producing Strowger switches, rather than newer crossbar or semi-electronic models. They made money supplying replacement parts for this older technology, as depreciation schedules were stretched out to interminable lengths by restrictions on Post Office spending. It was a situation that left British manufacturers well behind as technology changed rapidly after 1960.[38]

Co-ordination between the existing demands of the network and new technology was a long-standing issue in telecommunications. Economic advance in this industry required a structure to balance complex relationships between innovation, manufacturing, and service. Such a structure had to encourage improvements in

---

[37] Cable & Wireless Archives, London, UK (hereafter C&WL), 'Addresses and Discussion at the First Annual Conference of Managers and Officers', 1–5 July 1935, pp. 2–5.
[38] BT Post 70, Standing Telephone Advisory Committee (1967).

basic design, co-ordinate them with manufacturing skills and capacities, and spread new technology at minimum cost and disruption through the network.

America had long relied on vertical integration to provide such co-ordination. With Western Electric the captive (and sole) supplier of all Bell equipment, it was easy to adjust manufacturing processes in accordance with discoveries coming out of the research laboratories. Indeed, since the 1930s, Bell Labs had assumed more and more control over the entire Bell System. By the end of the Second World War, the American industry was largely technology driven. Steadily growing demand provided incentives to exploit economies of scale. Research was oriented toward finding new technologies that could increase system capacity without the expense of radical redesign or wholesale junking of existing equipment.

Britain maintained arm's-length relations between private equipment suppliers and publicly owned and operated networks. The idea here was to encourage competition and diversification in design. Sometimes it even worked. Competition between British suppliers at about the time of the First World War, for example, had led to an important modification of the Strowger automatic switch. In response to improvements coming out of Germany (Siemens) and America (International Western Electric), the British maker of Strowger technology, Automatic Telephone Manufacturing (ATM), developed the Director System. It employed some of the advanced features of competing technologies, while retaining the virtues that had made the Strowger design a favourite in Britain and elsewhere. The result was to extend the life of this durable technology, allowing automation of the British network to proceed at lower cost. By contrast, one could argue that the American structure failed in this case. Western Electric devised the panel switching system. Criticized as expensive and over engineered, it never caught on internationally.[39]

Supplier–network relations in Britain worked fairly well in the inter-war decades. By the 1920s, a cartel of equipment manufacturers was supplying the British Post Office. Britain relied on informal co-ordination between public and private entities in place of the more formal vertical structure Bell was building. Equipment was procured through bulk contracts, and the Post Office worked closely with suppliers in setting standards and scheduling equipment purchases. With government support, manufacturers formed the Telephone Equipment Manufacturers' Association in the 1920s. Similar associations were also formed for radio equipment and cables. Association reduced excess manufacturing capacity and encouraged large-scale production.

Bulk contracting tended to lock in existing suppliers, but there were methods of limiting opportunism. By working closely with private firms and carrying out

---

[39] BT Post 86/18, 'Director's System: Agreement between Post Master General and Automatic Telephone Company Limited, 1923'. Post Office Archives, London, UK, *The Post Office*, 1934. It remains undecided whether the Director or the panel system was in the long run superior. And it is well known that ATM was favoured in part because it was a British licensee of an American firm, and it planned to manufacture the switches on British soil.

their own research, Post Office engineers acquired the necessary knowledge to monitor and supervise performance. By setting common standards, they prevented lock-in by one firm. Research costs were shared among manufacturers, but the final design of any part or piece of equipment had to be interchangeable among all the contractors. In return, the Post Office assured each of the co-operating firms a slice of the manufacturing pie. These sorts of long-term relationship aimed to achieve a balance between central control and market competitiveness.[40]

When used properly, this structure could lower costs and encourage innovation. Bulk contracting allowed discounts on equipment when purchases went above a certain level. To take advantage of these economies, the Post Office accelerated modernization of the London exchanges in the 1930s. Co-operative improvements and rationalization reduced the costs of equipment, particularly for smaller exchanges and rural areas. Since Britain had a smaller universe of users than did the United States, it was important to produce equipment that was cost effective in smaller markets.[41]

Cumulatively, British telephone density (telephones in relation to population) grew by 194 per cent between 1923 and 1934, quadruple the rate of Germany, and twice that of Sweden. In the United States, density increased only 16 per cent, though this low figure reflected the prior high level of penetration established in the competitive era.[42] (See Figure 5.4.) The inter-war decades also witnessed rapid automation of switching in Britain, despite favouritism to domestic firms.[43] (Figure 5.5 plots this automation.) Even with the devastation of the Second World War, Britain managed to close some of the productivity gap with the United States by 1950.[44]

Though successful in the inter-war period, the British structure was not prepared to deal with national needs after the Second World War. The 1920s were a time of 'catching up', with Britain adopting many of the innovations coming out of America. Similar opportunities for catching up were also present after the

---

[40] 'Development of Standardization Policy', *Post Office Electric Engineers Journal* 28 (1935–6), 263; 'The Introduction of the Standard Telephone Relay', *Post Office Electric Engineers Journal* 27 (1934–5), 45–9.

[41] BT Post 86/65, 'London Automatic Committee Reports' (1930–4).

[42] BT Post 70, box 4/156, 'Post War Planning'. In part, this difference may have reflected the greater impact of the Depression across the Atlantic, as well as the high percentage of British telephones in cities, especially in the southern counties. These were not hit nearly as hard by the economic crisis as the countryside or the northern urban centres. In the United States, with its much larger rural population, farmers were priced out of the telephone market by the sudden fall in incomes.

[43] BT Post 70, box O, 'Telephone Charges Committee, Engineering Programme' (1935–6). Standardization through association provided an effective means of dealing with the frantic pace of development that came about in the 1920s. In the Depression-ridden 1930s, the bulk supply contracts also tended to mute price swings. Under the negotiated agreements with manufacturers, the Post Office could not radically lower prices. But this long-term policy also meant a steady market for manufacturers and continued reconstruction of the British network.

[44] Foreman-Peck and Millward, *Public and Private Ownership*, 312. Considering the great United States lead in productivity, even a minor relative improvement is noteworthy.

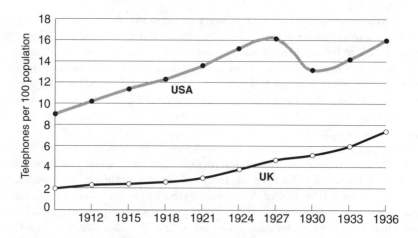

Figure 5.4. Telephone Density in the UK and the USA, 1909–1939

Source: B. R. Mitchell, *British Historical Statistics* (Cambridge: Cambridge University Press, 1988); United States Department of Commerce, Bureau of the Census, *Historical Statistics of the United States from Colonial Times to 1970* (Washington, DC: Government Printing Office, 1975).

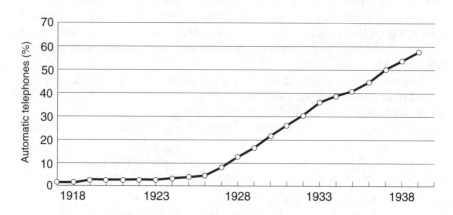

Figure 5.5. Automation of Telephone Switching in Britain, 1917–1939

Source: British Telecom, *Telephone Statistics*, Aug. 1981.

war. Telecommunications technology took a tremendous leap forward during the Second World War, and despite low levels of investment, Britain was adding transmission capacity with the use of coaxial, carrier, and microwave transmission systems. The key bottleneck was developing in switching. Britain was behind here, using equipment that dated back to the 1910s. Instead of narrowing, the gap with the rest of the world only widened.

In America, the Bell System led the way in developing new generations of switches, beginning with the crossbar models of the 1940s. Despite the flaws of the panel switch, it had proved a valuable learning experience, directing engineers' attention to the problem of designing large-scale switches for growing networks. The panel design had also provided invaluable experience with technologies to control these large and complex machines. Advances in hard wired logic circuits—the predecessor to programmable electronic switches—were a distinguishing feature of the several crossbar generations.[45]

After the war, both Labour and Conservative governments attacked the cosy relationships that had developed between the inside 'ring' of equipment manufacturers and the Post Office. This cartel, it was charged, limited competition and maximized the return from existing assets. The problem of collusion was especially acute in periods of slow network growth and limited public investment. It was considered unwise to let established firms go out of business, for fear that once the 'machine was stripped down, it would be hard to rebuild'.[46] Because capital was scarce, it seemed prudent to use existing capacity from incumbent manufacturers, rather than encourage new firms to create additional capacity. Large existing firms, such as General Electric (GEC), Standard Telephone and Cables (STC), Siemens (in Great Britain), and Automatic Telephone were also export oriented and had been supplying the colonial market. So they were better prepared to meet government mandates for exporting.[47] They had longer production runs, and it was believed this would allow them to exploit scale economies. New electronics firms possessing new skills, on the other hand, were excluded from telephone manufacturing.

In the 1950s, Conservatives passed stronger anti-cartel laws in an effort to boost manufacturing performance. Facing tight budgets, they pressured telecommunications suppliers to eliminate collective pricing and slash price lists, policies that inevitably placed weaker firms at risk. Since the law was tolerant of formal combination, mergers were the result. In 1961, Plessey acquired two old central office equipment suppliers, Automatic Telephone and British Ericsson. Encouraged by the Labour government in the 1960s, GEC acquired Associated Electric Industries (AEI) and English Electric.[48] Similar consolidation took place among cable manufacturers. Yet at the same time the government was promoting this 'rationalization' of the industry, it was also demanding competitive tendering in supply contracts. As a result, Britain lost its older structures for co-ordinating relations between technology, manufacturing and the network, but failed to create a competitive market alternative.

---

[45] On the crossbar, see Robert Chapuis, *100 Years of Telephone Switching, 1878–1978* (Amsterdam: North-Holland, 1982), 359–91. Amos Joel *et al.*, *A History of Engineering and Science in the Bell System: Switching Technology, 1925–1975* (New York: Bell Telephone Laboratories, 1982), 59–86.

[46] BT Post 70, Standing Telephone Advisory Committee (1950).

[47] BT Post 70, box 4/156, 'Post War Planning'.

[48] Hills, *Deregulating Telecoms*, 86. 'A Background History of the Ericsson/Plessey Factory at Beeston', undated typescript, Nottingham Industrial Museum, Nottingham, UK.

By the 1960s, Britain's telecommunications manufacturing industry was unprepared for the coming wave of electronics. As demand for telephone service picked up, so did orders to manufacturers. But there was a problem. Rapid modernization of the network was expected to overwhelm equipment-producing capacity. Yet no one wanted to encourage the manufacturers to expand, for fear that they would be left with obsolete facilities and excess capacity. Immediate reconstruction of the network to meet rising demand still required much of the older, electro-mechanical equipment. But the electronics revolution in telecommunications was at hand. By the 1970s, it was expected that electronic switches would be replacing Strowger models. Firms that expanded in 1965 would be ill prepared for this changeover.[49]

Though lacking experience with advanced switching systems, Britain attempted a rapid catch-up in this technology in the late 1950s by leapfrogging intermediate technologies. The result was a premature digital switch that was installed at Highgate Wood in 1962. The Highgate Wood experiment used electronic components for both the control mechanisms and the switching parts themselves. In 1962, however, electronic components of necessary quality were simply not available to support such a technology. It was possible to design a programmable electro-mechanical switch. Only the control unit had to be digital.

In 1965, AT&T showcased its 1ESS, the first switch to use 'Stored Program Control'. Computer control of switching operations, it turned out, substantially reduced costs, as well as allowing a whole host of new features for telephone users. The evolution of this technology in America followed the path marked by AT&T's early adoption of the controversial panel switch in the 1920s, and the well-regarded crossbar models in the 1940s. Both contrasted with the venerable old Strowger design by employing sophisticated, hard-wired control devices. A programmable switch was the next step. Lacking experience with these prior technologies, and facing the embarrassment of the Highgate Wood fiasco, Britain retreated to its existing switching technology, and missed the revolution in switching ushered in by Stored Program Control.[50]

In the face of AT&T's move to programmable switches, Britain developed a wired-logic reed relay system. This switch, which only reached the British market in quantity in 1973, used hard-wired control mechanisms, rather than software. It was completely out of step with the move to the more flexible, modular computer-controlled designs. At the same time, Britain also began to deploy some of the now dated crossbar technology. Thus Britain was forced to integrate several different types of switching systems—Strowger, reed relay, and

[49] BT Post 70, Standing Telephone Advisory Committee (1965), 'Investment Review and Industrial Inquiry'.

[50] Only two British firms had experience in crossbar technology. One was ATM, taken over by Plessey. The other was STC, a subsidiary of ITT. ITT competed in many markets and had connections to many networks, where crossbar and electronic technology was in demand. Amos Joel and Robert Chapuis, *Electronics, Computers and Telephone Switching* (New York: North–Holland, 1990), 197, 225–32. Pryke, *The Nationalised Industries*, 164–73.

crossbar—which reduced productivity gains. Finally, at the end of the 1960s, came another ill-advised attempt at rapid catch-up—the System X episode.

Between 1968 and 1976 a consortium of British manufacturers and the Post Office co-operated on an advanced electronic switch that was dramatically to modernize the British network. Work progressed extremely slowly, and it proved difficult to co-ordinate the numerous and complex tasks in an atmosphere stressing competition rather than co-operation. British manufacturers were trying to develop a product that would be globally competitive, but also meet rigid specifications set down by the Post Office. Post Office officials continued to debate the merits of alternative switching systems, as even GEC, STC, and Plessey were pouring resources into System X. As late as 1977, there were no firm domestic orders for this advanced design. Turning the Post Office into a public corporation in 1969 only further separated the network provider from its suppliers.[51] Trying to make the leap to the most advanced stage of technology, without significant experience from earlier stages, led to extraordinarily high development costs. Software forms a big part of modern telecommunication switches, yet the British consortium did not include a computer company. In the end, Britain had a product that, while advanced when it was first proposed, was no more than ordinary by the time it was finished. It has yet to find a large international market.[52]

Perhaps the most apt comparisons for these British manufacturing experiences come from Germany. Like Britain, Germany had a relatively small, publicly owned network supplied by a cartel of private manufacturers. It also had a strong commitment to an older switching technology, the rotary switch. But the German industry proved capable of continued improvement. After the Second World War, the Deutsche Bundespost introduced an upgraded version of its electro-mechanical system, the EMD (Edelmetall-Motor-Drehwähler). The switch performed exceptionally well and found a small export market. An alternative to the American crossbar design, it was a German answer to the need for a fast, reliable technology capable of automating long-distance service.[53] It met the needs of the German network until 1978. Despite the devastation of the Second World War, Germany ended up with a greater degree of automation than Britain.

One reason for this favourable performance may have been structural. Where Britain experimented with both competitive and co-operative arrangements in the post-war period, Germany stuck closely with the cartel model, incorporating some of the virtues of vertical integration. Siemens was the main German equipment supplier and it maintained close relations with the German Post Office. Other manufacturers were in essence junior partners in this enterprise. Siemens thus could invest in research without fear that new designs or improvements

[51] Hills, *Information Technology and Industrial Policy*, 120

[52] Peter Grindley, 'System X: The Failure of Procurement', London Business School Working Paper 29, Aug. 1987.

[53] Chapuis, *100 Years of Telephone Switching*, 227, 240–2.

would be rejected by the government in favour of a competing supplier. At the same time, close interdependence between Siemens and the Post Office gave the government a way of limiting opportunism. Commitment to network modernization in turn assured Siemens of a growing market for advancing technology.[54]

The German model laid heavy stress on standardization. Although this commitment worked against variety and radical innovation, it proved extremely effective at incremental improvements in basic designs, as with the EMD system. It also allowed manufacturing to take place on a large scale, with producers acquiring a high level of skill in improving production and design. The disadvantage, of course, was that Germany stuck with the electro-mechanical switch long after the dawn of electronics. So like Britain, it had difficulty moving into the electronic era before the late 1970s. But this fault was partially offset by an unexpected virtue. By waiting, Germany also avoided the pitfalls of moving too early or embracing the wrong technology at a moment of technological uncertainty. Given a relatively small domestic network, plodding consistency, leavened by a strong commitment to research, was apparently a superior strategy in telecommunications to ill-conceived experimentation.

Part of the problem British policy makers faced after the Second World War was deciding what, if any, structures they wanted to borrow from other nations. The success of the American telecommunications industry certainly loomed large in the minds of admirers. They particularly liked the high-quality research that came out of Bell Laboratories and the seamless co-ordination between the design, manufacture, and diffusion of new technologies. Critics also noted, however, that this monopoly structure resulted in equipment too expensive for a Britain on a tight budget, and inflexibility in design. So British policy alternated between efforts to streamline and rationalize telecommunications along American lines, and contrary efforts to promote competition. Comparisons suggest that strong user–producer interactions were key to innovative success, but British policy alienated manufacturers and the network from each other. Such divisions contributed to ill-timed moves into and out of electronic switching.

One solution, of course, might have been to simply imitate the American Bell System more thoroughly. But there is reason to suspect that the American model itself was already past its prime. Vertical integration, a near-monopoly position, dominance over questions of standards and design, and a substantial investment in research were certainly important reasons for Bell's success. Beneath the imposing edifice of the great Bell System, however, lay other sorts of links and connections, less formal, but no less important. These joined university departments, company scientists, production engineers and users of telecommunications technology.

---

[54] Although Germany did not adopt the crossbar or space division, computer-controlled electronic switches of the 1950s and 1960s in any quantity, it did continue to experiment with both designs, developing models that were never widely used. But this knowledge may have been important in preparing for the leap to digital in 1979. The experimental work was largely carried out by Siemens: Joel and Chapuis, *Electronics, Computers and Telephone Switching*, 204–17.

One of the great strengths of the pre-war Bell System had been its strong links between markets, science, and manufacturing. Research at AT&T was a model of the 'open' and 'porous' industrial laboratory. Some of the most important breakthroughs came at the hands of independent inventors. The crucial audion tube was invented outside AT&T, by Lee de Forest. The loading coil came from a Columbia University physicist, Michael Pupin. AT&T held only one of the several key patents for radio. It developed a system of sound motion pictures, but simultaneously with another competing system that actually proved superior.

These outside developments do not mean that corporate research within AT&T was a failure. Leonard Reich argues that before the First World War, AT&T's research was mainly defensive—to secure patents and fend off challenges from new technologies.[55] But I am drawn to a different conclusion. The proliferation of outside innovations coming into the Bell System reveals how well attuned to the plethora of scientific and technological breakthroughs going on around them company scientists really were. Most of the outside innovations had to be adapted to the needs of telecommunications, and this the corporate laboratories did extremely well. Using de Forest's tube, AT&T constructed electronic signal amplifiers for both cable and open wire. It devised carrier transmission systems, a whole series of them in the pre-transistor era, each offering greater signal-carrying capacity than the previous generation. This was unheroic work to be sure, but economically it was extremely valuable.

Bell Labs brought greater order to the innovation system. It helped to link innovations more firmly to manufacturing. Initially, however, it did not close off the company to the external environment. As David Mowery has argued, industrial research is at its most effective when it is learning based, when it serves as a conduit of knowledge and information between the firm and the outside world.[56] Before the Second World War, big science, military funding, and increasingly centralized, detached, and inward-focused industrial laboratories broke these important links to the outside world. Bell Labs, the model of post-war science, actually led the way, away from its own pre-war openness. Thus it came to serve as a model of industrial research at the very moment it had lost its historic sources of competitive strength.[57]

One nation that did emphasize openness and learning in telecommunications after the Second World War was Japan. The Japanese network operator, Nippon Telephone and Telegraph (NTT) was a separate, privately run public corporation that built connections to AT&T and other outside organizations.

[55] Leonard Reich, *The Making of American Industrial Research: Science and Business and GE and Bell, 1876–1926* (New York: Cambridge University Press, 1985).

[56] David Mowery, 'The Boundaries of the U.S. Firm in R&D', in Naomi R. Lamoreaux and Daniel M. G. Raff (eds.), *Coordination and Information: Historical Perspectives on the Organization of Enterprise* (Chicago: University of Chicago Press, 1995).

[57] To a degree, both the French national telecommunications laboratory, CNET, and a new International Telephone and Telegraph Laboratory located in England, STL, were modelled after Bell Labs. Both were formed after the Second World War.

In 1952, Japan decided to reconstruct its telephone exchanges using advanced crossbar technology, licensing this equipment from the United States and Sweden for home production. By sponsoring a rapidly growing domestic network, NTT assured suppliers of the growing market that they needed to build up skills in a changing telecommunications field. Over time, Japanese manufacturers moved from being licensees to being exporters of switches.[58]

Japan was also able to leverage its domestic network to support new ventures in electronic switching and computers, helping to build the skills needed to compete in these rapidly changing fields. Like most countries outside the United States, Japan moved slowly into the field of electronic switching. But in 1963, just before AT&T came out with its 1ESS switch, NTT and its suppliers began co-operating on a Japanese version, which became the D10 in 1973. NTT laboratories also worked with Japanese electronics firms to design a prototype fully digital switch in 1964. These experiences encouraged a flow of skill and knowledge between telecommunications and computers that allowed Japan to remain competitive in telephone switching. Using a structure that combined co-operation with limited competition, NTT was even more successful in supporting Japanese advances in computers and data communications.[59]

Given the strong export orientation of British policy makers, the Japanese road might well have been a logical one to travel. The Japanese pattern of 'controlled competition' seemed to be closest to what Britain was seeking.[60] As in other industries, such as automobiles, Japan captured the advantages of vertical integration through long-term relations with unintegrated suppliers. But it avoided the problems of opportunism by creating a more competitive environment within its cartels. Rather than pursue scale economies and rationalization, Britain could have forced constituents of the supplier 'ring' to compete with each other on projects whose outcome would benefit all cartel members. This policy would have provided a needed note of competition without breeding mistrust or abrogating the structures of co-operation between the network and the manufacturers, as the wholesale abandonment of bulk contracts seems to have done. But such bridges were never built. So perhaps it should not be surprising that the last great British effort at co-operative development—System X—had so many problems and delays.

Why Britain never embraced any of these options remains an open question. We can suggest some of the possible causes. Supporting British firms through the domestic telecommunications network ran against powerful political and cultural constraints on public investment. Under Conservative governments, the network had to pay for itself, indeed produce a healthy profit to compensate for

[58] Chapuis, *100 Years of Telephone Switching*, 452–3.

[59] Martin Fransman, *Japan's Computer and Communications Industry: The Evolution of Industrial Giants and Global Competitiveness* (Oxford: Oxford University Press, 1995).

[60] Martin Fransman, 'Controlled Competition in the Japanese Telecommunications Equipment Industry: The Case of Central Office Switches', in C. Antonelli (ed.), *The Economics of Information Networks* (Amsterdam: North–Holland, 1992), 253–75.

losses in mail delivery and telegraphs. There was talk of using public industries like the Post Office to support manufacturing exports, but little came of it, particularly when budget pressures forced retrenchment in spending. Controlled competition requires careful co-ordination, and painstakingly slow elaboration of standards and specifications. Both characterized Post Office sponsored joint development projects. But in reaction to high costs, long lead times, and cartel behaviour, Britain veered towards a policy of pure competition. This 'winner take all' approach encouraged consolidation among the manufacturers, while making them reluctant to acquire skills in emerging, but unproven new technologies.[61] The result was to stretch the crucial network–supplier relationships to the breaking point by the 1970s.

Experiences of other nations indicate that there were several possible paths to telecommunications modernization. All of them depended, however, on the key relationship between research, manufacturing, and the network. No manufacturer could hope to succeed if it did not have good relations with a network, preferably a rapidly growing and cutting-edge one. The American model offered one way of forging these connections, but it was hardly the only answer. Some nations succeeded through hybrid structures similar to Japan's. In Sweden, the Ericsson Company and the national network operator, Swedish Telecom, built a relationship on one measure of co-operation, and another of competition. The network procured equipment from both Ericsson and its own manufacturing subsidiary, Telegrafverket Verkstad. But it co-operated with the private corporation through joint development projects. Ericsson crafted cutting-edge switches for the home market and for export. Government factories turned out basic equipment and older-model switches still needed on the network. This division of labour allowed Ericsson to penetrate foreign markets by selling low-cost, high-quality, advanced designs. It overcame barriers to protected networks elsewhere through foreign subsidiaries, licence agreements, and intensive research that kept it at the cutting edge of technology.[62]

5.4 Conclusion

I have suggested that post-war British telecommunications policy was misguided from the start, and on several levels. Britain overestimated its strengths in manufacturing while neglecting its domestic network.[63] It misapprehended the true nature of the much-admired Bell System, how it worked, and why it seemed so successful. It ignored its own prior, successful experiences with international communications in the inter-war period.

---

[61] Fransman, 'Controlled Competition', 271–2. One result of that can be seen in the drastic reduction in R&D expenditure by GEC taken at this time: Hills, *Information Technology and Industrial Policy*, 134.

[62] Claes-Fredrick Helgesson, 'Coordination and Change in Telecommunications', Economic Research Institute, Stockholm Institute of Economics, Feb. 1994.

[63] Charles Feinstein, 'Success and Failure: British Economic Growth since 1948', in Floud and McCloskey, *Economic History of Britain*.

Particularly harmful was the failure to forge links between technology, manufacturing, and the network. Experiences with competition had caused America to evolve a telecommunications regime that emphasized universality and network growth. But Britain's own context of experience blinded it to this, the most obvious strength of the American model. Constraints on growth after the Second World War in turn limited the size and sophistication of the British telecommunications network. As a result, manufacturers were handicapped in responding to new opportunities at the technological frontier.

Though some constraints on telecommunications were external, both Labour and Conservative governments sought to limit public sector borrowing. British public debt was not large by historic or comparative standards. Yet limits on public investment certainly stifled modernization and improvement in telecommunications.[64] Such choices reflected in part a policy culture with strong assumptions about the value of communications and what sort of network the British public needed. Post-war decisions were structured by prior experiences that reflected different paths taken in the USA and the UK. Policy makers justified their choices by pointing to America, even where America offered in fact the opposite lesson. They sought to imitate an American model that itself was starting to experience problems. At the same time, they seemed to have missed important modifications of the American model evolving in places like Japan.

How significant were these decisions for Britain as a whole? Clearly no industry, even one as big as telecommunications, can by itself derail an entire economy. Even though information became more and more important in the decades after the Second World War, many growing, successful nations continued to labour with inferior, outdated, and inadequate telecommunications services. That roster includes not only Britain, but France, Germany, and to a degree even Japan.

These same nations, Britain included, rapidly and dramatically restructured or reinvested in telecommunications during the 1980s. This sudden turnaround suggests that a substantial underinvestment in facilities may have taken an increasing toll on economic growth as the years passed. Lack of attention to telecommunications may have been one of the more serious costs of misplaced investment, helping to lock Britain into what Charles Feinstein calls a vicious circle. Low initial investment in telecommunications contributed to a slower rate of productivity growth which caused trade imbalances, which led to retrenchment in public investment, especially for telecommunications.

The social costs of poor or inadequate service, spread out over the entire population of telephone users, may have been even greater than the impact on production. As Leslie Hannah and Steven Broadberry have found, British (and indeed European) productivity growth lagged further behind America in util-

---

[64] See conclusions of Foreman-Peck and Millward, *Public and Private Ownership*.

ities than in manufacturing.[65] As we now know, improvements in the quality of goods and services are not always fully captured by traditional productivity measures. Telephones and related communications devices are, like computers, one of those technologies whose economic impact is hard to measure, but which none the less contribute to well-being, more so now than in the past. Policy makers with free access to communications may not have felt the pinch, but ordinary citizens who had to manage their increasingly complex, time-stressed lives without adequate telephone service bore the full cost.

[65] Leslie Hannah, 'The American Miracle, 1875–1950, and After: A View from the European Mirror', *Business and Economic History* 24, 2 (1995), 197–220. Stephen Broadberry, 'Manufacturing and the Convergence Hypothesis: What the Long-Run Data Show', *Journal of Economic History* 53 (1993), 772–95.

Chapter 6

Creative Cross-Fertilization and Uneven Americanization of Swedish Industry: Sources of Innovation in Post-War Motor Vehicles and Electrical Manufacturing

HENRIK GLIMSTEDT

6.1 Introduction: Americanization and the 'Swedish Model': An Elective Affinity?

Since the late nineteenth century, prolific and suggestive images of mechanized production of standardized goods in large volumes for vast and stable markets—Americanization in short—have fascinated European industrialists. Widespread perceptions of an emerging gap between the USA and Europe in production methods thus drew a steady stream of industrial pilgrims across the Atlantic. Swedish industrialists and engineers, as many examples show, were no less mesmerized by the promises of mechanized mass production than many of their European colleagues. Much of the literature on the modernization of Sweden assumes that the Americanization of domestic industry proceeded rapidly and smoothly. But there is also widespread agreement that the country's industrial trajectory, and particularly the spread of US industrial practices in domestic industry, has to be seen in terms of a larger social and institutional context, the so-called Swedish model.[1]

The first version of this chapter was presented at the 1995 Business History Conference in Ft. Lauderdale, Florida. A more detailed version was discussed at the conference, 'Americanization and its Limits', in Madison, Wisconsin, in March 1997. The section on electrical engineering was tried out at the 1996 European Business History Conference in Göteborg. I thank for long-standing encouragement, thoughtful criticism, and useful advice: Bruce Kogut, David Soskice, Steven Tolliday, and, in particular, Jonathan Zeitlin. I equally owe thanks for comments on earlier versions to Steve Casper, Tomas Engström, Patrick Fridenson, Takahiro Fujimoto, Bob Hancké, Gary Herrigel, Alf Johansson, and Dan Raff. The usual disclaimers apply.

[1] For the general argument, see, in particular, Anders L. Johansson, *Tillväxt och klassamarbete* (Stockholm: Tidens förlag, 1989), or his 'Technological Optimism and the Swedish Model', Working paper, Swedish Centre For Working Life Studies, 1984. See also Bo Strååth,

This 'Swedish model' is broadly defined to comprise not only a structure of union–employer relations, but also a pattern of economic and social policy that conditioned the development of that structure. Together, they are held to have embodied a 'historical compromise' over the management of the economy. There is also wide agreement that the new Social Democratic élite which came to power in the 1930s laid the foundations for a novel progressive governance formula combining low levels of industrial conflict and support for rationalization with a modernized unemployment programme. After ultimately turning its back on a planned economy in the late 1940s, the Swedish labour movement based its main strategy for increased welfare on the promotion of economic efficiency.

On this view, industrial modernization, rather than full-fledged planning and Keynesian demand management, is central to the making of the modern Swedish welfare state. To maintain full employment while controlling inflationary pressures in the post-war economy, the Swedish Trade Union Confederation (LO) devised the well-known Rehn–Meidner (RM) model. Instead of pursuing expansionary macroeconomic spending programmes, this approach was based on a combination of restrictive fiscal and monetary policies, active labour market policies, and solidaristic wage policies within the framework of centralized collective bargaining. Tight integration of these elements was expected to decrease dependence on potentially inflationary domestic demand and expenditure. To achieve non-inflationary full employment, it was believed, public spending should be focused not on expansion of domestic demand but instead on enhancing Swedish industry's capacity to extract resources from global markets. Thus, centralized wage bargaining occupied centre stage in the RM model for promoting structural change. In centralized negotiations, average wages were thus set just above the level that the firms in the less advanced sectors could sustain, and below those of foreign competitors in the advanced, export-orientated industries. In other words, the solidaristic wage policy placed heavy burdens on firms operating in the less advanced sectors characterized by thin margins while keeping wage costs at a relatively low level in the advanced sectors of the economy. At the same time, however, state support for structural economic change required a complementary role for active labour market policies to lower the transaction costs of labour redeployment through subsidized mobility, training, and recruitment.

The foundation for these arguments about the role of centralized wage barganing is, of course, the oft-asserted 'Fordist' character of Swedish industry. The implementation of the RM model was accompanied by increased rigidity of the barganing system, which limited industrial strategy in the 1960s. Rather than move into new product markets—it is claimed—Swedish industry subsequently responded to increased international competition in the 1960s by

'Verkstadsklubbarna vid Volvo och Saab: facklig politik i två företagskulturer', in Bo Öhngren (ed.), *Metall 100 år: fem uppsatser* (Stockholm: Svenska Metallindustriarbetareförbundet, 1988), 83–106.

rationalizing the existing industrial structure through increased mechanization, plant closures, mergers, and vertical integration.[2]

SWEDISH INDUSTRY BETWEEN FLEXIBILITY AND MASS PRODUCTION

The attraction of this argument about the affinity between Americanization and the Swedish model lies in its apparent integration of cultural, social, and industrial modernization within a single theoretical framework. Proponents on all sides agree that a positive-sum bargaining process is a feasible vehicle for joint conflict resolution under conditions of economic growth. More problematic, however, are the underlying ideas about micro-level modernization. Those insisting on the Taylorist and Fordist cornerstones of the Swedish model subscribe to assumptions about economic development rooted in a simplistic understanding of the Chandlerian school of business history. Thus they share the view that modern economic growth can ultimately be attributed to the advent of mass production and scientific management as the core of a universal 'industrial best practice'. But the assumptions about the sources of international competitiveness on which this perspective rests are by no means unproblematic. To support the validity of the production cost–export performance relationship, it must also be shown that decreasing costs in production lead to rising sales and profit margins. In particular, it must be demonstrated that increased economies of scale through standardization and subdivision of labour lead to radical gains in market share, permitting in turn heavy investment in R&D, rapid erection of more rational manufacturing facilities, and more aggressive marketing to stay ahead of competitors soon able to imitate the original advantage in mass production.

Yet, much of the recent debate on contemporary and historical industrial divides reveals that few industries, let alone entire national economies, are easily categorized as either 'Fordist' or 'flexibly specialized'. Take Sweden's important pulp and paper cluster. First, Sweden's main industrial competitors undoubtedly saw her dominance in pulp and paper in the post-war era as the result of almost a century of gradual investments aimed at achieving larger scale and greater speed in production, vertical integration, and increased deskilling of labour in order to manufacture large volumes of standardized products at a low price. The successful implementation of sulphate-based methods in the late 1940s had a dramatic impact on the development of economies of scale in Swedish pulp, since it increased the range of timber types that could be utilized in its production. From the early 1960s onwards, furthermore, lowered trade barriers and increased competitive pressure from Finland and Canada set off intensified trends towards vertical integration of vast timber resources and large-scale pulp and paper industries. Huge fixed investment in machinery required not only vertical integration between pulp and paper, but perhaps even

---

[2] For the most elaborated version of this argument, see Jonas Pontusson, *The Limits of Social Democracy: Investment Policy in Sweden* (Ithaca: Cornell University Press, 1992).

more so control over the uninterrupted supply of the raw material. So great in fact were Sweden's advantages in mass production of standard qualities of paper that firms in countries like Norway consolidated their production of pulp around a high value-added strategy for integrated production of high-quality paper and cardboard to escape the resulting competitive pressures.[3] In the late 1940s and early 1950s, however, the production of machinery for the pulp indus-try remained largely untouched by preplanning of production, and use of semi-skilled labour remained very limited despite efforts to standardize, mechanize, and subdivide work during the preceding decades. A case in point is the Hedemora engineering works studied by Maths Isacson.[4] While the manage-ment tried to supplement the more complex production of machinery and tools with simpler products that allowed Taylorized work schemes, commercial real-ities limited these efforts. Purchasers of complex investment goods, such as the pulp industry, required tailored designs to fit their specific needs and previous structure of investments in machinery, the skills of the workers, and the profile of demand in their particular product markets. The workers' high skill levels and their independent judgement of the work situation as it existed constituted the backbone of this strategy; absolute cost cutting and adversarial piece-rate practices were rare and seldom successful.

A more clear-cut 'Fordist' example is the Swedish shipbuilding sector, which successfully took advantage of the soaring demand for large standardized oil tankers following the nationalization of Iranian petroleum resources and refiner-ies and the closure of the Suez Canal in the mid-1950s. The Swedish yards were well positioned to pioneer Fordist-style production methods in the 1950s and 1960s as the post-war course of political events concentrated international ship-ping demand on large tankers. By the 1930s, Swedish shipyards had gained ex-perience in product standardization, preplanned production, and all-welded construction of oil tankers for Norwegian shipowners, although craft methods still remained dominant in the production process. From the 1940s onwards, Swedish shipbuilders invested heavily in improving the existing yards and built new berths to accommodate the largest ships. As Jan Bohlin's recent study of Götaverken shows, productivity gains depended just as much on overall increase in production volume and the related learning in production planning as it did on longer series of standard ships with identical designs. Concentration on large tankers required, furthermore, advances in the welding process, since Lloyd's Register did not establish acceptance rules and quality control standards for welded construction until the mid-1950s. Norwegian shipowners thus worked closely with both the Swedish yards and their own national classification society,

---

[3] See *inter alia*, Svein Olav Hansen, 'Globalisering i treforedling: En historisk og aktuell nordisk utfodring', in Henrik Glimstedt and Even Lange (eds.), *Globalisering: Drivkrefter og konsekvenser* (Oslo: Fagboksforlaget, 1998), 48–60; Anders Melander, *Industrial Wisdom and Strategic Change: The Swedish Pulp and Paper Industry 1945–1990* (Jönköping: Jönköping International Business School Dissertation Series, No. 1, 1997).

[4] Maths Isacson, *Verkstadsarbete under 1900-talet: Hedemora verkstäder före 1950* (Lund: Arkiv, 1987).

Det norske Veritas, to win acceptance for structural engineering formulae. Replacing the old approved methods linked to practical experience of shipbuilding with theoretical models for the structural mechanics of materials, Det norske Veritas put shipbuilding on a scientific footing. The introduction of new process and product strategies yielded dramatic results: Sweden was the only shipbuilding nation that took full advantage of the emergence of standardized demand in post-war shipping by increasing output, capacity, and export market share without significant state subsidies. But the limitations of this approach are equally important to bear in mind: the Fordist strategy in shipbuilding ran into serious troubles as soon as the demand shifted from large standard tankers towards more diversified production during the mid-1970s. It was the much more diversified and less 'Americanized' Norwegian yards that proved capable of penetrating the offshore oil-equipment sector after the energy crises struck in the 1970s and the discovery of North Sea oil fuelled demand for rigs and related marine equipment.[5]

No single explanation has a clearer bearing on the hybrid character of the Swedish industry than the connection between fragmented markets and flexible manufacturing methods. The two cases briefly discussed above suggest that standardization might have appeared to many industrialists as an attractive strategy for adapting to international competition, but the diversity of market realities forced even the most enthusiastic proponents of Americanization carefully to revise if not completely abandon plans for achieving economies of scale through standardization of work, division of labour, and deskilling. If this observation is correct, we also need to rethink some of the conventional connections between politics and economic life in the standard narratives of Swedish development. The argument of this chapter is, then, constructed as follows. Section 6.2 tries to establish why managers in the Swedish motor vehicle industry redefined their strategic options in the direction of standardization during the post-war era after a ten-year period of successful implementation of flex-

---

[5] For the inter-war years, see Kent Olsson, 'Tankers and Technical Development in the Swedish Shipbuilding Industry', in Jan Kuuse and Anthony Slaven (eds.), 'Development Problems in Historical Perspective: Scottish and Scandinavian Shipbuilding Seminar', unpublished report, University of Glasgow, 1980, 55–72. For an excellent empirical analysis of the limited impact of product standardization on productivity in Swedish shipbuilding, see Jan Bohlin, *Svensk varvsindustri 1920–1975: Lönsamhet, finansiering och arbetsmarknad* (Göteborg: Meddelanden från Ekonomisk-historiska institutionen, 1989). For Norway, see in particular, H. W. Andersen, 'Producing Producers: Shippers, Shipyards and the Cooperative Infrastructure of the Norwegian Maritime Complex since 1850', in Charles Sabel and Jonathan Zeitlin (eds.), *World of Possibilities: Flexibility and Mass Production in Western Industrialization* (Cambridge: Cambridge University Press, 1997), 461–500. For the scientific approach to ship construction see: H. W. Andersen and J. P. Collet, *Anchor and Balance: Det Norske Veritas 1864–1989* (Oslo: Universitetsforlaget, 1989). For a comparative overview of market trends and production strategies, see Jonathan Zeitlin, *Between Flexibility and Mass Production: Strategic Debate and Industrial Reorganization in British Engineering, 1830–1990* (Oxford: Oxford University Press, forthcoming), chs. 13–14; for a study which makes the opposite argument that there was no survival in the shipbuilding sector at all after the mid-1970s, see Bo Stråth, *The Politics of De-industrialisation: The Contraction of the West European Shipbuilding Industry* (London: Croom Helm, 1987).

ible, craft-based production of advanced trucks. It distinguishes between 'indirect' societal effects of the political construction of markets on the one hand and 'direct' societal effects of labour market policies governing working conditions on the other. The thrust of the argument is that the drive for Fordism in the Swedish motor vehicle industry during the 1950s was not based on the direct societal effects of conscious political support for domestic volume production to build up the international standing of Swedish industry, but was instead a consequence of managers' strategic response to a complex market environment shaped by the indirect societal effects of state commercial and regulatory policies. To corroborate that argument, the second part of our story (section 6.3) contrasts the policies towards the motor vehicle sector with the political priority enjoyed by Swedish electrical engineering in conjunction with the energy sector in the post-war era. It shows how the political construction of the Swedish energy market drove ASEA, the major Swedish player in electrical equipment, from dependence on volume manufacture of simple products towards exports of complex, customized, and bulky equipment, such as power transformers, generator sets, and turnkey power plants.

While an attempt to extend this argument across the key sectors of the Swedish industrial economy would take us far beyond the scope of this chapter, it should nevertheless be said that this pattern of collaboration between public agencies and private firms seems to have typified the growth of Sweden's high-tech industries. While the Swedish auotomobile industry received little institutional support until the 1970s, other sectors, such as telecommunications and pharmaceuticals, have gained from tight linkages between public and private R&D, intricate user–producer relations, and extensive public demand due to heavy investments in the technologically advanced Swedish infrastructure.

6.2 Post-war Adaptation in the Swedish Motor Vehicle Industry

THE EVOLUTION OF PRODUCT AND PRODUCTION STRATEGIES, 1930–1970

In considering the history of foreign influences on Swedish motor vehicle manufacture, it is often suggested that the industry has evolved through three discrete periods, each based on a distinct set of principles or production paradigms. First, it is assumed that the 1930s saw a breakthrough in American production methods, although craft still persisted to a large extent. The British and American automobile industries kept their dominant positions in the Swedish passenger car market, while the truck market experienced rapid change. By 1938, it was obvious that expansion during the 1930s had changed the balance between foreign and domestic truck producers. In heavy trucks, Volvo had by the late 1930s established itself as the leading truck producer in the Swedish market. American firms, on the other hand, remained strong in the lighter end of the truck market. The structure of Sweden's transport regulations limited the demand for simple trucks while channelling demand towards fragmented but demanding transportation in the agrarian and forest sectors. Hence, Volvo focused its resources on sophisticated trucks, combining rapid product development with decentralized supply of components from the advanced Swedish craft-based engineering sector and internal flexibility in assembly—in other words: flexible specialization.[6]

The standard story of Volvo's adoption of new organizational models and production strategies in the post-war era thus seems to be that of full-fledged Americanization in the 1950s and early 1960s. Volvo's Torslanda plant outside Göteborg on the west coast of Sweden is offered as proof of the firm's commitment to standardization of products and work methods. Several arguments have been advanced to support this viewpoint, of which the first is concerned with product standardization. Given the liberalization of trade that opened the international market in conjunction with the soaring domestic demand after the mid-1950s, Volvo embarked upon a Fordist trajectory. In essence, most new investment in passenger cars was dedicated to the production of a single standardized model, the P 444, designed during the war. This new car was a rather elegant and yet sturdy design that would prove to be a successful competitor in the medium-sized segment. Subsequent models, the Amazon and the 140, introduced in 1958 and 1968 respectively, shared this same basic characteristic.

Second, the question of volume and process has been highlighted. As production runs ballooned from the few thousand cars typical of the inter-war and immediate post-war years to more than 50,000 units by the mid-1950s, Volvo introduced mechanized moving assembly lines, abandoning team work and collective piece-rates in favour of the well-known American Method-Time-Measurement (MTM) system. As a result, work became individualized and cycle times dropped dramatically from about 20 minutes to 2.5 minutes. Assembly times dropped from about 21.5 hours to about 17 hours in the assembly of the PV 444, driving down costs by 15 per cent between 1954 and 1956. In engine production, man-hours per unit fell from 25 hours to 9 hours during the same period.[7]

Third, automation is held to have played a key role. Until the late 1950s, the trimming of surplus steel from the pressed body parts was still carried out by hand, while assembly of these parts remained basically a hand-welding process. But by the mid-1960s, automatic welding and transfer machines were in operation, reducing production times from about 30 to 15 hours per unit. In addition,

[6] This brief account of Volvo production is based on Henrik Glimstedt, *Mellan teknik och samhälle—stat, marknad och produktion i svensk bilindustri 1930–1960* (Göteborg: Avhandlingar från Historiska institutionen i Göteborg, No. 5, 1993), chs. 6–7. For Scania, see Erik Giertz, *Människor i Scania under 100 år: Industri, arbetsliv och samhälle i förändring* (Stockholm: Nordstedts, 1991).

[7] For this and the overview below, see in particular Kajsa Ellegård, *Bilder av ett produktionssystem* (Göteborg: Meddelanden från Göteborgs universitets geografiska institutionen, 1983); Gustaf Luthman, Holger Bolin, and Alf Viklund (eds.), *MTM i Sverige 1959–1990* (Stockholm: Sveriges Rationaliseringsförbund, 1990).

dedicated machinery and standardized production methods were gradually developed in the manufacture of engines, gearboxes, and chassis, which also resulted in higher levels of productivity.

Fourth, emphasis has been placed on vertical integration and functional centralization. In production engineering, for example, groups which had previously specialized in trucks, various kinds of car designs, and components were all brought together under one roof during the mid-1950s to form a central production department.[8] Relations with subcontractors were also transformed, in so far as they were tied closer to Volvo through ownership and joint strategic planning.[9]

The relationship between workers' reactions to the transformation of production and the innovative responses of company management completes the story. The accumulated effects of rationalization drove down costs and speeded up production. In the years between 1955 and 1965, productivity levels soared from 1.8 cars to 6.5 cars per employed worker. Naturally, such a dramatic rise in productivity was accompanied by increased volume of output and employment. In assembly, the number of employed workers increased from about 1,000 in 1950 to about 2,500 in the early 1960s and then soared to about 6,500 at the end of the 1960s.[10] Most studies covering the 1970s and later agree that there was then a rapid shift towards a new type of work organization based on sociotechnical principles.[11] That break with decades of Fordist practices was motivated by productivity and quality problems causing social revolts, including high absenteeism and high rates of labour turnover in the late 1960s and early 1970s. More precisely, the emergence of the new factories—first Trollhättan (Saab) and Kalmar (Volvo), and then Uddevalla (Volvo) and Malmö (Saab)—represents conscious strategies to pursue efficiency through work groups, skilled labour, and alternatives to line production. This era, thus, represents a return to high value-added production combined with flexible, neo-craft production principles.

This standard story, however, raises a number of empirical questions. Those central to this chapter concern how deeply Americanization actually ran in the organization of work and to what extent flexible systems still prevailed throughout the 1950s and 1960s.

[8] Volvo also formed a central strategic group, the so-called Main Committee on Production Technology, with representatives from the important subsidiary companies.

[9] Thus Volvo became the majority shareholder of Köpings Mekaniska Verkstad, manufacturers of gearboxes and transmissions, and took a minority stake in Olofström, its main supplier of pressed steel parts.

[10] Demand for labour in the Swedish automobile industry had, since the beginning of the 1950s, to be satisfied in a tighter labour market. Thus, labour shortage stimulated Volvo to recruit immigrant workers and, by the late 1960s, we will see that these had become numerically dominant in the company's Göteborg plant, but that tendency was already evident in the late 1950s.

[11] Christian Berggren, 'Alternatives to Lean Production: Work Organization in the Swedish Auto Industry', *Cornell International Industrial and Labor Relations Report* 22 (Ithaca: ILR Press, 1992); Thomas Sandberg, 'Volvo Kalmar—Twice a Pioneer', in Åke Sandberg (ed.), *Enriching Production: Perspectives on Volvo's Uddevalla Plant as an Alternative to Lean Production* (Aldershot: Avebury, 1995), 87–101.

Post-War Market Regulation and Americanization

Post-war Americanization at Volvo should first be placed in a broader comparative context of indirect societal effects flowing from national patterns of market regulation and trade policy. In post-war Europe, the motor vehicle industry faced surging demand for both passenger cars and commercial vehicles. Post-war governments in the chief producing countries were also eager to exploit the industry's growth potential, which indeed constituted a prime mover towards mass production of motor vehicles. Yet the political motives that propelled the drive towards mass production none the less varied across Europe. Whereas the French authorities focused on technological modernization as a key to long-term growth and political stability, the British Labour government pressed for increased production volumes in pursuit of short-term balance of payments gains rather than long-term commercial advantages.[12]

By contrast, the Swedish government saw the domestic motor vehicle industry neither as a key to future techno-economic development, nor as a potential export commodity to be exploited, as in the British case. On the contrary, by March 1947 the Swedish government imposed tight regulations on all imports through a strict licensing system, which clearly resulted in a contraction of domestic automobile production. Volvo was deprived of access to vital components because of the import regulations, while the supply of imported materials for the automobile sector fell more than the industrial average.[13] In addition, allocation of steel for chassis was particularly restricted, directly limiting production volumes until the mid-1950s. Even more important, however, was the Swedish government's utilization of the Road Haulage Act to achieve goals related to the country's balance of trade. Just as the state had used this Act to protect the railways from competition in the previous decades, it now used the same framework of regulations to influence the trade balance. A telling statement is that of the Importberedningen (Foreign Trade Regulation Working Committee), which recommended that 'concerning the issue of truck transport licences, the current restrictive levels should be maintained in order to keep the demand for trucks at a low level'.[14] These policies implied slower than expected post-war growth in the truck industry. As Volvo's managing director, Assar Gabrielsson, concluded: 'the demand for trucks is what we under normal circumstances would call very weak, but since foreign competition is limited by the import barriers we will get a barely satisfactory share of the market anyway'.[15]

---

[12] For annual rates of export and the role of the car industry in Britain's balance of payments policy, see T. R. Whisler, 'The Outstanding Potential Market: The British Motor Industry and Europe, 1945–1975', *Journal of Transport History* 3, 15 (1994), 1–19. For the more general argument see Zeitlin, *Between Flexibility and Mass Production*.

[13] For a more detailed analysis of post-war trade policy and the automobile industry, see Glimstedt, *Mellan teknik och samhälle*, ch. 8.

[14] Handelskommissionen, Importberedningen, protokoll 4/6 (1947).

[15] AB Volvos historiska arkiv, Företagsnämndens protokoll 23 March 1949 (Göteborgs Stadsarkiv, Göteborg).

Although the Volvo management saw no direct threats to its home market position in trucks, they were confident that other sectors of the automotive industry would grow at a faster rate. This belief stemmed from the fact that the market for Volvo's main product line—trucks—which accounted for the lion's share of the company revenues (80 per cent in 1947), was for a few critical years subjected to political limitations which made output uncertain.[16] Given such worrying uncertainty, the company had to find other products, such as tractors for the agricultural sector, in order to compensate for the lost truck production. But for a long-term solution of this problem, Volvo began to explore the business opportunities present in the potentially expanding passenger car market. It was, however, the announcement by the Swedish government of its intention to join the GATT negotiations, that decided the company to redirect its resources to production of passenger cars in what became known as the 'Jönköping programme'. It was assumed by the Volvo managers that the state would not permit future expansion of the truck market, but they were equally convinced that the government was likely to open the Swedish passenger car market to foreign competition. If this were to be the case, Volvo concluded that the firm was left with no other option but to develop volume production equivalent to that of other European car companies.

Having little if any experience in real mass production, Volvo found the move to volume production a strenuous process. The challenges were quite formidable. The firm's pre-war production of passenger cars can best be characterized as marginal, although the company had successfully manufactured commercial vehicles (mainly taxis) in batches of a few thousand per year. Wartime production had further accentuated Volvo's heavy vehicle profile. Hence as the company emerged from the Second World War, passenger car capacity was limited. Given that its car production was still almost negligible in 1948 at fewer than 3,000 units, Volvo had to develop an industrial strategy and structure that would revolutionize the volume of production in the space of a few years to survive in the market. A target of 50,000 units was therefore set for 1952.

AMERICANIZATION AND THE ORGANIZED LABOUR MARKET, 1946–1955

Per Söderström, managing director of Volvo Penta works and responsible for the supply of engines, was among the first seriously to address the issues involved in the transition to volume production. Söderström, who has subsequently been hailed as perhaps the most progressive figure behind the introduction of American work methods and the MTM system in Swedish engineering, identified the piece-rate system as the key to future productivity levels.

In a 1947 memorandum, Söderström outlined his views on how to achieve higher productivity levels.[17] First, he stated that actual earnings in industry had

[16] For the extended argument, see Glimstedt, *Mellan teknik och samhälle*, ch. 8.

[17] For Söderström's reflections, see Verkstadsföreningens Historiska Arkiv, AB Volvo, Rapport av den 23 December 1947 (Per Söderström).

soared far beyond the levels allowed by the central wage agreements due to local wage drift. According to Söderström's estimates, there was a gap of nearly 30 per cent between actual hourly earnings, which rose by an average of 44 per cent between 1938 and 1946, and the 15 per cent increase envisaged by the central agreement. Söderström's explanation for this discrepancy emphasized the workings of the piece-rate system. Put simply, actual earnings resulted not so much from high hourly wages in local agreements as from overly generous piece-rate bonuses. In his reflection on these bonuses, he concluded that 'general rationalization of production was hardly ever followed up by corresponding piece-rate reductions to reflect the new organizational or technical preconditions in production'. This pattern had emerged in the 1920s and had been reinforced by military demand during the Second World War. To Söderström, the situation reflected a collective mismanagement of the piece-rate system, involving a deep-seated distrust between the two parties. Whatever the cause, the achievement of increased productivity levels and volume production required, in his view, a revision of the piece-rate system. In short, his vision of the late 1940s revolved around a new deal with the union to ensure high productivity and high wage levels.

Söderström's response was to maintain the local piece-work contract, with its direct relationship between output and earnings, but to do away with the old rates established through decades of local bargaining. To achieve this goal, he drafted a two-part agreement. Paragraphs 1–5 specified how elimination of restrictions on output would allow for higher productivity, while workers would also have to assume responsibility for continuity of production and the achievement of effective work methods. The company would assume full responsibility for conditions outside the worker's control, such as internal logistics and supply of materials. The second section protected workers against unfair reductions of piece-rates and established a 'floor' for future local wage negotiations, with guaranteed minimum earnings of 20 per cent above the central agreement for 1947.[18]

Both the local union and the employers' association reacted promptly. While the local union approved the proposed agreement, the Swedish Engineering Employers' Association (SEEA)—in a less than friendly tone—refused to accept, seeing it as a violation of the national agreement in terms of both wage levels and extension of employers' responsibilities. The chairman of the SEEA wrote that: 'the local agreement goes far beyond the national agreement . . . Therefore, we are surprised that you did not, according to § 23 of our statutes, submit the outline of the agreement to us before turning it over to the workers.'[19] In SEEA's judgement, the local agreement, drafted by Söderström and accepted by the workers' local union representatives, was completely illegitimate.

[18] The details of the local agreement are described in the subsequent exchange between Söderström and the Swedish Engineering Employers' Association. See: Verkstadsföreningens Arkiv, AB Volvo, Letter to Per Söderström, 22 Mar. 1948.
[19] Ibid.

It was against this background that Volvo's management began, in the late 1940s and early 1950s, to look for alternatives to the rejected local agreement. In an effort to alter the traditional piece-rate system, an American firm, Method Engineering Council, was contracted to implement a new payment system—the MTM system.[20] While the MTM system is complex and highly technical, its basic principles are quite simple. Its fundamental idea is that engineers shall, under laboratory conditions, identify the best way to carry out a certain task. Once they have established the desired method, the job is then broken down into several stages. A simple routine, by way of example, would go something like this: look to the left, grip screwdriver, look to the right, look at the screw, move arm, point screwdriver at screw, insert screwdriver, etc. Attached to each and every one of these physical movements are standardized Time Measurement Units (TMUs), which allow the production engineer to calculate the standard time or MTM 100 per cent needed for a specific job. To calculate the cost of a specific job, the TMUs are multiplied by a pay factor, resulting in an hourly wage. Actual pay then depends on an agreed pace of work—for example, whether an assembly line should move faster or slower than MTM 100 per cent—and on the outcome of central negotiations on the pay factor.

Compared to the traditional group piece-rate system which had dominated production before the 1950s, the MTM system threatened the workers' discretion on the job in several ways. In 1952, efforts to implement the MTM system on the Göteborg assembly lines met in the initial stages with severe unrest and a series of wildcat strikes, as the company tried to establish the labour intensity of the work process through a statistical analysis of lost production time. Faced with these unexpected strikes, the management reacted in a rather heavy-handed way, giving notice to seven workers and suspending no less than a quarter of the work-force.[21]

Once again, the SEEA intervened to establish whether or not the wildcats were organized by an isolated clique of Communist workers, as Volvo's management had claimed. The SEEA representative, Gustaf Toller, concluded, however, that the unrest was not so much a matter of successful Communist agitation but rather a severe conflict in which both the local union and the company were in breach of the national labour market agreement. As a result, Toller recommended to both parties that the conflicts should be referred to the newly formed Time and Motion Study Committee (TMSC), established by the central labour market organizations to explore and regulate future uses of this practice in the wake of the 1945 national metalworkers strike.[22] Since the MTM system and the statistical lost-time analysis which had constituted the preliminary phase of its introduction were virtually unknown to the representatives of both

[20] Luthman et al., MTM i Sverige; see also Per Sundgren, 'Införandet av MTM-metoden i svensk verkstandsindustrie', Arkiv 13–14 (1978).

[21] The following account of the conflicts in 1952 is, when not otherwise indicated, based on Glimstedt, Mellan teknik och samhälle, ch. 8.

[22] For the Time and Motion Study Committee, see Johansson, Tillväxt och klassamarbete.

the Swedish Metalworkers' Union (SMWU) and the employers, the initial decision of the TMSC were ambiguous. Volvo was instructed, on the one hand, to
end the suspensions and re-hire those workers previously fired. On the other
hand, however, the company was permitted to continue with its work studies
for two weeks so that representatives of the TMSC could become familiar with
these hitherto unknown methods.

Unsurprisingly, the representatives of the SMWU were hard pressed to con-
vince their opposite numbers in the TMSC that the practice was unjust and
highly unreliable. Though the employers' representative, Oscar Werne, seemed
to accept the rationale of the MTM system strictly for assembly work, he soon
arrived at a more negative conclusion. Werne's problem, it appears, was not
whether the system would work in a technical sense, but rather, that he saw a
misfit between what he conceived as good Swedish engineering practices and
Americanization. Conspicuously enough, in the subsequent negotiations, both
organizations arrived at the same conclusion: namely that the Swedish engineering sector should avoid the new methods. In his report to the SEEA, Werne
described a meeting with Erland Fägerskjöld, Volvo's leading production engineer who was at the time heading the implementation of the new system at
Volvo:

I gave him [Fägerskjöld] some examples of the unrest and confusion that US experts and
their systems had already caused in the Swedish labour market, for instance the strike
caused by the implementation of the Bedaux system at Alm's shoe factory in Göteborg in
1936.[23]

In subsequent communications Werne was even more explicit, allowing continued experiments with the aim of convincing Volvo, as well as other employers,
of the system's unreliability. His counterpart in the negotiations, Lennart
Eckerström, was equally frank in his report to the Swedish Metalworkers'
Union: 'I came to the conclusion that everything should be done to impede the
application of this system in the automobile sector as well as in the industry
more generally'.[24]

Meanwhile, however, the views held by these respective organizations did not
inhibit Volvo from introducing the MTM system into the Swedish labour market during the course of the 1950s. While both the union and the employers association were doubtful about what they saw as 'radical Americanization', Volvo
remained committed to MTM. Indeed, by 1953 Volvo had stepped up its efforts
to implement the new system although it clearly flew in the face of Sweden's
organized labour market. At Volvo's Pentaverken engine plant in Skövde, the
introduction of the system continued. Time formulae for specific work operations were developed between 1953 and 1955, beginning with those for toolmakers and advanced machinists. For the first time, engineers integrated the

---

[23] Verkstadsföreningens Arkiv, AB Volvo, report 29 Oct.–8 Nov. 1952 (Werne).
[24] Svenska Metallindustriarbetareförbundets historiska arkiv, AB Volvo, Report, 8 Dec. 1952
(Eckerström).

planning of factory layout, construction of machines, and tooling as well as maintenance practices with pre-designed work sequences and time analysis. Early experiments in the toolroom were followed by three major expansions in the foundry, in the assembly of engines in Skövde, and in the assembly of trucks in Göteborg.[25] But these initial introductions of the MTM system owed their peculiar temporary and experimental character to the fact that the system was operated within the framework of the engineering industry's central piece-work agreement. Any permanent piece-work system under Swedish labour market regulations had to be integrated into section 4 of the national labour market agreement, which specified the rules for fixing rates. Adopting MTM as a permanent basis for incentive payments presupposed a revision of the current agreement. Although the wildcat strike of 1952 was not repeated, many of the early experiments were accompanied by heated debates within the local unions in Skövde and Göteborg, since workers identified the non-recognition of MTM in the national agreement as a possible way of blocking the spread of the system.

The emerging pattern was quite clear. While the Communist majority of workers—encouraged by their recent hands-down victory over management—wanted the SMWU to take firm actions against the experiments, the national union itself was more ambivalent.[26] By and large, this was a difference between those who wanted a radical effort to abolish the MTM system altogether and those who were convinced that the employers were determined to implement the system come what may, and preferred to incorporate the system in the collective labour market agreement to achieve some leverage over its future operation. While the SMWU hesitated in 1952, a few years later it saw no alternative but to accept the system. In the preliminary negotiations before the 1954–5 collective bargaining round, the union faced demands from the employers association to revise the paragraphs in the national agreement covering piece-work and rate fixing. The employers argued that the introduction of the MTM system called for a total revision of these paragraphs, but they eventually arrived at a less radical proposal. Although the ensuing revisions were moderate, they none the less paved the way for the future use of the MTM system.

In summary, the suggestion here is that Volvo stands out as the radical innovator, seeking to solve the strategic dilemmas in truck production resulting from a complex political process, involving the political construction of the market together with short-term goals, while the state and the organized labour market were still in the process of defining a conceptual framework for a post-war industrial structure. There is no evidence to support the idea that the Swedish state or the national labour market organizations were already in the 1940s and early 1950s committed to a Fordist trajectory for the domestic automobile sector.

---

[25] Luthman *et al.*, *MTM i Sverige*.
[26] For the union debates on MTM in the years 1952–5, see Sundgren, 'Införandet av MTM-metoden'; see also Stråth, 'Verkstadsklubbarna vid Volvo och Saab'.

SELECTIVE ADAPTATION AND MANAGEMENT CHOICES FOR THE 1960S

Product strategy

In the midst of the political battles over the introduction of American production methods, Volvo managers looked to the European automobile industry to gain knowledge about the post-war implementation of American practices. Thus, in October 1954 Per Söderström toured car factories in Germany to study the development of rationalization at Opel and Volkswagen. Reporting back to his fellow managers, Söderström declared that 'it was almost a shocking experience to see the resources gathered in those plants'.[27] After his return to Volvo, he met with his colleagues to discuss how they would need 'to redesign our production lines with an advanced degree of rationalization, including a more intensive use of the MTM system, if we were to remain competitive in the future'. The main thrust of the broader discussion involved, nevertheless, Volvo's limited capacity to develop a cost-based competition strategy:

> These gigantic concerns, with vast domestic markets and high customs duty on cars, have better conditions than Volvo. Our competitiveness thus depends on our designs, on our quality and our costs. We should be able to compete successfully with design, construction and quality, since designing modern cars in high demand and producing them at good quality is not something only a large manufacturer can do.

Söderström planted his argument in fertile soil. Although Volvo relied on a technically quite simple and somewhat rugged design in its successful 444/544 model, Volvo carved out a market niche in 'safe and reliable quality cars'. To become an early mover in safety, Volvo initiated close collaboration in particular with Nils Bohlin of the Chalmers Institute of Technology, the inventor of the modern seat belt and other safety devices successfully commercialized by Volvo.

Quality and Industrial Relations: Towards Union Pragmatism

To realize its high value-added potential, however, Volvo had to focus on quality. To put it another way: the more Volvo marketed itself as a high-value product, the more customers expected from the firm in terms of quality. Although Volvo successfully explored the market for safe quality cars, the company was less successful in achieving the necessary production quality. It is quite clear that throughout the 1950s and 1960s production quality deteriorated, while numerous sources suggest that because adjustments to the cars at all stages of the production line slowed down output, Volvo was unable to keep up with demand.[28] According to a former Volvo production engineer:

---

[27] These and subsequent quotations from the minutes of Volvo's Main Committee on Production Technology are based on an annotated selection of transcripts made available to the present author by Torsten Hagenblad, director of AB Volvo.

[28] In particular, Volvo's many requests for increased overtime work reflects quality problems in production. Production breakdowns, lost bodies due to insufficient painting and assembly, are frequently cited as causes for overtime. Svenska Metallindustri-arbetareförbundets historiska arkiv, AB Volvo, Ansökningar om utökad övertid.

Quality was not something that we achieved on the assembly lines. It was something that we really achieved through rectification. Yes, I would go as far as to say that the outgoing quality was determined by the level of post-production rectification. A very expensive business, too.[29]

A series of wildcat strikes from the mid-1950s onwards demonstrated that there was a closely linked nexus of problems at Volvo. While quality could be reduced to technical and logistical processes, managers began to realize that pay and working conditions, including the local industrial relations system, also played a role. Protesting workers and illegal strikes in the paint shop in the winter of 1955–6 illustrate these connections.[30] It turned out to be a vicious circle. Technological problems and lack of co-ordination on the production line caused repeated interruptions to production. Problems in the body shop, in particular, resulted in stoppages to the subsequent production sequences, since bodies accumulated in the rework or paint sections as a result of poor incoming quality.[31] Under the MTM system, workers benefited from high projected wages, but when the production line stopped, workers were paid a low hourly compensation rate. This discrepancy between projected and actual wages caused numerous wildcat strikes and general unrest among the workers. In turn, these wildcat strikes aggravated the problems of technical co-ordination and quality since the production line became overstretched on these occasions. Moreover, it is highly probable that the frustration caused by the strains within the production situation and uncertainty about earnings can hardly have contributed to an improved standard of work. It soon became apparent to the management that the system of industrial relations under MTM lacked the capacity to deal with this situation. Although quantitative estimates of the causes and the number of such conflicts are far beyond the scope of this chapter, suffice it to say that a picture derived from the minutes of the negotiation suggests the increasingly bureaucratic nature of the relationship between workers and management.[32]

Not much changed during the first half of the 1960s. In looking back at the period from the mid-1950s to the early 1960s, Volvo managers started to reflect

[29] Interview with ex-production engineer Bertil Andersson, May 1997. For more dramatic glimpses of the resulting quality problem, see n. 35 below.

[30] My interpretation of the situation is based on the correspondence between the SMWU and the SEEA in connection with the illegal strikes in 1955–6 (Svenska Metallindustriarbetareförbundets historiska arkiv, AB Volvo, rapport från centrala förhandlingar, 8 Feb. 1956).

[31] The records indicate that about 15 per cent of the bodies were rejected on the ground of poor quality.

[32] While the number of piece-rate negotiations fell by some 15 per cent in absolute terms, the numbers of employed workers rose from about 1150 in the late 1940s to about 3,200 a decade later, suggesting a far more dramatic change in relative terms. Also, the nature of the issues that caused disputes changed, from complicated arguments concerning how new models and other changes affected the balance between work and pay, to more formal arguments about the time of work breaks and the replacement of workers under the MTM regime. By the early 1960s, issues like break times clearly dominated the local negotiations. (The estimates given are based on the records of proceedings of local negotiations in the years 1948–68 and on interviews with Svante Simonsson, director of Volvo's negotiation office, May 1985. The reports of the proceedings were made available to the present author by Svante Simonsson.)

upon the fact that wildcat disputes more or less typified the situation in car production. When faced with a new crop of illegal strikes in 1963–4, management
again requested central negotiations between the SMWU and the SEEA. The
records of that meeting reveal that the management argued that although 'there
was no need to dig deeply into the details of the past, we do need to find guidelines to prevent future illegal strikes'.[33] Holger Olson, the ombudsman representing the SMWU at these negotiations, suggested two basic solutions, which
in a way also clearly defined the problems. First, Olson asserted that the difference between hourly wages paid during stoppages in production due to technical hold-ups or quality problems and the MTM wages should be narrowed
through an increase in time rates. Thus, the basic problem and the actual cause
of wildcat strikes could be tackled. Second, he also expressed the hope that
Volvo would be 'more generous in its economic contributions to the local union
to enable it to become more efficient in its undertakings'. Among other things,
Olson concluded that 'management should pay the full wage of the local union
president and some of the local officials'.

Confronted with this proposition from the SMWU, the chairman of the local
union and Volvo's management representative withdrew from the meeting for
a personal consultation. Their deliberations resulted in an agreement on the
general guidelines for closing the gap between projected and actual earnings as
well as an acknowledgement of the union's need for economic support. Svante
Simonsson, one of Volvo's most experienced managers and director of the negotiation office, recalled that:

We tried to establish a continuous, professional and pragmatic relationship with the
local union representatives, which means that we tried to formulate common policy
goals through centralized management–union relations rather than simply responding to
conflicts as they arose.[34]

In their search for strategies to solve the emerging nexus of industrial relations
problems Volvo's management tried to build a pragmatic alliance with the local
union. The outcome of the 1964 agreement suggests, however, that the local system of industrial relations was being pushed towards bureaucratization.

Bringing Craft Back In: The First Steps

Experienced production managers became motivated to go beyond this formula
for union pragmatism. They did so on two interrelated grounds. First, poor
quality remained very much the Achilles heel of production throughout the
1960s. For example, a consultancy report on Volvo's reputation and customer
satisfaction, commissioned by the company's management, showed that Volvo
enjoyed a far worse reputation for production quality than its competitors in the

---

[33] For this and subsequent quotations from this meeting, see: Svenska Metallindustriarbetareförbundets historiska arkiv, AB Volvo, rapport från centrala förhandlingar, 29 Sept. 1964.
[34] Interview with Svante Simonsson, director, AB Volvo, 16 May 1985.

important US market.[35] Second, to reconcile efficiency and quality, the company needed to look no further than the segment of the Swedish motor vehicle industry that was less affected by the development of economies of scale: Volvo's own production of trucks. As was suggested at the start of this chapter, the design of the assembly system and working conditions in the truck division differed from those characteristic of the car division. Between 1945 and 1989, in fact, even the most successful truck models were produced in runs of no more than 40,000 units, compared to the 1.2 million 140 series cars Volvo turned out between 1966 and 1974. Thus, it is not surprising that Fordism had less impact in the production of trucks and buses than in passenger cars. Typical cycle times in truck production ranged from 30–40 minutes up to 10–12 hours, compared to the 1–2 minutes typical of traditional assembly work. Given these differences, the truck-building work-force faced a far more demanding situation, with much higher skill requirements. While absenteeism and labour turnover rates grew rapidly in passenger cars, moreover, the work-force in truck production was far more stable.[36] Internal quality reports indicated the situation in truck production was quite different from that for cars: 'the quality in the production of trucks is, in general, good, and the external quality of the products [is] thus satisfactory'.[37]

Management thus became aware that the production of trucks had developed along a different course, combining quality with a stable work-force and continuing to rely on workers' skills and craft ethos. The course followed in Volvo's successful truck arm thus appeared to represent a possible route to a reconciliation of quality and stability of the work-force in passenger cars. Bertil Andersson, a production engineer, recalled that:

In the mid-1960s, Bertil Darnfors, the leading production manager, started to bring back old production engineers, like Hugo Hansson, who had more or less been released from his duties in the late 1950s because he had too strong a commitment to what were seen as outdated, craft-based production methods. Thus, it was experienced production engineers, such as Hansson, who made the connection between the two realities which typified the production of trucks and passenger cars respectively.[38]

[35] The alarming rise in quality costs due to general disorganization of production, high absenteeism, and high labour turnover, reported in yet another confidential quality report to the management, caused the engineer behind the report, Per Åke Sörensson, to suggest to his superiors that there was a clear need for general improvements, including the formulation of a consistent quality policy. But before such measures could be implemented, Sörensson suggested, quite extreme polices needed to be put in place. In particular, Sörensson wanted to limit the elements that put strains on the production line, particularly: 'the number of modifications in construction should be restricted to such extent, that, in principle, only those product innovations that were motivated by changes in the legal framework, or for quality reasons, should be accepted'. The same report presented a quality cost analysis, indicating that the costs for internal rectification and external warranties were no less than 8 per cent of the total sales in 1970: AB Volvo, 'Kvalitetssituationen', Konfidentiell PM utfärdad av P.Å. Sörensson och tillställd Bengt Darnfors 12 Nov. 1973.

[36] See Berggren, 'Alternatives to Lean Production'.

[37] For the source of this quotation, see n. 35.

[38] More than a decade later Bertil Andersson emerged as one of the key advisers and sources of inspiration for Professor Tomas Engström, Chalmers Institute of Technology, in his experiments

INTEGRATING CRAFT AND UNION PRAGMATISM

Although the push towards union pragmatism had already yielded some significant results by the mid-1960s, these craft ideas remained inside the company as a partially suppressed resource. It was not until 1969, however, that the management tried to mould these ideas into a more general framework of craft-based production principles for the automobile industry. By then the familiar wildcat strikes over workloads and pay had forced managers at the Torslanda plant to take a more active stance. In essence, managers worried not so much about the strike itself as about how the highly mobile work-force would react if the strike lasted for longer than a week. Svante Simonsson's report to the top management on the strike negotiations indicated the nature of the real threat:

> The resulting settlement has to be seen against the background of the nature of the illegal conflicts and the nature of the particular prevailing manpower situation at Volvo in Göteborg, with 53 per cent immigrant workers. I estimate that a prolonged conflict would most likely have implied that only about 60 per cent of the work force would have returned to work after the conflict was over.[39]

Against this background it is hardly surprising to find that ideas of craftsmanship and individual commitment, together with union pragmatism, paved the way for the reformulation of production paradigms. It is often argued that the wildcat strikes of 1969 and 1970 were the motivating force behind the decision of Volvo managers to go beyond Fordism. However, it should be noted that internal management reports on alternative production strategies were already advanced and circulating among top managers by the autumn of 1969. By this time production engineers had already formulated their strategy as a 'Program for Industrial Democracy'. This strategy was based on Louis David's work on job redesign which he had developed from experience in production during the Second World War.[40]

It is also clear from deliberations within Volvo's top management that they themselves doubted their ability to deal competently with the new requirements. Erik Quistgaard, managing director of the Torslanda plant, for example, commissioned younger colleagues to draft reports on alternative production strategies and also hired experts on sociotechnical strategies.[41] Internal reports, which emphasized the need for autonomous work groups, quality of working life, job rotation, and industrial democracy, show that it was individual motivation that first attracted attention. Efforts to formulate a basic company policy

with parallel production lines and non-line-based production systems for the assembly of cars at Volvo's Uddevalla plant.

[39] AB Volvo, Produktionstekniska huvudkommitten, 19 Dec. 1969.

[40] AB Volvo, 'Programförslag för ökad industriell demokrati inom Volvo Göteborgsverken', Utfärdat av H Lenerius och tillställt B Danfors och E Quistgaard, 10 Nov. 1996 (Reg. no. 71000-204).

[41] For example Berth Jönsson, a social scientist trained in the USA, with a background in industrial sociology and psychology, who later emerged as one of the most trusted advisers to Volvo's leader, Peer Gyllenhammar.

in this area favoured concepts like motivation and job satisfaction, although the ideas were still rather crude. A committee on job redesign and industrial democracy, chaired by Quistgaard, concluded in a confidential report that: 'the initial goal should be to achieve increased individual stimulation in parallel with increased efficiency—that is industrial democracy'.[42]

At this point, the deliberations within management circles reintegrated the union pragmatism of the 1960s with strategies for sociotechnical job redesign. One of the early strategy documents from this period pinpoints the idea of autonomous work groups as the key to craft-based production principles.[43] What was required was a trade-off between quality, productivity and the individual worker's need for meaningful job tasks or job enlargement. In particular, the document stated that Volvo needed to deploy policies rapidly to 'develop procedures for consultation between management and the work-force concerning day-to-day co-operation on the shop floor', and to 'study and design a work organization that would satisfy the individual worker's job requirements'.

In exploring the institutional grounds for the implementation of neo-craft-based production principles, the same document referred back to the pragmatic management–union relations and to the various committees and procedures for union consultation that had developed from the mid-1960s. In addition, it conveys the general argument that management–union relations at Volvo developed into relatively advanced institutional forms in the 1960s. At least two different bodies for union consultation that were established under the push towards union pragmatism during the mid-1960s could, according to the report, directly contribute to the formation of the institutional basis for the development of new production strategies: the Frånvarogruppen, a joint group formed to reduce labour turnover and absenteeism, and the Personalkommitén, a joint group on general staffing policy.

By the early 1970s, as has been widely recognized, the search for a new factory regime gathered momentum with the arrival of Volvo's charismatic new leader, P. G. Gyllenhammar. One of the first major reports on the problems of working conditions commissioned by Gyllenhammar, 'Volvo Socialkalkyl' (Volvo Social Calculation),[44] reached the same conclusions as the reports mentioned earlier. Although the importance of immigrant workers was played down, the general conclusion was that Volvo's main problems stemmed from workers' low commitment to work, high absenteeism, and labour turnover. In those respects, Volvo's situation was far worse than the average Swedish engineering firm. In summary:

Worse working conditions at Volvo than in Swedish industry in general are believed to explain the differences in labour turnover between Volvo and the Swedish industry in

[42] Confidential report of the proceedings of the reference group on forms of union consultation and work organization, dated 15 Jan. 1970. The author would like to thank director Torsten Hagenblad of AB Volvo for making this document available to him.

[43] AB Volvo, 'Programförslag'.

[44] This project was monitored by a reference group appointed by the prime minister, Olof Palme.

general . . . The interviews showed that psychological effects, like suffering from repetitive tasks, lack of freedom at work and heavy workloads are the main reasons why Volvo's workers tend to quit.[45]

Poor quality could thus in turn be attributed to the negative effects of repetitive work on the assembly line. Thus, the consultants' reports to management on the quality problem echoed the insights already gained in the 1960s, although they conveyed increasingly advanced arguments concerning the advantages of sociotechnical-based production strategies, implying radical departures from the moving assembly line.

6.3   Electrical Engineering: The Political Construction of a Route from Scale to Scope

Another sector closely linked to growth of domestic consumption during the post-war era was electrical equipment. Between 1946 and 1954 Swedish electricity sales increased at an annual rate of more than 10 per cent, from 11.3 to 24.7 terawatt hours. Real output of electrical goods expanded accordingly. The domestic producers of heavy electrical equipment tripled their home market sales of generators, transformers, and distribution equipment between 1946 and 1960. Electrical engineering, however, constitutes an interesting contrast to the motor vehicle industry. While the government did not identify motor vehicles as a key industrial priority, electrical engineering received full attention and support during the critical post-war years. Electrical manufacturers faced less severe restrictions than the automotive sector which experienced, as we saw in the previous section, strict limitations on its access to foreign currency and steel allocations. Imports of materials and components for the production of heavy electrical goods remained surprisingly stable even during the critical period between 1947 and 1953. Much the same pattern can be observed in the allocation of credits during the period of restricted lending between 1952 and 1957. Thus in 1952, two sectors—building and power—were given credit while industry as a whole saw its access to borrowing facilities diminish. In the mid-1950s, for example, the government excluded all sectors except the power industry from the Swedish bond market.[46] But these tendencies were not just reflections of post-war concerns with the rapidly expanding demand for electric power. They mirror also the formation of the Swedish development bloc in electrical engineering that had emerged around the turn of the century.

ENERGY POLICY AND SUPPLY INDUSTRY

By the last decade of the nineteenth century, Sweden turned towards its internal hydro power resources to limit its dependence on British coal. To contemporary

[45] AB Volvo, 'Volvo Socialkalkyl. Delrapport IV. Försök till helhetssyn' (1973), 72.
[46] Ingemar Nygren, *Svensk kreditmarknad under freds- och beredskapsåren 1935–1945* (Göteborg: Meddelanden från ekonomisk-historiska institutionen vid Göteborgs universitet, 1974).

policy makers, it thus became apparent that both the energy-intensive exploitation of Swedish ore resources and especially the transportation of iron ore across the steep mountain region between Kiruna and the Norwegian Atlantic coast required modern electric railways. Thus, the Swedish state acquired at least 17 waterfalls from private owners, which were to constitute the energy source for the planned nationwide electrification of the public railway system.[47] But the potential exploitation of hydro resources soon developed into a clash between industrial and agricultural interests. Plans to harness waterfalls to exploit hydro power resources diminished, or prevented access to forest, river banks, and water for agriculture and fishing. After the turn of the century, this clash involved a clear struggle between rural interests and the early beneficiaries of electrification: industry in north and central Sweden together with the expanding population centres in the south. The ensuing political conflict gave rise to a coalition of Conservatives and Social Democrats, both favouring further industrialization, in opposition to farmers whose rural interests were at stake. The urban–rural clash peaked as industrial entrepreneurs sought to develop remote hydro resources to service towns and industries.[48]

The prolonged debate concerning the exploitation of the waterfall outside the town of Trollhättan (north-east of Göteborg) is generally taken as the turning-point at which the Swedish authorities emerged as a principal owner and operator of generation as well as distribution networks. Between 1899 and 1907, the supporters of electrical energy improved their legal standing considerably. First, they proved able radically to alter the legal framework regulating exploitation of waterfalls and erection of regional grids. Second, by 1907, the leadership of the public committee advanced the idea that the state—and not private owners—should both own and operate the planned utility in Trollhättan. The state should not, it was further suggested, run public utilities individually but rather integrate them into a single centralized agency. Hence in 1907, after parliamentary approval, the state acquired the rights to the Trollhättan waterfalls and established the Vattenfallsstyrelsen, or Swedish State Power Board (SSPB).[49]

Within a few years, the SSPB had become the major actor in energy affairs. By 1910 it was operating two respectable-sized power plants (Trollhättan and Porjus) and over the following two decades it became the cornerstone of Swedish electrification, running the national grid and controlling a stable 40 to

[47] Staffan Hansson, *Porjus: En vison för industriell utveckling i övre Norrland* (Luleå: Institutionen för industriell ekonomi och samhällsvetenskap, 1994); Eva Jakobsson, *Industrialisering av älvar: studier av svensk vattenkraftsutbyggnad 1900–1918* (Göteborg: Avhandligar från Historiska Institutionen, Göteborgs Universittet, 1996); Arne Kaijser, *I fäderens spår. Den svenska infrastrukturens historiska utveckling och dess framtida utmaningar* (Stockholm: Carlssons förlag, 1994); Lars Lundberg, *Energipolitik i Sverige 1890–1975* (Stockholm: Sekretariatet för framtidsstudier, 1978).

[48] For an account of the struggle between agrarian and industrial interests, see Nigel Lucas, *Western European Energy Policies* (Oxford: Oxford University Press, 1985), 107–10; Lundberg, *Energipolitik i Sverige*; Eva Jacobsson, 'Norsk och svensk vattenkraftsutbyggnad', *Polhem* 3 (1992), 226–64.

[49] Jakobsson, *Industrialisering av älvar*.

50 per cent of national power generation capacity. Together with the major private and municipal electric utilities, the SSPB in 1934 formed a joint committee—Centrala Driftsledningen (Central Operation Management)—that soon reduced private power producers' rights to own and operate heavy-power distribution links. Furthermore, in 1946 the state granted the SSPB exclusive rights to operate the national grid, which included the task of forging agreements with those private producers wanting to distribute power through it.[50]

In his comparative analysis of West European energy regimes, Nigel Lucas describes the Swedish system as a state-led energy production cartel. The construction of the national grid for long-distance power transmission constituted the centrepiece of the cartel structure. Although the production of energy was balanced between public, private, and municipal producers in terms of capacity, the national state had secured a dominant position in the system at the central level. The SSPB had achieved its unrivalled position mainly through its control over the distribution links between the producers of power and their customers.[51]

ENERGY REGIME AND INNOVATION

In relation to technology and innovation, three major consequences of this integrated energy regime stand out. First, Swedish politicians and civil servants alike based their go-ahead approach to electrification on the emerging domestic equipment manufacturing sector. Generators produced in 1907 by ASEA for Norway's still young fertilizer industry were particularly promising. Swedish electrical engineering had, by 1913, achieved a foothold in the international market as an exporter of electrical equipment. Second, the Swedish authorities concluded that despite the promising state of domestic electrical engineering, the state should not take a back seat in the development of new technological solutions. Although state actors seemed confident about the competence of Swedish industry, they were impatient in their efforts to find advanced solutions to national infrastructural problems—not hesitating to pursue the idea that domestic industry would be capable of developing technologies that were not yet available on the world market. Third, the state took an active part in shaping technology through joint R&D programmes that not only assured private industry about the relative security of a particular line of investments, but also provided the firms with full-scale facilities and access to the user's experience.

Jan Glete's carefully researched work on ASEA and Swedish electrical engineering, and Mats Fridlund's more recent studies of the relationships within the Swedish electrotechnical sector provide detailed accounts of the intimate relation between users and producers of heavy electrical power equipment under the Swedish energy regime.[52] Despite ASEA's early investments in the heavy seg-

[50] Kaijser, I fäderens spår.   [51] Lucas, Western European Energy Policies.
[52] Jan Glete, ASEA under hundra år 1883–1983 (Stockholm: Institutet för Ekonomisk historisk forskning vid handelshögskolan i Stockholm, 1983); id., Nätverk i näringslivet. Ägande ochindustriell

ment (generators and transformers based on its patent for polyphase distribution), the company's position in the Swedish home market was not uncontested; around the turn of the century several domestic firms took the crucial initial steps to become 'full line producers'. Yet there is little doubt that strong financial interests, represented by the Wallenberg family, conceived ASEA as the key to a major restructuring of Swedish electrical engineering aimed at creating an internationally respectable competitor in heavy products. Critical investments by the Wallenbergs around 1903–4 set ASEA apart as the indisputable leader in the sector. This successful move was soon followed by ASEA's takeover of independent firms that resulted in the concentration of generator technology in its Västerås plants and of power transformer and distribution systems in factories in Ludvika previously owned by Nya Förenade Electriska AB (NFEA). To gain a foothold in the expanding turbogenerator market, ASEA also sought to develop closer links with manufacturers, resulting in a merger with STAL, the licensed producers of Ljungström turbines, as well as an agreement with the Ljungström patent-holding company about further collaboration on turbine and generator development.[53]

Apart from these important moves to integrate and develop Sweden's heavy electrical engineering industry, the less well-known steps taken by ASEA after 1903 towards the modernization of light production (motors, meters, and other types of electrical apparatus for industrial use) proved to be equally significant. Motors were the key to this production, which soon expanded to cover a wide range of products from elevators to electrical capital goods riding on the global wave of industrial electrification. Again, this modernization strategy, which included careful assimilation of American methods for volume production, also involved mergers with some of the leading Swedish firms in light engineering.

ASEA was instrumental, moreover, in the organization of a Swedish electrical engineering cartel that linked the firms and constituted the domestic basis for protection of the home market against foreign penetration. This cartel was thus the 'Swedish end' of an agreement also involving AEG, Siemens, and Brown Boveri, which limited the degree of foreign penetration of the home market to 13 per cent.[54] Therefore, it seems safe to assume that cartelization supplemented the vertical and horizontal integration that occurred in the first two decades of the twentieth century. This means that the value of electrical goods sold by the cartel and/or by ASEA's single firm monopoly soared in the 1920s and 1930s, according to the calculations of the Swedish Department of Commerce. In fact,

omvandling i det mogna industrisamhället 1920–1990 (Stockholm: SNS Förlag, 1994); id., 'Swedish Managerial Capitalism: Did It Ever Become Ascendant?', *Business History* 35, 2 (1993), 99–110; Mats Fridlund, ' "En specifikt svensk virtuoskonst": Empiriska och teoretiska perspektiv på utvecklingsparet ASEA-Vattenfals historia', *Polhem* 12, 2 (1994), 106–31; id., 'The Development Pair as a Link between System Growth and Industrial Innovation: Cooperation between the Swedish State Power Board and the ASEA Company', TRITA HST Working Paper 93/5 (Stockholm: Department of History of Science and Technology, Royal Institute of Technology (KTH), 1993).

[53] Glete, *ASEA*.   [54] Ibid.

these calculations show that electrical engineering emerged, together with simple staple goods and textiles, as the centre of the Swedish cartel movement.[55]

Swedish legislation imposed few constraints in this field. While the Acts of 1925 empowered and required the government to investigate monopolies and cartels, the 1946 laws imposed public registration of cartels on the industry. In spite of large investments to advance Swedish heavy electrical engineering, ASEA still depended on light products as its core strategy. The idea was to exploit the firm's international reputation in light electrical machinery and meters to safeguard its cash flow, partly to insulate the company from the irregularities of the market for heavy products, and partly to generate revenues for R&D. Subsequent efforts to enhance ASEA's position in the market for advanced, heavy products depended on success in the 'bread-and-butter lines' involving standardization, large volumes, and rapid throughput—a sharp contrast to the realities of production in, say, generators. In fact, Glete writes, the composition of ASEA's export sales reveals that light equipment dominated the complex products by a ratio of approximately 7:3. Thus during the 1930s, ASEA was still relying on volume and standardization to fund future achievements in complex products.[56] By then, as already indicated, the Swedish state had taken the lead in domestic electrification and was on its way to assuming some of the burden of developing new technologies that it believed could link hydro resources with transport and industry. On the one hand, the state was about to open up a vast market for heavy products, but on the other, as will be emphasized here, it was also committed to the development of the products and production to supply this market.

Though the first joint projects involving ASEA and the SSPB were modest indeed, the programmes initiated in the 1930s addressed core problems of electrical engineering. Joint research on transformer technology, for example, was initiated by the state in the 1930s, when Swedish technology was still lagging far behind that of German, American, and Swiss firms. During this period, ASEA invested in modern production facilities and a test laboratory for power transformers at its Ludvika plant. As the need for the rapid development of long-distance distribution emerged during the Second World War, the SSPB initiated a series of joint ventures to solve the fundamental problems of High Voltage Direct Current (HVDC) distribution systems and related transformation and high voltage switching technologies. Thereafter, shortly before entering into the extreme complexities of nuclear technology, the SSPB launched joint programmes for 400 kW distribution lines and air-cooled breakers.[57]

Conspicuously enough, Fridlund's account of the joint SSPB and ASEA development projects reflects a list of priorities coinciding with ASEA's later market success. Thus ventures such as the transformer technology and breaker projects

[55] Hans Brems, 'Monopoly and Competition in Scandinavia', in Edward H. Chamberlin (ed.), *Monopoly and Competition and their Regulation* (London: Macmillan, 1986), 167–88.

[56] Glete, *ASEA*.

[57] Fridlund, 'En specifikt svensk virtuoskonst'; id., 'Ett kraftfullt utvecklingspar'.

in the 1920s and 1930s, or the HVDC and 400 kW line projects between 1945 and 1956, show that the state's engagement was targeted at key areas of electrical engineering. These development projects demonstrate the centrality of user–producer interaction: it became unambiguously clear to both parties that short-term commercial interests had to be deferred in favour of reaping the mutual benefits of meeting long-term needs. ASEA advanced to a leading position in the international market by mastering the techniques of the 400 kW Alternating Current (AC) distribution network, by taking the lead in improving the critical performance–weight ratio for power transformers, and by solving the key stability problems of HVDC. Advances in these areas put subsequent pressure on engineering expertise in related fields (such as cables and condensers). Breakthroughs in one area were not as valuable unless other parts of the system also made the leap to the new paradigm. Glete summarizes the result of this form of collaboration:

Concerning power generation and distribution technology, [ASEA] was, for the first time, beyond doubt, second to none among the heavy electrical companies. The course of events illustrates also the close interdependency between ASEA and SSPB [Vattenfall]. Given the circumstances, SSPB could not turn elsewhere for supply and ASEA had, likewise, no other customer to whom it could sell its most advanced technologies.[58]

Swedish patterns of collaborative learning provided the bedrock for the shift in ASEA's approach to the business that occurred gradually between the late 1930s and the 1950s. The importance of mass-produced goods as the bread-and-butter lines which would secure its cash flow diminished as the firm's technological capability increased with the accelerated learning process in the home market. According to Bela Balassa's statistical analysis of the trends in revealed competitive advantage in the world market for power equipment, there is clear evidence that the Swedish home market-led approach to R&D and learning soon spilled over into the international product market in a more general way. In fact, according to Balassa's ranking of world trends, this catching up was forceful enough to position Sweden's heavy electrical engineering in the same division as those of Germany and the USA.[59]

From the trade statistics analysed by Michael Porter to identify national clusters of strength, it is also evident that Sweden had acquired a position as one of the world leaders in heavy electrical engineering. Although Porter's analysis reveals some obvious pockets of strength, such as transformers, it does not reflect the dominant market positions of more narrowly defined key technologies. The development of HVDC technology stands out as one of the most illuminating cases of interaction between state and industry. It was the development of this technology that provided Swedish electrical engineering with its most significant inroad into international markets. Winning the contract for the

[58] Glete, ASEA, 174. [This author's translation.]

[59] Bela Balassa, Comparative Advantage, Trade Policy and Economic Development (New York: Harvester, 1989).

English Channel submarine link between Britain and France in 1957 was the most significant international breakthrough for Swedish industry, leading to subsequent orders for submarine links connecting the main islands of New Zealand, Denmark to the Swedish grid, Sardinia with the Italian mainland, and, most important in terms of market penetration, the Pacific Intertie that distributes hydroelectric power from generation areas in the north-western United States to consumers in California. Between 1955 and the late 1960s, ASEA was unrivalled in its technological leadership in this area.[60]

This story of collective learning is thus closely related to national trade performance. Although Sweden's share of the global market for electrical power equipment remained fairly stable at 2.5–3 per cent of world trade from the inter-war period to the 1970s according to most estimates, the technological underpinning of that share changed dramatically during those years from mass production of relatively simple products such as motors to the complex core products of heavy electrical engineering.

6.4 Conclusions

By emphasizing the rigid character of centralized bargaining, the students of Sweden's political economy end up by wrongly characterizing its national industry as 'Fordist'. By contrast, this chapter argues that the strategic responses to increased competitiveness varied across sectors. In motor vehicles and electrical engineering, the experience of diversity of market realities, heterogeneous institutions, and conflicting political goals forced even the most enthusiastic proponents of Americanization to modify or abandon their view on the fundamentals of industrial efficiency. Pressure from customers for innovation, small batch production, adaptation, and successive upgrading of existing products generally created obstacles to standardization. It was not until increased liberalization of world trade and competitive pressures began to make themselves felt in export as well as domestic markets that the ambiguities of Swedish industrial practices became manifest, spurring national industry to redefine its identity through a process of strategic debate and selective adaptation.

This conclusion raises three main questions: how widely diffused were Fordist responses to cost-based competition across different sectors of the Swedish industrial economy; what temporal variations can we identify; and how responsive was the institutional setting of Swedish industry to such international competitive pressures? As a first approximation, initial convergence on a basic Fordist model hinged as much on international pressures, or influences, as on the structure of the domestic market. It should thus be emphasized that international pressures did not have an unambiguous influence, and were not unaffected by internal politics. On the contrary, international economic influences were politically refracted as they entered the domestic market. If the struc-

---

[60] According to ASEA's market statistics, their world market share of HVDC installations was over 70 per cent between the 1950s and 1980.

ture of the market is important, no less significant is the role of the state, conscious and unconscious, in shaping those markets: it is not merely the fact that the domestic market was fragmented that is important for our story, but also that it was shaped in a particular way by state policies. Americanization developed as a strategic managerial response to a complex market environment shaped by the state's policies as well as by the ambiguous attitudes of both the employers association and the labour movement to the industrial relations consequences of the post-war productivity drive.

Meanwhile, the American production strategy was not transferred to the Swedish context without being significantly reshaped. In a long-term perspective, innovation and competitiveness hinged less on the capacity to imitate foreign models of industrial efficiency than on the hybridization that occurred as different production strategies were mixed and adapted. Before the Second World War, Volvo's innovative strength in trucks emerged out of the exceptional mix of craft orientation in Swedish engineering, flexible subcontracting strategies, and a locally adapted model of line-based assembly work. Examining the transfer and transformation process in the post-war period, including the adaptation of the MTM system, reveals that beyond this apparent Americanization, US methods were intermingled with Swedish traditions of high value-added production and sociotechnically inspired strategies. Once the MTM system was established, Volvo's top management quickly moved away from a cost-based competition strategy, resulting in early efforts to establish pragmatic, high-trust relations with the union in order to combine efficiency with quality in production. In the heyday of Swedish Fordism during the 1960s, key actors at various levels thus defined some of the fundamental principles that years later became the basis for the sociotechnical transformation involving pragmatic collaboration between union and company and strategies for creating individual incentive structures based on notions of skilled labour and group work. Hybridization thus paved the way for the product and process innovations that came to define the unique feature of Volvo cars and a new class of vehicle: the safe and relatively inexpensive station wagon.

By contrast, Swedish electrical engineering, as Section 6.3 showed, was integrated into a dynamic state-sponsored development bloc. After the Wallenbergs' successful consolidation of the Swedish electrical engineering sector around the turn of the century, the equally imaginative formation of a resourceful and technologically sophisticated state agency provided a dynamic environment for ASEA. Whereas Swedish electrical manufacturing was tied to an international cartel, which limited ASEA's freedom of action *vis-à-vis* its major German competitors, the surge in post-war national demand combined with national procurement polices stimulated a drift towards a focus on more and more complex products in Swedish electrical engineering. Unlike some of its most formidable competitors, such as Siemens, ASEA launched a product strategy that left less and less room for standardized goods. Instead of aggressively trying to explore the market for mass-produced electrical goods, ASEA took large steps towards increased dependence on advanced complex products. The formation of that bloc of competencies and innovatory activities was crucial for ASEA's subsequent success, as the company has continued to prosper through concentration on its historic technical strengths.

Taken together, these two cases suggest that the widespread interpretation of post-war Swedish development in terms of a Fordist modernization strategy is conceptually and empirically misleading. An additional problem with the claim that Swedish industrial actors played down product innovation in the post-war period while giving preference to reduction of unit costs and vertical integration is the fact that the 1950s saw the formation and tighter integration of new skill bases that subsequently contributed to Sweden's strong position in sectors such as electronic switching, mobile telecommunications, and pharmaceuticals. Perhaps the most crucial clustering of related skills and innovation processes since the late 1950s occurred in telecommunications, laying the foundations for Ericssons' and Televerket's successful launching of the digital AXE system. These innovative activities—some of them reaping significant returns after a time lag of two decades or more—were also part of Swedish industrial reality during the supposedly Fordist 1960s and 1970s.[61] By contrast, the sector that went relatively furthest towards the full-scale implementation of Fordism, the Swedish shipbuilding industry, collapsed completely after overcommitting itself to standardized tanker production. So diverse were domestic responses to international pressures in post-war Sweden.

---

[61] For the development of the AXE system, see Ove Granstrand and Jon Sigurdson, 'The Role of Public Procurement: Technological Innovation and Industrial Development in the Telecommunication Sector: The case of Sweden', in eid. (eds.), *Technological Innovation and Industrial Development in Telecommunications: The Role of Public Buying in the Telecommunication Sector in the Nordic Countries* (Stockholm: Nordic Co-operative Organization for Applied Research, 1985), 147–73; John Meurling and Richard Jeans, *A Switch in Time: AXE— Creating a Foundation for the Information Age* (London: Communications Week International, 1995). For the development of Astra-Hässle and the Losec drug, see Gunnar Eliasson and Åsa Eliasson, 'The Pharmaceutical and Biotechnological Competence Bloc and the Development of Losec', and Rikard Stankiewicz, 'The Development of Beta Blockers at Astra-Hässle and the Technological System of the Swedish Pharmaceutical Industry', both in Bo Carlsson (ed.), *Technological Systems and Industrial Dynamics* (Boston, MA: Kluwer Academic, 1997), 139–68 and 93–138 respectively.

Chapter 7

A *Slow and Difficult Process: The Americanization of the French Steel-Producing and Using Industries after the Second World War*

MATTHIAS KIPPING

7.1 Introduction

The French appear to have been among the most enthusiastic participants in the American-inspired and sponsored productivity drive from 1948 to 1958. The report which summarized the efforts and achievements in the different European countries stated that 'the technical-exchange aspects of the program were quickly and enthusiastically received by the French' and that they had developed 'the largest and most varied productivity program in Europe'.[1]

There is nevertheless considerable debate in the literature regarding the French reaction to American production technology, management methods, and model of market competition after the Second World War. Luc Boltanski has noted a fascination with American management in France from the late 1940s onwards, especially within the emerging class of middle managers, the so-called cadres.[2] And on the basis of the reports from the productivity missions to the United States, Vincent Guigueno has identified a 'profound modification of mentalities with regard to production', highlighting the fact that French engineers espoused shop-floor-oriented control techniques of American origin,

---

[1] International Cooperation Administration (ICA), *European Productivity and Technical Assistance Programs. A Summing Up (1948–1958)* (Paris: ICA, Technical Cooperation Division, 1958). For background on the US efforts, see Jacqueline McGlade, 'The Illusion of Consensus: American Business, Cold War Aid and the Reconstruction of Western Europe 1948–1958', unpublished Ph.D. dissertation (George Washington University, 1995); Anthony B. Carew, *Labour under the Marshall Plan: The Politics of Productivity and the Marketing of Management Science* (Manchester: Manchester University Press, 1987); and Charles S. Maier, 'The Politics of Productivity: Foundations of American International Economic Policy after World War II', *International Organization* 31 (1977), 607–33 (reprinted in several of his edited volumes).

[2] Luc Boltanski, 'America, America . . . Le Plan Marshall et l'importation du "management"', *Actes de la recherche en sciences sociales* 38 (May 1981), 19–41, and id., *The Making of a Class. Cadres in French Society* (Cambridge: Cambridge University Press, 1987), 97–143.

while leaving social regulation and matters of organization to management.[3] According to Marie-Laure Djelic, the Americanization of the 'French economic and corporate landscape' in the immediate post-war period was even more comprehensive and relatively smooth: 'Large production units, standardization, and mass production would remain the foundations of French industrial policy and for many years to come this choice was never questioned'. In her view, this 'radical' transformation was largely accomplished by a 'small group of French modernizers' around Jean Monnet 'who had the necessary means and tools to launch and implement their ambitious project' and also benefited from 'American intervention and support'.[4]

By contrast, other authors have expressed doubts about the extent of the alleged Americanization of French industry in the first post-war decade. Richard Kuisel, for instance, comes to a rather different conclusion based on his reading of the reports from the French productivity missions. He agrees that the visitors were in general impressed by what they discovered, including management as a method, 'a function of proper training and proper organization'. However, he also points out that a majority of members of French missions considered the application of the American experience to be unfeasible or even 'undesirable'.[5] Detailed analyses of the post-war attempts to reform French competition legislation, which had encouraged rather than prohibited cartel agreements, and of the efforts to introduce business-related training into the French universities and Grandes Écoles also paint a different picture. In both cases, Americanization proceeded only very slowly—at least until the mid-1950s—and encountered significant resistance from a variety of political and economic interest groups.[6]

As this chapter shows in detail, the Americanization of production technology, management methods, and market order in the French steel-producing and using industries after the Second World War was also a rather slow and difficult process. First of all, the question of whether the US example was appropriate and applicable to the different French circumstances was the subject of

---

[3] Vincent Guigueno, 'L'Éclipse de l'atelier. Les missions françaises de productivité aux Etats-Unis dans les années 1950', unpublished mémoire de Diplôme d'Etudes Approfondies (Ecole Nationale des Ponts et Chaussées, Université de Marne-La-Vallée, 1994), 15.

[4] Marie-Laure Djelic, Exporting the American Model. The Postwar Transformation of European Business (Oxford: Oxford University Press, 1998), 150, 157. It should be noted that Djelic relies almost exclusively on sources from the French Planning Agency (Commissariat général du plan) which might have led her to mistake plans for reality and overestimate the influence of Monnet and his American allies.

[5] Richard F. Kuisel, Seducing the French. The Dilemma of Americanization (Berkeley: University of California Press, 1993), 70–102, and quotations, 99–101. See also his articles, 'L'American Way of Life et les missions françaises de productivité', Vingtième siècle 17 (Jan.–Mar. 1988), 21–38; and 'The Marshall Plan in Action: Politics, Labor, Industry and the Program of Technical Assistance', in R. Girault and M. Lévy-Leboyer (eds.), Le Plan Marshall et le relèvement économique de l'Europe (Paris: Comité pour l'histoire économique et financière de la France, 1993), 335–58.

[6] Matthias Kipping, 'Concurrence et compétitivité. Les origines de la législation anti-trust française après 1945', Etudes et Documents VI (1994), 429–55; Matthias Kipping and Jean-Pierre Nioche, 'Politique de productivité et formations à la gestion en France (1945–1960): un essai non transformé', Entreprises et Histoire 14 (June 1997), 65–87.

considerable—and sometimes acrimonious—debate among the industrialists concerned, a debate which commenced before the end of the war. Second, the fact that those who argued in favour of the American model, as they perceived it, proved successful was not necessarily due to their superior arguments, but to the entrepreneurial initiative of a few key individuals, most of whom were active businessmen rather than 'technocrats' or government officials. Third, despite what appears like a 'breakthrough' in the immediate post-war period, the implementation of US-type technology, management, and competition took quite a long time, mainly because they had to contend with deeply rooted indigenous industrial structures, practices, and mentalities.

The steel-producing and using industries were chosen for in-depth analysis because they played a crucial role in the post-war reconstruction and the subsequent 'economic miracle' in France. They also illustrate the effects of the mutual interdependence between basic materials and their transformation, mass-production industry, and machinery suppliers on the Americanization process. The chapter addresses both industries in turn, and this examination is followed by an overall conclusion and a brief epilogue dealing with the influence of Americanization during the immediate post-war period on subsequent developments.

7.2 The Steel Industry

As in many other continental European countries, the debate about Americanization in the French steel industry centred on two major issues, the installation of modern strip mills and the European Coal and Steel Community (ECSC). Concerning the former, the debate demonstrates quite clearly that most of the French steel producers recognized the superiority of US production technology, but disagreed about the degree to which it could and should be applied in France. It also shows that the initiative for the installation of the new equipment came from the producers themselves rather than from French policy makers or the Americans. It was only slowed down by the lack of sufficient funding which was, however, alleviated by the Marshall Plan from 1948 onwards. Although the French were among the first in Europe to introduce US production technology, its potential was realized more slowly than in countries like Italy, because the French did not start with a more or less clean sheet. American influence had to contend with existing, locally embedded industrial structures, human resources, and management practices, all of which were modified only gradually. The debate about the ECSC reveals similar divisions and subsequent difficulties with regard to the creation of larger and more competitive markets in Europe, following the—perceived—American model.

The development of hot and cold strip mills in the United States from the 1920s constituted a major technological breakthrough for the production of flat-rolled steel. Unlike the previous, manually operated mills, these installations made it possible to produce steel sheet in large quantities and of high quality, in terms of both regularity and ease of further transformation. Steel sheet produced on these mills had so-called deep-drawing qualities which were of major importance for the production of automobile bodies and domestic appliances, such as refrigerators. In addition to differences in the width of the sheet produced, there were two basic types of mill in terms of the speed and the direction of the production flow. On a continuous mill, the steel ingot would make only one passage through several stands (usually five), sufficient to achieve the desired thickness. In the semi-continuous process, the mill had fewer stands and therefore two passages (in opposite directions) were required. The initial investment for the latter was lower, but so was its overall production capacity.[7]

The new technology spread rather slowly to Europe, where only two such mills were installed before the Second World War, one in Ebbw Vale in Britain and the other in Dinslaken in Germany. In 1945, the latter was dismantled and handed over to the Russians. While not prepared to install a modern rolling mill before 1939, the French steel producers apparently recognized the need to catch up with their American and, more importantly, European counterparts during the war. Under German occupation, they established a commission to investigate the question.[8] Shortly after the cessation of hostilities, in July 1945, a French steel delegation comprising both producers and representatives of the responsible government department went to the United Kingdom to study the strip mill in Ebbw Vale which, after the dismantling of the German installation, was the only one of its kind in Europe.[9]

While there seems to have been little doubt that the equipment had to be acquired in the United States, the French steel producers did not share the same views about the extent and speed of the necessary modernization. The industry's trade association, Chambre syndicale de la sidérurgie française (CSSF), suggested the installation of two semi-continuous strip mills. They were intended to replace the older, manual mills, without increasing the overall output. However, if the French were guaranteed a larger share of exports markets, so the document continued, 'the installation of a continuous mill could be envisaged, leading to an increase in its production capacity'.[10] This position reflects the deeply rooted cartel approach in French (and European) industrial circles, where market shares resulted from agreement, not competition. And it prefigured the debate about market size and organization following the proposal for the establishment of the ECSC in 1950.[11]

[7] For more details, see Robert Casey, 'Rolling Mills', in B. E. Seely (ed.), *Iron and Steel in the Twentieth Century* (n.p., 1994), 373–4, and United Nations, *La Sidérurgie européenne et le train continu à larges bandes. Etude sur l'évolution et les perspectives de la production et de la consommation des produits plats* (Geneva: UN, Economic Commission for Europe: Industry Division, May 1953).

[8] Philippe Mioche, 'La Sidérurgie et l'Etat en France des années quarante aux années soixante', unpublished thèse d'Etat, Université de Paris IV, 1992), 599–602.

[9] Roger Martin, *Patron de droit divin . . .* (Paris: Gallimard, 1984), 52.

[10] Archives de Pont-à-Mousson, Blois, PAM 70671, 11 May 1945.

[11] See Volker R. Berghahn, 'Montanunion und Wettbewerb', in Helmut Berding (ed.), *Wirtschaftliche und politische Integration in Europa im 19. und 20. Jahrhundert* (Göttingen:

But not all of the French steel producers adopted the same wait-and-see attitude. In February 1946, the two major companies situated in the north of France presented to the French administration a detailed project for the purchase of a continuous hot and cold strip mill in the United States.[12] These plans had been developed independently during the war, when these companies were subject to the German occupation authorities in Belgium and had thus been cut off from the rest of the French steel industry. The project foresaw the merger of the operating activities of the two companies into the Union sidérurgique du Nord de la France (Usinor). Funding was entirely private, comprising a large increase in capital, a bond issue, and medium-term bank loans. Government approval and help were only needed to obtain the necessary dollars. This initiative was based on an optimistic view of the market prospects for flat-rolled steel products which was not shared by many of the other producers at the time. Those situated in Lorraine, where about two-thirds of French crude steel was produced, seemed particularly reluctant. According to a government official, the head of the largest company, François de Wendel, reacted to the project by saying that 'the people in the North were crazy' and that 'such a machine was not suitable for the conditions of the French market'.[13]

It is important to note that this project was developed and presented before the presentation of the French Modernization and Equipment Plan, the Monnet plan, named after its initiator Jean Monnet. A Commissariat général du plan (CGP) had been created by a decree of 3 January 1946, but detailed proposals for the different sectors were developed only during the course of the year by the so-called modernization commissions, consisting of representatives from industry, labour, and the administration. Together with cement, agricultural machinery, transport equipment, energy supply, and coal, steel became one of six key branches considered crucial for the rapid recovery of the French economy and an increase in output. The deliberations of the steel industry's commission once again revealed significant differences in outlook among the French producers. Explicitly referring to the US example, its final report advocated the creation of production units with an annual output of one million metric tons, a suggestion which was, however, categorically rejected by representatives of the trade association, CSSF.[14]

Vandenhoeck & Ruprecht, 1984), 247–70; Matthias Kipping, *Zwischen Kartellen und Konkurrenz. Der Schuman-Plan und die Ursprünge der europäischen Einigung 1944–1952* (Berlin: Duncker & Humblot, 1996). A revised and extended French translation of this book will be published by the Comité pour l'histoire économique et financière de la France in 1999.

[12] Odette Hardy-Hémery, *De la croissance à la désindustrialisation. Un siècle dans le Valenciennois* (Paris: Presses de la Fondation nationale des sciences politiques, 1984) and, also for the following, Matthias Kipping, 'Competing for Dollars and Technology: The United States and the Modernization of the French and German Steel Industries after World War II', *Business and Economic History* 23, 1 (1994), 229–40.

[13] Quoted in Archives Nationales (hereafter AN), Ministère de l'Industrie, IND 11512, 'Historique et psychologie des trains à bandes', 22 Feb. 1949.

[14] *Premier Rapport de la Commission de Modernisation de la Sidérurgie* (Paris: CGP, 1946), 31–2 and 56–7. Compared to most other steel-producing countries, France had a much less concentrated

The commission also proposed the installation of a second continuous strip mill, to be built in Lorraine. This was part of a plan to increase the French steel output from a pre-war maximum of 9.7 million tons (achieved in 1929) to 12 million by 1951 and subsequently to 15 million. However, as noted above, most of the companies there showed little enthusiasm for the proposal. Given their doubts about the market prospects, they hesitated to commit themselves to the considerable investment necessary. Roger Martin, who at the time represented the Ministry of Industrial Production on the modernization commission, inter-preted this debate in his memoirs as a conflict between conservatives and mod-ernizers (*Anciens et Modernes*): 'The latter believed in the expansion of markets and a bright future (*les lendemains qui chantent*), the former didn't.' According to Martin, the modernizers were convinced that steel users would prefer better-quality products at lower prices.[15]

On the surface, their arguments seem to have carried the day. Thus, in October 1947, a group comprising most of the steel producers from the east of France, which later became the Société lorraine de laminage continu (Sollac), asked for permission and financial help to install a second strip mill. There are many reasons for this change in attitude, beginning with the initiative taken by the rival producers in the north. In addition, there was a credible threat by one of the major users, Renault's President and CEO, Pierre Lefaucheux, to acquire a semi-continuous mill for the company's own steel subsidiary, SAFE. Lefaucheux also believed in the future expansion of the car market and, conse-quently, the need for sufficient high-quality and cheap flat-rolled steel.[16] Finally, the opportunities to obtain the necessary funding had improved considerably with the announcement of the European Recovery Program by US Secretary of State George C. Marshall in June 1947. The Sollac project indeed received not only American technology, but also massive financial assistance. Out of a total cost of $147 million the Economic Cooperation Administration (ECA) financed directly $58 million, making it the single most important US-funded project in Western Europe. In addition, the ECA authorized the use of more than $80 mil-lion in counterpart funds and took over part of the financing of Usinor's strip mill.[17]

As a result of these investments, in terms of continuous rolling mills the steel industry in France was the best equipped on the European Continent by the mid-

steel industry. At the end of the war, there were 177 production units, 23 of which carried out the whole transformation process from blast furnaces to rolling mills. For details of the debates and especially the opposition from the CSSF, in detail Philippe Mioche, *Le Plan Monnet. Genèse et élab-oration 1941–1947* (Paris: Publications de la Sorbonne, 1987), ch. XI and id., 'La Sidérurgie et l'Etat'.

[15] Martin, *Patron de droit divin*, 52. Martin soon thereafter joined Pont-à-Mousson, a major French iron and steel producer, where he became CEO in 1964, and, after its merger with the glass producer Saint-Gobain, disposed of all steel-making facilities.

[16] Kipping, *Zwischen Kartellen und Konkurrenz*, 43–7.

[17] $13m. out of 61m.; for details see Kipping, 'Competing for Dollars and Technology'. Part of the reason for the sudden 'enthusiasm' of French steel producers and government representatives might also have been their—so far as we know unfounded—concern that the Americans would oth-erwise authorize and finance the installation of a modern strip mill in West Germany.

1950s. Producers in most other countries opted for the semi-continuous variety instead, which had a lower capacity and not quite the same potential for cost reductions, but was also less risky.[18] Despite the relatively early and large-scale adoption of US technology, the implementation of related elements of the American model proceeded much more slowly and with considerable difficulty in the French steel industry.

This concerns first and foremost a rationalization of the industry structure, dispersed in terms of both ownership and geography. As seen above, during the deliberations of the modernization commission, steel industry representatives opposed the concentration of activities into a limited number of production units with a minimum efficient scale. Their fear of nationalization, which would have been facilitated by further concentration, might partially explain this attitude.[19] But the desire of family concerns, like de Wendel, to maintain their position and influence probably played at least as important a role.[20] In this respect, it is important to note that the installation of the second strip mill did not result in a merger of the companies involved. Sollac was a technical co-operative, with shares distributed according to the crude steel-making capacity of each participating producer. But even in the case of Usinor, the merger did not lead to a rationalization of production activities and sites. Hot and cold rolling equipment were actually installed at two different plants, Denain and Montataire, over a hundred kilometres apart. This resulted in considerable diseconomies, because the coils had to be reheated.[21]

More important in the short term was the full exploitation of the production potential of the new strip mills. As a French productivity mission to the United States in February and March 1951 stressed, the solution for the French productivity gap 'cannot only be found in the construction of new installations'. The major difference between the two steel industries was the utilization of the rolling mill capacity, of crucial importance because of the amount of capital tied up in these installations. In France, on the whole, capacity utilization appeared much lower due to an 'imbalance' in the production process, namely the insufficient input of crude steel, and the lower speed of throughput, resulting partially from a lack of mechanical handling.[22] In addition, it appears that the original capacity specifications for both French strip mills were based on US standards from the 1920s.[23]

[18] See for details United Nations, *La Sidérurgie européenne* and the contributions of Ruggero Ranieri and Gary Herrigel in this volume.

[19] Mioche, 'La Sidérurgie et l'Etat', 593–6.

[20] Jean-Pierre Daviet, 'Some features of concentration in France', in H. Pohl (ed.), *The Concentration Process in the Entrepreneurial Economy since the late 19th Century* (Wiesbaden: Franz Steiner, 1988), 67–89.

[21] See for details Eric Godelier, 'De la stratégie locale à la stratégie globale: la formation d'une identité de groupe chez Usinor (1948–1986)', unpublished Ph.D. thesis (EHESS, Paris, 1995).

[22] *Etude par la mission de la sidérurgie française aux Etats-Unis (février–mars 1951)* (Paris, 1952), 84–9.

[23] This is highlighted by Anthony Daley, *Steel, State, and Labor. Mobilization and Adjustment in France* (Pittsburgh, 1996), 58, on the basis of a French research report from the 1970s.

In the case of Usinor, where the cold mill at Montataire had become operational at the beginning of 1950 and the hot mill at Denain in March 1951, it took
a long time to address these imbalances and to optimize the overall output.
During the 1950s, the company improved the productivity of the existing furnaces, added new steel-making facilities as well as a second slabbing mill, and
increased the capacity for the reheating of the steel ingots. Moreover, the organization of work was improved, the passage of steel through the mill increasingly automated, and the electric power at the stands (and thus the speed of
reduction in thickness) augmented. Estimated originally at 700,000 metric tons
per annum, the output of flat-rolled steel at Denain was 823,000 tons in 1955 and
then increased to 1,048,000 in 1958, 1,461,000 in 1960, and finally reached over
2 million tons in 1962.[24]

The above-mentioned mission from the steel industry also highlighted other
ways to help improve productivity in France, following the US example, namely
specialization of production, reduction in the variety of products, and an
increase in research efforts. Regarding the latter, the French steel industry had
made considerable progress with the establishment of a joint research institute,
the Institut de recherches de la sidérurgie (IRSID), in the immediate post-war
period.[25] But, as the mission noted in the conclusion of its report, 'our management of human relations (*technique sociale*) in each company is much less
advanced than our steel-making technology'. Once again, the example of Usinor
is instructive. Compared to the old manual installations, the new continuous
strip mills shifted the required skill from the production workers to schedul-
ing/planning and maintenance functions. But this transition was far from
smooth and provoked considerable resistance from the shop-floor to the
deskilling and the reduction in status.[26]

Most of these transition problems resulted from the different context in
which French steel firms operated compared to their American counterparts,
namely with respect to the skilled and 'localized' work-force. Another major
difference, highlighted by the productivity mission, concerned the organization
of steel markets: 'The Americans are, as a matter of principle, hostile towards
cartels, but the conditions prevalent in their country are so different from ours
that they can hardly judge based on a knowledge of the facts'.[27] The French steel
industry had been more or less cartelized since the inter-war period, a situation
reinforced by wartime controls. Orders were centralized by the sales syndicate
Comptoir des produits sidérurgiques (CPS). The government retained ultimate
control over basic steel prices. But the body responsible relied on information
provided by the trade association, CSSF, which apparently supplied the cost

---

[24] Godelier, 'De la stratégie', 24 and 64–9. The author would like to thank Eric Godelier for providing additional information and clarification regarding the technical and organizational developments at Denain.

[25] For details see Mioche, 'La Sidérurgie et l'Etat'.

[26] Godelier, 'De la stratégie locale', chs. 3–5. See also Daley, *Steel, State, and Labor*, ch. 4.

[27] *Etude par la mission de la sidérurgie française aux Etats-Unis*, 176, 186–7.

structure of the least efficient works in order to ensure their survival. In general, French legislation and practice had long distinguished between good and bad cartels, with the former seen as introducing stability and promoting increased specialization.[28]

Internal French discussions about restrictive business practices exercised by the trade associations (*dirigisme professionnel*) and their effect on the French economy began in 1948. They reached a peak during the debates about the ECSC. The French Foreign Minister, Robert Schuman, proposed, on 9 May 1950 to 'pool' the coal and steel resources of France and Germany. This proposal had been drafted by Jean Monnet and owed much to the US example, mainly as regards economies of scale and competition policy.

With his modernization plan Monnet had initially aimed at an increase of production in the 'basic' industries in order to remove the bottlenecks for an overall recovery. However, when the shortages of raw materials began to disappear in 1948 and the government started to relax its controls, it became apparent that the expansion of the French economy and the competitiveness of its industries were hampered by the cartel-like practices described above, which resulted in higher prices for the downstream industries. In July 1948, Monnet therefore placed renewed emphasis on an improvement of productivity. It is not clear whether the Americans, who launched their productivity drive at about the same time, were the source of this major shift in Monnet's policy orientation. A small group in the CGP headed by Jean Fourastié had been working on productivity issues since 1946. Monnet put Fourastié in charge of a working party which met for the first time on 28 July 1948. It was to outline a French productivity programme and suggest the necessary organizational structures.[29] The American example played an important role in the working party's deliberations. It blamed the small size of the French market for the low utilization of production capacity and advocated the creation of a larger economic system and a removal of trade barriers. Some members of the working party also insisted on the need to abolish cartels, which hampered free trade and economic progress as much as tariffs or quotas. On the cartel question however, opinions diverged, with some, in the French tradition, highlighting their beneficial aspects, namely as a way to promote specialization.[30]

Following the report of Fourastié's working party, the French government established a provisional productivity committee headed by Monnet in May

[28] For more details on steel prices and sales practices, see Jean Sallot, 'Le contrôle des prix et la sidérurgie française 1937–1974', unpublished Ph.D. thesis (University of Paris I, 1993); for the French cartel tradition, see Kipping, 'Concurrence et compétitivité'.

[29] Djelic, *Exporting the American Model*, 143–51 highlights Monnet's earlier interest in the concept of productivity and links his later efforts to the US initiative. But there are some reasons to believe that domestic considerations and pressure from the using industries might have motivated his shift in emphasis in 1948: Kipping, *Zwischen Kartellen und Konkurrenz*, esp. 68–75.

[30] See, for details of this point and the following discussion, Kipping, 'Concurrence et compétitivité'. By contrast, Djelic completely ignores both the different views within the working party and Monnet's early efforts to promote an anti-trust law.

1949. It co-ordinated the French efforts in this area and organized the French productivity missions to the United States which began in mid-1949.[31] Apparently, Monnet's Planning Agency also proposed an anti-trust law modelled along the American lines which explicitly outlawed cartels. But his proposal found little support in the French administration and government and was also opposed by a majority of industrialists. Despite his excellent contacts with the US authorities, Monnet failed to enlist their support for his law. Aware of the resistance within the French government to such a strict measure and, more importantly, the French reluctance to yield to any overt US pressure, the Americans feared that any intervention in favour of tough anti-trust legislation would be the 'kiss of death' even for the more modest draft prepared by the French Finance Ministry.

As seen above, steel was among those sectors where restrictive business practices hampered an increase in productivity and an improvement in international competitiveness. Because of its crucial position at the beginning of the supply chain, the consequences of the *dirigisme professionnel* exercised by the CPS and the CSSF were felt by a large number of steel users. From 1949 onwards, their representatives complained vociferously about the high prices and the irregular quality of French steel products, which seriously endangered their position *vis-à-vis* the resurgent German competition. But the centralized allocation of production quotas and the control of exports also found critics within the steel industry, especially among those, like Usinor's CEO, René Damien, who had expanded and modernized their facilities. To make full use of the potential of the new strip mills, especially in terms of cost savings, it was necessary to increase their market share, both in France and abroad.

The socialist politician André Philip was one of those who recognized and expressed this logic in the most lucid way. Minister of Finance and the National Economy in 1946/47, he had also represented France at the GATT negotiations and at the UN Economic Commission for Europe in Geneva. He is often portrayed in the literature as an economic interventionist, but several of his contemporaries have stressed that he was far from rejecting liberal ideas. An important figure in the European movement, he had as his main interest the so-called Ruhr problem, namely the future of Germany's heavy industries. Initially an advocate of internationalization, he later on saw a clear link between this question and French recovery and modernization. Most of his suggestions were inspired by the US example, as is clearly demonstrated in his article on 'The Economic Unification of Europe' published in March 1949.[32] He highlighted

---

[31] One year later, Monnet lost control over the productivity drive when the Comité national de la productivité was placed under the authority of the Finance Ministry. The situation changed again when an independent commissariat was created in 1953: see Kipping and Nioche, 'Politique de productivité', 72.

[32] Originally published in French in *Cahiers du monde nouveau* 5, 3 (1949), 32–8. An English translation can be found in Walter Lipgens (ed.), *Documents on European Integration*, microfiche, 4 vols. (Berlin: de Gruyter, 1985–90), 88–94. He expressed similar views in a number of other publications and public speeches; for details see Kipping, *Zwischen Kartellen und Konkurrenz*, 149–56.

that only the creation of a large market, similar to the United States, would enable European industry to achieve economies of scale, increase productivity, lower production costs, and become internationally competitive. This was especially true for basic industries with high fixed costs. But he also stressed the need for anti-cartel legislation and a European agency for its enforcement, equivalent to the Federal Trade Commission.

Many of Philip's suggestions found their way into the Schuman Plan. But the proposed ECSC also addressed a number of other problems, especially the US pressure for Franco-German reconciliation in the interest of Western European integration against the Soviet bloc. And, by focusing on coal and steel, it overcame the resistance of other industries to earlier integration schemes. The steel users in particular had opposed plans to liberalize trade in their products without prior opening up of the European steel markets for competition. The treaty negotiations and the debate about its ratification in France focused on the same issues and highlighted again the different views about the best market organization, with the US example occupying centre stage in these discussions.

Not surprisingly, the French steel users favoured the Schuman Plan, because it responded to their claims for an opening up of European steel markets. Jean Constant, Secretary General of the trade association of the mechanical engineering industries, the Syndicat général des industries mécaniques et transformatrices des métaux (SGIMTM; it was renamed Fédération in 1954), had been one of the most outspoken critics of the steel industry. He also became one of the Plan's most fervent supporters. In two of his widely read editorials in the association's monthly journal, he highlighted the need for a large *and* competitive European market to achieve the necessary capacity utilization in high fixed-cost industries and thus reduce production costs.[33] André Philip expressed very similar views in the Socialist newspaper Le Populaire and in a report about the ECSC to the Parliamentary Committee on Foreign Affairs on 12 April 1951:[34]

Since the end of the war, we have realized more and more that national borders and trade restrictions hamper economic development. To improve our balance of payments, we have to develop new industries capable of exporting to foreign markets, which necessitate considerable investments. These will only be profitable if their cost can be spread over a large number of units produced. In this way, we experience a revolution of productive forces against national barriers. Europe can only survive by developing scale intensive manufacturing and mass production methods. This requires the creation of integrated European markets.

[33] 'Productivité' and 'Le Problème de l'acier', Les Industries Mécaniques, Dec. 1950 and Feb. 1951 respectively. For a detailed overview of these criticisms see Matthias Kipping, 'Les Tôles avant les casseroles. La compétitivité de l'industrie française et les origines de la construction européenne', Entreprises et Histoire, 5 (June 1994), 73–93. The title of this article is based on an expression frequently used by Constant to highlight his claim that the markets for steel should be opened to competition before further trade liberalization in the using industries, which he did not oppose per se.

[34] 'Rapport d'information de M. André Philip sur le projet de pool du charbon et de l'acier', 12 April 1951 [translation by Matthias Kipping]. It can be found in the private papers of his son Loïc, Aix-en-Provence.

Those in the French administration and the steel industry who rejected the Schuman Plan did not necessarily oppose the integration of European markets. But they would have preferred a solution where producers negotiated specialization agreements and cartels instead of being subjected to competition. These views were reflected in their position towards the American model. According to a group of steel industrialists, the European proponents of anti-trust laws followed an 'artificial culture of competition corresponding to certain mathematical theories, but with little relation to reality'. They had elevated the US example to a 'myth', whereas the respective legislation was actually applied rather pragmatically in the United States, where industries had even been forced to conclude cartel agreements during the New Deal.[35] Thus, rather than reject the US example completely, the advocates of cartelization offered a different interpretation. But they clearly opposed a direct transfer of American legislation and proposed its adaptation to the specific circumstances in Europe.

But these suggestions to conclude agreements between European producers, based on national steel cartels, were not only opposed by Monnet, Philip, and influential representatives of the steel using industries. They also failed to find sufficient support within the French steel industry itself. Between March and August 1950, the producers engaged in protracted negotiations over a domestic cartel which was to replace the organizational structures established during the war and maintained immediately afterwards to deal with the shortages of raw materials, namely the sales syndicate CPS. While the leadership of the trade association saw such an agreement not only as a domestic imperative, but also as a basis for any international collaboration, like the Schuman Plan, many producers opposed it as a 'straitjacket' for their modernized and expanding production.[36]

In the end, largely as a result of Monnet's superior coalition-building skills and US pressure (mainly on the reluctant German government), the Schuman Plan preserved its rather liberal design and was ratified in France, somewhat against the odds. Its impact on the market order of the French steel industry was quite significant. The ECSC, together with the still rather rudimentary domestic competition legislation led to a reform of the steel sales syndicate, CPS, towards which the using industries had directed much of their criticisms. The prospect of a more open and competitive market prompted a further concentration of the French steel industry, especially the creation of Sidélor, which combined the steel-making facilities of three producers in Lorraine. In addition, to sweeten the bitter pill of competition, the French government also granted a number of additional advantages to its steel producers, making new funds available for investment in new plant and lowering the interest rates on outstanding loans from the Treasury.[37]

[35] 'Plan Schuman', *ACADI Bulletin*, 45 (Feb. 1951), 54–75. This anonymous article was elaborated by a working party from the steel industry.

[36] Not surprisingly, Usinor's René Damien was the staunchest opponent of the proposed quota regime, together with André Grandpierre from Pont-à-Mousson: Kipping, *Zwischen Kartellen und Konkurrenz*, 197–203.

[37] For all of the above, ibid., part IV.

But once again, Americanization was far from complete and European traditions in terms of cartels survived for some time.[38] As John Gillingham and others have shown, European steel producers had reached an agreement on prices for exports outside the ECSC before the opening of the common steel market in 1953.[39] But similar efforts were less successful for the Community itself, at least until the formation of crisis cartels in the 1970s, which was done with the approval of the European Commission.[40] More importantly, the ECSC constituted a first breakthrough in terms of American-style competition legislation in France and Western Europe. It was extended to other industries, albeit with somewhat weakened conditions, following the establishment of the European Economic Community (EEC) or Common Market with the Treaty of Rome in 1957.[41]

Thus, the initiative of a few producers, considerable pressure from the using industries, and the increasing competition within the framework of the ECSC, pushed the French steel industry down the path of Americanization, despite considerable doubts about the wisdom of such a move in terms of the market prospects. But while American production technology, mainly in the form of two continuous strip mills, was installed rapidly, other aspects of the US model took hold only slowly and with some considerable difficulty. Most of them conflicted with the prevalent structures and practices of the industry, including the reluctance of family-owned firms to be subject to further concentration, the resistance of workers to deskilling and loss of status, and the close ties of most steel plants with their local environment. Similar debates and difficulties may be observed in the downstream industries.

7.3   The Steel Users

The influence of the US example on the French steel-using industries is much more difficult to assess than for the steel producers. This is mainly due to the wide variety of activities subsumed under this general heading. On the one hand, they differ in terms of the degree of transformation, that is, the value added at this stage of production, which is reflected in the importance of steel as a percentage of costs. On the other hand, there are significant differences in the scale of operations, ranging from small artisan shops to automobile producers

[38] See in general Wendy Asbeek Brusse and Richard T. Griffith, 'L'"European Recovery Program" e I cartelli: una indagine preliminare', *Studi storici* 37, 1 (Jan.–Mar. 1996), 41–68.

[39] John Gillingham, *Coal, Steel, and the Rebirth of Europe, 1945–1955* (Cambridge: Cambridge University Press, 1991), esp. 310–12; Charles Barthel, 'De l'entente belgo-luxembourgeoise à la Convention de Bruxelles 1948–1954', in M. Dumoulin, R. Girault, and G. Trausch (eds.), *L'Europe du patronat de la guerre froide aux années '60* (Berne: Peter Lang, 1993), 29–62.

[40] See, for details, Dirk Spierenburg and Raymond Poidevin, *Histoire de la Haute Autorité de la Communauté Européenne du Charbon et de l'Acier. Une expérience supranationale* (Brussels: Bruylant, 1993).

[41] For the further evolution of French competition legislation see Hervé Dumez and Alain Jeunemaître, *La Concurrence en Europe. De nouvelles règles du jeu pour les entreprises* (Paris: Seuil, 1991), ch. 3.

with large factories. Overall, the using industries are dominated by small and medium-sized enterprises. Furthermore, because of their mutual interdependence, Americanization has different, sometimes apparently contradictory effects on these industries. Thus, for example, a move by car manufacturers to automation and large-scale production drives their suppliers in the same direction. But, at the same time, it forces companies producing the necessary machine tools to become highly specialized.[42]

With these caveats in mind, it seems nevertheless justified to say that the Americanization of the French steel users after the Second World War showed many similarities with the developments in the steel industry, namely concerning the debates and differences within the industry about the extent and the speed of the necessary modernization, the role of entrepreneurial initiative, the constraints imposed by the lack of sufficient funding for new investments, which was even more severely felt than in steel production, and the rather slow adoption of American-style production and management methods.

Like the steel producers, the French steel users did not wait for government or American pressure to develop initiatives for the post-war period. Under German occupation, both the metal-transforming industries and the automobile producers had critically examined the state of their equipment as well as their industrial structures and outlined the necessary steps for the future. Most of these deliberations took place in the so-called Comités d'organisation (organizing committees), which had been established by the Vichy government in 1940 in all sectors of the economy as a replacement for the disbanded trade associations and cartels, in order to co-ordinate and control industrial activities. At about the same time, the different departments in the newly created Ministry of Industrial Production undertook similar efforts and elaborated more or less detailed plans for the post-war reconstruction and expansion of French industry.[43]

Among both industrialists and government officials, there seems to have been little doubt about the fact that the equipment in the French steel-using industries was obsolete and that a greater specialization of production was necessary. Thus, the organizing committee of the automobile industry discussed the need for a reduction in the number of models and an increasing standardization of parts. During the German occupation, both Renault and Citroën secretly developed small and cheap models, to be built after the war in large numbers for an avid but impoverished population of would-be motor-car owners. At the same time, Peugeot introduced a standardization policy and even made attempts to

---

[42] This interdependence and its effect on the specialization of the machine tool industry is described by Arndt Sorge and Marc Maurice, 'The Societal Effect in the Strategies of French and German Machine-Tool Manufacturers', in B. Kogut (ed.), *Country Competitiveness. Technology and the Organizing of Work* (New York: Oxford University Press, 1993), 80.

[43] See, in general, Richard C. Vinen, *The Politics of French Business 1936–1945* (Cambridge: Cambridge University Press, 1991), ch. 10; Richard F. Kuisel, *Capitalism and the State in Modern France* (Cambridge: Cambridge University Press, 1981), chs. 5 and 6; Andrew Shennan, *Rethinking France. Plans for Renewal 1940–1946* (Oxford: Oxford University Press, 1989); and Mioche, *Le Plan Monnet*, chs. I–III, IX, X.

import machine tools from the USA.[44] The most 'extreme' expression of these attempts was contained in the five-year plan for the French automobile industry (1946–50), elaborated shortly after the liberation of France by Paul-Marie Pons, an official in the Ministry of Industrial Production. It foresaw a relatively high degree of specialization among the existing producers in terms of car and truck models as well as some concentration.[45]

Similarly, in 1943 a commission comprising representatives from most branches of the French metal-working industries—with the exception of the automobile producers—established an 'economic balance sheet' of the sector and made suggestions for possible improvements.[46] Their report stressed the high average age of installations and machines (most of which dated back to the First World War), the lack of specialization, and the almost complete absence of large-scale/mass production (*fabrications de série*).[47] Making a clear reference to the US example, the report advocated a vertical disintegration of the final producers, combined with a concentration and specialization at the level of companies producing inputs:

If we consider that a company executive is wrong to produce at the same time air heaters, pumps and diesel engines, it is not clear why we should consider him to be qualified for casting, stamping, wire drawing, or bolting. . . . In America, Mr. Ford buys his door hinges from a specialized enterprise which produces 40,000 of them per day. Maybe for this reason, his car doors don't squeak. . . . The standardization of different models has namely to lead to a standardization of their constituent elements. In order to allow a sensitive reduction of prices, the large-scale production of these elements has to occur in much more important quantities than the production runs necessary for the profitable assembly of the final goods.[48]

While there seems to have been a widespread consensus about the need for modernization of equipment, as well as more specialization and standardization, the actual departure from pre-war production policies and practices was not so easy, nor was there agreement about the best way to achieve the desired results. Thus, taking a similar approach to that followed in the simultaneous deliberations in the steel industry, the commission of the metal-working

[44] See for more details on specific firms and models Jean-Louis Loubet, *Citroën, Peugeot, Renault et les autres. Soixante ans de stratégies* (Paris: Le Monde Editions, 1995); id., *Automobiles Peugeot. Histoire d'une réussite industrielle, 1945–1974* (Paris: Economica, 1990); and Patrick Fridenson, *Histoire des usines Renault, Naissance de la grande entreprise 1898–1939*, I (Paris: Seuil, 1972) and II (Paris: Seuil, forthcoming).

[45] See for the Pons Plan, Jean-Louis Loubet, 'Les grands contructeurs privés et la reconstruction: Citroën et Peugeot 1944–1951', *Histoire, Economie et Société* (July–Sept. 1990), 441–69.

[46] *Bilan économique des industries transformatrices des métaux. Rapport général*. The commission elaborating this report consisted of Marcel Champin, Ernest Chamon, Jean Constant, Marcel Danbon, Auguste Detoeuf, Marcel-Edouard Lambert, Jacques Lenté, René Painvin, Pierre Ricard, Maurice Roy, Emile Taudière, and René Touchard. They synthesized earlier studies dealing with specific branches, commissioned by the French administration in 1942. The report was drafted by Jean Constant. The first edition in 1943 was limited and circulated only confidentially, but a second edition was published more widely by the SGIMTM after the Liberation. A copy can be found at the documentation centre of the Fédération des Industries Mécaniques (FIM) in Courbevoie, near Paris.

[47] Ibid., 14–17.   [48] Ibid., 50–1 [translation by Matthias Kipping].

industries warned against massive new investments in specialized machinery and favoured small adjustments to the existing multi-purpose machines instead, in order to improve their productivity. The majority of its members also showed very little confidence in market mechanisms or, as they called them, 'the fierce battles of competition (*luttes acharnées de concurrence*)'. Instead, following the French tradition, they advocated cartel agreements (*ententes*) between the producers as the best way to specialize production.[49]

But not everybody shared these views, either within the commission or outside. As in the steel industry, most of the changes were actually instigated by the entrepreneurial initiative of a few dynamic individuals. This was clearly the case for the nationalized car producer Renault, where the new president and chief executive, Pierre Lefaucheux, more or less single-handedly imposed the manufacture of the small and cheap model developed during the war, the so-called 4CV. In 1943, Louis Renault himself had actually decided against its production, an opinion probably shared by most of the company's top managers and engineers. Renault's decision should not be mistaken for a full-scale rejection of the US model. As his earlier interest in scientific management demonstrates, he was certainly open to the adoption of some of its elements. But he did not believe that there was a sufficient market for a small car in France.[50]

Lefaucheux, on the other hand, was convinced that an expansion of the market, both in France and abroad, was possible, so long as the industry managed to lower its production costs and sales prices. This optimistic view, supported by rudimentary market research, was certainly behind his decision to launch the 4CV and to construct the necessary specialized high-volume automated transfer machinery.[51] It also helps to explain his constant pressure for lower prices and better quality of steel sheets. In this respect, Lefaucheux clearly followed in the footsteps of his predecessor who had already aimed at controlling most elements of his costs and had therefore integrated backwards into steel production. Lefaucheux's attempts to obtain a semi-continuous strip mill for the company's steel subsidiary SAFE were not crowned by success (see also above), nor was his project to take over one of the sequestrated steel mills in the Saar. But he did, at least, secure a widening of the second continuous mill of Sollac from 66 to 80 inches, which was better suited for the needs of automobile production— against objection from the steel producers and the relevant officials in the

[49] *Bilan économique des industries transformatrices des métaux. Rapport général*, 41–2 and 59–67.

[50] Patrick Fridenson, 'La 4CV Renault', originally published in *L'Histoire*, now (re)printed in *Puissance et faiblesses de la France industrielle. XIXe–XXe siècle* (Paris: Seuil, 1997), 309–23; id., 'L'Industrie automobile: la primauté du marché', *Historiens et Géographes* 361 (March–April 1998), 227–42; for Renault's earlier reaction to the US example, id., *Histoire des usines Renault*, I.

[51] Fridenson, 'La 4CV Renault'. For a detailed description of this machinery, see Pierre Debos, 'Application des procédés d'usinage moderne à la fabrication des 4CV Renault', in *2ème Congrès international des fabrications mécaniques*, Paris, Sept. 1949, 13–35. A copy of these conference proceedings can be found at the archives of ORGALIME in Brussels (see below). For background on R&D at Renault, see Jean-Pierre Poitou, *Le cerveau de l'usine. Histoire des bureaux d'études Renault de l'origine à 1980* (Aix en Provence: Presses de l'Université de Provence, 1988).

Ministry of Industry. He also became a 'fervent supporter' of the Schuman Plan and the opening up of European steel markets to competition, as a result of which he hoped to obtain steel sheets at prices comparable to those paid by his foreign, mainly German, competitors—like Volkswagen.[52]

Without detracting from his personal role and achievements, it has to be said that Lefaucheux benefited from his excellent contacts with socialist as well as centrist politicians, and, until mid-1947, from the support of the communist trade union, CGT, in his efforts to improve productivity and increase output. In addition, as a nationalized company, Renault had easier access to funding than its private counterparts. It received, for example, $2.5 million of Marshall Plan aid for the purchase of machine tools in the USA. Between 1945 and 1958, the number of its machines increased from 16,900 to 22,168, while their average age dropped from 18.30 to 15.45 years.[53] As a result of these investments and the related changes in work organization, Renault became the most Americanized car producer in France.

The other manufacturers worked under less favourable circumstances and therefore had to make compromises. Citroën, for example, suffered—both financially and psychologically—from its disastrous investment policies during the inter-war period. André Citroën had expanded and modernized the company's main plant in Paris in the middle of the Great Depression. But the poor quality of the revolutionary new model with front wheel drive led to poor market performances and, combined with the financial weight of the investments, resulted in the company's bankruptcy. The new owners, the tyre producer Michelin, imposed a product policy which kept that model and added, at the lower end of the market, a very small vehicle, under development from the late 1930s. At the same time, they prohibited major new investments. When the new model, known as the 2CV, was finally launched in 1949, it was therefore not produced on an assembly line, but individually, in batches, albeit with highly standardized parts, themselves produced in large quantities.[54]

Similar debates and differences can be observed in the other steel-using industries. Traditionally, France had acquired the majority of its machine tools abroad, in the USA, Britain, and, especially from the late 1920s, in Germany.[55]

[52] See, for Renault's policy, Fridenson, *Histoire des usines Renault*, I, 134–7, 220–6; for Lefaucheux, Kipping, *Zwischen Kartellen und Konkurrenz*, 43–7, 91–4, 273–6.

[53] André Garanger, *Petite histoire d'une grande industrie* (Neuilly sur Seine, 1960), 257, n. 1. With $6 million, Renault received more than half of the total $11 million of Marshall Aid funds allocated to the French automobile industry between 1948 and 1951; see Patrick Fridenson, 'L'industrie automobile française et le plan Marshall', in R. Girault and M. Lévy-Leboyer (eds.), *Le Plan Marshall*, 283–9.

[54] The choice of production process clearly limited output, at least initially. Only from the mid-1950s did Citroën finally manage to satisfy the high demand for the 2CV: Patrick Fridenson, 'Genèse de l'innovation: la 2CV Citroën', *Revue française de gestion*, Sept.–Oct. 1988; Jérôme Thuez, 'La 2CV', originally published in *L'Histoire*, reprinted in *Puissances et faiblesses de la France industrielle*, 325–37.

[55] On the eve of the First World War, approximately four-fifths of the machines in France had been imported. This situation changed little during the 1920s, when the ratio of French exports to imports of machine tools was about one to six in value terms. Following the Great Depression,

There is no doubt that interest in foreign technology and machinery remained considerable, even during the Second World War. For example, in 1941 Albert-Roger Métral, professor of mechanical engineering at the Conservatoire National des Arts et Métiers (CNAM), undertook an unofficial study trip to the United States.[56] It also seems likely that American technical publications continued to arrive in France via Switzerland. Nevertheless, because of the war and the German occupation, these and other efforts to keep abreast of technological developments elsewhere were bound to remain limited. Therefore, after the cessation of hostilities, the French steel users were keen to learn about the progress made in other countries, and especially in the United States, since 1939. Thus, representatives from the major French car producers visited American factories in 1945 and 1946.[57]

The French also played an important role in re-establishing the contacts and facilitating the exchange of information between representatives of the mechanical engineering industries in Europe. Apparently, Jean Constant, the above-mentioned Secretary General of the SGIMTM, and his Belgian equivalent, G. Velter, took the initiative in this respect and organized a first meeting in Brussels on 25 June 1947. The representatives of the French, Belgian, British, Swedish, and Swiss associations decided to organize an International Mechanical Engineering Congress and established a permanent Secretariat, attached to the French association. The first congress was held in Paris in September 1948, and was followed by similar events in 1949, again in Paris, 1950 in Brussels, 1952 in Göteborg, 1953 in Turin, 1956 once again in Paris, and 1958 in The Hague.[58] The vast majority of contributions at these meetings were of a technical nature, but some of them allow us to assess the reaction of leading French industrialists to US technology and production management.

imports dropped drastically, but nevertheless still exceeded exports; see Garanger, *Petite histoire*, for details.

[56] François-Henri Raymond, 'Métral, Albert (1902–1962)', in C. Fontanon and A. Grelon (eds.), *Les professeurs du Conservatoire National des Arts et Métiers. Dictionnaire biographique 1794–1955*, 2 (Paris: Institut national de recherche pedagogique: Conservatoire national des arts et métiers, 1994), 261–4. As a result of his trip, he and his family had to flee to North Africa, where he founded a mechanical engineering company. It is not clear whether this was the Ateliers G.S.P., a small producer of machine tools in France, of which he was the chief executive after the Second World War.

[57] Fridenson, 'L'Industrie automobile française', 284.

[58] After 1958 it was decided to discontinue this congress because of competition from an increasing number of more specialized events. The organization of the congress also served as the nucleus for the creation of more permanent bodies, representing the mechanical engineering industries at a European level, namely the COLIME (Comité de liaison des industries métalliques européennes) in 1952, for the associations from the six ECSC member states, and the ORGALIME (Organisme de liaison etc.) in 1954, open to associations from all OEEC members. These bodies merged in 1960 and today also represent the electrical and electronics industries; see Nils Lundquist, 'The Historical Background of the Creation of Orgalime' (unpublished typescript, Sveriges Mekanförbund, 1971), available at the ORGALIME archives in Brussels, where copies of the proceedings of the international congress can also be found. The author would like to thank ORGALIME for granting him access to these materials.

The Paris congress in 1948 was devoted to the technical progress made since 1939. Of special interest in the present context is the comparison between the European and American 'concepts' of machine tool production, presented by Métral, now also chief executive of the machine tool producer Ateliers G.S.P.[59] He identified the level of automation and the degree of specialization as the major differences between the machine tools produced in the United States and Europe. In relation to the former, the Americans benefited from a wide range of accessory equipment, for example electronic controls, many of which were not (yet) available in France. In terms of the market for specialized, high-volume machine tools, Métral highlighted the fact that very few industries in France had sufficiently large production runs to justify the necessary capital investment:

If well adapted to the needs of the producer, the transfer machine has a capacity which exceeds the saturation levels even of an enlarged European market. And when the needs of customers change, it can also become obsolete very rapidly. In both cases, it is imperative to amortize it over a period of three to four years at most, compared with ten years for multi-purpose or modular machine tools with common elements. These considerations together with the high initial costs lead us to conclude that such . . . equipment is not of interest for the European countries', especially in the current state of the economy.

Thus, while transfer machines and long production runs made lower unit costs possible, the necessary amortization of the heavy investment reduced the company's flexibility to respond rapidly to changes in the market or to introduce new models. For these reasons, machines of this type had not yet been manufactured in France. Automobile producers therefore had to acquire them elsewhere or, as seen above in the case of Renault, construct their own. According to Métral, among the European countries, Britain came closest to the US model in terms of automation and specialization, partially as a result of low-skilled, often female labour. At the other end of the spectrum he placed Switzerland where universal machines with universal tools were the norm, reflecting the very high qualifications of the work force. Despite his caution, Métral indicated nevertheless that developments in France pointed more towards the Anglo-American model in the long run.

He was not the only one to question the appropriateness of US technology for the French market. For example, at a press conference in 1951, a representative of the producers of earth-moving equipment stressed that the giant US machinery was not adequate for highway construction and maintenance under the spatially more constrained European conditions. He also highlighted the differences resulting from the availability and the price of energy.[60] More generally,

---

[59] Albert-Roger Métral, 'Conception européenne des fabrications de machines-outils', in *1er Congrès international des fabrications mécaniques*, Paris, Sept. 1948, 495–509 (translation by Matthias Kipping). See also his more specific analysis of high-volume machines and the conditions for their installation, 'L'Adaptation des machines-outils à la fabrication de grande série', in *2ème Congrès*, Paris, 1949, 161–217.

[60] 'Le Matériel de Travaux Publics et de Génie Civil', Exposé de Monsieur Pommier, SGIMTM, Service de presse, 8 Feb. 1951, p. 5.

in his history of the machine-tool industry André Garanger repeatedly criticized the use of the United States as an example for the French producers to copy in terms of long production runs, specialization, and the much larger size of companies.[61] He stressed instead the predominance of small and medium-sized machine-tool producers in France and the fact that they had to offer increasingly specialized and customized machinery, if the consumer goods industry was to move towards large-scale production.

A compromise between the requirements of low-cost production and the existing industrial structure in France was proposed by Jean Constant, owner of the Manufacture d'estampage du Nord-Est, a forging company in the Ardennes, who had become Secretary General of the SGIMTM in 1944. In a contribution, under the theme of 'Productivity', at the second international congress in Paris in September 1949, he addressed the ways in which medium-sized firms could make improvements in this respect.[62] First of all, he underlined that there were a number of impediments to the concentration of production in ever larger units. They included the human factor, as regards both workers and employers, and increased congestion in urban areas. But he also highlighted the advantages of medium-sized firms, mainly their higher flexibility and lower overheads. In order for these firms to improve their productivity and remain competitive, he advocated subcontracting. Compared to vertical integration, this would require less investment and avoid having expensive machines left idle. The subcontractors, on the other hand, could specialize and thus lower their own production costs, and this would in turn benefit their clients. But Constant also made it clear that for such a subcontracting relationship to function properly, both parties had to collaborate very closely, especially in the technical field.

Overall, it appears that the development of the French machine-tool industry, and the metal-working industries in general, followed Constant's suggestions. The vast majority of mechanical engineering companies in France remained rather small. In 1970, only 2 per cent of them had more than 500 employees, whereas 88 per cent employed fewer than 100 people. Among the largest producers was, not surprisingly, Renault.[63] Overall, the production of the French machine-tool industry was highly specialized and more or less artisanal in nature. Before 1980, there was a large proportion of craft workers (about 50 per cent) and very few unskilled workers (around 10 per cent) in machine-tool production. Thus, workers' qualifications differed both from those of French industry as a whole and from those of the German machine-tool producers, where the proportions of unskilled workers were higher, approximately 40 and 25 per cent respectively. The French machine-tool industry was also highly successful dur-

---

[61] He quotes among others, an article in the *Revue économique et sociale* of May 1946 and another in *Le Figaro* of 26 May 1954; expressions of what Garanger calls the French 'xenomania': Garanger, *Petite histoire*, 243 and 261–3.

[62] Jean Constant, 'Les Entreprises moyennes et l'accroissement de la productivité', in *2ème Congrès*, Paris, Sept. 1949, 15–23.

[63] 'Les Industries mécaniques', *Géographie et industrie*, 81 (Jan. 1971), 1–7.

ing the post-war expansion until the mid-1970s. It grew faster than its German counterpart, albeit from much lower bases.[64]

But, as in steel production, Americanization cannot be limited to questions of technology and industrial structure. In his presentation at the first international mechanical engineering congress, Albert-Roger Métral identified a number of additional lessons from the US example which could be introduced without much delay or heavy investment. They related mainly to maintenance and the standardization of tools and dies, leading, in his opinion, to better-quality output and lower production costs. Just the replacement of tools at regular intervals, as practised in American industry, could produce considerable savings. Similarly, the French productivity missions, like their counterparts from most other European countries, had pinpointed materials handling as one of the areas where relatively minor investments and adjustments could lead to considerable improvements in the short term.[65]

However, as the reports of visits by US experts to French factories in the mechanical engineering industries in 1950–1 reveal, at that time only a few firms had actually implemented such measures.[66] There were a number of exceptions, among them, not surprisingly, the companies headed by Métral and Constant. For Ateliers G.S.P., the US visitor highlighted the 'very good' production control, cost system and plant layout, designating it overall as 'the best-run plant we have seen in Europe'. And at Manufacture d'estampage du Nord-Est, the visitors also found very little room for improvements. Both the machinery and the buildings were new, the production flow properly organized, the layout of the tool room 'splendid', and materials handling 'well thought out'. They concluded that there was 'broad gauged, progressive management alive to the possibilities of the latest manufacturing techniques'. The only downside was the apparent lack of orders, resulting in very low capacity utilization.

There was even the occasional plant which could almost rival its American equivalents. The Compagnie électro-mécanique, formerly a subsidiary of Brown, Boveri & Co., but now French owned, was such a case. Its factory, located at Le Bourget, produced steam turbines, electric generators, and related products. It was housed in modern buildings, equipped with modern machine tools of French, American, and Swiss make, and materials handling had 'been

[64] For details and additional references, see Sorge and Maurice, 'Societal Effect', 77–81.

[65] For the French missions and their suggestions on materials handling, see Claude Doucet, 'Fenwick 1884–1984: L'équipement industriel du négoce à la production', unpublished Ph.D. dissertation (University of Paris IV, 1997), ch. 10; for similar observations by the British missions, see Jim Tomlinson and Nick Tiratsoo, 'Americanisation beyond the Mass Production Paradigm: the Case of British Industry', in M. Kipping and O. Bjarnar (eds.), The Americanisation of European Business, 1948–1960: The Marshall Plan and the Transfer of US Management Models (London, 1998), 115–32.

[66] All of them can be found in National Archives and Records Administration (NARA), Washington, DC and College Park, Record Group 469, Office of the Special Representative in Europe, Office of the Deputy for Economic Affairs, Productivity and Technical Assistance Division, Subject files 1950–56, Box 27 of 52. These visits were of very short duration, lasting at best one day and usually only a few hours. Their conclusions therefore have to be read with considerable caution.

given attention'. The consultant compared it to the Elliott Company, 'the smallest in this line in the U.S., but even so . . . considerably larger than this French company'. However, they considered the latter to have a 'clever set-up' in some of the mechanical operations. Not surprisingly, on the other hand, the American firm had an advantage in much of the equipment, especially special-purpose machinery.

But most of the other firms seemed to lack even essential equipment, such as high-quality tools and dies, let alone modern machinery. They very clearly suffered from lack of funding, especially the absence of Marshall Plan grants. Financial constraints also deterred many companies from extending their current production site or relocating, even though the conditions were often described as 'cramped'. In a number of cases, this seems to have been caused by an inadequate plant layout or bad production scheduling, the latter resulting in excessive stock levels. Quite aware of the lack of financial resources, the consultants very often confined themselves to suggestions which could be implemented with little additional investment. More often than not these concerned materials handling, which could be considerably improved by the acquisition of a forklift truck or a few additional (rolling) pallets. They also suggested introducing incentive pay systems, such as piece-rates, in the frequent cases where they were not present. Overall, the managers in the companies visited seem to have been quite receptive to these suggestions. One executive even asked for 'practical examples where the methods of his company were not competitive with American methods'.

During the 1950s, many French companies seem to have implemented these suggestions and measures to some extent.[67] That such changes did take place is also demonstrated by the success of Fenwick. This company had imported American machinery into France from the end of the nineteenth century. Faced with increasing competition from domestic producers after the Second World War, it diversified into the production of mechanical-handling equipment, namely forklift trucks which it initially built under American licence. Production of this type of equipment rapidly developed into its major activity, with sales increasing more than threefold in absolute terms between 1948 and 1960 from 167 to 514 million francs (net of inflation), and their share of the company's total sales growing from 51 to 65 per cent during the same period.[68]

Finally, the steel-using industries played an important role in the introduction of more competition in France and Western Europe. As seen above, from 1945

[67] For the adoption of US-inspired standard costing techniques in French industry in general and at Renault in particular during the 1950s, see Henri Zimnovitch, 'Les Calculs du prix de revient dans la seconde industrialisation en France', unpublished Ph.D. dissertation (University of Poitiers, 1997), 166–80, 428–33; for statistical quality control see Patrick Fridenson, 'Fordism and Quality: the French Case, 1919–93', in H. Shiomi and K. Wada (eds.), *Fordism Transformed: The Development of Production Methods in the Automobile Industry* (Oxford: Oxford University Press, 1995), 166 and 170; and Vincent Guigueno, 'Productivité et contrôle statistique dans les années 1950', *Entreprises et Histoire* 13 (Dec. 1996), 149–51.

[68] Doucet, 'Fenwick', ch. 10, esp. 365, table. In order to exclude the effects of inflation, he has standardized sales to 1984 francs.

onwards Renault's Pierre Lefaucheux attacked the French steel industry and their lack of consideration for the needs of the steel users, especially with regard to the installation of modern strip mills, necessary for the production of body sheets for the automobile industry. The Secretary General of the SGIMTM, Jean Constant, also highlighted the difficult conditions under which the using industries had to compete internationally, especially with the reviving German producers. All of these factors, combined, amounted to a recipe for disaster:

Take expensive steel, heat it with onerous coal, mix it with unproductive labour seasoned with high social security benefits. Based on this, our industry will easily beat the foreign competition, because we are so intelligent and because we have modernized our production equipment with Marshall Plan grants all of which were actually allocated to the nationalized enterprises.[69]

Constant highlighted the lack of funding for the downstream industries (even though his assertion that *all* Marshall Plan funds went to the state-owned companies is not correct),[70] and the cost of raw materials and labour. But his major criticism, repeated from 1948 onwards in numerous public statements and letters to the relevant ministries, concerned the cartel-like behaviour of the French steel industry. He blamed the centralization of all orders by the sales syndicate, CPS, and the lack of direct contact between producers and users for the high prices and the irregular quality of French steel.[71] In his presentation at the 1948 congress, Métral had also identified the irregular and sometimes low quality of speciality steel as a major problem for the French machine-tool producers, attributing them 'to a lack of specialization in our steel industry'. He voiced similar complaints about the supply of cast iron, which was due to a lack of concentration in the foundry industry. This therefore remained at the craft production stage, which was 'particularly harmful for the quality of output'.[72]

These confrontations between steel producing and using industries in France were not new, but had a long history.[73] What was new in the period immediately after the Second World War was their increased acrimony, and, more importantly, the fact that the representatives of the using industries consciously

[69] Quoted from his editorial 'Les paliers du mensonge', *Les industries mécaniques*, Aug.–Sept. 1949, 1–4 [translation by Matthias Kipping].

[70] But they received indeed very little compared to the priority sectors, see Frances Lynch, *France and the International Economy: From Vichy to the Treaty of Rome* (London: Routledge, 1997), 69, 91.

[71] For more detail see Kipping, 'Les tôles avant les casseroles'.

[72] It should be noted that Métral, who had become president of the SGIMTM in 1950, initially pursued a less confrontational approach in his dealings with the French steel producers than Lefaucheux or Constant. He even forced the latter to resign in March 1951, after he had openly attacked the head of the French employers organization CNPF, Georges Villiers, and his critical attitude towards the Schuman Plan. But Métral quickly became disillusioned and, in the mid-1950s, even asked the members of his association to deduct a percentage from their payments to the steel producers in protest over the high prices: interview with one of Constant's successors as Secretary General, Georges Imbert, 13 Oct. 1992.

[73] For an overview of their relationship, see Matthias Kipping, 'Inter-Firm Relations and Industrial Policy: The French and German Steel Producers and Users in the Twentieth Century', *Business History* 38, 1 (Jan. 1996), 1–25.

and repeatedly referred to the USA as an example. Thus, in his speech to the second international congress of the mechanical engineering industries in 1949, Constant underlined the importance of the high quality and the regularity of raw material inputs which he 'personally considered to be an essential element of the high productivity of the American factories'. And in his testimonial before the commission of the French Economic Council investigating the ECSC Treaty, he explicitly identified the competition in the US steel industry as the major reason for its high productivity.[74]

As mentioned above, it is not important to assess the extent to which Constant's view of competition and productivity improvements in the American steel market corresponded to reality—probably very little.[75] But it is clear that the support of the steel users, backed up by their considerable and increasing economic importance, especially in terms of employment and exports, was one of the crucial elements for the success of the Schuman Plan and the—partial—opening up of European steel markets. But the French steel-using industries not only played an important role in the Americanization of the steel industry. They appeared also fairly Americanized themselves by 1960, albeit in different forms according to the type of activity. The car manufacturers, and among them first and foremost Renault, had clearly moved towards mass production of standardized models, such as the 4CV, or the 2CV in the case of Citroën. Their suppliers also developed in the same direction, specializing their output, without necessarily growing significantly in size. A similar specialization occurred in machine tools, where the industry structure, dominated by small, artisanal firms changed little, but output became customized in order to suit the needs of the consuming industries. As in the steel industry, the adoption of US methods of shop-floor organization initially progressed rather slowly. But mechanical handling and statistical quality control, for example, seem to have been more widely implemented in the French steel-using industries during the 1950s.

7.4   Conclusion and Epilogue

In the 1940s, the initial response of the French steel-producing and using industries towards the American model was rather cautious. Most of the industrialists in both industries hesitated to embrace high-volume production technology, because they were not sure about the market prospects in post-war Europe. But the move towards US-style large-scale production was precipitated by a few

[74] *La Communauté Européenne du Charbon et de l'Acier* (Paris, 1952). The Economic Council, on which all economic interest groups were represented, had only advisory powers. However, a negative vote on the Treaty would have seriously hampered its chances of ratification in the National Assembly.
[75] At about the same time, the pricing policy of US Steel was under renewed investigation and the academic George Stigler, for example, contended that 'the forces of competition in the steel industry are now not sufficiently strong to justify us to leave the industry alone'; see Paul A. Tiffany, *The Decline of American Steel. How Management, Labour and Government went Wrong* (New York: Oxford University Press, 1988), 89.

entrepreneurial individuals, like Usinor's René Damien and Renault's Pierre Lefaucheux, who had a much more optimistic view about the future development of demand, especially if quality could be improved and production costs reduced sufficiently. At the same time, political figures such as Jean Monnet and André Philip argued and strongly acted in favour of larger and more competitive European markets, as a condition for passing on lower costs to the consumers and thus becoming the basis for an increase in consumption and living standards.

In the immediate post-war period, the departure from cartel practices and stagnant markets, characteristic of France and other European countries during the inter-war period, was a hope, not a certainty. But through their joint and, in the case of the ECSC, concerted action, the above-mentioned individuals certainly contributed to making it happen in a kind of self-fulfilling prophecy, because they helped create the technological and market conditions necessary for the 'thirty glorious years' of post-war expansion of the French economy.[76] Most important was probably the fact that, as a consequence of these initiatives, the other producers, much more reluctant and cautious, had to move in the same direction. The steel producers were more or less forced to build a second strip mill and the mechanical engineering and machine-tool industries increased their specialization in order to supply the automobile producers with the necessary parts and machinery.

What was the role of the United States in the process? The Americans provided the necessary technology and, especially from 1948 onwards, much of the funding for the large-scale investment projects such as the continuous strip mills. At the same time, as the example of the anti-trust legislation very clearly shows, they refrained from intervening directly and openly in internal French debates. Probably most importantly, the United States acted as a kind of 'reference society'[77] which could be and was used by the above-mentioned group of individuals as a very strong argument in favour of large-scale production and competitive markets. That their depiction of the US example was not always an accurate reflection of American reality is only of secondary importance. Their perception and interpretation of American production technology, management methods, and market order shaped the ways in which they were adopted and implemented in the French context.[78]

---

[76] For the important role of renewed business strategies rather than 'enlightened technocrats', see also Patrick Fridenson, 'Who is Responsible for the French Economic Miracle (1945–1960)?', in M. Adcock, E. Chester, and J. Whiteman (eds.), *Revolution, Society, and the Politics of Memory* (Melbourne: Melbourne University Press, 1997), 309–13.

[77] This term was coined by Mauro F. Guillén, *Models of Management: Work, Authority, and Organization in a Comparative Perspective* (Chicago: University of Chicago Press, 1994), 290. For its application in this context, see also Ove Bjarnar and Matthias Kipping, 'The Marshall Plan and the transfer of US management models to Europe: an introductory framework', in Kipping and Bjarnar (eds.), *Americanisation of European Business*, 1–17, esp. 8.

[78] In the case of market order in the German steel industry, the *interpretation* of the American model played a similar role; see the contribution of Gary Herrigel in this volume.

This implementation, however, was a slow and arduous process. The 'break-through' achieved in the immediate post-war period by the adoption of the new technology, namely the continuous strip mills and transfer machinery, and the opening up of European steel markets to competition was far from complete. Usinor's hot and cold rolling mills reached their full potential capacity only ten years after they were first installed. Mechanical handling, for example, was introduced on a large scale only during the 1950s, as were the standard costing and statistical quality control procedures. Industrial structures were even slower to change. The French steel companies merged only reluctantly and belatedly and a significant rationalization of production sites did not occur until the 1980s. The opening up of markets to competition also happened only gradually, because it had to contend with the deeply rooted cartel thinking of many French (and European) industrialists. In this respect, the formation of the EEC in 1957 was the next important step.

Nevertheless, at the end of this rather drawn-out process, in the widely shared conclusion of Michael Piore and Charles Sabel, France was 'the country that went furthest towards the U.S. system'.[79] As these and many other authors have shown, the strict adherence to American technology and shop-floor management proved detrimental, when market conditions changed dramatically during the 1970s, and more flexibility was required. On the basis of the above analysis, it could also be argued that the slow and difficult Americanization of the French steel-producing and using industries in the 1940s and 1950s was one of the reasons why both found it so hard to extricate themselves from the US model during the 1980s.

The steel industry is an excellent example in this respect. Struggling to keep up with demand during most of the 1950s and 1960s, the industry finally embraced Americanization more fully with a series of mergers and the construction of very large-scale coastal plants in Dunkirk and especially in Fos-sur-Mer, near Marseilles, a region with no tradition in steel production.[80] Somewhat ironically, the latter came on stream shortly before the two oil shocks plunged the world steel markets into turmoil. It took a long time, the nationalization of the industry, and the injection of considerable public funds, to bring the French steel industry back from the brink of bankruptcy and extinction. During the 1980s, under government ownership, production sites were rationalized further and the remaining companies combined into a single entity, Usinor-Sacilor. Subsequently, the company focused increasingly on high value-added products and improved its relationship with the steel users. Having been privatized in 1995, Usinor-Sacilor is now one of the world's largest and most successful steel producers.[81]

[79] Michael J. Piore and Charles F. Sabel, *The Second Industrial Divide. Possibilities for Prosperity* (New York: Basic Books, 1984), quotation, 135.

[80] For an overview of the increasing concentration see Michel Freyssenet and Françoise Imbert, *La Centralisation du capital dans la sidérurgie 1945–1975* (Paris: Centre de Sociologie Urbaine, 1975); for the new investments see Godelier, 'De la stratégie locale', 295–7, 321–9; Daley, *Steel, State, and Labor*, ch. 5.

[81] For details of the difficult turnaround, see Daley, *Steel, State, and Labor*, chs. 6–9.

The reasons for the decline and the difficult resurgence of the French steel industry are manifold, and include government intervention through price controls, the rather belated rationalization of production sites, the strained relationship with the using industries, and the failure of the different companies to co-ordinate their investment plans.[82] But it might also have been that the steel producers were reluctant to give up technology and management methods which they had implemented with considerable difficulty and which also had proved so successful during most of the post-war period.

[82] Michel Freyssenet and Catherine Omnès, *La Crise de la sidérurgie française* (Paris: Hatier, 1982).

Chapter 8

Remodelling the Italian Steel Industry: Americanization, Modernization, and Mass Production

RUGGERO RANIERI

8.1 Introduction

This chapter discusses the impact of American industrial practices on the Italian steel industry in the post-war period. The area involved is by definition a fairly elusive one, and does not lend itself to clear-cut conclusions. It should be pointed out, however, that the Italian steel industry as a whole was largely dependent for its post-war reconstruction on new American technology. For a number of larger firms in the state-owned sector the influence of the American model extended deep into management and into organizational design. This was particularly true of Cornigliano's new integrated plant near Genoa based on a wide strip mill, which provides one of the best examples of an attempt at full Americanization by an Italian company.

The chapter is organized into three sections. Section 8.2 provides some basic background information on the structure and performance of the Italian post-war steel industry. Section 8.3 examines the impact of the Marshall Plan on post-war reconstruction investment. The central part of this section deals with the negotiations over the allocation of US funds to the state-owned sector, but there will also be brief accounts of Marshall Plan aid to other steel producers, particularly the Falck group and Fiat. The final section (8.4) looks at the attempts made by Italian managers, particularly in Cornigliano, to copy and follow the 'American model'. This examination begins with an account of how those managers viewed the American model, and what contribution they thought it could make in Italy, and then reviews some of the steps taken to import it into Italian companies. Finally it considers some of the limitations and subsequent failures of this attempt.[1]

---

[1] This chapter is based on records from the following archival sources:

- Washington National Record Centre (WNRC), RG. 469, Records of the US Foreign Assistance Agencies, 1948–1961.
- Washington, World Bank Group Archives.

8.2 Reconstruction and Expansion of the Italian Steel Industry, 1945–1970

Before the war Italian crude steel production peaked, in 1938, at about 2.4 million tonnes, with capacity estimated at around 3 million. Rolling-mill capacity was estimated at around 4 million tonnes, while the output of finished products did not exceed 2 million tonnes in 1935, the best pre-war year. This pointed to a low level of capacity utilization, due to insufficient demand and market fragmentation.[2]

In 1945 production of steel was down to 1.1 million tonnes, of which over 50 per cent was carried out in electric furnaces. Pig iron production had fallen even further, down to about 170,000 tonnes. Wartime destruction hit the industry's public sector holding company, Finsider, particularly severely. Finsider's blast furnaces at Portoferraio, Piombino, and Bagnoli had been destroyed and a number of important units had also been removed from the plants of Campi, Terni, and Apuania. Pig iron capacity was entirely knocked out, with smelting and rolling capacity also severely dented. The industry's private sector, on the other hand, suffered much less: its electrical furnaces emerged from the war relatively intact, and damage to its open-hearth and rolling-mills units was also contained. It was, thus, better placed to exploit the rising level of demand prevailing throughout the reconstruction years. For the state-owned sector, on the other hand, reconstruction was protracted and costly.

The public sector of the industry had been created in the early 1930s, when the state was forced to step in to rescue the large 'mixed' banks, particularly the Banca Commerciale and the Credito Italiano. Thus saddled with a number of important steel companies, the state proceeded to entrust them to the Industrial Reconstruction Institute (IRI), a new holding agency, which was created in 1933 and given permanent status in 1937. IRI was a majority shareholder in companies representing some 42 per cent of Italy's joint-stock capital, covering a wide variety of interests in the utilities, banking, and manufacturing, particularly in heavy industry. As far as the steel industry was concerned, IRI owned some 40 per cent of total capacity. The old company structure was left intact, but all steel assets were placed within Finsider, one of IRI's sectoral holding companies.

* Fondazione Luigi Einaudi, Torino, Carte Osti.
* Archivio Ansaldo, Genova, Archivio Storico Ilva.
* ACS, Roma, Archivio storico IRI.

Quotation of documents, however, is selective. Fuller references are to be found in the relevant secondary literature. In particular see R. Ranieri, 'Il Piano Marshall e la ricostruzione della siderurgia a ciclo integrale', *Studi storici* 1 (1996), 145–90 and G. L. Osti, *L'Industria di Stato dall'ascesa al degrato—Trent'anni nel gruppo Finsider. Conversazioni con Ruggero Ranieri* (Bologna: Il Mulino 1993).

² This section is drawn mainly from M. Balconi, *La siderurgia italiana 1945–1990—Tra controllo pubblico e incentivi di mercato* (Bologna: Il Mulino, 1991), and from R. Ranieri, 'Assessing the implications of mass production and European integration: the debate inside the Italian steel industry (1945–1960)', in M. Dumoulin, R. Girault, and G. Trausch (eds.), *L'Europe du patronat de la guerre froide aux années '60* (Berne: Peter Lang, 1993).

The public sector consisted of four large companies: Ilva, Siac, Terni, and Dalmine, of which Ilva, a vast and fairly shapeless conglomerate, was by far the largest. These four companies owned 21 steel plants, of which 5 were hot-melting shops, 10 were cold-melting shops, while the remaining 6 were re-rolling facilities.

After the war, Finsider's new president, Oscar Sinigaglia, set out to reorgan-ize the industry. He was no newcomer to the steel industry: he had, among other things, been involved in attempts to reorganize Ilva in the early 1930s. His new plan, therefore, which provided the blueprint for the reconstruction of the pub-lic sector, drew on his previous thinking about the need to modernize the indus-try, concentrate facilities, and develop large coastal integrated plants. Much of the reorganization he had in mind was to take place within the public sector. Three large coastal plants at Piombino, Bagnoli, and Cornigliano were to form the core of Finsider, each one specializing in a particular section of the market. Bagnoli was to supply merchant bar and rod to the engineering and construction industries. Piombino would specialize in rail and other heavy products, while Cornigliano, equipped with a wide strip mill of American design, would spe-cialize in thin flat products, turning out coils for the automobile industry, as well as tin plate and galvanized sheet. It was also proposed to upgrade and modern-ize tube-making facilities at Dalmine and special steel sheet production at Terni. For the rest the Sinigaglia Plan proposed to eliminate a number of cold-metal shops, turning them into mere re-rolling facilities or, in some cases, closing them altogether.[3]

By and large, the Sinigaglia Plan achieved its main objectives, despite the opposition met in many quarters, particularly, as we shall see, from the indus-try's private sector. By the early 1950s the public sector had been reorganized, the larger plants had been expanded and modernized, while Cornigliano entered into production in 1953. Finsider's production of steel had increased from about 150,000 tonnes in 1945 to over 1.5 million tonnes in 1952.

The group of executives surrounding Sinigaglia must share with him some of the credit for successfully carrying out such an ambitious strategy. Prominent among them were Mario Marchesi, Guido Vignuzzi, and Ernesto Manuelli. In the wings was a group of younger managers, such as Gian Lupo Osti, Enrico Redaelli Spreafico, and Alberto Capanna, who were to play a key role at a slightly later time. The older generation of Ilva managers included men such as Antonio Ernesto Rossi, Eranio Fidanza, and Giulio Pescatore. They were less happy with Sinigaglia's reforms, and often opposed them, although somewhat surreptitiously, given that they had lost some of the influence they had enjoyed in the inter-war period. Wedded to a more relaxed steel-making strategy, ba-sically reliant on government protection and industry-wide cartel arrangements, they disapproved of the construction of Cornigliano and entertained at best a

---

[3] For a short summary of the Sinigaglia Plan see O. Sinigaglia, 'The Future of the Italian Iron and Steel Industry', *Banca Nazionale del Lavoro Quarterly Review* 4 (1948), 240–5. See also Finsider, *Sistemazione della siderurgia italiana* (Rome: Finsider, 1948).

lukewarm commitment towards Americanization. The group most committed to Sinigaglia's reforms, therefore, tried to stay clear of Ilva's influence. Cornigliano was made into a separate company within Finsider and it remained so until the Italsider merger in 1961.

The private sector of the industry was composed of a mixture of firms, mostly located in north-western Italy. In the first rank there were a number of medium-sized independent steel makers, such as the Falck group, Caleotto, Bruzzo, La Magona, the Redaelli group, and a few others. They ran cold melting shops based both on open-hearth and electric furnaces, which made them heavily reliant on external scrap supplies, and they turned out a large variety of shapes, both in common grade and special steel. The Falck group, of which A. F. L. Falck (Acciaierie e Ferriere Lombarde Falck) was the main company, accounted in 1948 for a yearly output of over 200,000 tonnes of crude steel. Its steel-making capacity was estimated at around 10 per cent of the country's total, while for hot and cold-rolled products its share was about 15 per cent.

During the inter-war period a few large engineering companies such as Fiat and Breda had set up their own steel plants. They used most of the steel they produced for their own internal consumption. Fiat was by far the largest of this group and its capacity was of the same order as Falck's. Finally there was a group of very small firms, scattered across the pre-alpine valleys, only a few of which produced steel in electrical furnaces often on a seasonal basis, while many others—more than one hundred—were re-rollers. Dedicated to the supply of local markets, they were grouped in an association called ISA (Industrie Siderurgiche Associate).[4]

In 1952 private firms as a whole accounted for over half of the national capacity for both crude and finished steel. Their pig iron production capacity, however, consisting merely of a small number of electrical blast furnaces which had been set up during the war and continued into the reconstruction period mainly as a stop-gap measure, was much less important. A few years later, in 1957, once the public sector had completed its post-war investment, the relative importance of privately owned steel making and rolling mill capacity declined to below 50 per cent of the total. When considered in absolute terms, however, given that the country's aggregate steel capacity was undergoing very rapid expansion, it still meant that these companies had expanded.

The Falcks enjoyed a position of leadership among private producers. Their spokesmen often articulated the interests of most of the other firms, with the exception of Fiat, which followed an independent strategy and alone supported, as we shall see, Finsider's post-war rationalization plans. The literature has devoted much attention to the contrast of opinions that opened up between the

---

[4] ISA-affiliated companies were estimated, in 1950, to account for 12 per cent of the country's steel-making capacity. On the history of this section of the industry see M. Pozzobon, 'L'industria padana dell'acciaio nel primo trentennio del Novecento', in F. Bonelli (ed.), *Acciaio per l'industrializzazione—Contributi allo studio del problema siderurgico italiano* (Turin: Einaudi, 1982).

Falcks and Sinigaglia over post-war investment priorities. A very brief summary of that argument is therefore in order here.[5]

According to the Falcks, there was little need in Italy for large integrated steel plants. It was sufficient to have a range of medium-sized and small firms, operating cold-metal steel shops, primarily based on electrical furnaces. For such plants, labour was cheap and skilled, energy supplies were plentiful, the firms were well connected to their customers, and the production was of high quality. Mechanical engineering in northern Italy, the argument went, was a low-volume, high-quality business. It did not require large batches of standardized steel shapes, but rather a variety of products, customized and delivered according to what today would be called a 'just in time' schedule. Whenever, on the other hand, cheaper ranges of standardized shapes, particularly long and heavy products, were required—as for example by the state for major infrastructure projects—then, argued the Falcks, it was cheaper and more convenient to import them from abroad.

Sinigaglia retorted that small scale, and multiple production runs were a recipe for low productivity. What was required instead was standardization and specialization in conjunction with better marketing. To achieve this steel production should be based primarily on hot metal shops, in order to avoid being overdependent on imported scrap. Scrap prices, Sinigaglia pointed out, fluctuated in accordance with the economic cycle, peaking when the market was more buoyant. Furthermore it was essential that production runs should be rationalized so that rolling mills could be used closer to capacity. This required a small number of large specialized outfits to replace smaller multiple-product facilities. Smaller firms were to be restricted to special steel production and to the finishing end of the trade. Large firms, on the other hand, were not to restrict themselves to supplying semi-finished goods, a recipe for cartels and low profits. On the contrary they should enhance their finishing facilities to gain a foothold in the most profitable markets.

The differences between Sinigaglia and the Falcks touched upon the question of industrial expansion. Whereas Sinigaglia envisaged a role for Italy as a major steel and engineering producer and exporter, for the Falcks there seemed no reason to look beyond an economy based on a rather narrow industrial base, geographically concentrated in the North. It is true, therefore, that the Falcks, as well as most private industrialists, did not wish to entertain any particular industrial strategy. They ran fairly successful businesses, with traditional but well-proven methods, and wished to continue doing what they had done in the past. They therefore disliked the idea that Finsider—a public company—would

[5] The basic texts of the debate between Sinigaglia and his opposite number Giovanni Falck are in Ministero della Costituente, *Rapporto della Commissione Economica—Presentato all'Assemblea Costituente—II—Industria—ii—Appendice alla Relazione (Interrogatori)* (Rome, 1947) and in the columns of the magazine *Il Mondo*, in April–June 1949. For a good summary of the debate from the point of view of Finsider see Osti, *L'Industria*, 113–48; for the position of Fiat see P. Bairati, *Vittorio Valletta* (Turin: UTET, 1983),156ff.

compete with them in supplying steel users in northern Italy. Their condemnation of state intervention at one level, however, did not stop them from seeking from the government high protective tariffs as a way of holding on to their preferred markets in the face of foreign competition. On the other hand, Finsider promised that modernization would for the first time in Italy's history deliver internationally competitive steel prices.

The question of market expansion was also the key to the way Fiat, under the energetic leadership of Vittorio Valletta, viewed the post-war period. What Fiat was aiming for was mass production of a few ranges of cheap automobile models, which would be accessible to consumers on average incomes. To achieve this, large new modern plant equipped with the latest technology was required, allowing for greater standardization and lower costs. The meeting of minds between Valletta and Sinigaglia went further. In order to fulfil its plans, Fiat needed vast quantities of coils and just before the war it had commissioned a new strip mill from the German plant supplier DEMAG in Duisburg. Although ultimately, as we shall see, this piece of equipment did not prove to be successful, in the immediate post-war years it was very difficult to assess its value. One of Finsider's first concerns was to prevent Fiat from engaging directly in the production of coils. They insisted that the DEMAG order should be passed on to them so that they might install it in Cornigliano in return for a commitment to supply Fiat with all their requirement in coils. At the beginning of 1948 Fiat accepted the deal and the two companies forged a powerful alliance, which would have a lasting effect on the whole industry.[6]

In 1950 the Italian government chose to join the French initiative for pooling the European coal and steel industries, known as the Schuman Plan. At the time Italy's steel industry was much the smallest of the other participating countries: its output of crude steel was little over 2 million tonnes per year whereas France had an output of 8.5 million tonnes, Belgium of 3.7 million tonnes, and West Germany of 11.2 tonnes. Negotiations were to result one year later in the creation of the European Coal and Steel Community (ECSC). Italy's steel industry was cushioned by a number of special clauses and exceptions, particularly the fact that tariffs on most steel shapes within the Community would only come down gradually over a transitional period of five years. Although on this occasion public and private sector industrials co-operated to secure the best possible deal, it was Finsider that played the most active role in the talks: Italy, it was said, was still in a weak competitive position, and needed time to complete her modernization programme. Iron ore deliveries from French overseas territories were secured for Finsider's blast furnaces along the Tyrrhenian coast. At the same time as the Schuman Plan was being negotiated, the building of Cornigliano was clearing its last obstacles with the American authorities. The Italian government let it be understood that its playing a part in the new

---

[6] On the DEMAG mill see Ranieri, 'Il Piano Marshall'.

European arrangement was strictly dependent on the completion of Finsider's modernization.[7]

By the end of the 1950s, the competitive state of Italian steel had greatly improved. Whereas in 1952 prices for standard section and for thin sheet on the Italian market had been roughly 15 to 20 per cent above comparable prices in the rest of the ECSC, by 1958, at the end of the transitional period, there was hardly any difference. Large increases in productivity made these gains possible: Finsider, for example, employed 45,000 in 1945 and 47,500 in 1957 with output per head increasing over the period from 23 to 71 tonnes. By 1965 the group had raised the number of its employees by 18,000, but its output had more than doubled again.[8]

The general picture of the 1950s was one of high rates of economic growth: rates of GDP averaged over 5 per cent, and of industrial production over 8 per cent. Steel output grew by 12 per cent a year during the period 1951–8, and at a faster rate thereafter. In 1951 output of crude steel was little over 3 million tonnes, but by 1958 it had doubled to 6.3 million tonnes and in 1961 it exceeded 9 million tonnes. Italy achieved the fastest rate of growth within the ECSC. The growth of output, moreover, was greater than the growth of domestic consumption. Another important development was the growing importance of flat products, whose share in the market for rolled products grew from about 25 per cent in 1950 to 43 per cent in 1960, a vindication of Sinigaglia's post-war projections. Where Sinigaglia, on the other hand, was less accurate was in thinking that the time of cold-metal shops was coming to an end. Despite the expansion of integrated steel facilities, the ratio between pig iron and crude steel in Italy remained one of the lowest in the world, at about 30 per cent. Scrap remained abundant and cheap, its price kept down, among other things, by generous compensation schemes set up within the ECSC.

Finsider's role in the steel market grew enormously in importance. Whereas in 1952 the company had accounted for 44 per cent of crude steel production, in 1957 it accounted for over 50 per cent and by the mid-1960s for about 60 per cent. The same was true of hot rolled products, Finsider's share of national output of which climbed from 43 per cent in 1952 to 55 per cent in 1957. Furthermore, Finsider's share of flat products reached 70 per cent of the market by 1960 and was to increase thereafter. A considerable part of Finsider's expansion was accounted for by Cornigliano, which by 1960 was turning out 1.3 million tonnes per year of crude steel, as well as about half the national output for both hot and cold-rolled products. Cornigliano sold coils to Fiat, Falck, and La Magona, all of which were equipped with cold strip mills, but it also turned out

[7] On Italy and the Schuman Plan see R. Ranieri, 'La siderurgia italiana e gli inizi dell'integrazione europea', *Passato e presente* 7 (1985), 65–95; see also D. Spierenburg and R. Poidevin, *Histoire de la Haute Autorité de la Communauté Européenne du Charbon et de l'Acier. Une expérience supranationale* (Brussels: Bruylant, 1993).

[8] Statistics are taken from the following: Balconi, *La siderurgia italiana*; Assider, *La siderurgia italiana in cifre negli anni dal 1946 al 1960* (Milan: Assider, 1961); Assider, *Steel in Italy* (Milan: Assider, 1978).

large quantities of its own tin plate and galvanized sheet, which it sold both at home and abroad, supplying the fast booming consumer durable trades.

By the late 1950s further expansion of Italy's steel capacity was firmly on the agenda. A new discussion took place between the public and the private sector, based on arguments very similar to the ones that had been raised during the reconstruction period. Public sector executives advocated more concentration and standardization and less dependence on scrap, and they pointed to the fact that coastal locations enjoyed increasing advantages, for they could receive shipments of comparatively cheap and high-quality iron ore and coal from recently developed overseas mining sites. Plans for a new coastal integrated plant at Taranto, in southern Italy, were thus brought forward. They were conceived as part of the current drive to channel a large share of state-owned-sector investment into the industrialization of the South, with the benefit of special government subsidies. Private producers pointed out that the largest steel demand was in the North, a fact that would put a southern mill at a disadvantage and they insisted that there was a risk of generating over-capacity. While they recognized that there was a need for further primary steel-making capacity, they did not wish the state-owned sector to increase its share of rolled products. Their objections were brushed aside, however, and the Taranto plant, with a capacity of 4 million tonnes per year, was built between 1960 and 1964.

The construction of Taranto was accompanied by a major reorganization within Finsider. Cornigliano was merged with Ilva to give rise to a huge new corporation, Italsider, which was meant to amalgamate most existing public sector companies. The merger took place very much on Cornigliano's terms, and can be seen, therefore, as the ultimate fulfilment of Sinigaglia's wishes, although Sinigaglia himself had died in 1953. Standardization, new technology, specialization among the different plants, efforts at introducing up-to-date managerial and organizational practices modelled on American companies were very much part of the new Italsider ethos. The success of the operation, however, was far from unqualified: for one thing the southern location of Taranto brought with it its share of problems, especially since the original production targets were deliberately overshot and the capacity of the plant was doubled in the course of the late 1960s. Disagreements over priorities, pressures to dilute the organizational reforms, mounting levels of debt, and a breakdown in industrial relations all contributed to Finsider's declining performance during the late 1960s and 1970s.

It is worth following developments within the industry's private sector, which also expanded considerably. Fiat's and A. F. L. Falck's capacity for hot and cold-rolled finished products moved up respectively from 1.1 million and 800,000 tonnes per year in 1959 to 1.9 and 1.6 million tonnes per year in 1970. The same was true for a number of other producers. These companies nevertheless found it increasingly difficult to compete with Finsider in the market for standardized, common-grade steel goods, particularly thin flats, and they were therefore pushed to concentrate on long products and on special steel grades.

At the same time extraordinary new developments were taking place among the smaller electrical producers and re-rollers, which had traditionally constituted the lower tier of the private steel makers. Encouraged by low scrap prices and a booming demand for steel by the construction industry, these firms stepped up their output of long products, particularly concrete reinforcing bars. Traditional scrap-reheating techniques were gradually replaced by integrated production chains, whereby electrical steel makers supplied re-rollers with semi-finished products. A host of new firms entered the trade: for example former Milanese scrap merchant, Emilio Riva, opened his own steel shop in 1957. Their overheads were very low, business was kept within the family, unions were banned, and labour was employed at very low wage rates. Long products did not require elaborate technology and technical economies of scale were not an important factor. On the other hand continuous casting technology first introduced during the 1960s lowered barriers to entry and pushed costs further down, boosting the mini-mills' share of the market. Between 1959 and 1970 mini-mills increased their production of hot-rolled products by a factor of five, reaching an output of 3.8 million tonnes per year.[9]

In 1970 steel consumption in Italy exceeded the 20 million tonne mark which, on a per capita basis, was equivalent to about 400 kilograms, only slightly below the equivalent figure for France, Belgium, and the UK. This was a remarkable progress from the bare 65 kilograms per capita obtaining in 1950 and it heralded Italy's transition to a fully industrialized economy. Output of steel, at 17.3 million tonnes, was only slightly behind France's 24 million, although still far behind West Germany's 45 million.

8.3   The Impact of the Marshall Plan

During the reconstruction period, the modernization of the industry benefited greatly from an injection of Marshall Plan funds, particularly through the Industrial Project loan scheme, set up in November 1948, to encourage the purchase of new industrial equipment. Bidding companies had to justify their applications by detailing the kind of machinery they requested and the name of US producers they would buy from, as well as stating whether they had tried alternative non-dollar sources. They were also required to spell out the impact the investment would have on their levels of output and productivity. The loans were granted on a 25-year basis, with an interest rate of 5.5 per cent, which, in Italy, was about 2.5 points below the current market rate. Further sums were loaned at a comparable rate by the Italian government, against the ERP Counterpart Funds.

The sums allocated to the Italian steel industry totalled a little less than $82 million (post-war value), of which $54 million went to Finsider and the rest to

---

[9] On mini-mills see Balconi, *La siderurgia italiana*, 155–66; ead., *Riva 1954–1994* (Milan: Casagrande, 1995); ead., 'The Notion of Industry and Knowledge Bases: The Evidence of Steel and Mini-mills', *Industrial and Corporate Change* 3 (1993), 471–507.

private steel firms. Within Finsider, Cornigliano benefited from loans equivalent to some $35 million. Among private producers the Falck group received loans equivalent to about $9 million while Fiat's steel plants received over $7 million. Smaller sums went to Sisma, Redaelli, La Magona, and Ilssa Viola. Further amounts, not part of the Industrial Project scheme, were disbursed as loans to smaller firms.[10]

It is important to assess the relative importance of this aid. The first wave of post-war investment is estimated to have amounted to roughly 200 billion lire (post-war value), equivalent to $300 million. ERP and other smaller US loans, such as the Eximbank one of 1947, accounted for slightly more than one-third of the figure. The most expensive industrial project to be carried out was the rebuilding of Cornigliano, which was alone worth around $90 billion lira, equal to about $150 million. In total, investment in the state-owned sector amounted to about $215 million, whereas private firms with a capacity of 50,000 tonnes or more received about $70 million.

The loan received by Cornigliano was instrumental in the purchase of a US-made wide-strip mill as well as a cold-strip mill; on the other hand, it made up only about one-quarter of the cost of the new plant. It is fair to say, therefore, that US loans were comparatively more important for private company investment than they were for Finsider. Private companies, in fact, did not enjoy access to additional external credit in the same way as the public sector, and their non-ERP investment funds came mainly from the ploughing back of profits. New developments in thin flats,[11] in particular, were greatly facilitated by European Recovery Programme (ERP) money. Not only did the Americans finance the hot and cold-strip mills at Cornigliano, they also granted considerable sums to Fiat, A. F. L. Falck, CMI (Cantieri Metallurgici Italiani), and Terni to upgrade their existing units or install cold-strip mills as well as finishing and coating equipment.

[10] Figures for ERP loans are drawn from R. Ranieri, 'The Marshall Plan and the Reconstruction of the Italian Steel industry (1947–1954)', in R. Girault and M. Lévy-Leboyer (eds.), *Le Plan Marshall et le relèvement économique de l'Europe* (Paris: Comité pour l'histoire économique et financière de la France, 1993), 384–5; for an assessment of post-war investment in the Italian steel industry, see WNRC, RG 469, Mti, OFD, S.F. 48-57/48-54, Box 54: 'Small Business, Loan Program for Small Steel Firms', 21 Aug. 1953.

[11] New thin-flat technology developed in the inter-war period in the USA included wide strip mills (continuous or semi-continuous); continuous cold reduction mills, referred to as cold strip mills; and continuous electrolytic lines for tin plate and galvanized sheet (zinc grip mills). For a full account see T. J. Misa, *A Nation of Steel. The Making of Modern America 1865–1925* (Baltimore: Johns Hopkins University Press, 1995), 241ff.; the most important contemporary source is T. J. Ess (ed.), *The Modern Strip Mill. A Record of the Continuous Wide Strip Mill Installations and Practices in the United States* (Pittsburgh: Association of Iron and Steel Engineers, 1941). A good Italian account is A. Scortecci, 'Rapporto sul progresso tecnico nell'industria siderurgica italiana', in Centro Nazionale di Prevenzione e Difesa Sociale, *Il Progresso Tecnologico e la Societa' Italiana—Effetti economici del Progresso Tecnologico sull'economia industriale italiana (1938–1958)*, iii, *Industrie Varie* (Milan: Giuffrè, 1961).

The Allocation of Marshall Plan Aid to the
State-Owned Sector

The application by Finsider for Cornigliano's wide strip mill was at the same time the most important and the most controversial and it therefore deserves particular attention. At first, in early 1948, Finsider applied for a loan to the World Bank as part of a wider bid sponsored by the Italian government covering different industrial sectors. Fifty million dollars were requested for the steel industry in the form of two investment projects, one by Finsider and one by Fiat, tied together by the fact that Cornigliano's coils would feed a new cold-strip mill in Fiat's steel plant in Turin. The World Bank had just extended large loans to France for the Usinor project, also based on a new wide strip mill, as well as to Belgium.[12] Their main steel expert was Wayne Rembert, an engineer who had also been a consultant in Italy first for AMG (Allied Military Government) and then for the United Nations Relief and Rehabilitation Administration (UNRRA) Mission. Rembert travelled to Italy, consulted widely, and produced a long report questioning a number of Finsider's assumptions. Thereupon, acting on this report, the World Bank authorities decided to offer Finsider a loan half the size of the one requested.[13]

Impatient with the World Bank, in late 1948 Finsider addressed their application to the ERP. The reception they received from the European Cooperation Administration (ECA) was, however, very mixed. The ECA mission to Rome, headed by David Zellerbach, initially decided to reject their request altogether. After some reflection, however, they decided to call in Wayne Rembert to act as consultant, this time on their behalf. Not surprisingly Rembert set out to restrict allocations to Finsider while allowing greater scope for loans to a number of private companies. In the latter regard he was indeed successful. Finsider, however, refused to give up and took its case to the ECA's Washington headquarters. The Washington people were having their own differences with Zellerbach at the time and were better disposed towards mass production and expanding markets, so they gave Finsider's case a somewhat more sympathetic hearing.

In the early summer of 1949 the ECA in Washington had approved Fiat's reconstruction plans, which were predicated on receiving coils from Finsider. Finsider, in the meantime, was pressing its case very hard, relying on the backing and the advice of two influential American companies, the engineering consultants Arthur McKee of Cleveland and the steel makers ARMCO (American Rolling Mill Corporation) of Middletown, Ohio. The issues at stake were basically two. Should Finsider be allowed to install a wide-strip mill and which kind of strip mill and, second, if affirmative, where should the mill be installed, at Cornigliano as Finsider were asking or at Piombino, which would have meant

---

[12] International Bank for Reconstruction and Development (IBRO), *International Bank for Reconstruction and Development 1946–1953* (Baltimore: Johns Hopkins University Press, 1954).

[13] A detailed account of negotiations surrounding the Finsider loan can be found in Ranieri, 'Il Piano Marshall'.

that Italy's coastal integrated plants would be restricted to two? The talks were resolved in the autumn of 1949 to Finsider's entire satisfaction: a new modern wide-strip mill was earmarked for Cornigliano.

The main positions to emerge during what was a very intricate negotiation process were three. First, there was the total opposition to the Sinigaglia Plan, inspired by the private industrialists and supported to some extent by the Rome ECA Mission and by Zellerbach. Second, there was the intermediate view held by Rembert, who favoured a reduced version of Finsider's plans, and finally there was Finsider's position, reflecting Sinigaglia's modernizing agenda but also increasingly shaped, in so far as practical modalities were concerned, by ARMCO.

Radically opposed to Sinigaglia's plans was the view that broadly reflected the arguments laid out by the Falck group. It was a view inspired by a strong bias against the state-owned sector, accused of being cushioned by state orders and unduly favoured by government policies. Such an argument fitted well with the free trade inclinations of a number of American officials. More fundamentally it disputed the very fact that Italy could successfully manufacture blast furnace steel, given her lack of raw materials and high import costs:

The Italian experience has not been particularly satisfactory with integrated plants. It was through a series of financial crises and bankruptcies that the Government intervened and secured many of the steel plants that are now in the Finsider group.

Letting Finsider proceed with its plans would almost certainly result in 'high tariff protection' and cartelization—(an accusation which Sinigaglia rejected, claiming that it was his opposite numbers in the private sector, as well as in Ilva, who had always acted against liberalization and competition). On the other hand there was much to be said for encouraging the private sector:

there are sixty odd companies in Italy making and/or rolling steel. Obviously these concerns vary in size from small, obsolete, special process plants to the larger, well organised ones. A small number of companies produce a certain amount of pig iron via electric furnaces and iron pyrites. The principal tonnage however, is the production of steel through cold charge in open hearth and electric furnaces. Some of these companies are well organized and make a very definite and worth while contribution to the Italian economy. They are long established and have over a period of years made substantial progress. In some instances they have their own hydro-electric development for power and high tension transmission lines bring this electricity to their steel plants . . . Italian rolling mill equipment is by American standards outmoded. Its condition is poor, which reflects upon the both the quality and quantity of production. It should be pointed out, however, that the type of equipment very generally lends itself to the pattern of Italian requirements. The facilities are flexible, which is essential to the varying and small quantities of Italian industry.

The quotations above are taken from a long report drawn up by Charlie Moore of the ECA Rome Mission.[14] Somewhat awkwardly this report tried to

[14] WNRC, RG 469, I.D. Of.D., S.f. 1948–51, Box 17, 'Steel Branch 1949', Memorandum by C. E. Moore, 'Finsider Reconstruction and Modernization Plan', 10 Jan. 1949.

project a favourable image of private steel producers, while at the same time acknowledging the deficiencies in their equipment and the shortcomings of their production, which were all too apparent in American eyes. Small here was not beautiful, but then Italy should restrict her industrial ambitions firmly within a kind of 'second tier'. The recommendations following this analysis were just as explicit: Finsider's plans were too ambitious and should be dropped. As an alternative, US aid should be channelled to a variety of rolling mill plants:

For example presently existing semi-continuous rolling mills could, by only adding some new stands, be made into continuous mills with a very important saving of capital . . . There should be an expansion of existing steel works, with particular consideration to electric furnaces producing quality steel, and their subsidiary departments. These include cold rolling, wire drawing, bolts and nuts, etc. This has been a specialty of independent steel companies. Their products include less material and more labor and are products in which Italy can successfully compete. They are most essential and useful for the Italian economy . . . The scarcity of capital clearly indicates the benefits of aid to the scrap processing plants. These small and independent companies are more elastic. They are more accustomed to the requirements of Italy and it is felt that they would do a far better job . . . than would be done under the proposed new expansion. When a modest crude steel production proves insufficient for her needs, Italy could supplement this with ingots, billets, blooms and even finished rolled products from other OEEC countries and America.

Extreme in its conclusions and at the same time very naïve—the Falcks themselves would not have gone so far—this position met with strong opposition within the ECA itself. Wayne Rembert pointed out that an entire nation's steel industry could not be based on scrap and that Italy's requirements of finished steel shapes were much more varied than Moore's report suggested. Rembert believed that Finsider's plans were fundamentally correct in their desire to modernize coastal integrated plants, working them to full capacity, and endowing them with new rolling mills of a continuous type. This is how a modern steel industry should be run. The problem with Finsider's plans, according to Rembert, was that they were too ambitious and would have led to overcapacity. He also thought that they were too expensive and would have required a degree of managerial skill that a public company was unlikely ever to acquire.[15]

Rembert's prescription therefore sought to address these points. First, he insisted that Finsider's plans should be based on a national output not exceeding 2.5 million tonnes per year. This was a very low ceiling, considering that apparent consumption in Italy had already exceeded that figure in the late 1930s. On the other hand, it could not entirely be dismissed if one took into account the unstable prospects of the later 1940s, when demand was flagging after the cyclical reconstruction boom. Finsider themselves, although much more buoyant (and rightly as it turned out) on the long-term prospects of the market, were

---

[15] The Rembert Report is in WNRC, RG 469, Osr/E, Ird, GF 1950–53, Italy Box 15, IBRD, 'Report on Specific Projects in Italian Loan Application. Part II, Finsider Steel Mill Projects', prepared by Wayne Rembert, 19 Aug. 1948.

not without a certain amount of caution over the possibility of over-capacity in the short and medium term. Second, Rembert wanted Finsider to scrap their plans for Cornigliano and to concentrate modernization in Bagnoli and Piombino, the latter to be equipped with the wide-strip mill originally meant for Cornigliano. The cost of building Cornigliano was going to be huge, Rembert argued, and production costs there would still be too high to meet international standards. Third, Finsider should be discouraged from installing cold-rolling, tinplating, and other coating units, which should be left instead to private companies such as Fiat, La Magona, and Falck. Only private company managers could be trusted with these plants, which required a good feel for the market.

Rembert's prescriptions were perhaps the most insidious from Sinigaglia's point of view, for they also endorsed mass production and standardization, but projected them in such a way as to undermine Finsider's ambitions. Significantly, for example, Rembert rejected the idea that Bagnoli should install a narrow-strip mill, a technology which had been used in the inter-war period and which some European firms still seemed to prefer. Narrow strip should, Rembert said, be obtained by cutting the product of the wide-strip mill. Sinigaglia agreed with him, although Ilva's executives did not and, while their request for an ERP-funded narrow-strip mill was rejected, they still managed to install one some years later. Also many of the points that Rembert raised about relative costs at Cornigliano and Piombino were probably not far off the mark. Furthermore, Rembert's views were shared by many US officials in the World Bank and the ECA as well as by members of the business community. In the autumn of 1948, for example, a number of American steel industrialists, including Clarence Randall of Inland Steel Corporation, Mr Wysor of Republic Steel, and Mr Ascarelli of US Steel, were consulted on the Italian loan applications and endorsed his recommendations.

The core of Rembert's thinking was that Finsider's two large plants were to be seen as 'base load' plants. In other words, they would be large enough to satisfy domestic demand when the market was slack, whereas in more buoyant times their output would have to be supplemented by imports of semi-finished steel. The straitjacket around Finsider therefore would have been twofold. On one side, not being able to fulfil the needs of the domestic steel market, they would have to give ground to private firms, allowing them to re-process imports. Second, being unable to expand their finishing facilities to cover the most profitable trades, they again would have lost out to private producers. The benefits of their partial Americanization were to be reaped by others. What Rembert wanted to avoid at all costs was that Finsider should become a profitable, successful company able to outmanœuvre its private competitors: he did not believe that they could do it, nor did he want them to try. Despite, therefore, his attempt to couch it in merely technical and financial terms, Rembert's argument rested essentially on a prejudice against Finsider—a fact that became clearer in the following months as his position weakened, and he was forced to come down to his bottom line.

The strength of Finsider's argument lay first in the fact that their plan had become official state policy; second, in their links with Fiat; and third, in the dogged commitment and ability of Sinigaglia and his team. Technically speaking the idea of integrated coastal plants was a sound one, and one that was to prove immensely successful around the world. Sinigaglia had been on record for some time about the gains stemming from rationalization and mass production. Well before the prospect of American aid, he had been an Italian prophet for Americanization:

One of our main faults, derived from the German practice—reads a memo of 1946 certainly inspired by Sinigaglia himself—is for each plant to want to produce its own steel to feed its rolling mills. It would be much better to follow the American model, and use finishing mills to roll billets of the right size, adequate to the final shape which is required . . . These plants . . . should be located next to the steel users, so that they can enjoy cheap transport costs and fulfill market requirements. On the other hands it is necessary to concentrate billet production in large integrated plants, employing specialized mills of high capacity.[16]

Contacts with American producers were instrumental in refining and articulating these concepts. The first drafts of Finsider's plans contained a preference for Bessemer basic (Thomas) converters in both Bagnoli and Cornigliano. Only later, under the influence of ARMCO, were open-hearth furnaces prescribed for Cornigliano. Similarly, ARMCO advised on the layout of the new plant there and on the characteristics of the wide-strip mill to be installed. In operating a continuous mill throughput was essential, so that the right balance was required between the melting shop and the hot and cold rolling units.

Finsider were fast learners. Their own contribution lay essentially in the strength of their conviction about future market growth. Their case rested on the fact that per capita consumption of steel was comparatively still very low in Italy. Moreover the country's industrial apparatus, particularly the engineering industry, had been expanded and partly modernized during the late 1930s and the war years. There was no reason why that drive should not be carried forward, nor why it should not extend to trades requiring thin flat products, such as domestic appliances (for thin sheet), the automobile industry (for sheet and coils), the canning industry (for tin plate), or the construction industry (for galvanized sheet).

The attraction of the American model for Finsider went further, however. It provided an outlet for the deep antipathy nurtured by Sinigaglia for restrictive cartels as well as for the shady deals between the private and public sector, which, in his view, had dominated the industry in the inter-war period and which he blamed essentially on Ilva. All this could now be finally replaced by a competitive drive for low prices, provided the new Finsider company at Cornigliano were allowed to service end-users in the market. On this point,

---

[16] ACS, Archivio Storico IRI, 'Vitalita' della siderurgia nazionale', Rome, 10 May 1946, edited by Guido Vignuzzi.

there was a full meeting of minds with ARMCO. In their report for the ECA presented in the summer of 1949, ARMCO stated that:

with the existing plant reconstructed along modern lines, with modern rolling equipment and the high type of operating personnel and efficient management this location can maintain, Cornigliano should be the most economical quality steel producer in Italy . . . We would recommend all products produced from the hot and cold strip mills be finished to the most complete stages within this plant, including tin plate. This method will allow the utmost economy and lowest production costs.[17]

Installations at the new plant should be comprehensive: they were to include a new 80-inch semi-continuous wide-strip mill, pickling units, a cold-strip mill, annealing equipment, zinc grip, and tinplate lines. It was to be modelled on the Irvin Works near Pittsburgh, a greenfield site developed just before the war by Carnegie–Illinois Steel Corporation, which belonged to US Steel. While the initial capacity would be around 500,000 tonnes, the layout would be such that, at a later date, it would be possible to double it by adding one or more stands turning the mill from semi-continuous to continuous. ARMCO's recommendations seem to have been followed to the letter.

It is interesting, finally, to consider the fate of the DEMAG mill, the order for which had been passed over to Finsider from Fiat. The mill had been originally commissioned by Fiat before the war, and reflected the current efforts of European producers to adapt new American wide-strip mill technology to smaller production runs and narrower markets. The output of the mill was supposed to reach 250,000 tonnes, a good 30 to 50 per cent below the level of the smallest semi-continuous US strip mill on offer. Based on a modified version of a United Engineering & Foundry patent, it fell short of providing the wider variety of coils and sheet demanded by European car-body producers. The spacing between the stands of the mill, moreover, was very conservative and there were only four finishing stands instead of the standard six. This meant that the gauge of the coils would be above the norm, therefore requiring a larger amount of expensive cold rolling.

All these points emerged from the detailed report drawn up by consultants Arthur McKee for the benefit of the ECA. Much of the DEMAG mill remained to be done, and the cost would be substantial. The real question, however, was another one; if the mill were to be finished it would have to meet US standards and specifications. How could this be achieved, given that no US company was prepared to hand over its latest designs to a German producer, and given that DEMAG itself did not have enough experience in wide-strip mill plant to proceed on its own? In the event, Finsider was all too happy to conclude that the DEMAG mill was not a viable option and to switch over to demanding a new wide-strip mill of genuine US make and design. It finally secured it from plant suppliers Mesta Machine Co. of Pittsburgh.

[17] WNRC, RG 469, Osr/E, Ird, Iss, GF 1950–53, Italy, Box 16, 'Report on Iron and Steel Manufacture for Societa' Finanziaria Siderurgica per Azioni', by Armco International Corporation, Middletown Ohio, Sept. 1949.

MARSHALL PLAN AID TO PRIVATE PRODUCERS

Interesting observations about Americanization and the adaptation of intermediate technologies can be also gleaned by examining the use made of Marshall Plan loans by the two largest private producers, Fiat and Falck.

Fiat's steel production was carried out in two plants: the main one was at Ferriere, located next to the firm's automobile factory in Turin; it had a smelting, refining, and finishing capacity. The other one was at Avigliana, which specialized in the finishing end of the business. At the end of 1948 Fiat's capacity was estimated at 46,000 tonnes of pig iron (produced in electric furnaces) and 250,000 tonnes of steel. Fiat used its steel to cater for the requirements of its large and growing production of buses, truck, cars, diesel engine, and aircraft. In fact the steel loans, although treated by the ECA as separate applications, were linked to the larger loans designed to renovate Fiat's mechanical equipment. The first instalment of Fiat's steel loan, for example, amounting to over $4 million, cleared ECA authorization in August 1949, one month after the approval of a $14 million loan for its automobile plants.[18]

Fiat was one of the main Italian beneficiaries of the ERP programme. This is not surprising since the company was highly rated by the Americans. Its management, part of which had been trained outside Italy and enjoyed good international connections, was considered of high quality, and Vittorio Valletta, the company's President, enjoyed the confidence of many US automobile producers. Contacts in both the technical and the financial field stretched back into the inter-war period: for example Fiat had been the beneficiary of a $10 million loan through J. P. Morgan in 1926. In terms of Fiat's steel investment strategy, however, Americanization was not unqualified. In 1939 the company had commissioned a large amount of equipment from the German plant supplier, DEMAG. Apart from the strip mill considered above, other items in the same order were a blooming mill, a universal roughing mill and an 80-inch cold-strip mill. The orders were not discontinued after the war; rather, Fiat sought to re-jig this equipment by combining it with a number of US-made items: for example the cold-rolling mill was supplemented with strip reels and coil-handling equipment from ARMCO and United Engineering. Although the DEMAG order was invoiced in dollars, Fiat received no assistance for it either from the ECA, or from the World Bank. It would have therefore been advantageous for the company to switch its orders to a US manufacturer and the fact that they did not do so suggests that they found the German equipment attractive. On the other hand, just as Finsider had done, they called in ARMCO as consultants to assist in the purchase and installation of the machinery and in the handling of the cold strip mill.

---

[18] For Fiat's investment programme see WNRC, RG 469, SRE, Ird, Iss, Geog F. 1950–53, Italy, Box 15; see also D. Velo, *La strategia Fiat nel settore siderurgico* (Turin: Gruppo Editoriale Forma, 1983); on overall company strategy see Bigazzi's chapter in this volume.

Most of the remainder of Fiat's steel modernization was dependent on new US equipment: new electrical and open-hearth furnaces, wire drawing plant, new mechanized plant for bolt and nut production. Despite the deal with Cornigliano that was meant to supply most the bulk of its steel requirements, Fiat, fully endorsed by the ECA, did nothing to scale down its own steel-making capacity. On the contrary, new electrical and open-hearth equipment was installed. As a result a crude steel surplus emerged and the company sought new market outlets. In 1954 they purchased a Sendzmir Mill patented by ARMCO for cold reduction of high-grade stainless steel sheet with a capacity of 72,000 tonnes per year. This opened up new market opportunities, allowing Fiat to compete with established Swedish and French producers.

A. F. L. Falck's position on the steel market was much more vulnerable than Fiat's. Initially the company had hoped to stop the implementation of the Sinigaglia Plan. Their first application to the ECA in early 1949, therefore, had been a very limited one, based on the request for a few new electrical furnaces and miscellaneous accessory equipment. There was no drive towards specialization and the company in fact seemed proud to restate its commitment to churning out, using traditional methods, the widest possible variety of shapes of all sizes. In the light of what was to emerge on the state of their plant, their initial lack of attention to new technology was indeed remarkable.

The company's main plant was the Unione one, at Sesto San Giovanni, in the vicinity of Milan. It was provided with electrical blast furnaces and electric and open-hearth steel capacity. Also near Milan were most of its other plants, which were engaged in a variety of finishing trades or in ferro-alloy production. Another member of the Falck group, although organized as a separate company, was CMI (Cantieri Metallurgici Italiani), a tin-plate producer situated in the neighbourhood of Naples. The group's strength lay in its high output, its proximity to steel users, and its vast self-generated electrical supply. The work-force, numbering 12,700 workers and over 2,000 clerical staff in 1950, had a reputation for being skilled and committed. The company's top management, composed, among others, of members of the Falck family, was influential in political circles and also well known among European steel industrialists, having been, during the inter-war years, the spokesmen of the Italian steel association.[19]

When the battle was joined, however, between 1948 and 1949, the Falcks proved unable to stop Finsider from going ahead with their plans. Indeed Sinigaglia was sufficiently magnanimous to offer them a stake in Cornigliano, which they refused. Instead the mood of the company's executives turned to bitterness, particularly towards the ECA, which they blamed for having given in to the powerful Finsider–Fiat combination.

---

[19] On A. F. L. Falck's investments see WNCR, as above and Carte Osti, Box 5, 'Acciaierie & Ferriere Lombarde Falck', 2nd ERP, Year 1949–1950. See also M. Pozzobon, 'La siderurgia milanese nella ricostruzione (1945–1952). Ristrutturazioni produttive, imprenditori, classe operaia', *Ricerche Storiche* 1 (1978), 277–305.

The company's application for a second instalment of ERP loans, handed out in March 1950, marked a decisive U-turn and reflected a real sense of urgency. It was based on extensive purchases of US rolling-mill equipment, including a new four-high heavy plate mill, a new cold-strip mill patented by United Engineering and ARMCO, new equipment for seamless tube rolling, and machinery to ensure automation of the rolling passes in the wire and rod mill. The new plate mill was scheduled to increase the rolling speed by four times, replacing an older mill allegedly in such poor condition as to require 47 passes to secure the required gauge. Equally the new cold-strip mill, of the narrower 48-inch kind, would take the place of a number of old, one-stand, hand mills.

The company had increased its hot-rolling capacity for thin flats by purchasing in 1942 a narrow-strip mill of German design, which, at the time, was considered the most modern facility for thin flats in the country. This narrow-strip mill's capacity was estimated at around 60,000 tonnes, so that it was thought that, in order to meet further demand and keep the cold-rolling mill fully employed, its output would have to be supplemented by a certain amount of external inputs of sheet and coil. The mill, moreover, although mechanized, was not of the fully continuous type and its productivity was in no way comparable to that of the American models. Also, the fact that it was of the narrow type, only suited for tin-plate production, cut Falck out of the profitable market for auto-body sheet. In other words Falck's thin flat capacity, although later bolstered by a new American cold-strip mill, was tied down by the company's former commitment to the intermediate technology developed by European plant suppliers in the inter-war period in response to US wide-strip mills.

A more radical shift to American technology took place at CMI, where hand-pack mills were replaced with a new American cold-strip mill and other finishing and coating facilities. The company's steel-smelting and rolling facilities were thus rendered obsolete at one stroke. The most likely source for the coils was obviously Cornigliano, which could have shipped them down along the Tyrrhenian coast. Faithful to their opposition to the Sinigaglia Plan, the Falcks, however, refused to accept Finsider's deliveries, preferring to purchase US supplies. Eventually Italian government officials forced them to comply, by raising the tariff on imported coils, thus proving that Falck's fears of a public sector monopoly were not wholly misplaced.[20]

Speaking to ECA officials in June 1950, A. F. L. Falck's Director General pointed to the fact that in order to compete with Finsider, Falck would be forced to concentrate on special steel, particularly sheet and plate, provided they could complete the replacement of their older rolling mills.[21] However this was not a fair reflection of the company's strategy. The modernization of their rolling mills was extensive and it increased their capacity by over 50 per cent across a

---

[20] See M. Lungonelli, *La Magona d'Italia—Impresa, lavoro e tecnologie in un secolo di siderurgia toscana (1865–1975)* (Bologna: Il Mulino, 1991), 119–22.

[21] See Plans and Problems of S. A. Falck Steel Company, 29 May 1950 in WNRC, RG 489, DQO, OEO, I.D, D. F. 1948–1954, Box 17, 8.51 'Iron and Steel'.

wide spectrum of finished shapes. The financial effort required was also considerable. In other words it does not appear that Falck was carving itself a niche in the market, rather it was chasing the game by injecting new US technology into its large, albeit rather outmoded, rolling-mill plant.

Falck sent technical missions to America, as well as to West Germany, Belgium, and Britain. They found, however, that European plant suppliers had very long delivery times, and offered installations that were not equal to the American ones. Nevertheless Falck's commitment to Americanization remained very tentative. Among other things, and in contrast to Finsider and Fiat, they made no use of US know-how in either selecting or operating their new units.

8.4 Importing the 'American Model': From Cornigliano to Italsider

PERCEPTIONS OF THE 'AMERICAN MODEL'

On balance one is struck by the amount of new plant of American design that was introduced into the Italian steel industry. This was true for steel-making units, particularly electric furnaces of the Lectromelt kind, for open-hearth furnaces, and, even more, for rolling mills. In most cases it was a case of introducing new plant of a continuous type, of which there had been little previous experience in Italy. This was nearly always the case for flat products, as in Cornigliano, Falck, Fiat, and also in the case of the hot-rolling mill for electrical sheet at Terni. New continuous-mill American technology also affected long products, as for example the new rod mills at the Rogoredo plant of the Redealli group, at Sisma, and at Ilva's Bagnoli plant. New blooming and roughing mills were also purchased in the USA, such as the slab and billet mill by Ilssa Viola, a small firm specializing in special steel sheet, and the large semi-continuous roughing mill installed at Bagnoli. Everywhere pre-existing rolling-mill capacity seemed to be largely outdated, with low levels of mechanization. The hope that A. F. L. Falck seem to have briefly entertained, of muddling through with this kind of equipment, speaks for itself about the company's aims and strategy.

A different question is whether the injection of new US plant succeeded in bringing about standardization and rationalization. The answer here is partly affirmative, at least for thin flats. When the coils from Cornigliano hit the market, a number of former producers were unable to compete and had to leave the trade. This was true, for example, for the Bruzzo firm near Genoa, as well as for Terni, where production of common grade sheet was discontinued and replaced by electric sheet. The fact, however, that there was still room for Falck to hang on to its narrow-strip mill and for Bagnoli to install another one a few years later, shows that the rationalization was far from complete and alternative technological and market strategies could still be entertained.

On a slightly different level were the attempts to combine intermediate technologies developed in Europe with new larger-scale American ones, as in the case of Fiat's DEMAG orders. In the field of both narrow and cold-strip mills progress among European plant suppliers in the inter-war years had been considerable, although this was not the case for wide-strip mills. Another good example of the different plant layouts available in Germany and the USA is provided by alternative bids to supply Redaelli with a new rod mill. The equipment offered by SIEMAG entailed the output of smaller-section ingots; the mill was not wholly continuous; for some operations such as looping it would still have required a good amount of manual work, which could only be accomplished with skilled labour. Finally, there would have been more stands, not all of them in a straight continuous line, however, as in the US-designed mill.[22]

Although they usually opted for the US technology, Italian producers remained somewhat cautious. In the case of Cornigliano for example, at least initially, there were doubts as to whether the market would be able to sustain large output levels. The wide-strip mill chosen was of a semi-continuous kind, and the company executives hedged their bets further, by ensuring that the mill's single reversing roughing stand could also be used to turn out plates, for delivery to Campi, a nearby specialized steel plant supplying the shipyards. In the event, however, this proved hardly necessary and the demand for sheet and coil picked up together with the country's industrial boom.[23]

In other sections of the trade, there is little evidence of rationalization: in fact there appears to have been much patch-up investment. Eventually private companies, with the exception of mini-mills, tended to specialize in high-grade steel, where they did not have to face the challenge of the integrated Finsider, but their conversion was a very lengthy one. On the whole US loans helped to push through a large and untidy expansion, well above the fictitious targets with which Wayne Rembert had tried to give Finsider a thrashing.

American plant suppliers clearly enjoyed considerable success in Italy during the early 1950s, and the number of companies involved was very large. The largest orders were taken up by United Engineering & Foundry, Mesta Machine Co., Lectromelt (for electric furnaces), and a few others. The most complete attempt at Americanization in the industry, with access to US patents and know-how as well as technical assistance, was the one carried out in Cornigliano. After this came Fiat's new departure in cold reduction of auto-body sheet, Terni's new electrical sheet department, and a few others.

Sinigaglia was one of the most committed Americanizers in the field. He, and his group, clearly thought in terms of introducing the American model to replace what they considered the outdated and inefficient 'German' or 'North European' model, prevalent during the inter-war period. Undoubtedly there was a degree of simplification in this paradigm but some of the points were well

---

[22] A detailed report on the Redaelli mills in 'Soc. p. Az. Giuseppe & Fratelli Redaelli, Milano', 23 Jan. 1950, WNRC, RG 489, SRE, IRD, ISS, G.F. 1950–53, Italy, Box 15.

[23] Fears about over-capacity were expressed by Marchesi and others at an important meeting about the construction of Cornigliano—see Archivio Storico Ilva, Fondo Redaelli, 007-4062, F. Petit a Mario Marchesi, Rome, 8 Feb. 1951, with minutes of Riunione presso la Finsider sul programma di impianti e di finanziamenti dello stabilimento di Cornigliano, 2 Feb. 1951.

taken. In the German model the emphasis had been on the skill of the workforce and the technical personnel. Steel companies were often part of very large conglomerates, which had come together as the result of both horizontal and vertical integration strategies. They were conceived and organized as large workshops, turning out a number of different shapes in small batches of different size and quality. Many Italian firms, like Falck, Terni, Breda, and Ansaldo had indeed modelled themselves on this pattern. In other large and more basic producers, such as Ilva, reliance on Thomas converters encouraged production of large quantities of cheap semi-finished goods, which were supplied to rerollers.

Adopting the American model, on the other hand, meant opting for economies of scale and designing new plant geared for continuous throughput. Process innovation would be accompanied by product innovation at the finishing end. According to Sinigaglia, standardization and specialization were to go hand in hand with a greater involvement in marketing, directly approaching prospective customers, in an attempt not just to supply, but to 'make' the market.

How the 'American Model' was Introduced

The adoption of American managerial and organizational techniques during the 1950s and 1960s was largely a consequence of the importing of new technologies based on continuous throughput and mass production, analysed above. The evidence that will be presented here focuses on Cornigliano and Italsider and does not, therefore, take into consideration other cases such as Terni or Fiat, that might have offered comparative perspectives, nor does it discuss firms in which the American model was not adopted. Nevertheless the companies considered were among the largest and most successful. Their adoption of the American model was not an isolated phenomenon; rather it reflected wider trends in Italy's industrial culture. During the 1960s the influence of the American model declined, to be partly replaced by other influences, particularly the Japanese one.[24]

The critique of Finsider's modernizers extended to the way most Italian steel companies were organized and run. It was believed that most companies were sheltered by cartel arrangements, or by local or national monopolies. Large conglomerates such as Ilva had developed as large bureaucracies rather than as

[24] This section draws upon Osti, *L'Industria*, as well on as F. Amatori, 'Cicli produttivi, tecnologie, organizzazione del lavoro—La siderurgia a ciclo continuo integrale dal piano autarchico alla Italsider (1937–1961)', *Ricerche storiche* 3 (1980), 57–611; further discussion of the impact of the Productivity Missions and the Marshall Plan in Italy can be found in P. P. D'Attorre, 'Anche noi possiamo essere prosperi: aiuti ERP e politiche della produttivita' negli anni Cinquanta', *Quaderni storici* 20, 58 (1985), 55–93 and L. Segreto, 'Americanizzare o modernizzare l'economia? Progetti americani e risposte italiane negli anni Cinquanta e Sessanta', *Passato e presente* 14, 37 (1996), 55–86; see also R. Ranieri, 'Learning from America. The remodelling of Italy's public sector steel industry in the 1950s and 1960s', in M. Kipping and O. Bjarnar (eds.), *The Americanisation of European Business. The Marshall Plan and the Transfer of US Management Methods* (London: Routledge, 1998), 208–28.

centres for entrepreneurial or managerial innovation. Throughout the industry there were no formal training programmes, accounting techniques were primitive, decision making was highly hierarchical, and industrial relations were poor, low pay being matched by low productivity.[25]

Clearly the Sinigaglia Plan called for the reform of such practices, and particularly so in Cornigliano. In Cornigliano the entire production process required careful planning and monitoring from the centre. As a result managers were to take a greater role at all levels of the firm, while the work-force needed to conform to stringent standards of automation and regularity. Costing of the plant's operation on one side, and marketing of final output on the other, were bound to play an important role if the company was to be profitable. There was an idea that Cornigliano, like other public-sector companies, might eventually break away from state control, attracting a larger share of private capital and becoming an independent public company.[26]

Missions to the United States

A number of trips by leading public-sector executives to the Unites States prepared the ground for the intense process of assimilation that was to follow. An important mission, for example, was the one carried out in October 1952 by four top Ilva executives, with visits, among others, to the headquarters of US Steel, Bethlehem, ARMCO, Inland Steel, and Kaiser Steel Corporation. It led to a long and detailed report which highlighted managerial and technical differences between the US and Italian steel industries. It was couched in a sober and rather neutral tone, devoid of any particular enthusiasm towards the American model.[27]

Among the aspects of US corporate organization which the Ilva team singled out was the amount of statistical detail produced, and the way statistical information was disseminated throughout the company. This allowed extensive and detailed planning to be conducted within the operating departments. It also allowed budgeting and costing at all levels, with each unit being assigned a standard cost. Another difference with Italian companies lay in the importance American corporations assigned to their sales department. What particularly struck the Italian executives were Market Development Offices staffed by specialists of the using industries. Also unknown in Italy were Public Relations Offices.

[25] On the backwardness of Italy's industrial culture see, among others, M. Martinoli (ed.), *La formazione del Lavoro* (Bari: Laterza, 1964); for a Finsider view see G. L. Osti, *L'Industria*, 146ff. See also G. Sapelli, 'Gli organizzatori della produzione. Tra struttura d'impresa e modelli culturali', in *Storia d'Italia, Annali 4* (Turin: Einaudi, 1981).

[26] Sinigaglia had always been particularly insistent on the need for training, or as he put it, to raise the 'cultural level' of the technical and managerial staff. He soon convinced himself of the need to draw on superior American experience in management, strategy, and organization. Notes on Sinigaglia's life and personality can be found, among others, in P. Rugafiori, 'I gruppi dirigenti della siderurgia "pubblica" tra gli anni Trenta e gli anni Sessanta', in Bonelli, *Acciaio*, 335–68.

[27] Archivio Storico Ilva, Fondo Redaelli, 027/4089, Box 3, 'Relazione sull'industria siderurgica Americana', 5 Nov. 1952.

The Ilva team was impressed with the sophistication of the commercial sector. US steel makers, seeking to confine themselves to supplying bulk standardized orders, quoted large 'extras' for small orders and non-standard specifications. Stockholders, however, possessed a large array of very sophisticated cutting and finishing equipment in order to deal with small batches. In Italy extras were not much used and were in any case much smaller, whereas the number of standard sizes was larger than in the USA. Stockholders, on the other hand, did not have the same profit margins and were much less well equipped.

Industrial relations in the US steel industry did not strike the Italian managers as being very successful, despite the high wages and good living standards of the work-force. In fact they found much concern among US managers about trade union militancy, frequent wildcat strikes, and persistent industrial action, well beyond the periods in which new contracts were negotiated. The pay structure seemed more rational and fair than that in Italy, there was less politicization among the work-force, and personal contacts were open and frank, but the picture was not wholly reassuring.

The message that the Ilva team drew from their visit was tinged with a degree of scepticism: they recognized that US corporations enjoyed some advantages, but to what extent were they simply a consequence of the larger markets and superior scale of output prevailing in that country? It was pointed out, for example, that large standard orders were a luxury for a small market like the Italian one, and that planning and budgeting were all very well when demand was sustained, raw materials plentiful, and prices stable; but how feasible were they in the much more volatile Italian conditions? Nevertheless the American methodical approach to organizational questions, their reliance on facts and figures, and their exposure to a competitive environment were seen as potentially stimulating for Italian companies.

More enthusiastic was the message drawn by another important mission carried out one year later by five top managers from IRI and Finsider, including Enrico Redaelli, one of Cornigliano's top executives. By 1953 Cornigliano had already established its own American links, with about seventy employees having completed a training stage at ARMCO. The significance of this particular mission, however, was of a broader nature: originally it was meant to study 'budgetary' techniques in US firms, but it was soon extended to cover most aspects of corporate management. The final report was a very thorough and prescriptive document. Most of its recommendations were in fact taken up, in the following years, particularly at Cornigliano.[28]

The mission was organized in conjunction with management consultants Booz Allen & Hamilton of New York. It started at their headquarters in New York with ten different lectures on various aspects of management science. This was followed by visits of one week each to ARMCO headquarters in

---

[28] Archivio Storico Ilva, Fondo Redaelli, 018/4080, Box 2, 'Concezioni aziendali e metodi direzionali negli Stati Uniti' (Missione dirigenti amministrativi Finsider negli Stati Uniti, marzo–aprile maggio, 1953).

Middletown and Inland Steel in Chicago and by shorter stays with Republic Steel, Globe Tube and Steel, and Pittsburgh Steel.

The mission's report recognized how US steel companies, even the better managed ones, which were thought to be ARMCO and Republic Steel, were not at the forefront of US corporate culture. The market for steel had been so buoyant for such a long time and profits so generous, that a degree of sharpness seemed to have been lost. Nevertheless there was much to learn and many of the American methods and techniques could be fruitfully transplanted to Italy. Some of them were indeed already well known, but they had not been applied with the same thoroughness. In particular what was distinctive in US corporations was the attempt to develop a company 'ethos', with which all the staff could identify.

Every aspect of corporate management was analysed. Attention was given to organizational hierarchies and distinctions between staff and line workers, to long- and short-term planning functions and 'profit goals', to budgeting and standard costing techniques, to the collection and transmission of statistical data and budgeting practices, to committee structure, job description, personnel training and motivation techniques, and finally to industrial engineering and marketing. Out of this detailed analysis a number of prescriptions were retained. According to the IRI–Finsider report, it was not good enough, as the Ilva executives had done one year earlier, to attribute superior American practices to the natural advantages which US producers enjoyed. Undoubtedly Italy's public sector steel had been largely reliant in the past on government aid and was thus less receptive to the profit motive, its factories were over-manned; planning was difficult because of price and market instability; personnel choice and personnel morale had been negatively affected by the war. All this, however, should not be taken as a justification to stifle reform, rather as further incentive to pursue it.

Urgent corrective action should be taken to increase the level of training for existing managers, by means of more missions abroad, more contact with research and university facilities, more formal instruction within the companies. Moreover, to better monitor the training requirements of all the staff, suitable job descriptions for each post should be drawn up. Company organization should be radically reformed in the direction of decentralization, there should be a clear separation between line and staff, and middle managers should be empowered in their particular sphere of responsibility. Here, it was said, Italian firms had much to learn, since in Italy too much power was exercised by too few people, leading to fewer controls, slower decision making, and insufficient managerial talent. Following the American example, on the other hand, executives were to be confined to executive functions, and not permitted to interfere unduly in the company's day-to-day management.

Long-term planning should also be gradually introduced, while annual company budgets should become an essential concern of the Board. Equally it was important to introduce standard cost procedures, and this would be greatly facilitated by improving industrial engineering, quality controls, and production plans. Furthermore the wage structure should be made fairer and more predictable by the application of job analysis and evaluation. More effort should be put into involving workers in the firm, providing them with information, calling them to participate in regular meetings with the staff, discussing their individual careers.

Finsider's report concluded by advocating a radical reform of the Italian steel industry on American lines. It went far beyond what the Ilva managers had observed one year earlier. In some cases, in fact, it took a different view. As we shall see, the managers of Cornigliano were the most assiduous in following up on some of its recommendations. Two separate agreements had been struck by Cornigliano with ARMCO, one giving them a general consultancy and another providing for staff training to be carried out at ARMCO's headquarters in Middletown. Training was offered to workers in line for promotion as foremen, to technical cadres, and to young clerical staff in line to be promoted as managers. Such training affected hundreds of people throughout the 1950s.[29]

ARMCO was not the only company to establish a relationship with companies in the Finsider group, although it remained a very important partner well into the Italsider phase. At a slightly later stage US Steel was brought into the picture to act as consultants. Also important was the agreement with Jones & Laughlin for the training of operatives for the oxygen converters that were installed at Taranto.

Organizational Changes at Cornigliano

In the following years, Cornigliano modelled itself along strictly functional lines, placing managers under the authority of a central office.[30] Great care was taken to draw up job descriptions for all positions and to draw a clear division between staff and line employees. The General Secretary co-ordinated Staff and Personnel and acted as a kind of administrative supremo. He also controlled the sales department. The production and sales departments were closely integrated and a strong emphasis was put on marketing.

Organizational problems became more complex in the case of Italsider. Italsider took over most of Finsider's plants, including the new one under construction at Taranto. The merger was designed as a gigantic effort in centralization, which effectively ruled out the option of creating a multidivisional corporation. It was decided, instead, to pursue the functional model, albeit allowing some scope for flexibility by creating separate divisions for specific tasks. Italsider's executives formulated their plans after having looked at a number of American corporations, but the model they followed more closely was that of US Steel, which also operated over two-thirds of its facilities in the

---

[29] Exchanges with ARMCO are fully described in Osti, *L'Industria*, 144–8.

[30] For more details and bibliography of the innovations introduced at Cornigliano and Italsider see R. Ranieri, 'La grande siderurgia in Italia. Dalla scommessa sul mercato all'industria dei partiti', in Osti, *L'Industria*.

functional mode. Thus Italsider concentrated all its staff operatives at its head-quarters, so that each plant retained only a production and a sales office. Finsider, on the other hand, was supposed to restrict its tasks to the financial field.

Organizational reforms increasing the scope of planning and marketing were bound to lead to a large number of new managerial appointments. University graduates, often with a background in the social sciences and the humanities, were increasingly chosen to fill a number of positions. Personnel offices were entrusted to carry out extensive training programmes. Cornigliano initially relied on the 'training within industry' programme initiated by the Marshall Plan, with separate sessions for directors, middle managers, supervisors, and foremen.

Further organizational changes, modelled on American methods, included the introduction of a General Contractor, standard costs, and management incentive plans. The General Contractor was a firm that specialized in co-ordinating plant extension and renovation, thereby relieving plants of the need to do their own construction work. The General Contractor was introduced as a result of the massive constructional effort required at Cornigliano. Mario Marchesi, the company's Chief Executive Officer, decided that the staff involved in building the plant should be taken on by the company Innocenti. Further amalgamation led to the creation of Italimpianti, which acted as a specialized contractor for the whole of IRI.

The practice of applying 'standard costs' involved turning each production unit into a budgeting centre, thus generating separate accounting departments and forcing each of them to link up with the corresponding technical unit. A standard cost was set for each operation, so that the relevant unit's performance could be appraised. Fiat, which had also done its homework in the USA, had adopted the system in 1928. Cornigliano applied it from the very start and seemed to operate it effectively, speed of appraisal being supplemented by flexibility in the implementation of corrective measures. In the late 1950s the monitoring was made more effective by a new computerized central system, patented by Univac—the first of its kind in Italy.[31]

A further step towards empowering management was the management incentive plan (MIP), attempted at Italsider on the model of US Steel. The MIP consisted of involving managers in seeking efficiency gains leading to a reduction of standard costs within their units, against the award of a share of the company's attendant financial gains. This process was supposed to extend beyond managers to clerical staff and workers, creating pay differentials among the company's departments. By following the method employed at US Steel, Italsider

[31] See A. Baldisserra, G. L. Bravo, A. Luciano, A. Pichierri, and E. Saccomanni, 'Sistemi informativi e trasformazioni organizzative in un grande stabilimento metallurgico', *Quaderni di Sociologia* 21, 3 (1972), 249–346; on standard costs at Fiat, 'Fiat. L'applicazione del controllo budgetario', Rome, 16 Feb. 1953, Archivio Storico Ilva, F. Redaelli, 018, 4080, Box 2.

hoped to use MIP to mitigate the centralization with which their new company had been designed.

Another notable reform in Cornigliano was the introduction of job analysis and evaluation. Previous experiences with Taylorism in Italian industry had been very patchy, involving regulation of piece-work but no fundamental restructuring of the production line. In Cornigliano, on the other hand, every job was analysed and classified according to a scale of points. Points were assigned for each task, taking into account a number of different parameters, the aim being to define a value-free objective scale of remuneration. Clearly this was a sharp break with the craft system, which put a premium on the skill and experience of the individual worker. The condition that made this new system possible was the fact that workers for Cornigliano's new plant were recruited among unskilled operatives with no previous experience in the steel industry—many of them chosen among those employed in the plant's construction.[32]

The system proved fairly successful. One of the reasons for this was that wages were kept at a higher level, relative to the rest of the industry. Moreover management granted bonuses and productivity-linked pay rises to sections of the work-force. Jobs within point bands were brought under specified categories, which were then referred to the current pay scales—allowing trade union involvement. Initially applied only to manual workers, gradually job analysis and evaluation was extended to the white-collar force. Difficulties emerged when the system was carried over from Cornigliano to the rest of Italsider, a tenfold increase in the number of employees covered. Applying the system in the large integrated plants, based on continuous throughput, such as Piombino and Taranto, was relatively simple, despite the presence in Piombino of a militant Communist-inspired trade union. Difficulties arose, however, in some of the other Italsider plants, such as those at Trieste, Lovere, Savona, which were more in the nature of large engineering workshops, essentially based on skilled craftsmanship.

Cornigliano's management established a very close relationship with CISL, the Catholic-inspired trade union confederation. CISL (Confederazione italiana dei sindicati lavoratori), which had originated in 1948 as a splinter organization of the largely Communist Confederazione generale italiana del lavoro (CGIL), adopted an ideology based on efficiency and productivity-linked wage settlements. To some extent CISL's ideology was influenced by American thinking, which it combined with its own Catholic social doctrine. Attempting to increase its membership and seeking a wider legitimization among the work-force, it

---

[32] On the nature and origins of job evaluation in the USA see M. Quaid, *Job Evaluation. The Myth of Equitable Assessment* (Toronto: University of Toronto Press, 1993), and J. Stieber, *The Steel Industry Wage Structure. A Study of the Joint Union Management Job Evaluation Program in the Basic Steel Industry* (Cambridge, MA: Harvard University Press, 1959); on its application in Italy see Amatori, 'Cicli', 602–3, and F. Cai, 'L'esperienza italiana sulla job evaluation. Il caso Italsider', in *Ascesa e crisi del riformismo in fabbrica. Le qualifiche in Italia dalla job evaluation all'inquadramento unico* (Bari: De Donato, 1976). See also A. Fantoli, *Ricordi di un imprenditore pubblico* (Turin: Rosenberg & Sellier, 1995), 55ff.

sought to promote an essentially plant-based bargaining strategy.[33] The application of job analysis and evaluation, calling for very minute contractual agreements, favoured its approach. By 1962 decentralized bargaining at the plant level had gained the support of the other trade unions. Italsider was one of the first companies to switch over to the so-called *contrattazione articolata*, or articulated bargaining, whereby significant elements of the pay structure and working conditions were agreed upon at company and plant level.[34]

Other measures tending to Americanize relations between managers and workers should also be mentioned. At Cornigliano in particular there was a great effort to forge an inclusive culture, by bringing in innovations such as a suggestion box, by providing amenities and cultural facilities, by offering workers company shares, and by individual staff–worker contact. There was also an attempt to project the company's image: an Arts Director was appointed and the company promoted art shows and festivals, as well as financing new infrastructure around the city of Genoa.

LIMITS OF AMERICANIZATION

Cornigliano's reforms made a considerable impact on Italy's industries. Many other companies, in both the private and the public sector, followed its example. Reforms were carried out at Olivetti, Fiat, at Mattei's ENI, and at a number of other companies. By the time of the Italsider merger, a constituency of middle managers had developed, who pressed for the continuation and development of the reforms in the steel industry as well as in other sectors. During the later 1960s and 1970s, however, the momentum was lost. Old organizational patterns were reinstated; poor management and over-expansion led to a fall in profitability and a growing burden of debt. Among the developments affecting the American model at this later stage were the emergence of new technological trends, the partial failure of Italsider's organizational reforms, the deterioration of industrial relations, and finally the fading away of the drive for modernization as a distinctive feature of the public sector.

As far as technology was concerned the American model began to give way during the 1960s to the Japanese model. The US steel industry did not perform particularly well during the 1950s, losing its technological edge and its commercial dynamism to the point that many see those years as the beginning of the serious crisis which was to overcome it later.[35] Albeit belatedly, this fact began to

---

[33] On CISL see S. Sciarra, 'L'influenza del sindacalismo americano alla Cisl', in G. Baglioni (ed.), *Analisi della Cisl* (Rome: Edizioni del Lavoro, 1980); M. Maraffi, *Politica ed economia in Italia. La vicenda dell'impresa pubblica dagli anni Trenta agli anni Cinquanta* (Bologna: Il Mulino, 1991), 205ff.

[34] G. Giugni, 'Recent Developments in Collective Bargaining in Italy', *International Labour Review* 91, 4 (April 1965).

[35] P. Tiffany, *The Decline of American Steel. How Management, Labour and Government went Wrong* (Oxford: Oxford University Press, 1988); J. P. Hoerr, *And the Wolf Finally Came. The Decline of the American Steel Industry* (Pittsburgh: University of Pittsburgh Press, 1988).

dawn on Italian public-sector steel leaders. Nevertheless in the early 1960s important new partnerships were struck with a host of American companies: for example Terni and US Steel set up Terninoss to develop mass production of stainless steel flats. As consultants for the Taranto steelworks, however, Nippon Steel were chosen.

Taranto was a watershed for public-sector steel. Its capacity was doubled after the example of the huge Japanese greenfield plants and against the best advice of many of the most experienced Italsider executives. The adoption of the Japanese model was also problematic. The Japanese ethos of extreme loyalty to the company never seemed to fit in with the Italian ways, nor was there the same amount of cross-fertilization that had been allowed to develop with American practices. Moreover the adoption was selective: large-scale plants required flexible organizational structures and human resource policies to enhance teamwork and flexibility, whereas Italsider's system remained centralized and rigid.[36]

A second problem related to the structure of the steel market in Italy was the weakness of Italsider's commercial organization. Admittedly the marketing of steel from Taranto was made more difficult by the lack of flexibility of existing pricing rules which discouraged relocation of steel users next to the plant. Also steel demand remained, on the whole, fragmented and diverse—a fact which enabled independent stockholders to establish a strong grip on the market. In some cases stockholders were happy to undercut Italsider's prices by relying on cheaper imports. The commercial organization of Italsider proved inadequate to the task. The lesson of the American model pointed to the need for well-organized stockholders, able to service the small batch market with adequate finishing equipment (shearing, extra coating etc.). In Italy, given the prominence of the retail market there, Italsider would have needed to carry this further and secure a number of captive stockholders.[37]

Many have seen this failure as proof that Sinigaglia's intended reforms were unrealistic and that public-sector steel by going for the large scale was bound to lose touch with end-users. It seems, in retrospect, that public-sector modernizers at Cornigliano were not sufficiently aware of the fact that their commercial success was largely abetted by the market oligopoly they enjoyed. Their attempts at improving marketing were nevertheless genuine but, in order for them to carry them over to Taranto, an even greater reforming drive would have been required. After all Taranto lacked one of the key features of Cornigliano—

[36] H. Nakamura, *Il paese del sole calante* (Milan: Sperling & Kupfer, 1993). The author, a Japanese steel manager with long Italian experience, was appointed briefly as Chief Executive Officer of Ilva in 1993. See also pertinent observations by Balconi, *La siderurgia italiana*, 24–5. For an investigation of the work of Japanese consultants at Taranto during the 1970s and 1980s, see Anthony Masi, 'Nuova Italsider-Taranto and the Steel Crisis: Problems, Innovations and Prospects', in Yves Meny and Vincent Wright (eds.), *The Politics of Steel: Western Europe and the Steel Industry in the Crisis Years (1974–1984)* (Berlin: de Gruyter, 1987), 476–501.

[37] This aspect is dealt with in both Osti, *L'industria*, 141, and Balconi, *La siderurgia italiana*, 24–5.

vicinity to final users. Instead, the failure to have in place a good commercial organization brought back the practice of sharing the market among a number of private firms, by supplying them with semi-finished goods.

Moulding Italsider as a large corporation organized along functional lines proved much more difficult than expected. There was much resistance both on the part of the management of the old companies and, more importantly, on the part of Finsider itself, which did not want its influence reduced. A merger into a single-divisional firm of a number of separate companies, sometimes hundreds of miles apart, was probably asking too much. What happened was that tasks were duplicated rather than rationalized: Finsider, for example, retained a large layer of administration, in Finance as well as in Personnel. It was able to draw on the support of managers in peripheral firms, who felt sidelined by the reforms. Gradually public-sector steel began to resemble an unwieldy, disorganized conglomerate.[38]

Equally some of the other organizational reforms attempted with some success at Cornigliano ran into trouble when they were extended to the whole of Italsider. Standard costs, for example, required flexibility and speed in the application of corrective measures. Within Italsider, however, growing delays in remedial action made the exercise much less rewarding. A comparison with Thyssen carried out at the time highlighted the problem. Thyssen also applied standard costing, and the company's bureaucracy appears to have been slow in spotting the problem areas. However once they had been spotted remedial action was fast and effective. At Italsider, on the other hand, problem solving was bogged down by a Byzantine decision-making procedure, characterized by overlapping committees and divided responsibilities. The obstacles to MIP proved even greater, to the point that it never really took off. Italsider managers had hoped to use MIP in order to mitigate the effects of centralization, but they soon discovered that Finsider officials resisted its introduction, on the ground that it would have curtailed their powers over promotion and staff benefits.[39]

Job evaluation proved ill suited to many parts of Italsider. It was applied without the necessary flexibility. Initially industrial relations at Italsider, based on 'articulated bargaining', high wages, and good management–union relations were seen as an example for the rest of manufacturing industry. Soon, however, the picture changed. First, the smooth application of job evaluation depended on a kind of informal partnership with a compliant union, enabling management to set the terms. In the course of the 1960s, trade unions at Italsider, reflecting the national mood, became increasingly militant and confrontational and the CISL, in particular, turned away from its previous commitment to efficiency and productivity. Eventually job evaluation was vehemently rejected by the unions and formally dropped.

[38] An inside view of Italsider's organizational decline is Osti, *L'Industria*, 191ff.

[39] A critical view on the relevance of standard costing for US corporate management is expressed in H. T. Johnson and R. S. Kaplan, *Relevance Lost. The Rise and Fall of Management Accounting* (Boston: Harvard Business School Press, 1987).

Finally the whole ethos of public-sector steel was transformed. The post-war reformers were essentially modernizers, committed to industrial expansion. They were part of a wider coalition of business leaders and technocrats committed to rapid industrialization, the development of the South, the creation of a large domestic market to raise the standard of living of the population and foster the emergence of a new middle class. By the early 1960s, with Italy well on course in her 'economic miracle', most of these objectives seemed close to being fulfilled. The debate had moved on, and the more progressive wing of reformers had embraced a new body of ideas known as 'neo-capitalism'. Public and private-sector corporations were thus seen as agents of reform, prime movers in the creation of new infrastructure and adequate social services. Committed to a sustained increase in per capita incomes they would act in conjunction with the local and central government to smooth economic fluctuations, plan investment, and build up the stock of social and human capital. Professional managers were at the core of this doctrine and they were supposed to act in accordance with the ethos and objectives of the modern, progressive corporation.[40]

The neo-capitalist reformers, however, did not enjoy the same success as their post-war forerunners. Initially the new Centre–Left coalition, which gained power in the early 1960s, did seem to fulfil their expectations, since it was committed to raising social expenditure and to indicative planning. On the corporate side, moreover, many large firms were setting up their marketing and staff services, and the power of managers seemed to be increasing. New large investment schemes, such as Taranto, were indeed transforming the nation's economy. Soon, however, it appeared that this had been achieved at a price and old, unresolved problems resurfaced. For one thing the idea of attracting more private capital was never allowed to materialize. It went furthest at Cornigliano, but there again IRI–Finsider retained the majority of the company's stock. In fact large investment schemes required new injections of public money. Political party leaders and government ministers who in the past had largely been content to entrust responsibility to chief executives, now became much more intrusive. They acquired the habit of interfering with company appointments, often through the intermediary of IRI, a Christian Democratic fief. [41]

By the beginning of the 1970s the state-owned sector of the steel industry had lost in quality what it had gained in quantity. Italsider was a huge company, with its Taranto plant one of the most modern in Europe. But with steel output

[40] On this wave of neo-capitalist thinking see, among others, G. Berta, *Le idee al potere. Adriano Olivetti fra la fabbrica e la Comunita'* (Milan: Comunita', 1980); A. Salsano, 'Il neo-capitalismo. Progetti e ideologia', in *Storia d'Italia Einaudi, vol. 5. I documenti,* I (Turin: Einaudi, 1973).

[41] There is a large literature on the crisis of Italy's public-sector industry. See, for example, E. Gerelli and G. Bognetti (eds.), *La crisi delle partecipazioni statali: motivi e prospettive* (Milan: Angeli, 1981) or P. McCarthy, *The Crisis of the Italian State: From the Origins of the Cold War to the Fall of Berlusconi* (London: Macmillan, 1995), 81–101. The case of steel is dealt with by Osti, *L'Industria,* and Balconi, *La siderurgia italiana.*

reaching new heights, performance was increasingly deteriorating. Profits and productivity were declining, managers were being replaced by political place-men. Reforms that had been so successful during the earlier years were either taken for granted or, in many cases, resisted and overturned.

Chapter 9

Mass Production or 'Organized Craftsmanship'? The Post-War Italian Automobile Industry

DUCCIO BIGAZZI

9.1 The American Model in Italy between the Wars

In 1945 the Italian industrial sector was fairly well prepared to measure itself against the American model. In the engineering sector, direct contacts between technical experts on the two sides had been established at least since the turn of the century, and had been intensified during the First World War. The war had given impulse to serial production and had convinced many large enterprises of the need to introduce working methods based on standardization and specialization. However, the 'titanic energy' inspiring the ambitious plans of vertical integration crumbled before the harsh realities of post-war reconstruction, which cut down to size the aspirations of making a 'religion' out of standardization, as expressed by Ansaldo's managing director, Mario Perrone.[1]

During the Fascist period, the diffusion of American methods met a stumbling block in the traditionalist approach to technology and organization of most Italian industrialists, who preferred low wages and a strongly authoritarian form of paternalism.[2] None the less, the most innovative managers and entrepreneurs continued to look up to America as the source of the most efficient technical and productive model.

The regime's relationship with the USA was unstable and ambivalent. The press, by highlighting the conquests of American industry and technology, sought to uphold Fascism's self-representation as a progress-oriented movement, but at the same time denounced the social consequences of modernity: 'mechanization' and the lack of spirituality. From the standpoint of a country

---

[1] Ferdinando Fasce, L' Ansaldo in America (1915–1921)', in id., *Tra due sponde. Lavoro, affari e cultura tra Italia e Stati Uniti nell'età della grande emigrazione* (Genoa: Graphos, 1993), 111–30. An account of the first trips to America is given in Camillo Olivetti, *Lettere americane* (Milan: Edizioni di Communità, 1968); Ing. Ernesto Breda and Co., *Le locomotive in America e in Europa. Osservazioni e confronti* (Milan: Breda, 1900).

[2] Giulio Sapelli, *Organizzazione, lavoro e innovazione industriale nell'Italia tra le due guerre* (Turin: Rosenberg & Sellier, 1978), 111–28.

forced to seek relief in autarkic restriction of consumption, the USA was dis-
quieting because of its waste and excess, all the more so because these were supposedly accompanied by a lack of values and stable traditions: such abundance
could not bode well for a society in which adoration for 'the almighty dollar'
translated into 'the slavery of high wages'.[3] In this climate, even the many entrepreneurs and technicians who reported on their frequent trips to the USA had to
temper their enthusiastic descriptions of the 'wonderful improvement in the living standard of the masses' and 'the passage from an economy for the privileged
to a mass economy' with more or less sincere appeals against the risk of loss of
the 'sense of values' and of a levelling of intellectual individuality.[4]

But the managers of the major firms looked to America mainly for operational methods and concrete suggestions. In the automotive industry, American
influence was symbolized by the Fiat Lingotto plant, planned in 1916, in which
the rationalism of production engineers merged with a futurist and metaphysi-
cal vision, contrasting starkly with the traditional balance of Turin's urban
architecture: production proceeded from the lowest to the highest of five floors,
until the finished car reached the testing track on the roof. The archetype, Ford's
Highland Park plant, had indeed been carried to an extreme in the shape and
dimensions. Fiat, usually more sober in its self-representation than in its practical realizations, succumbed in this case to the temptation to affirm a debatable
international primacy.[5]

Opened in 1923, the 'imagined' Lingotto had to come quickly to terms with a
social and economic reality which would not easily adjust to mass production.
Even from a technical and organizational standpoint the plant, designed as a
single, huge machine, could not guarantee an acceptable level of co-ordination.
In 1926 a long study trip to the United States by the managers of the plant made
it clear that the inefficiencies did not originate in equipment, machinery, or even
in an insufficient application of line production: the literal imitation of the
Fordist model advocated by some, would indeed have had disastrous consequences.

This and other American visits rather suggested the need to rethink from a
methodological standpoint the approach required by large-scale, complex
organizations, a lesson summarized by the Fiat mission in the slogan, 'research,
order, method'. Adriano Olivetti, who had travelled to the United States in
1925, reported his impressions in similar terms ('the secret was not in the men

[3] Michela Nacci, L'antiamericanismo in Italia negli anni trenta (Turin: Bollatio Boringhieri,
1989), esp. 28–31.
    [4] See particularly the opinions of Alberto Pirelli and Gaetano Ciocca, quoted in Nacci, L'anti-
americanismo in Italia, 54–5 and 65; also Antonio Alessio, 'Il divenire dell'industria nel moto
dell'insegnamento scientifico e professionale', in Lezioni di tecnica industriale (Bologna, 1943), 18
('a high-wage policy meant to increase consumption [by the masses] can distract from every con-
tentment and indeed from every cultural betterment').
    [5] Duccio Bigazzi, 'Strutture della produzione. Il Lingotto, l'America, l'Europa', in Carlo Olmo
(ed.), Il Lingotto 1915–1939. L'architettura, l'immagine, il lavoro (Turin: Umberto Allemandi,
1994), 281–336.

. . . but in the structure, in the organization and in the rigor of the methods') as did, a few years later, the general manager of the Officine Reggiane Antonio Alessio ('assembly processes are much less a technical problem than one of discipline, method, order and movement').[6] The American model had to be adapted to Italian reality. Even the managers at Magneti Marelli in Milan, among the most enthusiastic advocates of 'rational organization methods' learned in the United States, remarked that blind emulation 'would be simply an unnecessary luxury for us'.[7]

The imitation of the American model, however limited, could only be carried out by a restricted group of large enterprises. An attempt was made to extend the principles of scientific management beyond the few producers of standardizable consumer goods, through the creation in 1927 of the Ente nazionale italiano per l'organizzazione scientifica del lavoro (ENIOS), but this was interrupted by the Depression. After a period of few contacts, new occasions for rejuvenating relations with America and for profiting from agreements centred on technical cooperation (licensed production, technology, and equipment transfers) were offered by the Chicago (1933) and New York (1939) World's Fairs, but shortly afterwards Italy's entry into the war alongside Germany cut the channels just reopened. Relations with large German enterprises, however, while strongly influencing the production choices of the Italian firms, enlarged their technical perspectives and partially replaced the flow of technology that previously came from the United States.[8] They also focused attention on the theme of organization, the backward state of which was among the basic causes of the failure of war production in Italy: the managers of the most important firms had to admit that organization was 'our problem'.[9]

9.2 Heterodox Visions: Specialization, Flexibility, Decentralization

Behind the apparent vagueness of this formulation stood a broader debate which had been taking place among managers and technicians during the war

---

[6] Bruno Caizzi, *Camillo e Adriano Olivetti* (Turin: UTET, 1962), 153; Antonio Alessio, *Prospettive della industria meccanica* (Rome: Federazione nazionale fascista dirigenti aziende industriali, 1942), 18. For statements by French industrialists utilizing the same terms see Patrick Fridenson, *Histoire des usines Renault. Naissance de la grande entreprise 1898–1939* (Paris: Seuil, 1972), 195.

[7] Perry R. Willson, *The Clockwork Factory. Women and Work in Fascist Italy* (Oxford: Oxford University Press, 1993), 42–62.

[8] As Paul N. Rosenstein-Rodan had observed in 'Technical Progress and Post-War Rate of Growth in Italy', in *Il progresso tecnologico e la società italiana. Effetti economici del progresso tecnologico sull'economia industriale italiana 1938–1958* (Milan: Giuffrè, 1962), 162; post-war observers had insisted, on the contrary, on the negative effects on Italian industry of isolation from American technology: see for instance Vittorio Zignoli, *Aspetti tecnici della crisi del Piemonte* (Turin: Camera di commercio, industria ed agricoltura, 1947), 91; Attilio Jacoboni, *L'industria meccanica italiana* (Rome: Istituto poligrafico dello Stato, 1949), 31.

[9] Alessio, *Prospettive della industria meccanica*, 8; also Agostino Rocca, 'Parole dell'amministratore delegato al personale dirigente dell'Ansaldo il 13 novembre 1943', in Archivio A. Rocca, Fondazione L. Einaudi, Turin, b. 16, f. 92.

years. Within a framework of shared technocratic aspirations, involving a variety of perspectives from Fascist and Catholic corporatism to liberal and socialist planning, these discussions covered a wide range of issues from productive specialization and inter-firm co-operation to plant flexibility, industrial decentralization, and the need for a new model of labour relations based on real collaboration. Here, too, an important inspiration came from the American industrial environment. Francesco Mauro, founder of ENIOS and an eager popularizer of scientific management practices, focused attention on the decentralized structure of General Motors. The standardization of components and tools which gave GM flexibility was contrasted with the rigid 'intensive specialization' adopted by Ford and others. Mauro was convinced that the most desirable development would have been the disappearance of the big 'dinosaurs' and their replacement with smaller, more specialized plants, in an idyllic rural setting which would have allowed the labour force to integrate their jobs with 'regenerative' agricultural work.[10]

Even within Ansaldo, a typical example of vertical structure devoid of any clear specialization, the economists who made up the core of the influential 'ufficio studi' were quick to point out that in the USA 'the current tendency [is] to avoid huge plants in favour of smaller ones', not bigger than a thousand workers.[11] More important, a few concrete plans for decentralization and specialization were being formulated; among others, by Alessio of Reggiane, who proposed in 1944 to 'decongest' the firm's main plant in Reggio Emilia by moving out all intermediate stages and gradually limiting its functions to initial and final operations (casting, forging, assembling, repairs) in order to keep the entire structure running smoothly.[12] Alessio's localization plan was part of a paradigm of industrial development focusing on product quality and skilled labour. In his view, the Italian engineering industry's ideal models should have been Switzerland, Sweden, and Germany rather than the USA: in those countries medium-sized firms, 'organized on the basis of series production criteria', had been able to reconcile efficiency and innovation.[13] This model would allow Italian firms to make the most of a wealth of 'abundant, high-quality labour'. Incidentally, it was that very labour force which, leaving the Reggiane spontaneously or because of the post-war reconstruction crisis, set up their own businesses, thereby establishing the conditions for the emergence of small and medium-sized flexible production systems around Reggio and in other neighbouring areas.[14]

The dilemma evoked by Alessio was anything but new: since the beginning of Italy's industrialization, a number of economists and technical experts had sin-

[10] Francesco Mauro, *Teratismi dell'industria. Anomalie e squilibri* (Milan: Hoepli, 1942), 121–2, 177–98, and 388–9.

[11] Alberto Campolongo, *Ricostruzione economica dell'Italia* (Milan: Giuffrè, 1946), 167.

[12] Antonio Alessio, *Luci ed ombre della rinascita. Libro bianco* (Reggio Emilia, 1945), 21–2.

[13] Alessio, *Prospettive della industria meccanica*, 8–10.

[14] See on this Giulio Sapelli et al., *Terra di imprese. Lo sviluppo industriale di Reggio Emilia dal dopoguerra ad oggi* (Parma: Pratiche, 1995).

gled out Switzerland and other regions lacking in natural resources (particularly Alsace) as appropriate examples for the country to follow. Now, however, this course was recommended in direct and stark contrast not only to the over-ambitious rearmament policy pursued by the Fascist regime ('the "infatuation" with the "colossal" '),[15] but also to those who attributed the wartime successes of American industry exclusively to giantism: big plants, huge production volumes, hence large series production. What Italy needed was rather paths to modernization which could preserve the human resources readily available in the country: serial production did not necessarily mean an impoverishment of the range or quality of products.

Perhaps the most lucid conception of the relationship between quality and serial production was that proposed by Adriano Olivetti and his managerial associates. Giovanni Enriques, for one, undertook in 1937 to analyse in depth the features of what he called 'complex mass production': that is, large-scale series production applied to consumer products made of a great number of components. Enriques insisted on the central role of jigs and fixtures in avoiding the need to have recourse to single-purpose machines, a solution implying the availability of a substantial pool of highly skilled workers. The over-simplifications dear to the 'fetishists of Fordism', who 'for a time made the world believe that increasing beyond reason . . . the gap . . . between two classes of workers, the highly skilled in charge of the equipment, and the others assigned to the mere harvest, not different in that from farmhands in agriculture' would usher in 'an era of ever-increasing prosperity', should be avoided. On the contrary, mass production involved serious risks, pre-eminent among which were those connected with 'inertia': 'Time and habit have a way, even more than in smaller firms, of making the whole system dull.' Those who tied traditionalism and conservatism exclusively to artisanal and small-batch production were therefore sadly mistaken.[16]

Ugo Gobbato, general manager of Alfa Romeo, had advanced with respect to the opportunities for post-war reconstruction of the automotive industry in Lombardy (Alfa Romeo, Isotta Fraschini, Bianchi), an approach to specialization in large companies which avoided a purely artisanal perspective, endeavouring on the contrary to maintain and even enhance an existing orientation towards serial production. Gobbato identified as a crucial need of the times the affirmation of 'a spirit of standardization and unification made necessary by the narrowness of the Italian market itself'. His proposal echoed the hypotheses concerning the 'rationalization' of the automotive industry, and more generally, of the various sectors of war production, which had been much discussed, although to no great effect, in the previous years. But it combined attention to efficiency issues with a territorial and social dimension. The new firm would be divided into four separate productive units (casting, forging and presses;

[15] Scrutator, 'L'avvenire dell'industria metalmeccanica italiana', *Macchine* (1946), 4.
[16] Giovanni Enriques, *Caratteristiche dell'industria complessa di massa* (i.c.di m.), (Rome, 1937).

engineering; body works; design and auxiliary shops) arranged in succession so that they could exploit most efficiently the economies of transportation derived from the use of waterways. Each unit would have made for a 'self-sufficient industrial entity', including, not far from the plant, the blue- and white-collar workers' lodgings. Following a neo-paternalist model already in place at the Pomigliano d'Arco plant near Naples, each village would boast a complete range of social services: 'Mother and Child Unit, a small centre for the assistance of the sick, a meeting and recreation place, a vocational school, and shops'.[17] For Gobbato, as well as most Italian advocates of rationalization, the principles of Fordism were translated and adapted in a context of social relations founded—if not directly on nineteenth-century agro-industrialism—then on familism, authority, and company values.

In the post-war period, however, the debate became simplified and impoverished as a stark contrast between two extreme, alternative positions replaced the more articulated visions just presented, in which scale, specialization, decentralization, and localization were all closely connected elements. On the one hand, there was large-scale production, radical centralization, ever-increasing fixed capital investments, and a production strategy targeted towards consumption growth. On the other, there was an approach based on product quality alone, which insisted on the unavoidability of maintaining or even returning the country's industrial activities within the limits of a huge and well-developed artisanal sector. Typical in this respect are the opposite standpoints endorsed at the beginning of 1946 by Vittorio Valletta for Fiat, and by Pasquale Gallo, then interim CEO at Alfa Romeo, on the subject of the automotive industry, whose opportunities for growth represented, together with those of the steel industries, the focus of the debate in the reconstruction years.[18]

Valletta's strategy was no rehashing of Fordism nor a prefiguration of what would happen in Turin from the 1960s onwards. The company's plan, while aiming at 'doubling and even tripling production', was rightly defined as 'not too daring': it mainly sought to put the plants back on track, utilizing them, at last, according to the productive targets and scale for which they had been planned. Valletta based his faith in Fiat's future on the belief that the USA had good reasons for wanting the recovery of Europe's automotive industry: in Italy's case, besides the volatile social situation and the already critical unemployment level, there was the peculiar nature of Fiat's production, mainly

---

[17] Archivio storico Alfa Romeo (ASAR), Arese (Milan), Direzione generale e Segreteria generale, f. 368, 'Promemoria per il Sig. Bugatti', 14 Jan. 1945. On Pomigliano, see Augusto De Benedetti, *La via dell'industria. L'IRI e lo sviluppo del Mezzogiorno, 1933–1943* (Catanzaro: Meridiana Libri, 1996), 123–31.

[18] Ministero per la Costituente, *Rapporto della Commissione economica presentato all'Assemblea Costituente. II. Industria, II. Appendice alla Relazione (Interrogatori)* (Rome, 1946), 125–37 and 345–55 (hereafter *Interrogatori*). See also Piero Bairati, *Vittorio Valletta* (Turin: UTET, 1983), 158–60 and Giulio Sapelli, 'L'organizzazione del lavoro all'Alfa Romeo 1930–1951. Contraddizioni e superamento del "modello svizzero" ' in id., *L'impresa come soggetto storico* (Milan: Il Saggiatore, 1990), 316–32.

focused on small cars. Fiat, for its part, had a clear understanding of the constraints dictated by the international context and by the available resources: 'We do not seek to produce large and medium-sized cars simply because we do not have proper plants and we would have to spend more than we can afford in order to build them.' The contacts already initiated during the German occupation gave Valletta every confidence that the relations with the big US corporations, which had always been 'magnificent', would resume on a basis of assistance rather than competition.

Gallo's point of view was different and much gloomier. In his view the Italian market could not absorb even the three hundred cars a day which Fiat normally produced. And besides, 'we think we have in Fiat a big corporation, and we only have a small American firm': better to leave mass production to those capable of handling it, such as Ford and General Motors. Quality production, on the other hand, was a wholly different matter: Alfa Romeo 'has an excellent, classy product, and that is craftsmanship's niche. It produces the 110 mph car, the car that picks up easily, the car, in short, for the happy few. The Americans are not interested in small clienteles, but in big markets. Alfa Romeo can therefore be saved.'

Here Gallo contrasted quality and series production in a way which could only marginalize Italy's automotive industry. Valletta therefore had an easy time criticizing the 'Italian notion (widespread even among intelligent people) that the standardized product is of necessity second-rate, while custom production is by definition excellent'. Worse still, Alfa Romeo's CEO spoke not only about the automotive industry: in his opinion the Italian industrial sector would forever be limited to niche or secondary production by an international division of labour that had to be assumed, at that point, as a given. Gallo's position was derived from a certain Italian liberal line of thought supporting the development of industries which 'could naturally fit in Italy': cotton textiles, agricultural machinery, shipbuilding, non-ferrous metals, and such like. The criteria for this selection hinged on the characteristics of the product as a commodity and on a very narrow consideration of the available external economies, particularly the transportation costs of raw materials, while issues which would prove all-important for the recovery of the Italian industrial sector, such as the overcoming of technological and organizational backwardness and the search for new ways to serial production, remained in the background.

9.3   Italian Managers and American Factories: The Ambiguous Lessons of Post-War Visits

The short-sightedness of Gallo's vision opened the door for the triumph of Valletta's expansionary perspective. Non-mainstream positions such as those examined earlier—both IRI technocrats (Rocca, Gobbato, Alessio) and Olivetti and his associates (the latter much more complex and sensitive to the need for new social relations)—lost ground. However, the mass production model adopted by Fiat was recast, at least during the first post-war decade, in a very sober and subdued fashion.

The major firms, above all Fiat, privately anticipated the institutional mechanism of study trips to the United States, even before opportunity for such was granted through official channels: from 1946 they began sending specialized technical delegations to the most prominent American factories on a regular basis. A Chrysler representative met Valletta in Turin at the end of 1945 and the company's plants were the destination of the first post-war trip to America by Fiat's technical experts in December 1946. It was a short visit, intended solely as a means to renew the ties with the American corporations.

This understanding was put on a more formal footing in April 1947 through a 'technical co-operation' agreement signed by Valletta during a trip to the United States which had important political and financial connotations. Unlike Ford (with which the first contacts had come to nothing) or General Motors (which was loath to try its luck in Europe again after the Opel disaster), Chrysler was ready to grant 'the desired assistance, without forcing us to pay for it, or to accept it with meek gratitude'.

Fiat's general manager insisted on those features of the agreement which emphasized reciprocity and highlighted 'those activities in which we too are up to date and proficient'. Chrysler's interest had moreover stimulated a similar attitude on the part of General Motors. Although the Chrysler–Fiat agreement included an exclusivity clause, two missions by Fiat in 1947 gained access to all the main GM plants.[19] The relations established in the pre-war period with plant and equipment suppliers were also restored, first and foremost with Budd of Detroit which had, since 1933, granted patents and technical assistance for dies and bodywork equipment.[20]

The willingness of American firms to cultivate these connections was made even clearer in the meticulous and detailed reports written by the Fiat men on their visits. These were no 'industrial pilgrims' travelling to modernity's 'promised land'.[21] No inferiority complex nor fascination with large-scale production prevented them from looking at the American automotive industry in a clear, analytical way, with the aim of translating the technological and organizational language of Detroit into the Mirafiori context. The American industrial model, though successful, was by no means free of problems: the earlier reports, in particular, included a variety of criticisms, mainly about solutions deemed unnecessarily expensive, a management style often appearing to be

[19] Bairati, *Vittorio Valletta*, 151–2 and 177–83; Dante Giacosa, *Progetti alla Fiat prima del computer* (Milan, 1988), 113 and 127–9; Archivio storico Fiat (ASF), Turin, Verbali del Comitato direttivo (VCD), 7 May 1947; Fondo Divisione affari internazionali (DAI), b. 107/1 and 107/2.

[20] ASF, Rapporti presidenza, 23 Jul. 1947; Giacosa, *Progetti alla Fiat*, 128; Bigazzi, 'Strutture della produzione', 321.

[21] This term is taken from James P. Womack, Daniel T. Jones, and Daniel P. Roos, *The Machine that Changed the World* (New York: Rawson Associates, 1990), 231; see also Wayne Lewchuck, *American Technology and the British Motor Vehicle Industry* (Cambridge: Cambridge University Press, 1987), 113.

focused on 'moving the metal' at all costs, and industrial relations that seemed at least as thorny as they were in Italy.

From this point of view the report written in July 1947 by Alessandro Genero, a former worker turned plant manager first of Lingotto, then of Mirafiori, can be considered typical. First, Genero made a comparison with his previous visit, made in 1936 in preparation for the construction of Mirafiori: 'With a few exceptions . . . The machines, equipment and plants are basically the same and, being older, make a poorer impression'.[22] On the other hand, the factory environment had decidedly changed for the worse, especially as far as discipline and control were concerned:

Even then most factories left much to be desired in the matter of tidiness and cleanliness, and today it is still worse; Ford itself, once an example for others, cannot be cited as such any longer. Discipline is less complied with by both blue and white-collar workers. In some plants we have seen workers quit their jobs up to 15 minutes early and line up in front of the clock waiting for the departure time in order to leave as soon as possible. At the starting time, the same thing happens: the workers . . . loiter instead of beginning work immediately, then let the machines idle for a while but in some cases true production only begins half an hour later. Smoking is tolerated everywhere, even at Ford and at Pratt-Whitney, where, before the war, the ban was strictly observed.

This was not merely the *cri du coeur* of a notoriously stern manager confronted with the uneasiness and crises of post-war American labour.[23] Nor was the issue at stake here only 'tidiness', although this element was duly noted in all the reports as a necessary precondition of efficiency and, at the same time, as direct evidence of the good organizational and disciplinary standing of a plant.[24] Rather, Genero had been able to capture, if not to analyse fully, the deep changes in social relations that the war and its aftermath had brought about in Detroit, the Fordist factory-town to which Turin had constantly looked up: 'All this turmoil is especially noticeable in big-city factories and particularly in Detroit . . . In almost all the small-town plants we have seen a more regular, quicker pace of work than in Detroit.' It was at these scattered and specialized plants, and in particular at the GM foundry at Saginaw and at other smaller

[22] ASF, b. 107/2, 'Impressioni riportate nel viaggio compiuto negli USA dal 23 maggio al 18 luglio 1947'; similar comments are in A. Fiorelli, 'Relazione viaggio in America 16–20 dicembre 1946' and in Rapporti presidenza, 23 Jul. 1947.

[23] See Nelson Lichtenstein, *Labor's War at Home: The CIO in World War II* (Cambridge: Cambridge University Press, 1982), 221–32; David A. Hounshell, 'Planning and Executing "Automation" at Ford Motor Company, 1945–65: The Cleveland Engine Plant and Its Consequences', in Haruhito Shiomi and Kazuo Wada (eds.), *Fordism Transformed. The Development of Production Methods in the Automobile Industry* (Oxford: Oxford University Press, 1995), 55–6; David Brody, *Workers in Industrial America. Essays on the Twentieth-Century Struggle* (Oxford: Oxford University Press, 1980), 173 ff.

[24] On the '*maladie de la propreté*' widespread also among other European manufacturers see Olivier Cinqualbre and Yves Cohen, 'Les Usines dans l'action d'un grand industriel: Citroën, Quai de Javel' (Paris, 1984, mimeo.), 215–16 and also Yves Cohen, 'Quand un homme de Peugeot visite Renault (janvier 1939)', *Renault Histoire* (Nov. 1989), 25. On the 'aesthetic values' of Taylorism, Judith A. Merkle, *Management and Ideology. The Legacy of the International Scientific Management Movement* (Berkeley: University of California Press, 1980), 51.

foundries in Michigan and Indiana, that it was possible to obtain the best results.

These comments did not, however, call into question the whole territorial and social organization of the automotive industry. Genero also pointed to a few excellent examples of efficiency found in Detroit itself, such as Chevrolet's and Dodge's forges, Budd's press shops, and Ford's engine shops at River Rouge. General manager Bruschi, who considered the Saginaw foundry the 'best plant [we] visited', praised Ford: 'These are, even now, the most interesting plants in the world . . . Speed and precision are extraordinary: they build 4200 motors a day on a line 100 yards long.'[25] As for the big plants of the Detroit area, Fiat's men remained fascinated by the transfer equipment used in the production of cylinder blocks at the Buick plant in Flint and by the 'iron hands' and automated pincers which moved sheets along the press lines at the Fisher and Briggs plants, although they realized that these methods were only suitable for 'continuous productions' and had, for the moment, 'only a documentary interest'.[26]

As a general rule, for Fiat's 'modernization programme' it was necessary to 'analyse in detail every purchase request' in order to make sure that it made economic sense. After all, Fiat's technicians reported that, 'in a number of plants, even among the best ones, we have seen older machinery work satisfactorily alongside newer and more up-to-date equipment'. Older installations were replaced only when maintenance placed too severe a burden on production costs.

As for a more general assessment on the American style of production, a recurrent theme in these reports was surprise at the 'tremendous mass of materials' visible in the machining departments and bodyshops. This situation at first struck Fiat's men as 'abnormal' and likely to get out of control, but this evaluation was modified by the subsequent trip in October–November 1947. This change of heart is the first signal of a gradual adjustment to a different logic, a shift the consequences of which would become clear only years later, in the era of Fordism's triumph. The urgent need for a return to acceptable productivity levels pushed Fiat's management to reconsider the priorities according to which the production cycle had been organized up until then. The general manager, Gaudenzio Bono, remarked that: 'First and foremost comes the attainment of the production target; then cost considerations; and last the problem of waste. Production is so important that when, for whatever reason, it slows down or stops altogether, all other issues fade into the background.' For his part, Valletta took it for granted that: 'for us too, as for the Americans, production comes first'. During his trip to the United States, he had been fascinated, like all European visitors, by an atmosphere of wealth and opulence, which was in stark contrast with the difficulties of Italian reconstruction. The situation called for a new and different outlook: 'We need to refute the notion that there is a cult of waste in America. There is, instead, a cult of production in order to

[25] ASF, Rapporti presidenza, 23 Jul. 1947.
[26] ASF, Dai, b. 107/1, 'Note sul movimento materiali'.

create jobs and therefore, naturally, the urge to see that what is produced can be absorbed through consumption.'[27] The 'American spirit', still so distant from the Italian post-war situation, highlighted the possibility of joining mass production and mass consumption: waste was a by-product, to a certain extent unavoidable, of a social model focused on production.

But there were also technical reasons for what had appeared at first to be an unjustifiable excess of materials on the shop-floor: first, the elimination of warehouses and their replacement with stocks arranged along the production lines; then the Sloanist tendency towards product differentiation, which had spread throughout the industry and implied a much greater variety of possible combinations. Plymouth, Fiat's technicians reported in 1946, assembled 8 body types on the same engine, in addition to 14 interior finishes, 20 colours, and 7 sets of mechanical components (including steering, suspension, and so on): no car was exactly identical to the next.[28] As for the materials needed to guarantee the continuity of the production cycle, Plymouth's statement that 'they had to be sufficient for a period of time from a day and a half to six days' was met with scepticism: 'We think the assessment fair only for the most important groups and components.' Mirafiori's experience left these men apparently doubtful about the possibility of achieving comparable levels of flexibility with stocks which at the time seemed extremely light.

Fiat's technicians were only then discovering the pivotal role of the movement of materials in American automotive plants. Not that, even from this standpoint, there was anything completely new ('the basic organizational principles are already known to all'). The visitors were, however, struck by the use of moving equipment, much more generous than at Mirafiori. The American plants were continuously criss-crossed by electric trucks, while aerial conveyors represented the 'arteries' of the system: here, it was remarked in a not too covert allusion to what instead happened at Fiat, conveyors 'are really utilized to capacity, and they do not run empty', so that 'everything moves quickly between shops, from shop to assembly lines, and from the latter to the warehouses'. Higher costs for plant and for space taken up in the shops were compensated by economies and organizational advantages: 'One can easily see how enormous are the weights involved in the movement of materials and how much work it requires. According to American data an arrangement which allows for moving in total only 50 times the weight of the finished product may already be considered efficient.'[29] The movement of materials therefore made an impact on the

[27] ASF, VCD, 7 May 1947; Rapporti presidenza, 14 and 23 Jul. 1947.

[28] ASF, DAI, b. 107/2, 'Relazione viaggio in America 16–20 dicembre 1946' and 'Relazione viaggio in America ottobre–novembre 1947'. The complexity of this product mix is generally understated by the supporters of the absolute novelty of the Toyotist model. See for instance Haruhito Shiomi, 'The Formation of Assembler Networks in the Automobile Industry: the Case of Toyota Motor Company (1955–80)', in Shiomi and Wada, *Fordism Transformed*, 38.

[29] These were Fiorelli's impressions of Chrysler's assembly plant at Windsor, ASF, DAI, b. 107/2, 'Relazione viaggio in America ottobre–novembre 1947'; see also b. 107/1, 'Note sul movimento materiali'; and Eiji Toyoda's comments on his visit to River Rouge in 1950 (Shiomi, 'The Formation of Assembler Networks', 35).

•

generally prudent attitude towards the purchase of plant and machinery: 'It is pivotal that everybody becomes conscious of the movement issue . . . Capital investment . . . here more than anywhere else must satisfy the principle of spending in order to make savings possible.' On the whole, however, the strategy chosen was classically Fiat: 'No radical solutions, rather a logical and economical middle course which will allow us to reconcile the product price with the amortization costs of the equipment.'[30]

Many of these observations could be repeated for Alfa Romeo.[31] Its general manager, Alessio, in a presentation given on his return from a trip with a Finmeccanica mission in June–July 1949, showed the same preoccupation with discipline and tidiness: at Pullman, Chicago, and at Hudson, Detroit, for instance, he had had 'an impression of great disorder' and had noted 'the slow work pace' and the 'poor and old machinery'. Even the visit to River Rouge, though 'amounting for a technician to what for a tourist entering the Eternal City is a visit to Saint Peter's or to the Forum', stirred ambivalent feelings: the foundry appeared to him a 'pit of Hell' and the machinery in the engine shop looked 'not up to date'. On the other hand, 'the pace of the machines and the synchronization of the workers with them' in the press shop was 'striking', and at the engine shop 'all movements are mechanized: thus endless belts, and chains and hooks are everywhere'.

The only plant which received unconditional praise was that of Oldsmobile (GM) in Lansing: 'a very modern-looking plant with buildings grouped in the most rational way . . . The machinery is spectacular and its arrangement exemplary.' Here Alessio could detect a degree of specialization and, above all, an 'operational automatism' not reached as yet by others: each worker 'becomes the minder of a completely automated machine—the same as in weaving'. Above all, Lansing allayed the perplexities stirred once again by the 'remarkable' stock of materials kept at the final assembly plant at River Rouge: this appeared now as the 'backbone on which the smooth and efficient working of large-series production rested'.

Paradoxically, the fascination with automated processes and large scale struck the manager from the niche firm, Alessio, more than it did those from Fiat.[32] But Alessio had not forgotten the plans made during his stay at the Reggiane. The production strategy he had seen in action at the GM plant had indeed reinforced his convictions regarding the greater efficiency of decentralized and specialized organizations: unlike Ford, GM 'has made no effort to produce every single part of a car, and has instead concentrated the production of semi-finished sheets, castings and forgings, differentiating only the specific

---

[30] ASF, DAI, b. 107/2, 'Relazione viaggio in America ottobre–novembre 1947'.

[31] ASAR, 'Relazione riassuntiva sulla missione agli Stati Uniti d'America giugno–luglio 1949'.

[32] See also Giuseppe Bianchi, managing director of Bianchi: 'We sent one of our managers to visit an American plant which turned out 10,000 chassis a day. On his return he told me he had seen the most wonderful sight of his whole life . . . It had been fantastic to see all the components coming in on a flying carpet, and coming together on a device without the presence of a single worker' (Interrogatori, 283–4).

features of the engines and the final assembly and subcontracting the rest to external suppliers which it has then gradually taken over'. This sort of production structure could represent the solution to the difficulties which threatened the survival of smaller firms in Italy: 'During our trip it had occurred to us that perhaps it would be possible to build a similar organization in Italy, merging the forces of Alfa Romeo, Lancia, Isotta, Bianchi and making an agreement with a famous body manufacturer as in the case of GM and Fisher.' The visit to the Budd press works in Detroit and Philadelphia even added to these ambitions a European dimension. The operations which required the costliest investments and which at the same time had too high a minimum efficiency scale could be reserved for a few specialized producers; the car manufacturers could then concentrate their resources on design innovation and high-quality production: 'What would the European automotive industry be if there existed a Budd, a Briggs, a Fisher?'[33] All these dreams were to fade away shortly but Alfa Romeo based its reorganization on a policy inspired, as we will see later, by the conclusions drawn from this visit to America.

9.4   Marshall Aid and Technology Transfer

Marshall Aid undoubtedly had an important role in the technological updating of Italian industry. The size of the loans obtained by Fiat was exceptional both in national terms (it amounted to more than one-fourth of the total sum allotted to iron and steel and engineering firms)[34] and at a European level (see Table 9.1).

But the figures given in Table 9.1 only partially explain the diverse outcomes of the reorganization plans carried out by the Italian automotive firms: in the period from 1945 to 1951 foreign loans represented little more than a quarter of the investments in plant and machinery made by Fiat (17.7 billion lire out of 65.6); as for Alfa Romeo, government support through IRI and Finmeccanica, intended to offset the losses and allow new investments reached, at the end of 1953, 18 billion lire.[35] ERP aid thus formed part of firms' larger technical modernization projects, but the limited modernization of Lancia and other smaller companies was related more to the lack of domestic support than to the modest size of American loans.[36]

In Fiat's case, the financial requests presented to the ECA were based on what had been elaborated during the missions to America and reinforced the pattern

---

[33] Alessio referred to the attempt made by the German-American joint venture Ambi–Budd at the end of the 1920s.

[34] Comitato interministeriale per la ricostruzione (CIR), 'Lo sviluppo dell'economia italiana nel quadro della ricostruzione e della cooperazione europea' (Rome, 1952), 370–1; see also Marco Doria, 'Note sull'industria meccanica italiana nella Ricostruzione', *Rivista di storia economica* 4 (1987), 53.

[35] ASF, Verbali del Consiglio di amministrazione (VCA), 10 Mar. 1952; ASAR, VCA, 5 Oct. 1953.

[36] Franco Amatori, *Impresa e mercato. Lancia 1906–1969* (Bologna: Il Mulino, 1996), 147–55; Amatori, however, insists on the importance of ECA's financial support (or lack thereof).

Table 9.1. ERP Loans Granted to the Most Important Automotive Firms in Italy and France ($)

Italy		France	
Fiat	30,943,000	Renault	6,000,000
Alfa Romeo	2,150,000	Simca	1,850,000
Bianchi	970,000	Peugeot	1,174,000
Lancia	800,000	Citroën	528,000

Sources: Tre anni di ERP in Italia (Rome: Missione Americana per l'E.R.P. in Italia—Divisione Informazione—Uff. Stampa [Information Division—Press Office of the US ERP Mission in Italy], 1951), 143, 145, and 191; Patrick Fridenson, 'L'Industrie automobile français et le Plan Marshall', in René Girault and Maurice Lévy-Leboyer (eds.), Le Plan Marshall et le relèvement économique de l'Europe (Paris: Comité pour l'histoire économique et financière de la France, 1993), 285.

of vertical integration pursued by the firm since its beginnings. Car production took on a central role with respect to other fields such as railway construction and aircraft, but a substantial share of the desired funds (more than $10 million) had been assigned to the group's steel plants.[37] As for Mirafiori, expenditures of $13.2 million were envisaged; but much of this money would go to overhauling the cast-iron foundry and, above all, to improving the stamping operations: the 1400 model marked the shift to the unitary-construction body, requiring new power presses for deep-drawing larger panels; the construction of the dies and the assembly equipment, ordered almost entirely from Budd, cost an additional $3 million. The extent of the press shop renovation dwarfed the efforts of the lesser firms which had, however, to utilize most of their resources for the purchase of a few large presses (of 800 and 1,000 metric tons).[38]

At any rate, the most substantial investments were made in the machining departments, where Fiat replaced 20–25 per cent of the equipment in this period. Half of these machines (usually the costlier and more complex ones) were bought with ERP funds. The renovation carried out at Alfa Romeo was less significant: although since 1945 15 per cent of the machinery had been replaced, in 1953 the firm's equipment was on average 14 years old. At Lancia, similarly, only a third of the machine tools were less than 10 years old in 1951. As for the features of the equipment requested through the ECA, Lancia was above all

[37] ASF, VCA, 19 Nov. 1948; Bairati, Vittorio Valletta, 213 and 410.
[38] For orders of machinery through ERP by Fiat and others, see Archivio centrale dello Stato (ACS), Ministero dell'Industria e del Commercio, Direzione generale della produzione industriale, Finanziamenti ERP, b. 17, f. 269 and b. 18, f. 286; Notiziario ERP, 1948–1953; Duccio Bigazzi and Giancarlo Subbrero, 'Tecnologia e organizzazione produttiva alla Lancia (1906–1969)', in Franco Amatori et al., Storia della Lancia. Impresa, tecnologie, mercati 1906–1969 (Milan: Fabbri, 1992), 242–7.

interested in high-output machines for volume production, whereas Alfa Romeo seemed more interested in devices which could enhance the quality of the product: gear cutters, grinders, precision boring machines, superfinishers.[39]

Fiat's plan satisfied both requirements: the need for greater productivity while employing unskilled labour inspired the requests for finishing and precision machinery. But the plan also included single-purpose machines equipped with automatic devices and mechanisms allowing multiple operations for series production. The prevailing trend, in all fields, was that of integrating previously distinct stages. Hence the introduction, at least from 1948,[40] of a few large 'complexes' with turning platforms: machines with multiple heads capable of performing subsequent operations on single pieces located on a revolving table. Also through the ECA Fiat sought to obtain a Greenlee boring-milling machine with 8 chucks which allowed the firm to 'perform automatically three subsequent operations on two brake shoes at a time' and to 'minimize the time needed for piece loading . . . , reaching a pace of 70–80 pieces per hour'.

These machines had initially perplexed the Italian observers, who thought 'that Americans were embracing too enthusiastically the use of "complicated" machine tools'. They had soon realized, however, that this was the way to overcome the artisanal style of national automotive production: Italy must follow the 'Anglo-Saxon spirit' which tended 'to "standardize" any job'. The complexity of this kind of equipment was in any case offset by the simplicity of their construction: they were made of basic elements which could be recombined, should a change in the product features require it.[41] Incidentally, the engineering division of Olivetti (OMO) acted on just such a strategy: it produced 'combined' machines which utilized unified elements (bases, columns, etc.) and standard units for the various operations (drilling, boring, tapping, etc.). These machines had been adopted, for instance, by the firms that produced motor scooters.[42]

Thus, Fiat's purchases of specialized multi-head 'complexes' were augmented by those of multiple drilling and boring heads (produced by Natco, Ex-Cell-O, or Heald) to be mounted on standard structures. It is likely that these purchases, uncharacteristically small in number by comparison with those made by Alfa and Lancia, were aimed at studying the most recent technological advances in order to update the firm's in-house production: in 1951 the creation of the Stabilimento produzioni ausiliarie (Auxiliary Equipment Establishment), with a capacity of about 400 metric tons a year, replacing a department which had previously formed part of Mirafiori, made it possible to manufacture special

---

[39] Fiat, 'Assemblea annuale degli azionisti, 31 Mar. 1950'; Bigazzi and Subbrero, 'Tecnologia e organizzazione produttiva', 242–3.

[40] Olinto Mario Sassi, 'Considerazioni sul progresso tecnologico alla Fiat nella produzione automobilistica', in *Il progresso tecnologico e la società italiana*, vol. 2, 205–6.

[41] E. Vandone, 'Considerazioni e notizie sulla seconda Esposizione europea della macchina utensile ad Hannover', *Rivista di meccanica*, 29 Nov. 1952, 4; id., 'Artigianato', ibid., 17 Jan. 1953, 3; id., '343 Miliardi di dollari', ibid., 31 Jan. 1953, 3. Vandone, a former Fiat and IRI manager, had visited the USA in 1949: cf. id., *America tecnica* (Turin, 1951).

[42] *Macchine*, Sept. 1953, 970.

machinery for serial production.[43] Two other plants (Grandi Motori in Turin and Fiat Modena) made machinery both for the firm and for the external market. At Fiat reverse engineering had always existed alongside the acquisition of reproduction licences: since 1937, for instance, there had existed an agreement with Sciaky of Chicago allowing construction in thousands of copies of the various models of welding machines designed by the American firm.[44]

ERP, therefore, made possible a more complex and lasting kind of technology transfer than that implied by the sheer purchase of equipment.[45] It was based on cultivating reciprocal trust-based relationships with plant and machinery producers who played an important brokerage role between automotive firms, facilitating the visits to the factories and often joining in them. A similar part was played by the consulting firms which co-ordinated the construction of complex plants involving a large number of suppliers: thus Fiat had developed a long-lasting relationship with Giffels & Vallet of Detroit, which managed the reorganization of both the casting and painting departments.

None the less the Marshall Plan could not stimulate a complete technological overhaul; this took place only a few years later. The perception that contemporary observers had of its role in Fiat's 'technical and productive revolution'[46] seems to have been inspired by the accelerated arrival of new machines, by the expectations they carried with them, and by the associated modernization message, more than by an objective consideration of their quantitative and qualitative features. But obviously the importance of this kind of assistance should be evaluated taking into account its more general effects. As has already been argued in relation to the French automotive industry, the ERP legitimated goals which would otherwise have appeared too ambitious: only by counting on this support could Fiat unabashedly pursue a development strategy based on mass production of cars.[47]

9.5   An Evolving Vision of Automation

To Fiat the new ' "philosophy" of production' advocated by John Diebold and other prophets of the forthcoming 'advent of the automated factory' was completely alien. Fiat's managers did not think that the increasingly frequent intro-

[43] *Macchine*, Oct. 1954, 1123–4.

[44] Bigazzi, 'Strutture della produzione', 324. Sciaky also co-operated with Alfa and Lancia, see Bigazzi and Subbrero, 'Tecnologia e organizzazione produttiva', 245.

[45] Nathan Rosenberg, 'The International Transfer of Technology: Implications for the Industrialized Countries', in id., *Inside the Black Box: Technology and Economics* (Cambridge: Cambridge University Press, 1982), 272; see also Richard Nelson, 'The Role of Firms in Technical Advance: A Perspective from Evolutionary Theory', in Giovanni Dosi, Renato Giannetti, and Pier Angelo Toninelli (eds.), *Technology and Enterprise in a Historical Perspective* (Oxford: Oxford University Press, 1992), 164–83.

[46] See Adalberto Minucci and Saverio Vertone, *Il grattacielo nel deserto* (Rome: Editori Riuniti, 1960), 57.

[47] Patrick Fridenson, 'L'Industrie automobile français', in Girault and Lévy-Leboyer, *Le Plan Marshall*, 289.

duction of automation implied 'a conceptual revolution with the same impact as Henry Ford's idea of the assembly line'.[48] Olinto Mario Sassi, speaking on behalf of the firm at an important meeting on 'Technological Progress and Italian Society' (1960) stated that: 'The replacement of conventional with automated machines has brought about no revolution. It has been a continuous but gradual trend, meant to make possible an increase in the scale of production.'[49] Even in the perception of the leaders of the movement towards technological change an event of great importance such as the introduction of transfer equipment could not alter the substance of the technological paradigm defined for the automotive industry fifty years before in Detroit. The concept of 'flow' was now supported by more efficient devices allowing conditions of 'automaticity' at a number of stages in the cycle, but these did not represent a new economic or socio-technical model. There was a confirmation of a practice made up of day-to-day adaptations, a kind of 'incremental reorganization of productive technology': the configuration of the lines changed constantly but the fundamental criteria on which the production cycle was based remained the same.[50]

During the 1950s, in any case, process automation at Mirafiori was very limited. This was especially true in bodywork, where there was piecemeal introduction of loading and unloading devices at the presses, electrostatic welding and then painting of the wheel rims, and electrostatic application of the first coat to bodies. In machining, by contrast, the use of multi-head automatics was increasingly common, and at the end of 1953 the first transfer machines were introduced on the cylinder block of the 1100/103.

But even in these two last cases, Fiat's technicians confirmed their evolutionary approach to technology. In their opinion, there was an obvious continuity between multi-head automatics and transfer machines: Sassi presented the transfer equipment as a result of a gradual replacement of conventional machines. Another manager, Neri Torretta, remarked how multi-head automatics and transfer machines were based on the 'same production concept', thereby representing 'equivalents' with 'a specific application field'.[51] Transfer

---

[48] John Diebold, *Applied Automation. A Practical Approach* (New York, 1955) quoted in Friedrich Pollock, *Automazione. Consequenze economiche e sociali* (Turin: Einaudi, 1970), 17; see also Diebold's testimony to the US Congress Joint Economic Committee, *Automation and Technological Change. Hearings before the Subcommittee on Economic Stabilization of the Joint Committee on the Economic Report* (Washington, DC, 1955), 8–9.

[49] Sassi, 'Considerazioni sul progresso tecnologico alla Fiat', 196. For Olivetti's evolutionary approach on automation see Luciano Gallino, 'Lo sviluppo dell'automazione in un'azienda processiva', in id., *Indagini di sociologia economica e industriale* (Milan, 1972), 137.

[50] See on this James R. Bright, *Automation and Management* (Boston: Division of Research, Graduate School of Business Administration, Harvard University, 1958), 84; and on a historiographical level, Stephen Meyer, 'The Persistence of Fordism: Workers and Technology in the American Automobile Industry, 1900–1960', in Nelson Lichtenstein and Stephen Meyer (eds.), *On the Line. Essays in the History of Auto Work* (Urbana and Chicago: University of Illinois Press, 1989), 74–5. For the quotation, Nelson Lichtenstein, 'Auto Worker Militancy and the Structure of Factory Life, 1937–1955', *Journal of American History* 67 (1980), 352.

[51] For these comments, see Sassi, 'Considerazioni sul progresso tecnologico alla Fiat', 196 and N. Torretta, 'Gli stabilimenti di Mirafiori e la lavorazione della Fiat 600', *Ingegneria meccanica*,

machines could only be developed along a linear route which had no spatial restrictions on the movement of components through the various work stages, as was true of multi-head automatics. From this narrowly technical perspective, common characteristics were emphasized over differences, thereby undervaluing the novelty of the *transfer* element: as Touraine has underlined, the name itself evokes the pre-eminence of organization and movement over and above the specific job performed, giving rise to complex problems of planning and balancing between the transfer mechanism and the other phases of the cycle.[52]

Fiat used transfer machinery less systematically than did other European automotive enterprises like Renault.[53] Even in machining, the share of totally or partially automated processes was significant but still a minority: according to the firm's data 20 per cent for the 1100/103 and, later, 30 per cent for the 600.[54] In most cases multi-head automatics, generally designed and produced inside the group, were still preferred: for the 600 series alone, Fiat manufactured 76 of these machines, some of which (like that for the steering column) were close to the maximum dimensional limits for the application of this type of equipment.[55] As for the transfer machines, the longest (123 yards) and most complex line was that dedicated to the 600 cylinder block: on its 108 stations, hundreds of automatic operations, three partial tests, and a manual assembly stage were performed. A total of 42 operating units were arranged alongside it, but most of the work was performed by three Sunstrand and Natco transfer machines; its total capacity was 50 pieces per hour.

These installations, impressive as they were in a European context, could not enable Fiat to compete, on size terms, with the American companies. Technical journals reported that at Pontiac and Plymouth, the cylinder block lines were 400 yards in length and had a theoretical capacity of 140 pieces per hour.[56] On the other hand, the introduction of transfer lines at Mirafiori did not simply imply the purchase of these machines in the United States. In 1953–4, combin-

Feb. 1956, 23. A similar view was held also by other technicians, as is shown by M. Chalvet, 'Macchine utensili a trasferimento e macchine affini', *Macchine*, Oct. 1952, 987–1001 (transl. from *Ingénieurs et Techniciens*); Hellmut Goebel, 'Le linee a trasferimento', *Macchine*, July 1956, 840, remarks moreover that, in a number of respects, transfer machines permitted more convenient technical solutions than did multi-head automatics.

[52] Alain Touraine, *L'Évolution du travail ouvrier aux Usines Renault* (Paris: CNRS, 1955), 33–5.

[53] In addition to Touraine, see Marius Hammer, *Vergleichende Morphologie der Arbeit in der Europäischen Automobilindustrie: die Entwicklung zur Automation* (Basel and Tübingen: Kyklos/J. C. B. Mohr (Paul Siebeck), 1959) and Jean-Louis Loubet, *Citroën, Peugeot, Renault et les autres. Soixante ans de stratégies* (Paris: Le Monde Editions, 1995), 51–4. On Austin, see Jonathan Zeitlin, 'Reconciling Automation and Flexibility? Technology and Production in the Postwar British Motor Vehicle Industry', *Enterprise and Society* 1 (2000).

[54] Olinto Mario Sassi, 'Limiti di convenienza economica di applicazione dell'automazione in alcune produzioni industriali', in Consiglio nazionale delle ricerche, *Convegno internazionale sui problemi dell'automatismo. Milano, 8–13 aprile 1956* (Rome, 1958), 892 and id., 'Considerazioni sul progresso tecnologico alla Fiat', 197.

[55] Torretta, 'Gli stabilimenti di Mirafiori', 28.

[56] Ralph H. Eshelman, 'Le nuove linee di lavorazione della Pontiac', *Tecnica ed organizzazione*, 24 (Nov.–Dec. 1955), 70 (transl. from *Tool Engineering*, December 1954); ASF, DAI, b. 107/6, 'Relazione tecnica visita a stabilimenti Usa 1955. Meccanica', 35 and 178.

ing the experience acquired on multi-head automatics with long experience in producing specialized machinery, the Stabilimento produzioni ausiliarie was able to manufacture the transfer equipment for the 1100/103.[57] The level of know-how rapidly reached in this field by Fiat is further demonstrated by the sophisticated transfer machine that performed certain stages in the processing of the camshaft for the 600. This extended automation to the loading, unloading, and testing of the component, and included a bench for the collection and adjustment of the tools, with lights indicating to the operator when the wear limit had been reached and the tool should be replaced. Fiat had installed these benches on the American transfer machines as well, solving one of the most troubling problems that affected process flow and product quality.[58]

Constraints on further introduction of transfer machinery were not, therefore, strictly technical. They had perhaps more to do with essentially conservative assessments of production levels and with the long commercial life of Fiat models in this period: the 600 remained in production from 1955 to 1970, with a total of 2,695,000 units being produced; the Nuova 500 was produced from 1957 to 1975, reaching a total of 3,678,000 units.[59] This conservative attitude was totally consistent with the company's evolutionary perspective. Further reductions in times could be obtained through new approaches to design and construction; automation itself led sometimes to a return to multi-head automatics or at any rate to shorter transfer lines. In the case of camshafts, the cycle time was nearly halved between 1948 and 1956, falling from the 58 minutes needed for the 500 Topolino to less than 30 minutes for the 600; for the former, conventional machines were used, apart from a multi-head drilling automatic, while for the latter a Fiat transfer machine with 26 stations was used. With the Nuova 500, the cycle time decreased again, to just over 23 minutes. This further progress had nothing to do with automation, originating instead in a simplification of the design and in the replacement of steel with spheroid cast iron, which allowed a greater throughput speed; the machinery included, in addition to conventional equipment, only two multi-head automatics and a transfer machine with three stations and five operating units.[60]

Fiat was therefore far from the 'push-button factory' utopia evoked by American publicists at the time when Ford's Cleveland engine plant had just opened. The Italian firm's technicians visited the Cleveland plant in 1951 and 1955, fully grasping the novelty represented by its systematical use of transport

---

[57] Torretta, 'Gli stabilimenti di Mirafiori', 26; the most remarkable of these transfer machines, 12.5 yards long, included 20 stations.

[58] Ibid., 28–9. On the issue of timely tool replacement see the testimony by Ralph E. Cross to US Congress Joint Economic Committee, *Automation and Technological Change. Hearings*, 503 (also quoted in Pollock, *Automazione*, 87–8) and Goebel, 'Le linee a trasferimento', 838.

[59] Robert Boyer and Michel Freyssenet, 'Emergence de nouveaux modèles industriels', *Actes du Gerpisa*, 15 (July 1995), 63; ASF, *Fiat: le fasi della crescita. Tempi e cifre dello sviluppo aziendale* (Turin: Scriptorum, 1996), 122.

[60] Sassi, 'Considerazioni sul progresso tecnologico alla Fiat', 204–25; see also id., 'Limiti di convenienza economica', 903–6.

and linkage devices between transfer machines, but they still preferred simpler and more manageable production lines.[61] The American experts themselves were beginning to register the effects of the increasing rigidity of production lines, and began to appreciate 'sectionized automation' or 'unitized automation' (according to whether GM's or Ford's jargon was used).[62] By 1960 Fiat also took for granted the use of 'recombinable elements which would allow changes in the type of production relatively easily and cheaply'. On the new cylinder block line, for instance, the passage from the 4-cylinder to the 6-cylinder version was quick and simple.[63]

The prevailing trend was towards the integration of processes, and Fiat also followed this course, if prudently and gradually. By 1960 the throughput cycle of the engine block included 126 stations and 58 operating units (mostly concentrated in five large transfer machines). The line was 200 yards long, one-third of which was made up of links and connections. These also served as buffer stocks through which the firm hoped to solve the problems of 'continuity of production . . . made more complicated and indeed exacerbated' by large transfer machines.[64] The use of these machines still gave rise to some second thoughts, in particular concerning some components for the 124 and 125, but with the 128 (1969) and later with the 127 (1971) they came into more general use. In 1966 there were 60 transfer machines in the engine shop alone, while the 127 gearbox line was deemed the largest in Europe. Its capacity was 150 pieces per hour whereas, according to the company's experts, American and Japanese factories turned out 120 per hour. A report by the Sezione produzioni ausiliarie written some years earlier had conceded that American machines represented a constant source of inspiration and at the same time had remarked with pride that 'Fiat's transfer machines are by no means inferior to their American counterparts'.[65]

9.6   The Mirafiori 'System' and the Centrality of the Conveyor

Since its foundation Mirafiori had been represented as a 'colossus', the 'largest plant in Italy'. The imperial drama of the factory's opening, in which Mussolini

[61] Bright, *Automation and Management*, 59 ff.; Hounshell, 'Planning and Executing "Automation"', 64 ff.; ASF, DAI, b. 107/4, 'Visita a stabilimenti USA. Autunno 1951. Relazione tecnica III parte', 207–48; b. 107/6, 'Relazione tecnica visita a stabilimenti USA 1955. Meccanica', 8–15.

[62] Eshelman, 'Le nuove linee di lavorazione della Pontiac', 68; Cross, in US Congress Joint Economic Committee, *Automation and Technological Change. Hearings*, 503; Hounshell, 'Planning and Executing "Automation"', 76–7.

[63] Sassi, 'Limiti di convenienza economica', 890; id., 'Considerazioni sul progresso tecnologico alla Fiat', 197.

[64] This line is described in detail in Neri Torretta, 'Osservazioni sui moderni sistemi di lavorazione meccanica dei getti. Memoria presentata al 6 Convegno Assofond' (Turin, 1961), 1–5.

[65] ASF, Stabilimenti Fiat, b. 3, 'La Fiat Mirafiori', Jan. 1966; DAI, b. 107/11, 'Visita in USA a costruttori ed utilizzatori di macchine a trasferta, 28 settembre–18 ottobre 1968'; b. 51, 'Visita in Giappone. Industrie automobilistiche. Fabbriche di macchine utensili. Novembre 1968'; Alberto Imazio and Carlo Costa, *L'organizzazione del lavoro alla Fiat* (Venice: Marsilio, 1975), 67–75. According to Lawrence J. White, *The Automobile Industry since 1945* (Cambridge, MA: Harvard University Press, 1971), 24, a gross production of 90 to 95 blocks per hour was considered satisfactory in America.

himself had taken part, reinforced this image. Then the labour movement had made the Turin plant its symbolic focus: beginning with the Resistance struggles, the organizational roots maintained in the factory testified to its general strength and representativeness.[66] In the 1950s the company further emphasized Mirafiori's centrality within its production structure, as an example of its technical achievements and a projection of its outlook for the future. Visits to the plant invariably produced a kind of mystified admiration for the complexity of the production process. In 1955 an anonymous engineer remarked: 'There is nothing there that doesn't have a reason, nobody without a job to do . . . Nobody can do without anybody else; nobody can know all this.'[67]

Alessandro Pizzorno remarked that in these years when the industrialization process was being completed, there existed in Italy a strong propensity to the 'ideological valorization and cultural dramatization of the technological phenomenon'.[68] Mirafiori was no exception to this imperative, which focused not on the individual machines (the transfer equipment or, as in the past, the huge presses), but rather on the *system*. Fiat's technicians insisted on this point, maintaining that its key element was materials handling. The greatest innovation in this field was the overhead conveyors: 'It can be said [wrote Torretta] that they are the essential feature of the plant . . . A different system for moving materials could not be adopted without upsetting the plant structure.' The increase in the adoption of these devices became the best indicator of the growth of the plants. In 1946, the overhead conveyor network had only 7 segments, 3.1 miles long, and carried only a few components (tyres, batteries, radiators, nuts and bolts). Between 1948 and 1950 the first storage conveyors, attached to the presses, were introduced, and the number of units increased to 23, approximately 6.9 miles in length. During subsequent years the company sought to utilize the plant's basement while increasing the number of storage conveyors: in 1956 there were 160 units, over 22 miles long in total. The transportation network grew in proportion to the plants' physical expansion: 69 miles in 1962 and 140 miles in 1968.[69]

Conveyors performed a double function: on the one hand, they simplified the movement of huge quantities of components that would otherwise jam traffic on the shop-floor (in 1955 it was calculated that they could move 230,000 components at once, a weight of 1,500 metric tons); on the other, they made it possible to eliminate stationary storage for batch processing, for instance at the

[66] Luisa Passerini, *Fascism in Popular Memory. The Cultural Experience of the Turin Working Class* (Cambridge: Cambridge University Press, 1987), 183–200; Tim Mason, 'Gli scioperi di Torino del marzo 1943', in Istituto nazionale per la storia del movimento di liberazione in Italia, *L'Italia nella seconda guerra mondiale e nella resistenza* (Milan: F. Angeli, 1988), 399–422.

[67] This quotation, dated 8 June 1955, is in a notebook collecting impressions from visits to the factory between 1954 and 1956, organized by the Associazione meccanica italiana (AMI). This notebook is now the property of this author.

[68] Alessandro Pizzorno, 'Fare teoria in una disciplina in sviluppo', *Sociologia del lavoro* (1984), 163.

[69] ASF, Stabilimenti Fiat, b. 7, Servizio impianti e sistemazioni, 'Relazione sui lavori eseguiti negli anni dal 1946 al 1955'; b. 3, Direzione stampa e propaganda, 'Cenni storici', 1962; b. 3, 'La Fiat Mirafiori', 1968; Torretta, 'Gli stabilimenti di Mirafiori', 18.

presses. Extensive use of conveyors was essential for the constant transformation of equipment and for the growth of production capacity: they had the all-important function of allowing 'the almost complete dissociation of the factory's topography from the sequence of production'.[70]

Since the 1920s, Fiat had drawn inspiration from the Fordist philosophy of 'keeping everything moving', as Lingotto's morphology itself testified. With Mirafiori, the outside appearance of the plant became anonymous: it was nothing but a container designed to host an ever-changing process. Thanks to a layout free from structural constraints, to the modular design of the departments, and to the availability of land nearby, it was possible constantly to reorganize the production lines and to adjust the plant for ever-higher volumes: from 300 cars a day in the immediate post-war period, to 1,000 in 1955. In 1956 the core of Mirafiori South was built, extending the roofed area from 31,500 to 45,000 square feet; this allowed 2,000 cars a day to be turned out by 1960. In the subsequent years, a seemingly endless growth took place. By 1967 the roofed area had reached 108,000 sq. ft. and production 5,000 cars a day.[71]

The focus on the movement of materials and components thus became a feature of the plant. On this subject, Fiat's technicians had had some differences with their American consultants, who in 1947 had, for instance, advocated the use of trolleys rather than hanging hooks on painting lines. The Italians insisted on the latter—still experimental in the USA—for reasons involving installation and maintenance costs, product quality, and process flow. At this stage they seemed less interested in the advantages trolleys presented from the standpoint of flexibility and product mix.[72] Likewise, in the following years, in all of the assembly phases the traditional ground lines were replaced by hook conveyors of the Webb type. During one of the first visits to Japan (in 1968), Fiat's men were struck by the relatively limited use of conveyors, especially in bodywork production.[73] But the same was true also for the United States, where in 1959 it had been noted that in the most recent Ford and GM plants the use of overhead conveyors was much rarer than at Mirafiori; particularly in the machining departments electric fork-lift trucks prevailed.[74]

Fiat's primacy in this field was questionable. Admittedly, it met the need to expand rapidly and to a certain extent suddenly, with fixed structures based on a constant flow but at the same time exposed to continuous change. The extreme focus on materials movement created contradictions and was doomed

[70] Torretta, 'Gli stabilimenti di Mirafiori', 19.

[71] ASF, VCA, 27 Jan. 1956, 30 Jan. 1960; Fiat, Assemblea annuale degli azionisti, 27 Apr. 1956, 27 Apr. 1962, 29 Apr. 1966, 28 Apr. 1967. See also Marco Revelli, Lavorare in Fiat (Milan: Garzanti, 1989), 31–3.

[72] ASF, DAI, b. 107/2 'Relazione viaggio in America ottobre–novembre 1947'; Bruno Bartalucci and G. Veraart, 'Linee di verniciatura per carrozzerie di automobili', Ingegneria meccanica (Nov. 1959), 43–4.

[73] ASF, DAI, b. 51, 'Visita in Giappone', Nov. 1968.

[74] ASF, DSI, b. 107/8, Fiat. Sezione costruzioni ed impianti, 'Relazione sopraluogo negli USA dal 15 al 29 novembre 1959'.

in the long run to exacerbate the problems it was meant to solve. Conveyors necessitated meticulous and constant control of the production process from the centre. This was perfectly consistent with an organizational logic traditionally based on a rigid, militaristic hierarchy: the command structures and industrial relations within the company failed to evolve, due also to an employment policy based on unskilled labour assigned to repetitive jobs and subjected to an increasingly intensive pace of work.

The company thus suffered the consequences of its choice to integrate processes, and to forgo decentralization and plant specialization, which had been pursued, apparently without qualms,[75] throughout the whole post-war expansion. The Turin plant complex, conceived from the start as a function of car production, became more and more skewed towards Mirafiori, which in turn had become an enormously complex machine, with a labour force of 54,000.[76]

Soon indiscriminate expansion was no longer possible.[77] But the company's solution to the problems admittedly posed by this logic was simply to duplicate Mirafiori at the new Rivalta complex, thereby exacerbating, among other things, the crisis of Turin as a city which had seen its population increase by 65 per cent between 1951 and 1971. Fiat accounted for most of this growth, as it brought in labour from the depressed South. The new workers had to adjust to a factory perceived as a frightening and chaotic Moloch and to living conditions and lodgings which left much to be desired: their uneasiness grew to the point where it gave rise, from 1969 and for the next decade, to serious social conflict.[78]

9.7   An Alternative Technological Style: Alfa Romeo and Italian Light Engineering

This gradual alignment of technologies and organization with the concept of mass production and specifically with 'Detroit automation' revealed a difficulty in maintaining and developing the productive style that had been Fiat's trade mark at least until the end of the 1950s. In that period the expansion of production had been dealt with through recourse to technical and professional resources collected over the years and left underutilized. ERP's role had been significant but Fiat's men did not go to America in search of technological or organizational revelations. As Paul Rosenstein-Rodan rightly noted: 'The management of Fiat . . . already knew in the 1930s everything about American

---

[75] This topic is treated only incidentally in Vittorio Bonadè Bottino, 'Criteri di impostazione delle costruzioni industriali', *Atti e rassegna tecnica* (1951), 10.

[76] ASF, *Fiat: le fasi della crescita*, 144.

[77] Fiat, Assemblea annuale degli azionisti, 28 Apr. 1967.

[78] Revelli, *Lavorare in Fiat*; Giovanni Contini, 'The Rise and Fall of Shop-Floor Bargaining at Fiat 1945–1980', in Steven Tolliday and Jonathan Zeitlin, *Between Fordism and Flexibility. The Automobile Industry and its Workers* (2nd edn., Oxford: Berg, 1992), 144–67; Stefano Musso, 'Production Methods and Industrial Relations at Fiat (1930–1990)', in Shiomi and Wada, *Fordism Transformed*, 252 and 258–9.

methods of automobile production. They did not introduce it, because the size of the Italian market did not justify mass production. When the Italian economy grew into a sufficient size, the previously known methods were introduced.'[79]

Rosenstein-Rodan, who knew Italian reality well enough, understood correctly the sophistication acquired by Fiat's management in the inter-war period, ascribed to a significantly earlier date what we can term their 'apprenticeship', and put into perspective the effects of the phase starting in 1945, in which contacts with the USA were more intensive and regular. This chapter has, however, sought to show that the transfer of the Fordist technologies and organizational methods took much more time, and was much more complex and contradictory than that author seems to have assumed. At least in the early stages, Fiat was able selectively to adapt American industrial culture to its national reality, deploying automation prudently and sparingly, utilizing internal technological capabilities, and above all avoiding an excessive rigidity in the production flow. In other respects, as in materials movement, the company went even beyond the practices more generally adopted in US plants.

Rosenstein-Rodan's approach may be especially misleading in that it establishes a direct, almost automatic relationship between market size and application of mass production techniques. Elsewhere I have sought to show how American methods could be adapted even to the smallish production scale of the inter-war Italian automotive industry. Lingotto's layout was obviously designed for continuous flow of materials and products, although, in reality, this concept had to be mitigated through the introduction of areas of batch production. At Lancia, where operations were on a much more limited scale, a team of top-notch toolmakers devised simple but creative jigs and fixtures which permitted a more efficient use of high-quality universal machines: multiple pieces could thus be processed at once, while turning tables allowed loading while the machine was working.[80] After the war production volumes were much higher, but success still did not depend on a reasonably rapid and complete adoption of mass-production methods. Nor did the achievement of a minimum (and very much debated) efficient size[81] represent the sole parameter for assessing cost reduction and therefore the capacity for survival of the lesser firms.

Alfa Romeo, in particular, which abandoned the purely artisanal approach advocated by Gallo was able to profit handsomely, in the 1950s and 1960s, from the opportunities opening in the sector of medium-sized cars: the Giulietta, which

---

[79] Rosenstein-Rodan, 'Technical Progress', p. 163, quoted in Vera Zamagni, *Dalla periferia al centro: la seconda rinascita economica dell'Italia 1861–1990* (Bologna: Il Mulino, 1990), 434; Duccio Bigazzi, 'The Production of Armaments in Italy, 1940–45', in Jun Sakudo and Takao Shiba (eds.), *World War II and the Transformation of Business Systems* (Tokyo: Tokyo University Press, 1994), 197; Francesca Fauri, 'The Role of Fiat in the Development of the Italian Car Industry in the 1950s', *Business History Review* 70 (Summer 1996), 179–80.

[80] Bigazzi, 'Strutture della produzione. Il Lingotto'; Bigazzi and Subbrero, 'Tecnologia e organizzazione produttiva', 223–7; see also Musso, 'Production Methods and Industrial Relations at Fiat', 255–8.

[81] A number of views on this issue are presented in White, *The Automobile Industry since 1945*, 50–3.

marked the decisive shift to series production, came to be associated by the public with the firm's glorious sporting tradition. Its overwhelming success was helped in no small measure by the technical capabilities acquired in manufacturing aero engines. Even before the war the recruitment of a number of bright, young engineers just out of the Milan and Turin schools allowed the creation of a top-notch, extremely closely-knit design staff. Admittedly, in the production departments the superior culture of this group, whose acknowledged leader was Orazio Satta, had to confront the authority of the technical personnel on the shop-floor, which was based on practical knowledge and experience. But these difficulties were overcome, at first thanks to Gobbato's charismatic personality and then, after 1951, with the advent of a number of capable organizers who had had previous experiences in serial production: a former manager at Pirelli, Franco Quaroni, Rudolf Hruska, an Austrian who had worked with Ferdinand Porsche's team during the completion of the Volkswagen project and, during the war, on the production of the Tiger 70 tank. The new management could therefore capitalize on the abundant designing resources already concentrated at the Portello plant, adapting a robust tradition to new needs.[82] Thus, first of all, the high construction quality standards demanded by aircraft engines were transferred to cars. Second, the company was able to exploit fully the wartime experience with materials such as aluminium and electron, a magnesium alloy. The legacy of aircraft production at Alfa Romeo included a modern foundry for light alloys and a remarkable extrusion plant. This equipment allowed the firm to secure significant reductions in the weight of cars: the Giulietta weighed 820 kilograms as compared to 1,004 for the 1900. Besides the obvious advantages in terms of performance, substantial economies in materials and processing costs could thus be obtained. The structure of the unitary body was streamlined through elimination of all unnecessary parts, and light alloys were utilized for a great number of mechanical components, among others the engine block, the gearbox, and the differential case; with regard to the latter, in particular, the company had to overcome the resistance of Gleason, which supplied the gear-cutting machines. Using the same logic, the company sought to replace machine tools in processing, replacing metal with mould casting or pressing processes.[83] Moreover, the production departments were able to benefit from the high quality and great professional versatility of the labour force. For instance, Alfa's toolmakers were able to manufacture in a few months at the Portello plant the dies for the 1900 which had been previously ordered from Budd. They therefore avoided a delay in the start of production which could have compromised the viability of the whole company.[84] Not everything had to be 'home-made',

[82] Angelo Tito Anselmi and Lorenzo Boscarelli, *Alfa Romeo Giulietta* (Milan: G. Nada, 1985); Griffith Borgeson, *Alfa Romeo: I creatori della leggenda* (Milan: G. Nada, 1990).

[83] Anselmi and Boscarelli, *Alfa Romeo Giulietta*, 35–7, 41–2; interview with Ivo Colucci, 21 Jan. 1986; ASAR, 'Piano di razionalizzazione e sviluppo delle officine "Alfa Romeo". III tranche ERP', vol. I; see also, on a more general level, Luciano Gallino, 'L'automazione, processo globale', in id., *Indagini di sociologia*, 168.

[84] ASAR, VCA, 23 Jul. 1951; interview with Ivo Colucci, 21 Jan. 1986.

though: many components such as brakes, shock-absorbers, and valves were purchased from specialized producers, both Italian and foreign (especially German).[85]

The available financial resources were used to update the equipment, and to improve the press and body works. The first assembly lines, belatedly established at the end of 1951 for the production of the 1900, were gradually expanded so that from the 20 units forecast by the initial daily plan for the production of the Giulietta, they reached an output of 100 units a day in 1958–9. To the standard bodies manufactured at Portello the company added two elegant and diversified lines, Sprint and Spider, contracted out to specialized body works such as Bertone, Pininfarina, and Zagato, which were largely responsible for the success of this car. Thanks to the Giulietta, Alfa Romeo steadily exceeded Lancia's performance both in production and in domestic market share.[86]

From at least 1952, Alfa Romeo's design teams had worked on the idea of a very small car that could challenge Fiat on its own ground. The realization of this project was, however, postponed a number of times because of the hesitations of Finmeccanica (the state holding company which controlled Alfa), as well as Fiat's natural opposition. In 1954, just before the Fiat 600 entered the market, Alfa Romeo forecast a daily production of 400 units for such a car. This theoretical target was increased to 1,000 units for an 800–900cc model but yet again the company had to relent and to confine its efforts to the production first of the Dauphine (1959) and then of the R4 (1962) under a licence agreement with Renault.[87]

The strategy of pursuing a lower market segment had poor chances of success, and represented a loss of company identity. In the 1960s, the management gradually became persuaded that imitation of the technological and organizational choices of larger-scale producers (namely Fiat) was unavoidable. The opening in 1964 of the new Arese plant, extending over an area of more than two million square metres, reflected an alignment with mass-production orthodoxy. This choice was made even clearer in 1968 at the new plant of Pomigliano d'Arco near Naples. This plant did, however, represent the first significant move away from the traditional hubs of the automotive industry, Turin and Milan.

Increasing market and financial difficulties eventually brought Alfa Romeo into the Fiat group in 1986. Lancia, unable to count on the state as a backer willing to finance its budget deficits on a regular basis, had met a similar fate 15 years before (in 1969). The result in both cases was a net loss in the design originality and technological style which had sustained the two companies in

[85] ASAR, VCA, 17 Jul. 1950; interviews with Giampaolo Garcea (13 Feb. 1985) and Giuseppe Busso (16 Apr. 1986).

[86] ASAR, VCA, 27 Apr. 1956; Amatori, *Impresa e mercato*, 174–5 and 255–6; ASF, *Fiat: le fasi della crescita*, 120–1.

[87] Giuseppe Luraghi, 'Alfasud mezzogiorno di fuoco', supplement to *Espansione* 64 (Feb. 1975), iv–v; Loubet, *Citroën, Peugeot, Renault*, 153–4.

their respective golden ages. This outcome, as well as the decline of Fiat's innovative capabilities during its Fordist period, was disappointing; however, it should not overshadow the peculiar characteristics and conditions which had made possible the remarkable post-war development of the Italian automotive industry.

The Fiat 600 and the Alfa Romeo Giulietta belonged to distinct market segments and were very different in their underlying concepts and strategies. None the less, they were representative of a more general technological style, which can be discerned even more clearly in the concurrent development of other branches of engineering. One thinks, on the one hand, of relatively complex products with a long tradition such as typewriters (Olivetti) or sewing machines (Necchi, Borletti), which made the transition from exclusively professional to widespread popular use during the 1950s; on the other, of certain mass consumer goods which contributed significantly to modifying national social patterns and ways of life during the same period: motor scooters such as those produced by Piaggio and Innocenti, domestic electrical appliances manufactured by Candy, Ignis, Zanussi, etc. What all of these products had in common, which determined their success on the domestic market as well as abroad, was a peculiar mixture of sobriety, simplicity, lightness, and elegance of line. Energy constraints, scarcity of raw materials, and the need for strict economies in running costs drove domestic firms to turn out goods of striking formal design, devoid of any hint of penury or austerity, which seemed to herald a new era of abundance for the whole society.

The transition to fully industrial production on a large scale did not stifle the need for product differentiation: frequent adjustments in designs, a wide range of colours, and a variety of shapes were entirely compatible with serial production.[88] In this spirit the new general manager at Necchi, Gino Martinoli, who had been trained at Olivetti, completely overhauled working methods, production ranges, and sewing machine designs. He recalled that a comparison of American and Italian production cultures did not favour the former:

In 1950–51 I went to visit Singer, and my hair stood on end seeing how primitive their ways still were. Years later I saw them . . . the President of Singer . . . came to meet me and told me: I have to thank you because you made us understand how to use the machines in a wholly different way. And indeed they had completely overhauled their [production] lines, they had included different colors, they had softened out their lines.[89]

[88] Stanislaw H. Wellisz, 'Studies in the Italian Light Mechanical Industry. I: The Motorcycle Industry' and 'II: The Sewing Machine Industry', *Rivista internazionale di scienze economiche e commerciali* (Nov. 1957), 1015–61 and (Dec. 1957), 1161–82; Vittorio Gregotti, *Il disegno del prodotto industriale: Italia 1860–1980* (Milan, 1982), 269 ff. On the long-term technological constraints see Renato Giannetti, 'Mutamento tecnico e sviluppo (1880–1980)', in Pier Luigi Ciocca (ed.), *Il progresso economico dell'Italia. Permanenze, discontinuità, limiti* (Bologna, 1994), 47–79.

[89] Excerpt published in 'Memorie di un decennio di modernizzazione. Archivio delle fonti orali sugli anni '50', *Censis. Note e commenti* 30, 4 (1994), 106. But see above all Gino Martinoli, 'Come un'azienda italiana ha migliorato la sua efficienza', *Produttività* (Oct. 1951), 897–902.

The artificial antagonism between mass production, low in costs and poor in quality, and a highly sophisticated and elitist form of craftsmanship dissolved, if only temporarily, into a happy co-operation. A number of factors contributed to this result. First of all, reinforcement of the engineering sector during the war had left behind capabilities that could for the first time be used to turn out consumer goods; these human resources could build on a wealth of new materials and military technologies which could be adapted to peacetime production. The most striking instance of the opportunities thus made available is perhaps the motor scooter industry, where the two most successful companies, Piaggio and Innocenti, turned out the same kind of goods, though starting with different technical and plant endowments. Corradino D'Ascanio, who had been a pioneer in helicopter design, adopted for the Vespa Piaggio an unusual approach, closely related to his previous experience: in order to create a nimble, convenient vehicle, whose mechanical parts need not be exposed, he designed a self-supporting body made of pressed sheet and other light materials. The Lambretta Innocenti, on the other hand, also designed by two aeronautical engineers (Cesare Pallavicino and Pier Luigi Torre), featured a load-bearing structure made of a single large tube. This was consistent with the company's wartime production, which consisted of bombs and ammunition made from tubular components.[90]

The conversion of these and other important firms to peacetime production relied on a network of specialized machine and component producers, previously absent. During the post-war crisis of the engineering industry, a wealth of professional and technical capabilities formerly clustered in the larger companies became available and moved on, to radically renovate production methods in the smaller plants. This phenomenon affected the periphery as well as the major industrial areas, since many workers and technicians who had been involved in war production returned home and started their own businesses: this is particularly evident in studies of Emilia-Romagna,[91] but holds as well for a great number of local production systems and Marshallian industrial districts in Lombardy, Piedmont, the Veneto, Tuscany, and the Marche.

This confluence of favourable circumstances implied an approach to the American industrial environment which, instead of looking exclusively to large-

[90] Anty Pansera (ed.), *L'anima dell'industria. Un secolo di disegno industriale nel Milanese* (Milan, 1996), 156–8; Tommaso Fanfani, *Una leggenda verso il futuro. I centodieci anni di storia della Piaggio* (Pontedera, 1994), 95–6; Giuseppe Lauro, 'Produttività nell'industria del motorscooter', *Produttività* (Feb. 1952), 157–61; Corradino D'Ascanio, 'Come è nata la Vespa' and Renato Tassinari, 'L'alveare delle Vespe', *Rivista della produzione e dell'organizzazione Piaggio* (Jan. 1949), 3 and 9–11.

[91] Pier Paolo D'Attorre and Vera Zamagni (eds.), *Distretti, imprese, classe operaia. L'industrializzazione dell'Emilia Romagna* (Bologna: Il Mulino, 1992); Vittorio Capecchi, 'Una storia della specializzazione flessibile e dei distretti industriali in Emilia-Romagna', in Frank Pyke, Giacomo Becattini, and Werner Sengerberger (eds.), *Distretti industriali e cooperazione fra imprese in Italia* (Florence: Banca Toscana, 1991), 35–50; also, for a good example, see Giovanni Solinas, 'Competenze, grandi imprese e distretti industriali. Il caso Magneti Marelli', *Rivista di storia economica* (1993), 79–111.

scale companies, appeared to bring together simultaneously all of the evolutionary stages of the American system of manufacturing. In post-war Italy, a number of goods which had become symbols of the development of American productive technologies (sewing machines, typewriters, two- and four-wheeled vehicles) jointly acquired full industrial maturity: in the space of a few years almost a hundred years' worth of catching up was completed.

Chapter 10

The Long Shadow of Americanization: The German Rubber Industry and the Radial Tyre Revolution

PAUL ERKER

10.1 Introduction

The history of the 'Americanization' of German industry reveals itself upon closer examination to be not a one-sided penetration, but rather a complex history of relations in which those involved had changing roles.[1] The fact that the attempt to identify the economic and technological influence of the United States on the German industrial economy yields no simple pattern is due primarily to three variables. First, the picture varies with the temporal perspective. The picture presented by the 1950s and 1960s will show variations in nuances and in fundamental features depending on whether one takes into consideration continuities with earlier 'periods of Americanization' before the First World War or during the Weimar Republic, whether one includes intervening and competing influences such as those during the time of National Socialism and the Second World War, or whether one distinguishes between indirect, informal influences (such as the orientation towards America of German entrepreneurs) as opposed to direct, formal, and intentional influencing from American sources. Second, depending on the industry, but also in different companies in the same industry, varying degrees of the economic and technological influence emanating from the USA can be ascertained, whereby the frequent conception of two 'types of Americanization'—German heavy industry on the one hand and in the modern chemical, electrical, and automobile industries on the other—is an oversimplification. Third, there is the question of the short-, middle-, and long-range harbingers of this influence. This chapter examines not only the American factor in the strategic, organizational, and technological roots of the success of German industry in the post-war period, but also the 'American share' in the seeds of economic misadventure, as well as technological failure.

[1] See Paul Erker, ' "Amerikanisierung" der westdeutschen Wirtschaft? Stand und Perspektiven der Forschung', in Konrad Jarausch and Hannes Siegrist (eds.), *Amerikanisierung und Sowjetisierung in Deutschland 1945–1970* (Frankfurt: Campus, 1997), 137–45.

Taking the example of the tyre industry—which more than almost any other industry was influenced macroeconomically by the economic developments in the United States, as well as microeconomically by American companies—this chapter traces various areas and phases of influence in German industry. The starting-point and fundamental perspective are provided initially by the 1950s and 1960s. The processes of Americanization will be examined by means of a comparison of two competing companies: Phoenix-Gummiwerke AG (Hamburg-Harburg), and Continental-Gummiwerke AG (Hanover). What will emerge are two 'Americanization models', not only with respect to the ways and means of the American influence but also with respect to the perception of the German companies. The results obtained will then be put into context—first by means of a look back at the 1920s and 1930s, and then by means of a 'preview' of the 1970s and 1980s—in order to be able to specify, in a diachronic comparison, assessment criteria for the degree of 'Americanization', as well as the degree of ambivalence resulting from the basis of success and secondary damage. We shall see how markedly the contours of these two pictures differ from one another.[2]

10.2 American Tyre Technology at Continental and Phoenix, 1945–1965: Implementation and 'Careful Examination'

After the Second World War the big American rubber corporations were among the first US industries to re-enter Germany. Their motives were clear: on the American domestic market cutthroat competition with excess capacities had continued to obtain, even in wartime. This situation made Europe, where motorization was just getting started, appear as a highly promising market—especially Germany. From the very beginning co-operation rather than domination prevailed, for the two largest German tyre companies, Phoenix and Continental, appeared to possess interesting know-how, and furthermore, as future partners, they saved the Americans the expense of costly investment in plants of their own in Germany. Before the end of 1945 the major US tyre corporations had therefore, using the resources of the engineering corps of the occupying power, which for the most part was filled with their own experts, begun to get a precise picture of the development and state of tyre know-how in Germany. The Americans were aware of their own basic technological lead, and the comprehensive data in the FIAT (Field Information Agency, Technical) and BIOS (British Intelligence Objectives Sub-Committee) reports on compound formulae, process technologies, and machine designs provided few insights.[3]

---

[2] This research is largely based on the USA-Reiseberichte (American travel reports) between 1920 and 1964 in the Continental Archive (CA), to which all file numbers in subsequent notes refer. See as regards the industry as a whole, Paul Erker, *Competition and Growth. The Continental AG Story* (Düsseldorf: Econ, 1997).

[3] See BIOS Report, 'Survey of Manufacturing Methods and Equipment, Continental Gummiwerke AG, 1947' and BIOS Overall Report 7, 'The Rubber Industry in Germany during the Period 1939–45', 1948.

But in terms of the analysis of and test methods for synthetic rubber, discoveries had been made in the German laboratories concerning 'various innovations . . . that generated great enthusiasm among American and British observers alike. It might very well be the case that no fundamentally new methods have been developed, and Germany does not seem to have come as far in comparison with Anglo-American advances, but there is an abundance of individual innovative items nevertheless.'[4] And in the processing methods for Buna (synthetic rubber) nothing fundamentally new was ascertained, 'but since at present for lack of a good theory implementation is of decisive importance, the reports on German praxis doubtlessly possess great value'.[5] It was mainly because of the particular complications of tyre technology, that here, unlike many other industries, there was an absence among both the Allied engineering officers and the Germans of that 'incomplete understanding of science and technology . . . which made the Allies think that science and technology could be channeled, controlled, taken away from here and implemented in their own countries'.[6] It was clear to the American tyre corporations that often only the Germans possessed the technological know-how to make the acquired knowledge productive. Thus, even in the late 1940s the West German rubber companies possessed enough valuable bargaining chips to start negotiations with Anglo-American companies about know-how and patent exchange.[7]

On the basis of initial, laboriously acquired information on tyre development and the tyre market in the United States, the Germans were of the opinion that 'during the war years no revolutionary inventions and innovations [had] been made' there.[8] At the same time Continental's tyre engineers were well aware that with respect to productivity and tyre-product technology the American corporations were clearly in the lead, particularly as regards compounds and tyre construction. Now for the first time it became painfully clear to what extent the German tyre industry had been cut off from every kind of information on developments in the United States. 'They apparently', concluded the BIOS officers as a result of their inquiries, 'had no idea at all as to how American tires are made'.[9] Thus, when the rapidly recovering German automotive industry approached the tyre companies even before the end of 1946 with the demand 'to come up with modern ideas in the tyre area which reflect an orientation to the developments in the major American companies', the tyre companies recog-

---

[4] BIOS Overall Report 7, 'The Rubber Industry in Germany during the Period 1939–45', 1948, 1 ff.

[5] Ibid.

[6] See Raymond G. Stokes, 'Assessing the Damages: Forced Technology Transfer and the German Chemical Industry', in Matthias Judt and Burghard Ciesla (eds.), *Technology Transfer Out of Germany After 1945* (Amsterdam: Harwood Academie, 1996), 81ff., as well as his 'Technology and the West German Wirtschaftswunder', *Technology and Culture* 32 (1991), 1–22.

[7] See as regards the electrical industry Paul Erker, 'The Politics of Ambiguity: Reparations, Business Relations, Denazification and the Allied Transfer of Technology', in Judt and Ciesla, *Technology Transfer*, 131–44; and CIOS (Combined Intelligence Objectives Sub-Committee) Report 23-1, 'Continental-Gummiwerke AG Hannover', Apr. 1947.

[8] CA, note, 16 May 1947, 6525 1/56, A25.   [9] BIOS Report, 29.

nized the scale of the problems they faced.[10] Although Continental had its own ideas about the new 'post-war tyre', the company nevertheless opted not to 'approach the automobile industry before we know what the United States is doing . . . We do not wish under any circumstances to pursue courses of our own, but rather orient ourselves to the trails blazed in the tire sector in the United States.'[11] Continental and Phoenix pursued different courses in their efforts to obtain the necessary information and close the yawning technological gap.

Before the middle of 1948, Continental had inaugurated a series of trips to the American tyre mecca in Akron, Ohio. These visits were to continue at intervals until the end of the 1960s. From these trips it is possible to put together an itinerary, so to speak, of the learning process at Continental. Facilitated by long-standing business contacts and with assistance from former officers from the Allied rubber control board, the first visit was focused on leading institutions outside the company in the areas of tyre chemistry (Smithers Laboratories) and tyre-making machines (National Rubber Machinery Co.), as well as Goodyear, General Tire, and Goodrich, which along with Firestone and US Rubber made up the 'Big Five' companies of the American rubber industry. The on-site inspections shocked Continental's engineers, because the technological gap was far greater than had been suspected. 'There is no doubt', stated their report, 'that we will be able to hold our ground only [if] we take part in the development in the American rubber industry with respect to technology and chemistry and to this end constantly maintain the contact with America.'[12] A mutually beneficial co-operation agreement with General Tire was soon in place. At the same time it became apparent that what made German companies so interesting for the Americans (in addition to the penetration of the market in Germany and in Europe in general) was that the US corporations, which had traditionally specialized primarily in tyres, especially Firestone, Goodyear, and General Tire, were right in the middle of revising their business strategy and beginning to diversify into the industrial products (IP) sector. And it was in this area that Continental (like Phoenix) had long had comprehensive know-how, which had scarcely become dated even during wartime. This gave Continental the opportunity to become familiar with the latest American methods of tyre production and tyre chemistry, and at the same time to avail itself of General Tire's commercial organization for purchasing and selling abroad.[13] Other reasons for the Continental board's choice of General Tire as the best partner was that it was the fastest growing American rubber company, and apparently had a lead over Goodrich and Goodyear in production methods.

After brief negotiations Continental and General Tire concluded a co-operation agreement on 22 March 1949 for an initial period of five years. The Germans made available their experience, formulae, and patents in the area of

---

[10] CA, letter, 24 Jan. 1947, 6525 1/56, A25.   [11] Ibid., 3 ff.
[12] CA, Hörisch report, Aug. 1948, p. 12, 6714 verschiedene Zugänge, A67, 1.
[13] CA, report of visit to General Tire, 6 Aug. 1948, 6610 1/85, A4.

industrial products, in exchange for which the Americans provided access to their production know-how, their world-wide sales network, as well as raw materials at cost price. Not only did Continental thereby save on high R&D costs; the necessary modernization of its tyre-fabrication facilities—the first of the new 'tyre automatons' were set up two months later in Hanover—also took place shortly afterwards. A final point was that in the opinion of the Continental board there was 'in the association with this company not the danger of being too strongly influenced from the outside'.[14] The intensive exchange of information and experience, which started even before the conclusion of the contract, was centred on production technology. Continental's engineers and chemists studied every last detail of the mixing procedures, heating methods, and automated tyre-building processes. With a plethora of blueprints and formulae in their pockets the Continental contingent returned to Hanover and set to work to implement the new information. Nevertheless, the new 'post-war tyre' that was now produced at Continental was not an 'American tyre'. The limited applicability of many improvements in details soon became apparent, because each change in the method of tyre construction or the type and quality of rubber mixture could not be implemented without creating problems of incompatibility with the existing tyre-production processes. The copying of technical details did not deliver identical results at the end of the production line. It was therefore not technological scepticism or a 'not-invented-here attitude' which led to difficulties in the technical adaptation but above all the peculiarity of the complicated procedures of rubber processing and tyre production. Furthermore, the situation in Germany was very different from that in the USA—not only with respect to automotive technology but also with respect to road surfaces and speed limits. To Continental's engineers this meant 'that our preferred qualitative construction of the tyre has proven itself in practice and [therefore] the know-how of General Tire cannot be approved without careful examination'.[15] In other words, there were technological limits to the 'Americanization' of German tyres.

By early 1950 the number and nature of the US trips had changed noticeably. The fact-finding visits became fewer, and questions concerning future technological development were more prominent: rim width, wire cord or nylon cord as a reinforcing component, tyre standardization, tubeless tyres, and tread design—these were now the topics of interest.[16] The focus of co-operation with General Tire shifted increasingly from production technology to central strategic decision-making aids. However, it is noticeable that information and ex-

[14] CA, Report of visit to General Tire, Aug. 1948, 6610 1/85, A4, 6.

[15] CA, report of visit to General Tire, 31 May–10 June 1949, Dr. Bickel, 6500/3 ZG 1/56, A8, 2, p. 6, 'The important European concern to build tyres that do not skid on the treacherous European roads even in wet weather does not exist in America', reads a somewhat laconic visitor's report. 'In America practically every tyre doesn't skid because the potential for skidding has been reduced through the improvement of the quality of the roads.' See Hübener report, June 1949, 6500/3 ZG 1/56 A8, 2, p. 12.

[16] See Continental–General Tire correspondence, 6714 ZG 1/71, A1.

perience transfer nevertheless remained restricted to R&D and production. Questions concerning company organization were not discussed at all; an analysis done in 1950 of General Tire's budgeting system explicitly revealed the non-existence of anything to emulate, and, although the thorough report on the marketing methods did acknowledge their efficiency, because of the differently structured channels of distribution, as well as the fact of the 'dealer system', there was nothing for Continental to adopt.[17]

At about the time that Continental and General Tire concluded their co-operation agreement Phoenix was just beginning to search for a strong American partner. Negotiations dragged on for two years, which gave Continental a considerable competitive edge at the start. The initiator and driving force behind Phoenix's search for support from one of the major US tyre corporations was Otto A. Friedrich. Friedrich, who had trained in the American tyre industry, assumed the chairmanship of the board on 1 April 1949. He had lobbied vehemently after 1945 for an 'Americanization' of German industry,[18] finally concentrating his efforts on the second largest of the American 'Big Five': the Firestone Tire & Rubber Company. The agreement that Phoenix and Firestone concluded towards the end of November 1951 and which took effect in mid-1952, established a type of association different from that between Continental and General Tire in a number of respects: the period of the agreement lasted 20 years, and during this time Phoenix not only based its production 'largely on Firestone patents', but also instituted under the brand name 'Firestone-Phoenix' a joint arrangement for export and marketing. A further element was a 25 per cent capital participation in the Harburg company on the part of Firestone. From the beginning Friedrich had conceived this contract as 'nothing less than a model contract for further contractual agreements between US and West German companies' and had therefore involved both the Marshall Plan administration and the West German government.[19]

With this arrangement Phoenix had tied itself far more strongly and lastingly to Firestone than Continental had to General Tire and had also made itself dependent on the American company not only financially, but also technologically and commercially. In a formal sense the independence of Phoenix was guaranteed, nevertheless the basis of co-operation was clearly more one-sided at the expense of the Harburg company, since Firestone—in contrast to General Tire—displayed to all appearances only a modicum of interest in the IP know-how of the Germans and much more in strategic advantages for the conquest of the German and European tyre market. Although Phoenix did not keep any notes on the US visits or reports on know-how transfer, there are nevertheless a

---

[17] See several reports in CA, 6500 ZG 1/63, A1 and 6714 verschiedene Zugänge, A67, 1.

[18] See Volker Berghahn, 'Otto A. Friedrich: Politischer Unternehmer aus der Gummiindustrie und das amerikanische Modell', in Paul Erker and Toni Pierenkemper (eds.), Deutsche Unternehmer zwischen Rüstungswirtschaft und Wiederaufbau. Studien zur Erfahrungsbildung von Industrieeliten, forthcoming.

[19] See several documents in the Phoenix Archive (unclassified) and CA, 6600 ZG 1/68, A5.

number of indications that a tyre was subsequently produced at Phoenix that was developed, produced, and marketed in accordance with American methods and principles, as well as being subject to the relevant 'American' cost-and-profit calculations.

In contrast to Phoenix, Continental, on the expiry of the contract with General Tire in the mid-1950s, made a strategic switch of partners. On 28 July 1954, just a few months after the contract ended, Continental concluded a co-operation agreement with Goodyear. The Germans' motives were primarily technological; apparently Continental considered the advantages of the know-how transfer from General Tire to be exhausted, whereas a co-operation agreement with Goodyear, the world market leader, opened up new opportunities for Continental in product technology and production technology. Furthermore, by that time General Tire was conducting negotiations for a co-operation agreement in the IP sector with Metzeler AG, one of Continental's main competitors in the German domestic market. The contract with Goodyear, which ran for ten years, led to a marked 'Americanization effect' at Continental. The crux of the agreement was the off-take production of Goodyear tyres by the Hanover company. Some of the production facilities in Continental's tyre plants—ranging from compound formulae, tyre-building, and lay-up machines, up to and including vulcanization moulds—were now converted entirely to the American method. Not only did Continental thereby combine access to Goodyear's tyre know-how with improved capacity utilization and considerable additional earnings; at the same time the company pursued—and this was the main thing—a defensive strategy as well: Continental wanted to prevent the American tyre company from setting up factories of its own in Germany. For Continental, therefore, the co-operation agreement was prompted by strategic competitive considerations as much as by technological motives.[20]

The co-operation between the German and American tyre companies was, by and large, marked with difficulties at first and characterized on both sides by initial mistrust. Continental was apprehensive—more so than with General Tire—of being dominated by the far larger corporation, whereas Goodyear was well aware that it was dealing with a competitor that had by then revived and constituted, at the least, a potential rival. Nevertheless, in the mid-1950s a second wave of visits to America commenced at Continental. In close succession board members, engineers, and plant managers travelled to Akron to study the situation at Goodyear. One of the main items on the initial agenda was the implementation of American production technology for off-take tyres, which did not meet with the satisfaction of the Americans until after a long and involved negotiation over production figures and production quality. The second primary interest of Continental's managers had to do with efficiency planning and profitability. In spite of the lessons learned from the General Tire phase, with respect

---

[20] See various notes and Continental–Goodyear correspondence, CA, 6600 ZG 1/68, A2 and 6714 verschiedene Zugänge, A78.

to production time and production costs Continental continued to exhibit a clear deficiency compared with the American company: in the USA an average of 39.6 minutes was required for the production of one passenger-vehicle tyre, whereas at Continental 84 minutes were needed. In addition, in spite of the lower German wages (but with far higher material costs, as well as a greater tyre weight), the unit production cost at Continental was almost twice that in the American plants: DM 47.10 against (the equivalent of) DM 28.25.[21] By the time Continental had increased its speed of production, aided by the experience of the American company, new planned-efficiency measures were already making themselves felt in the American plants. Here, too, the Continental engineers believed 'that we should not simply imitate Goodyear in an obtuse fashion in order to implement the requisite simplifications'.[22] To a far greater extent than previously, however, efficiency planning was now understood as a comprehensive process and as having relevance both for organizing work and for generating profit. Hence in the course of studying Goodyear's production methods—which were based on simplification, cost reduction, and a simultaneous increase in quality—attention was also directed at organizational structures, cost accounting systems, and industrial labour relations. As regards the purely mechanical set-up of the plant, Continental had had occasion to observe 'that we are not at all less modern; on the contrary . . . in some things we are even ahead'.[23] With respect to uniform plant management and collaboration between R&D and production, however, a number of things at Continental were in a sorry state. A visit report stated:

The full utilization of our recently introduced direct processing from the Banbury [mixing machine] to four cylindrical calenders is aborted by the lack of openness to changes which are just simply necessary for this economical fabrication method . . . In contrast to the manufacturing methods at Goodyear, traditional adherence to conventional methods of manufacture plays a large role with us, which is a circumstance that has a negative effect, particularly with respect to a great number of details.[24]

As a result, Continental's managers came back from Akron, particularly in 1955 and 1956, with a long list of detailed measures which the Hanover company promptly set about putting into effect.[25] With growing self-confidence, however, they set limits to the adoption of American methods, and even drew attention to Continental's own superiority. In particular with respect to tyre marketing the company continued to be sceptical about a changeover to the American methods: 'We already knew from General Tire and have now made the same observation in one of Goodyear's own facilities that both tyre manufacturers get automobile dealers to take the standard tyres off new cars, even those of competitors, and replace them with a higher quality of their own make',

[21] See correspondence notes, CA, 6500/3 ZG 1/56, A8, 2 Hübener Report, late 1949.
[22] CA, Beckadolph report, Oct. 1955, p. 4, 6575 ZG 1/71, A2.   [23] Ibid., 3.
[24] Ibid., 4.
[25] See Bähr/Richter/Warnecke report, CA, 17 Aug.–17 Sept. 1956, visit to Goodyear, 6575 ZG 1/71, A3.

stated a visit report for April and May 1955. 'The additional costs, on which both the car dealer and the tyre dealer make even more money, go to the expense of the consumer . . . If we wanted to introduce something similar in Germany we would no doubt stir up a storm of indignation on account of unfair market practices.'[26] The conclusion was that American marketing methods, as well as the dealership system in existence there (sole agency), were just as much products of organic growth as was the structure of the tyre business in Germany (as well as in the rest of Europe), with its sale of various brands, which had developed over a number of decades. 'For Continental, therefore, there are no discernible advantages that could occasion us to assume the risks of a thoroughgoing change in our system of sales for no specific reason.'[27] When in 1957 a group of Continental managers visited the United States for several months to study business organization and its implementation, as well as to assess its relevance for Continental, their report, of almost 100 pages, was similarly negative. Although the primary destination of the trip was Goodyear, they also visited Ford (to study the wage system), Sears, Roebuck & Co. (to study its delivery organization), as well as Harvard University and MIT (to study the educational system). It was reported that Continental's decentralized organization was better than the centralized form at Goodyear; further that although the American accounting system with its budget system was logically and simply constructed, ' the absence of planned-cost calculation—with the result that an accurate calculation of production costs is not possible—must be seen as a substantial drawback'. The apprenticeship for master craftsmen ('supervised training') was in some respects exemplary, according to the report, but was comparable with Continental's own company training programme. The report concluded:

If we wanted to give a short answer to the question, 'What have you brought back of real substance from America? What new procedures, methods and perspectives should now be adopted by us?', it would not be easy to provide an answer. First of all, the answer is negative. We have been told that in America you will find the country of uniformity and standardization. There planning is a pleasure. Our finding: the opposite is the case. Of the much-praised standardization there is nothing left but remnants. No one can stop this development. We have been told that in America even the horses run faster. Our finding: the rate of work is by no means faster than ours. The development is clearly moving in the direction of a social welfare state. We have been told that in America you will everywhere find an exemplary form of organization and clear cost calculations. Our finding: the organization is indeed in existence and covers the smallest details; but it is not in touch with plant operations, as well as being too costly. We have been told that America has the best educational system. Our finding: the educational system is extremely well maintained, but it is too expensive for us. The main positive result of the trip is the confirmation of the fact that we are on the right track with our own methods. We have seen an extraordinary amount of very interesting things and have received a great number of stimulating impressions.[28]

  [26] See Hoppmann /Niemeyer report, CA, 14 Apr.–14 May 1955, 3ff., 6575 ZG 1/64, A1.
  [27] Ibid., 16.
  [28] See Birn/Hahne/Otto report, CA, 15 Aug.–16 Sept. 1957, 6500 ZG 2/69, A4, 2.

Continental's managers indicated that they had been positively impressed not least by the attitude to work of the company employees, thereby putting into words a central mental barrier to the 'Americanization' of German industry.

Even if this or that in American and German working life are similar in kind, there is nevertheless an unmistakable fundamental contrast. In the final analysis the American gets more done . . . We frequently measure the economic performance of plant operations in terms of the ratio of white- to blue-collar workers; the opinion still prevalent in Germany is that the smaller the quota comprised by the former employee group the better the plant. This attitude constitutes a considerable obstacle for German efficiency planning, which by and large has to be effected by office workers, who have to make it prevail over the 'inborn' conservatism of the industrial worker, which in our plants is much more pronounced than it is in the New World. It is clearly the case that there every job instruction that involves a change is easier to put into effect than it is here. The American is more flexible in this respect, he regulates the labour input and the work itself with a view to the end result, whereby the employment of mental work connected with a programme or an organization, coupled with technological energy, brings him better results than the blanket application of costly muscle power.[29]

A third point of emphasis in the visits to Goodyear was the interest of Continental's engineers in the tyre technology of the future, and the likely primacy of steel-cord tyres or nylon-cord tyres. The impulse for the new reinforcing-component technology had been provided by Michelin towards the end of the 1930s, at which time a search was being made for improvements on account of the short service life of tyres on automobiles with a front-wheel drive system. Continental had also occupied itself during this period with the development of steel cord—but mainly in connection with the search for substitute materials. In 1948 the French tyre company was the first to launch a marketable product. The 'wire-cord tyre' then became a topic of the consultations between General Tire and Continental. There was agreement that the new tyre could become important and that R&D efforts would have to be undertaken in this direction. But the Americans had pointed to the development of fully synthetic fibre, and the likelihood that consequently the nylon tyre would soon supplant the wire-cord tire. And so it was that tyre research on the American side at the beginning of the 1950s was weighted increasingly in favour of the nylon tyre, and Continental followed suit with this reorientation of R&D efforts. The slow but steady advance of Michelin steel-belted tyres did create uncertainty among the Germans as well as the Americans, however. But at a conference of Goodyear and Continental managers in September 1955 both sides viewed the future of the steel-cord tyre once again as favourable.[30]

In 1957–8 Goodyear acquired a licence from Michelin for the relevant tyre technology, which was protected on all sides by patents, and Continental saw an opportunity to get hold of the new tyre know-how by a circuitous route. Goodyear began production of its 'Uni steel-cord tyres' in 1958. The technological advantage

[29] Ibid.   [30] See the minute in CA, 6600 ZG 1/68, A2.

of the new tyre was that, as a result of the radial instead of the cross-ply con-struction of the carcass, in combination with the use of steel cord, the service life of the product was more than twice as long as that of the traditional tyre. At the same time, the tyre also gave improved driving performance and greater load-carrying capacity. Goodyear's engineers made it clear to the Germans, however, that the new tyre would at best gain acceptance for trucks. In view of the prior-ity given by American car owners to driving comfort, the engineers regarded the steel-belted radial, with its appreciably harder driving performance, as having no future as a passenger tyre. They said that the automobile manufacturers in Detroit would have categorically refused to make any kind of change in con-struction design (suspension) to accommodate the tyres.[31] Continental's engi-neers for their part were well aware that, 'with respect to driving comfort European automobile manufacturers and drivers [are] far less sensitive, and far more concerned about value for money'.[32] But despite the Germans' critical atti-tude towards American work methods and production principles, in this area they had a good deal of respect for the judgement of the American world mar-ket leader. 'The study of the manufacturing problems resulting from the use of nylon in tyre construction, as well as the study of the production method for the belted tyre produced under licence from Michelin'—this was how they described their investigations in the USA in the following months and years.[33] As far as nylon-cord tyres were concerned, Continental proceeded with the uti-lization of American experience and know-how; as far as steel-belted tyres were concerned, the company kept an eye on developments and waited to see what the Americans would do.

By way of an interim assessment it can be said that by the early 1960s there were clear signs of the influence of the German tyre industry's orientation towards America. This manifested itself in a concentrated form in the make-up of the competitive landscape. The two 'Americanized' companies, Phoenix and Continental, despite the varying modalities of this influence, were pulling far ahead—with the help of the American corporations—of the other German tyre companies such as Metzeler, Fulda, Veith, and even the German branch of Dunlop—not only in productivity but also in profitability. It was characteristic of the situation that in the mid-1950s only Phoenix and Continental were in favour of the introduction of the tubeless tyre, which then quickly became stan-dard, whereas all the other German tyre companies were against it and tried to prevent the corresponding changeover in tyre technology.

The process of 'Americanization' had also had a profound effect on the com-petitive relationship between Continental and Phoenix. Otto A. Friedrich's strategic calculation on signing the agreement with Firestone was that this would enable Phoenix to break Continental's traditional predominance. This calculation seemed to be paying off. 'Is the technological assistance from

---

[31] See minutes of the Goodyear–Continental meeting, 15 May 1958, CA, 6500 ZG 2/69, A4, 2.
[32] Ibid., 3.   [33] See Warnecke/Richter/Heimberg report, CA, 20 Feb.–22 Mar. 1960, ibid.

Firestone,' Continental's board members were already asking themselves with alarm in the early 1950s, 'which is so emphasized by Phoenix, far more effective than that which Continental is getting from General Tire? Is a significant expansion of Phoenix exports to be expected as a result of the use of the double name Firestone-Phoenix, as well as the fact that Phoenix can make use of Firestone's outlets and general offices as if they were its own?'[34] Everything that Continental was able to glean in the following years from a comparison of the balance sheets tended to confirm its apprehensions: Phoenix had far greater liquidity and more available capital than Continental, had rapidly pulled even with respect to productivity, and surpassed the German market leader with respect to per capita profitability. Moreover, under the protective cover of Firestone, Phoenix had adopted a more aggressive market stance, with the result that the Harburg company was gaining growing market shares at home and abroad—to the detriment of Continental.[35] Even in its traditional position as technological leader Continental now saw itself under attack from Phoenix on the domestic market. Despite the fact that here as in other areas the Hanover company had followed the American example and introduced white-wall tyres, Continental had considerable problems in fabricating flawless Goodyear white walls, whereas with the help of Firestone know-how Phoenix was producing tyres of far better quality. In the intense struggle for the lion's share of original-equipment orders from VW, Continental was in danger of losing out. As far as the future development of tyres was concerned, however, it turned out that Continental held a substantial advantage. For although Firestone had also acquired a licence from Michelin for steel-belted radial tyres, Phoenix, unlike Continental, had scarcely noticed or capitalized on this, with the result that the Harburg company had no technical information on the 'wire-cord tyre'. Nevertheless, by the end of the 1950s and the beginning of the 1960s Phoenix was on the verge of wresting the leadership position—both in the market and in terms of technology—away from Continental.

10.3   Flashback: German Tyre Companies and 'Americanization' between the Wars

During the 1920s, as in the 1950s, Phoenix and Continental had each put a different 'Americanization model' into practice. At that time, however, it was Phoenix that was more reserved about US influences and which had turned down capital participation or other forms of co-operation with the American rubber corporations. As a result of disengagement from its Viennese partner, Phoenix had emerged greatly weakened from the years of inflation and, in the face of Continental's takeover attempts towards the end of the 1920s, had had difficulty defending its hard-won independence. It was not until the boom engendered by the military build-up under National Socialism and during the

[34] CA, note, 21 July 1952, 6600 ZG 1/68, A5.
[35] See Erker, *Competition and Growth*, 47.

Second World War that the company finally stabilized itself and, under the new arrangements of the war economy, as well as on the basis of the specifically German Buna technology, managed to initiate a revival.

Continental, on the other hand, had been seeking an American partner with which to sign a co-operation agreement immediately after the end of the First World War. It had chosen the B. F. Goodrich Company, which, because of the fierce competitive struggle in the American tyre market, was attempting to gain a position with potential in the European market and, in view of the decline of the German currency, was hoping also to be able to make a lucrative capital investment for itself. The contractual arrangements that Continental entered into with Goodrich between 1920 and 1929 comprised not only a close financial involvement, by virtue of a 25 per cent capital participation on the part of the Americans, but also a far-reaching technological, organizational, and marketing-strategy orientation to the American corporation on the part of the Germans. Here, as later, Continental's engineers were primarily interested, on visits to the USA, in the superior fabrication methods, at this time those pertaining to the then new balloon tyre. They studied carefully the Goodrich method of assembly-line production and put it into effect in Hanover almost down to the last detail.[36] Continental's board thereby submitted itself to what was in essence a process of 'business re-engineering' on an American foundation.[37] After the new production technology was introduced (in the early 1920s), the primary component in this process was the Bedaux system (adopted in the second half of the 1920s). This was a combination of a new type of job organization and wage arrangement, and a comprehensive plant-control system.[38] Thus Continental was far more highly 'Americanized' at that time than in the 1950s. Continental's rapid rise to international competitiveness again, in the period between the two world wars, along with its gilt-edged balance sheets, made its strategy appear to be a highly successful model in every respect. It was this Continental/Goodrich co-operation model that Otto A. Friedrich, who was executive board chairman at Phoenix after 1945, had in mind for his collaboration with Firestone!

The fact-finding visits and the reciprocal exchange of experience between Continental and the American tyre industry continued, even after the official departure of Goodrich in 1929, almost up to the outbreak of war in 1939. At the latest by the middle of the 1930s, however, Continental observed that—as a result of increasing government intervention with respect to the work-force, the wage system, and the transfer of capital and foreign exchange—the company was slowly but surely losing ground at the international level. In the course of

---

[36] See Bobeth report, July/Aug. 1920, 'Neuerungen in der amerikanischen Gummiindustrie'; and Simon report, 'Bemerkenswerte Fabrikationsmethoden bei Goodrich', Apr./May 1921, and several reports of visits, 1922–8, 6500/1 ZG 1/58, A1, 1–A1, 2.

[37] See Erker, *Competition and Growth*, 23 ff.

[38] See Paul Erker, 'Das Bedaux-System. Neue Aspekte der historischen Rationalisierungsforschung', *Zeitschrift für Unternehmensgeschichte* 41 (1996), 139–58.

the war the 'Americanization effects' became increasingly more tenuous, if not negated entirely. One reason was the need, dictated by the Nazi regime, to develop a specifically German model—not only for reasons of efficiency planning (using forced labour) but also for tyre technology (using substitute raw materials). Another was that a notable learning process of a different nature took place at Continental during this period: the initial successes of the war effort suddenly and unexpectedly gave Continental access to valuable Michelin know-how on both production and product technology, particularly the developments in the area of the 'wire-cord tyre'. However, the principles of the 'steel-cord philosophy', to which Michelin continued to adhere in the post-war period, were as we have seen displaced at Continental after 1945 by the 'nylon-cord philosophy' favoured by the Americans.

10.4   The Radial Revolution: The Decline of 'Americanized' Tyre Technology in the 1970s

The cycle of 'Americanization effects' returned—albeit in other circumstances and in a changed economic world order—in the 1970s. In the first half of the 1960s both Phoenix and Continental underwent thoroughgoing changes. In the United States the competition between the rubber corporations had intensified drastically as a result of the tough race to modernize, as well as their increasing capacities. The major corporations found themselves caught up in a wage–price spiral, and the growing resistance of the unions to streamlining measures had compelled them to set up—at considerable cost—new, state-of-the-art facilities outside Akron.[39] Furthermore, after the development of nylon-belted tyres in the 1950s, the rubber corporations put less of their profits into R&D, but rather pursued a diversification strategy, involving high initial costs, and invested in plastics, as well as in marketing battles for larger shares of the tyre market. By contrast Europe, where use of cars had grown rapidly, was a lucrative market unspoiled by the price wars that raged at home, and Germany became the major battleground of the American tyre corporations. The competition forced the small and medium-sized European manufacturers to seek 'protection' under the wings of larger corporations, or to switch to new lines of production to compensate for their shrinking tyre business. Whereas in Phoenix Firestone already had a secure bastion in Germany, Goodyear was positioning itself in late 1961 for the purchase and takeover of Fulda AG. Thereupon Continental announced the immediate termination of the co-operation agreement—a decision which surprised Goodyear just as much as the earlier termination had surprised General Tire. The vexation felt by the Continental board about Goodyear's move was deep, for now the smaller competitors also enjoyed the privilege of American technology and production methods, and Continental was

---

[39] See also Mansel Blackford and Austin Kerr, *BF Goodrich: Tradition and Transformation, 1970–1995* (Columbus, OH: Ohio State University Press, 1996), 274 ff.

confronted in effect with an 'American production facility' within the German domestic market.

Continental went its own way, and the separation from Goodyear was amicable, like that from General Tire. Hence the doors to the American tyre laboratories and production facilities remained open. In the early 1960s a third round of US trips by Continental board members and managers began, the purpose being to reduce the company's persistently large productivity gaps. Unlike Phoenix, which remained tied to Firestone and its methods, Continental now could obtain information from all the tyre corporations. Around March 1963, production and R&D specialists from Continental undertook an extensive trip to eight tyre plants, where they studied the production methods of the three major corporations—Goodyear, Goodrich, and General Tire. Their overall conclusion was that Continental's production methods were satisfactory but that 'the Americans [had] succeeded in raising the rate of work even higher and further reducing tyre production costs'.[40]

Meanwhile, however, Michelin had was capturing an increasing share of the European tyre market with a version of the steel-belted radial that had been further perfected in terms of production technology and product engineering. Whereas in the USA textile cross-ply, or breaker-strip, tyres still dominated the market, in Europe textile cross-ply tyres and steel-belted radial tyres were engaged in an open competitive struggle. But Continental's board was putting all its money on 'American technology'. '[We] have', it was explicitly stated in the annual report for 1965,

decided in favour of the fabric belt; for here as well the saying holds true that the better is the enemy of the good. The new Continental breaker-strip tyre, with its belt of CRG cord—which is a completely new product developed in conjunction with leading fibre manufacturers—combines the good mileage of the steel-belted tyre with the driving comfort of the conventional fabric-belted tyre. An entirely new kind of tread has also been developed for this tyre, which further improves its characteristics with respect to skid resistance and tracking ability. The production capacities for these tyres are already partly ready for operation or are being improved.[41]

Nevertheless, Continental saw itself in danger of being crushed between the American corporations on the one hand and Michelin on the other and in 1966 therefore took stock of its own strengths. It tried, together with Dunlop and Pirelli, to forge a grand European tyre alliance and also to undertake—admittedly somewhat half-heartedly—R&D with respect to steel-belt technology, which differed entirely from the traditional tyre technology. These attempts failed. Unsettled by the increasing signs of Michelin's advance, in the late 1960s the board at Continental looked around for new directions. The first step was a round trip to the German automobile manufacturers. They reported that the

[40] CA, Freundlieb/Junghänel and Braudorn/Mauck reports, 6500 ZG 2/69, A 4, 3 and 6575 ZG 1/71, A2–A4.
[41] Continental AG, Annual Report 1965, 13.

harder and therefore less comfortable steel-belted radials were not acceptable for German cars. And in fact a switch to the new tyres would have necessitated changes in the technology of German motor vehicles—particularly a more stable chassis—which no one in the German automotive industry was prepared to make. Without interest from the car companies, yet another costly conversion of the production facilities made no sense for the tyre industry. The board returned to Hanover reassured by the news. A consumer survey commissioned by the board provided a further all-clear sign: Adolf Niemeyer, executive board member in charge of sales, and also board spokesman, presented a study on the replacement market which indicated that the vast majority of consumers considered security to be the most important of the various tyre factors—security, service life, comfort, etc. Hardly any of those surveyed named the factors that were nevertheless soon to become decisive for consumer behaviour: price and mileage. No one surveyed—whether competitors, original equippers, consumers, or suppliers of reinforcing materials—saw any future for the steel-belted radial tyre. What finally decided the issue, however, was a trip made to Goodyear by Continental's board. It was not until Goodyear reiterated that the French tyre would have no chance against the technological and marketing power of the American corporations that Continental concentrated its R&D efforts on the further improvement and more economical production of the fabric-breaker tyre.

But there was no stopping the triumphant march of the Michelin tyre. Within a short time the market share of all the German tyre manufacturers combined slumped from 75 per cent to a scant 30 per cent. Continental, which as market leader in 1969 supplied about 45 per cent of passenger vehicle tyres and around 40 per cent of truck tyres, saw its market share plummet in 1972 to between 20 and 25 per cent. The initial sales losses occurred in the tyre-replacement market, for it was there—in contrast to what had happened with previous innovations—that the steel-belted tyre of the French tyre company had started its march to victory. The French tyre had twice the service life (60,000 km) and a higher load capacity than the bias-ply tyre. Furthermore, because by then Michelin had perfected its production technology, as well as having its own steel-cord factory, the French competitor was able to sell its tyre at a price just above the production cost for the fabric-breaker tyre. Almost helplessly the traditional tyre manufacturers in Germany, with Continental at the fore, were forced to stand by and watch as the French competitor continued to relieve them of their shares of the replacement market. When finally German car makers also succumbed to customer pressure and switched to steel-belted tyres and began to put stronger engines in front-wheel drive vehicles—which caused the performance of crossply tyres to drop rapidly from 15,000–20,000 km to 5,000–8,000 km—the share of the German tyre manufacturers in the original equipment market also dropped suddenly. Overnight, splendid profits turned into enormous losses.

On the German side and, after some delay, on the American side as well, a period of imitating and copying began, a time of dead-ends and detours, and a technological pursuit unprecedented in the history of the rubber industry.[42] A closer look at the nature of the R&D process in tyre technology makes it clear how great Michelin's lead was. Long development periods, high development costs, and a sizeable tie-up of capital were and are distinguishing features of the product-development process. It was these long innovation cycles that had shaped the thinking in the rubber industry for more than one hundred years. Whereas Continental now set to work on an intensive appropriation of the new steel-belt-radial technology, the American tyre industry continued to put its money on a development of its own: the 'bias-belted', or 'breaker-strip', tyre— a cross between the bias-ply or cross-ply tyre and the radial tyre, the sole advantage of which was that it could be built on the conventional lay-up machines and vulcanized in the old moulds. By the time the Americans finally switched over to the steel-belted radial as well, the European tyre manufacturers had attained a substantial technological lead. The main reason for the German ability to adapt to the Michelin challenge, which the Americans proved unable to do, was not greater flexibility in work organization but the time lag in market pressure. Europe was first to be hit by the marketing attack of the new French tyre but therefore was also forced to start reacting to the competition. Furthermore, the motorization boom of the 1960s and 1970s reduced the effects. The American tyre industry was hardest hit because until the mid-1970s the home market preferred the old type of tyre. Thus the first response of the American tyre companies—with the exception of Goodyear, which was the only American tyre corporation to invest offensively in the new technology (and hence in the end the only one to survive)—in the early 1970s was to retreat from Europe: Goodrich gave up its tyre plant in Coblenz, which had been built just a few years previously; Uniroyal searched desperately for a buyer for its entire European tyre division; and before the end of December 1970 Firestone had terminated its cooperation agreement with Phoenix. But in the mid-1970s radialization became dominant on the US market also, and this development coincided with the great slump in the automobile industry, triggered by the oil crisis. The American tyre companies paid the price for having removed the technical experts from its management committees.

At this point the 'American orientation' of the two German tyre companies— different in degree but still in evidence—threatened to backfire. Continental was the first to feel the consequences. From as early as the mid-1970s, when Continental had learned with great effort to master French radial technology, the major problem was that its production costs remained 20–30 per cent above those of Michelin. Further difficulties stemmed from the high initial investment needed for converting production facilities, and a simultaneous reduction in profits, as a result of which the company plunged deep into the red. Phoenix was in better shape at first thanks to its good capital structure. However, its technological dependence on Firestone meant that the company suddenly lacked new

[42] See Erker, *Competition and Growth*, 53 ff.; Julius Peter, *Dehnbare Erinnerungen: 1000 Stunden Gummiindustrie* (Ratingen, 1993), 107 ff.

tyre know-how. It seemed just a question of time before both companies—not to mention the remaining independent German tyre companies—disappeared from the market.[43] It appeared that only an amalgamation of the two companies could secure their survival. Despite repeated overtures, however, plans for a German tyre union miscarried.[44] In the end both companies did survive the crisis, but in the late 1970s Phoenix paid for its failings by being forced to leave the tyre market. It emerged from this period of radical change, permanently weakened, as a pure IP company. And it took Continental ten years of hard going before it, the sole surviving German tyre company, succeeded in catching up again with the leaders on the world market—Michelin, Bridgestone, and Goodyear. By the 1980s, the transformation of the industry's competitive landscape was such that the German company was now the technological leader, confidently confronting the reeling American corporations and holding the upper hand in the process of intercorporate rearrangement: Continental bought Uniroyal's European tyre plants in 1979, concluded a 'Technology Exchange and Co-operation Agreement' with General Tire in 1982, and finally, in 1987, took over General Tire entirely.

10.5 Conclusion

The turbulent history of the tyre industry constitutes a special case with respect to the effects of 'Americanization' on German industry. This is all the more so given the fact that ultimately the American methods and technologies backfired so drastically in this branch of industry—not only against the 'Americanized' German companies, but also against the Americans themselves. It is only a slight if polemical exaggeration to say that the Americanization of the German rubber and tyre industry after 1945 began with a 'Germanization' of the American tyre industry and ended, in the 1980s, with a further 'Germanization'.

---

[43] For detailed information see Erker, *Competition and Growth*, 51 ff.   [44] Ibid., 71 ff.

Chapter 11

The Evolution of the 'Japanese Production System': Indigenous Influences and American Impact

KAZUO WADA and TAKAO SHIBA

11.1 Introduction

During the 1990s Japan ran into economic difficulties, which she proved unable to overcome quickly. Then many people began to criticize her economic system. Some argued that the Japanese economic system should be thoroughly transformed: the model which had been well suited to the high-growth period by allocating scarce resources towards the most promising sectors was no longer appropriate for the new era of 'globalization'. Such critics traced the origins of the post-war Japanese model back to the war economy, particularly as it had developed during the 1940s: most of the key features of the Japanese model such as the main bank system, they argued, originated in these years.

Economists and historians debate the origins of the post-war Japanese economic system, but generally ignore the development of production methods. This is striking if one remembers the prevailing discussion of the Japanese economy during the 1970s and 1980s. After the oil shocks of the 1970s, the Japanese economy was soon back on its feet while the Western economies were still struggling to deal with the blows dealt by the rise in energy prices. This contrast understandably sparked a great deal of interest in the production systems of Japanese manufacturers. The secret of Japan's renewed growth was believed to be found on the shop-floor, as suggested by the popularization of the phrase, 'Japanese production system'. Nevertheless, many believed that this system was formed after the Second World War. This impression was further strengthened as the 'Toyota production system' became increasingly synonymous with the 'Japanese production system' more generally.

Oddly enough, Japanese economic and business historians also did not pay a great deal of attention to the development of production methods in Japan, and

This chapter was originally published as 'The Formation of the Japanese Production System' in Japanese in 1995, but has been extensively revised and expanded by Kazuo Wada. We should like to thank Jonathan Zeitlin for his stimulating comments on an earlier draft.

have written relatively little about it. As a result, scholars have often neglected the indigenous development of production methods in Japan before the Second World War, and have placed greater emphasis on the influence of American ideas after 1945. But 'Japanese management in the second half of the 1920s was under the influence of Americanization',[1] as Takenori Saito has rightly pointed out in dealing with the impact of American ideas on Japanese management in the post-war years.

Many observers have regarded this 'Japanese production system' as a totally different system from its Western counterparts, saying for example that: 'The success of the Japanese car industry is based on a system different in almost every feature from Detroit's mass-production system'.[2] Yet this production system evolved through efforts to catch up with, or to learn, imitate, and modify American methods before the Second World War. After the war, the fact that the country was defeated and occupied by the American forces led the Japanese to appreciate Detroit's mass-production system all the more. Even if the Japanese production system had seemed to be completely different from the American one, it evolved under the strong influence of US methods. This does not mean that most Japanese firms simply imitated and implemented American production methods in their plants. Some domestic engineers, believing that the strikingly different conditions in the two countries would hinder the implementation of American methods, sought to refine and modify indigenous techniques. Their goal was simply to find a way of achieving a similar level of efficiency under Japanese circumstances. In order to achieve this goal, they did not hesitate to experiment with new ideas or elements in their plants.

This chapter examines the American impact on the evolution of the 'Japanese production system', focusing on the fabrication or assembly industries in which US management ideas and practices were most influential.

11.2   The Japanese Production System before the Early 1930s

Japan, having begun its industrialization late in comparison with Western countries, was forced to play a game of technological catch-up. Consequently, the first priority of Japanese firms was to figure out how to make products that could compete with those produced by their international rivals. In the shipbuilding industry, for example, which was the first modern manufacturing sector in Japan, the most substantial hurdle faced by domestic shipbuilders in the first decade of the twentieth century was how to produce ships as big as those of

---

[1] Takenori Saito, 'Americanization and Postwar Japanese Management: A Bibliographic Approach', *Japanese Yearbook on Business History* 12 (1995), 16. He also rightly says that, before the war 'scientific management, with its symbol of the American management of that age (its practical orientation), was *not accepted* in Japanese academic circles' [italics added], although management consultants or practitioners welcomed it. Ibid., 10.

[2] Michael L. Dertouzos, Richard K. Lester, Robert M. Solow, and The MIT Commission on Industrial Productivity, *Made in America : Regaining the Productive Edge* (Cambridge, MA: MIT Press/New York: HarperPerennial, 1989), 48.

the Western powers, especially Britain. Their efforts paid off as 'world class' ships began to be turned out by the premier Japanese yards from around 1910.[3] In contrast to these efforts, however, the operational details of production were left primarily in the hands of skilled foremen. Firms did not have the necessary control over production management, or over employment, dismissal, promotion, and wage levels, these being handled by the skilled workers.[4] Only when the firms had finally weakened the power of the skilled foremen could they set out to establish direct control over their yards. The fact that the skilled foremen could not keep up with the level of technological precision required for modern shipbuilding, as well as the fact that the size of the vessels demanded was rapidly increasing, facilitated this shift. Accordingly, advanced shipbuilders responded to demands for high-precision technology by initiating their own training programmes for developing skilled workers. The first to do so was Mitsubishi's Nagasaki Shipyard, then the largest privately held shipbuilding operation in Japan, which had already begun such a training programme by 1889.[5] In the case of shipbuilding, then, firms moved to exert greater control over the details of production in response to requirements for improved technology. The 1907 financial crisis that followed the Russo-Japanese war provided further impetus. By the early 1910s, shipbuilding and almost all other Japanese manufacturing industries were withering after years of shrinking demand. Facing up to this situation, firms generally tried to curb losses by cutting costs, a move that led to the expansion of direct control over production. Mitsubishi's Nagasaki Shipyard took steps to tighten control over production facilities as early as 1908, including the implementation of incentive payment systems and improved cost controls. Similar moves can be observed in other industries, such as the introduction of a cost control system at Sumitomo's casting facility in 1912.[6]

Although awareness within Japanese manufacturing industry of the importance of production management appeared early on, its essential aims were limited to the mastery of technology and the reduction of costs. There was not yet room to consider the issue of how to promote efficient overall production. Japan's manufacturing companies were not yet operating on a scale that would require that type of awareness. This was particularly true of shipbuilding, as builders had to rely on intermittent orders for merchant vessels from a small number of state-subsidized lines. Even the Kawasaki Shipyard and Mitsubishi's

[3] Yoichiro Inoue, *Nihon kindai zosen-gyo no hatten* [The Development of the Modern Japanese Shipbuilding Industry] (Kyoto: Mineruva Shobo, 1990), 127.

[4] Tsutomu Hyodo, *Nihon ni okeru roshi-kankei no hatten* [The Development of Labour Relations in Japan] (Tokyo:Tokyo University Press, 1971), 234.

[5] Ibid.; see also Yutaka Nishinarita, 'Nichiro-senso-go ni okeru zaibatsu zosen kigyo no keiei kozo to roshi-kankei' ['Management Structures and Labour Relations at Zaibatsu Shipbuilding Firms Following the Russo-Japanese War'], *Ryukoku Daigaku keizai keiei ronshu* [Ryukoku University Review of Economics and Management] 18, 1 (1978), 53–5.

[6] Hyodo, *Nihon ni okeru roshi-kankei no hatten*, 247–8; Nishinarita, 'Nichiro-senso-go ni okeru zaibatsu zosen kigyo no keiei kozo to roshi-kankei', 55–8.

Nagasaki Shipyard, recognized as the leading Japanese shipbuilders, were building only a few vessels each year. Almost all Japanese shipbuilders designed and built vessels individually as orders were received from shipping lines. As each vessel was custom-designed, production processes had to be rigged up anew for each ship built. Maritime engineers were quick to point out that this system made for high costs. But despite calls for standardization in ship design,[7] this was not achieved at all before the war. Japanese shipbuilders concentrated their efforts on meeting the technical specifications presented by their customers, rather than on developing more efficient processes.

The outbreak of the First World War changed this situation: shipbuilders became increasingly aware of production control. The war completely changed market conditions for Japanese shipbuilders. Japanese shipping lines were all at once busier than ever before, causing a major increase in the demand for ships, and the lines found themselves no longer able to purchase second-hand vessels from Europe. As a result, Japanese builders were suddenly inundated with orders from tramp operators. The old production methods were now inadequate, and the shipbuilders were faced with the possibility of losing out on huge profits. In order to meet demand, each firm standardized its ship designs and produced vessels of the same design repetitively. In addition to this, leading firms tried to improve their production processes to allow the fastest possible repetition, because the speed with which ships could be constructed had a direct bearing on profits. 'The time periods required for shipbuilding processes were dramatically shortened' to the extent that a Japanese yard 'set a new world record of 30 days for a completed vessel'.[8]

Shipbuilders' positive attitude toward improved production processes during the wartime period did not last long. Demand for ships declined dramatically with the end of the war, undermining the incentive to continue with production control for large-scale output. In fact, the industry faced a depressed market awash with the excess shipping capacity that had been created during the boom years of the war, and its focus consequently reverted to the question of how to rein in costs. The situation was essentially the same for other industries as well.

Shipbuilding was not the only industry to undergo tremendous expansion in production volume during the war. With the end of the war, however, the industries that had been supported by the wartime boom faced a rapid market contraction. They responded first by scaling back production, and second by attempting managerial rationalization. It was this pressing need for rationalization or cost cutting that began to concentrate firms' attention back on the shopfloor, as they realized that the elimination of waste in production and the extent to which efficiency could be increased were directly related to financial results.

---

[7] Yukio Yamashita, 'Dai-ichiji-sekai-taisen ni sai shite no waga kuni zosen kakusha no seihin seisaku' ['The Product Policies of Japanese Shipbuilders During the FirstWorld War'], *Chuo Daigaku Shogaku ronsan* [Chuo University Commerce Review] 7, 2 (1965), 188–95.

[8] Kawasaki Heavy Industries Co., Ltd., *Kawasaki Jukogyo Kabushiki-gaisha shashi* [A History of Kawasaki Heavy Industries Co., Ltd.] (1959), 82.

The 'efficiency movement' that Japan experienced following the war took place in this context. The Ministry of Agriculture and Commerce established its Efficiency Section in 1920, and the Osaka Chamber of Commerce sponsored a training course on factory management the following year. An 'efficiency exposition' was also held in Osaka in 1924 with the goal of 'spreading efficiency to the general public'. It was against this background that Yoichi Ueno and other important figures advocated a radical rethinking of the production processes employed in manufacturing industries.[9]

The efficiency movement of this period was by no means primitive,[10] but it was limited by the fact that firms could not yet envision the creation of a flow which would encompass the entire production process and would reduce total costs. This was because most firms were still producing on a jobbing basis and had no real need for flow-type thinking, and because rationalization was being undertaken simply as a means of avoiding losses. Their main aim, therefore, was to eliminate waste in individual production processes.

While awareness of process control temporarily blossomed in the shipbuilding industry during the First World War, and while rationalization received significant attention throughout the subsequent years of economic hardship, serious re-evaluation of the entire manufacturing process did not occur until the 1930s. Firms were still not particularly focused on the production process, and the same can be said of their approach to materials purchasing. Hardly any firms in assembly industries can be found before the 1930s which had formed organizational relationships with outside suppliers in order to foster smooth materials procurement.

Issues concerning smaller firms in Japanese manufacturing industries have been widely discussed, particularly the inferior position of subcontracting firms.[11] Japanese manufacturers were already utilizing subcontracting during this period. A 1927 survey shows, for example, that Kawasaki's shipyard had 43 subcontractors working for it.[12] A Ministry of Commerce and Industry survey taken in 1932 concluded that the rate of dependence on external orders in the machinery industries in eight urban prefectures was 11.2 per cent.[13] The extent

---

[9] Hiroshi Hazama (ed.), Nihon romu-kanri-shi shiryo dai 2 ki dai 8 kan [Collected Documents on Japanese Labour Management; 2nd series, vol. 8] (Tokyo: Gozan-do Shoten, 1989), 10; Kenichi Okuda, Hito to keiei: Nihon keiei kanri kenkyu-shi [People and Management: A History of Japanese Management Studies] (Tokyo: Manejimento-sha, 1985), ch. 5.

[10] See Rinji sangyo gori-kyoku seisan kanri iinkai [Production Control Committee of the Temporary Industrial Rationalization Bureau] (ed.), Sagyo kotei kanri no kaizen [Improvement of Operational Process Control] (1933). On the efficiency movement and its evaluation, see Saito, 'Americanization and Postwar Japanese Management'.

[11] On this subject, see the following two excellent pioneering works published in those years: Takuji Komiyama, Nihon chusho-kogyo kenkyu [A Study of Japanese Small and Medium-Sized Industry] (Tokyo: Chuo Koron Sha, 1941); Keizo Fujita (ed.), Shitauke-sei kogyo [Subcontracting Industries] (Tokyo: Yuhikaku, 1943).

[12] Nohori Hashimoto, 'Kawasaki Zosensho shokko kaiko oyobi shitsugyo kyusai jigyo (jo)' ['Relief Practices for Workers Fired and Laid Off at Kawasaki Shipyards, Part I'], Shakai seisaku jiho [Journal of Social Policy] 85 (1927), 115.

[13] Juro Hashimoto, Dai-kyoko-ki no Nihon shihon-shugi [Japanese Capitalism During the Great Depression] (Tokyo: University of Tokyo Press, 1984), 268.

of subcontracting can be guessed from these figures, although external orders cannot be fully equated with subcontracting. During the 1920s, however, most firms had not yet reached the point at which they could effectively integrate subcontractors into their own production systems. Manufacturers 'were too concerned with their own survival to have the wherewithal to worry about organizing subcontractor factories'.[14] Therefore, they did not even try to create a level 'flow' throughout the total production process, including subcontractors' manufacturing process.

11.3 New Production Methods: Searching for a Path to High Volume

Firms' awareness of production processes significantly changed after the mid-1930s. The war with China required vast amounts of weapons and transport equipment. Manufacturing firms were faced with the need to boost output dramatically. As they set about expanding their plant and equipment, they had to reconsider the entire system of production. The inspiration that came to mind was 'the Detroit mass-production system'. An aeronautical engineer affirmed, for example, that the aircraft industry should adopt the 'essence' of the automobile production method in order to increase output.[15] Engineers knew that they could not implement Detroit methods in their entirety. Therefore, they analysed these methods thoroughly, and tried to find a way of achieving high-volume production in the Japanese situation.

Before the Second World War, most Japanese shipbuilders had designed and built vessels individually to customer specifications. As a result, they paid less attention to the methods of producing many vessels. But after the war broke out, domestic shipbuilders took the first step towards high-volume production by standardizing designs. Accordingly, all shipyards began converting to the newly established wartime standard designs from 1942. As the war worsened for Japan, however, shipping losses rapidly mounted much higher than had been anticipated, and further simplification of ship designs and functions was introduced in order to facilitate even greater output. At this point almost no attention was yet paid to the production process itself, and the increase in output remained unsatisfactory. Then, in 1943, the even more basic 'E' type design was introduced, and high-volume production methods were adopted. Because existing shipyards had been set up on the basis of custom orders, they were not suited to the requirements of high-volume production, and four completely new yards were established: the Fukahori Yard of Kawaminami Shipbuilding, the Mitsubishi Wakamatsu Shipyard, the Matsunoura Works of Harima Shipbuilding, and Tokyo Shipyards. Each of the four experimented with somewhat different high-volume production methods and three yards, but not Tokyo

[14] Komiyama, *Nihon chusho-kogyo kenkyu*, 35.

[15] Gakuji Moriya, 'Kokuki no taryo seisan' ['High-Volume Production of Aircraft'], in Y. Kobayashi (ed.), *Taryo seisan kenkyu* [Studies on High-volume Production] (Tokyo: Heiki Kokuki Kogyo Shimbun Shuppanbu, 1944), 50.

Shipyards, adopted the basic method of building entire sections of the ships on land, and then transporting them over to the slip for final assembly.[16] The new assembly method allowed work on different parts of the ship to progress simultaneously and consequently reduced the time required for production. However, unlike in the aircraft industry, no thought appears to have been given to breaking down design drawings from the outset in order to provide the necessary documentation for each individual process. Had this been done, it would have allowed the implementation of flow production techniques. It was only after the war that this kind of thinking was introduced and full-scale block construction methods were adopted.

In contrast to the shipbuilders, the Japanese automobile firms tried to establish a high-volume production system even before the war, either by purchasing US equipment and directly transplanting American production techniques, as in the case of Nissan, or by adapting Detroit methods to the Japanese situation.[17] Toyota took the latter course. This case encapsulates the difficulty of implementing high-volume car production systems in Japan. Toyota's main factory, the Koromo Plant, began operations in 1938. This was the first large-scale automobile factory completely designed by Japanese engineers.[18] Toyota aimed to produce 500 passenger cars and 1,500 trucks per month, far below the production levels of its American counterparts. Nevertheless, Kiichiro Toyoda, the founder of Toyota, directed Takatoshi Kan, responsible for formulating the overall design for this plant, to make it easy to change car models. Kan chose to install 'adjustable' special-purpose machines rather than many types of conventional dedicated equipment. He designed special-purpose machines with a unique mechanism: for example, a six-spindle boring machine for the cylinder block with a mechanism for changing the distance between spindles, which could thus accommodate design changes. Further, he placed machines not directly on the ground, but on wooden floorboards. This facilitated layout changes because machines could be easily unbolted from the wooden floor and moved to other locations.[19]

---

[16] See Ichiro Onozuka, *Senji zosen-shi: Taiheiyo senso to keikaku zosen* [The History of Wartime Shipbuilding: The Pacific War and Planned Shipbuilding] (Tokyo: Nihon Kaiji Shinko-kai, 1962); Shin Goto, 'Senji-ki Nihon zosen-gyo no seisan gijutsu ni kan suru kosatsu' ['Observations on Production Technology in Japan's Wartime Shipbuilding Industry'], *Kokusai Keiei Ronshu* (Kanagawa Daigaku) [International Management Review (Kanagawa University)] 3 (1992).

[17] See Michael A. Cusumano, *The Japanese Automobile Industry: Technology and Management at Nissan and Toyota* (Cambridge, MA: Harvard University Press, 1985), xviii.

[18] See Takatoshi Kan, 'Toyota Jidosha Kogyo Kabushiki-gaisha Koromo kojo no kensetsu' ['Construction of the Koromo Plant of the Toyota Motor Corporation'], in Toyota Motor Company History Editorial Committee, *Toyota Jidosha 20 nen-shi* [The First Twenty Years of Toyota Motor Company] (Koromo-shi: privately published, 1958). See Eisuke Daito, 'Automation and Production Organization of Japanese Automobile Manufacturers in the 1950s', *Enterprise and Society* 1 (2000).

[19] Even in the 1930s and 1940s, engineers discussed the inflexibility of machine layouts because many machines in those days were still driven by belts and shafts. As most machines at Toyota had individual motors, the layout could be changed relatively freely without the need to consider the position of belts and shafts. See Kazuo Wada, 'Nihon ni okeru "nagare-sagyo" hoshiki no tenkai'

In light of the lower demand for cars in Japan as well Toyota's more limited supply of capital, Kiichiro Toyoda and his engineers chose to mechanize operations less and utilized more hand labour than US manufacturers. Even in materials handling, the use of mechanical conveyors was limited at Toyota because of a lack of funds.[20] As a result, Toyota tried to synchronize diverse operations for smaller-scale production, relying not on mechanical devices, but instead on production management techniques, namely, the 'just-in-time' production method. Although this method has attracted the attention of many scholars, its origins and evolution remain obscure.[21] Kiichiro publicly used the term 'just-in-time' only once, when the Koromo plant came into operation in 1938.[22] He placed a strong emphasis on synchronizing diverse operations including materials and parts for smaller-scale production. The company history explains how 'just-in-time' production was achieved as follows:

[A] batch number was allocated to every 10 vehicles produced, and these numbers were used to keep track of everything related to those vehicles right down to the parts from which they were assembled. The progress of manufacturing could thus be monitored by checking the number assigned to the parts in each process at any given time. Flow production was employed, whereby the progress of each phase of manufacture was carried out in synchronization with the final stage; this was aiming toward realization of the Just-in-Time System.[23]

If this account explains how Toyota achieved 'just-in-time' production, there is a remarkable resemblance to the methods adopted at aircraft companies during the war and diffused to many Japanese manufacturers after 1945. Even if Kiichiro Toyoda had his own original splendid ideas about production control, he could not pursue them at his plant because of the army's tight control during the war, and he gradually lost his enthusiasm for company management.[24] The

['The Development of "Flow Production" Methods in Japan'], *Keizaigaku ronshu* [The Journal of Economics] (University of Tokyo) 61, 3 (1995), 30–1.

[20] Kan, 'Construction of the Koromo Plant', 608.

[21] Eiji Toyoda, the present honorary chairman of Toyota who once worked closely worked with Kiichiro, testifies as follows: 'Kiichiro's manual was impressive. A full four inches thick, it described in meticulous detail the flow production system we were to set up. This is the text that I and the other instructors used to teach the new system to the workers': Eiji Toyoda, *Toyota: Fifty Years in Motion: An Autobiography by the Chairman* (Tokyo: Kodansha International, 1987), 58. But unfortunately the manuals have not, to date, been found.

[22] 'After the facilities are built and the workers are able to use them, the materials, parts and other related items will be integrated through what I call "organic communication". In other words, each facility shall introduce the Just-in-Time system': (Kiichiro Toyoda, 'On the Completion of the Koromo plant', *Ryusenkei* [The Streamline Shape], Nov. 1938, 1.

[23] Toyota Motor Corporation, *Toyota: A History of the First 50 years* (Toyota City: privately published, 1988), 71–2.

[24] See Hanji Umehara (Takeshi Umehara, ed.), *Heibon no naka not heibon* [The most ordinary of ordinary people] (Kosei Shuppansha, 1990). Mr Umehara, then a lecturer at Tohoku Imperial University, joined the automobile division of Toyoda Automatic Loom as a commission engineer in 1936, and became a director of Toyota in 1950. Hence, he was in a position to know Kiichiro Toyoda's views. For the documents written by Kiichiro Toyoda, see Kazuo Wada (ed.), *Toyoda Kiichiro monjo shusei* [Corpus of Kiichiro Toyoda's documents] (Nagoya: University of Nagoya Press, 1999), which includes an introductory note by the editor in English, 'The Founding of the Automobile Business by Kiichiro Toyoda: From Conception to Realization'.

establishment of the 'just-in-time' production method at Toyota was thus hampered by the war, even if it was tried for a short period.

The war also had a serious effect on the aircraft industry and its production methods. After the conflict with Britain and the USA began in 1941, the government sought to increase aircraft production as much as possible, giving planes priority over trucks. Japanese aircraft engineers consciously tried to emulate the principles of mass production during the war, although the volume of plane production was relatively low.[25] It was particularly in the aircraft companies that Japanese engineers tried out and implemented many ideas in order to increase production.

Aircraft began to be produced in Japan shortly after the First World War. Aircraft makers initially called in experts from the USA and Europe, building planes under their direction and absorbing the fundamentals of aircraft technology and design. Domestic planes were developed thereafter, but production runs were limited and emphasis was placed on turning out prototypes. At this stage, assuming that a prototype passed military standards testing, subsequent production was carried out in the original prototype factory. Workers in these factories were 'non-specialized . . . [and they had to] fabricate parts, assemble, trim, and perform maintenance duties afterwards'.[26] Later, as increased production came to be required, factories specialized in producing certain sections, such as stabilizers or fuselages. Assembly areas were also set up within these factories for the manufacture of necessary components in order to increase production further. Still, no fundamental re-evaluation of the overall manufacturing process was undertaken, even when substantially greater output was subsequently demanded; factories were simply expanded, and more workers added. While this strategy did increase output in absolute terms, 'there was no increase in the volume of production per worker'.[27] This was to be expected, since the flow of production within the factories was left essentially untouched. 'Production control was in a state of chaos that clearly required attention,' and, 'given these conditions, the engineers attempted to solve the problem by establishing and expanding the production control department and introducing

---

[25] Kiyoshi Yamamoto, *Nihon ni okeru shokuba no gijutsu/rodo-shi 1854–1990* [The History of Technology and Labor on the Japanese Shop-Floor] (Tokyo: University of Tokyo Press, 1994), 232. The total volume of aircraft production from 1941 to 1944 was 58,822 in Japan, 92,656 in Germany, and 261,826 in the USA. Although the overall volume of production in Japan was low, two major aircraft companies dramatically raised their production after the outbreak of the war: Nakajima Aircraft and Mitsubishi Heavy Industries turned out 1,085 and 1,697 aircraft in 1941, but increased output to 7,943 and 3,628 in 1944 respectively. But in 1945, both companies rapidly decreased production: 2,275 aircraft at Nakajima and 563 at Mitsubishi. See Nobuhisa Fujita, 'Kokuki-bumon no keiei' [Business of the Aircraft Division] in Yasuo Mishima, Yasuaki Naganuma, Takao Shiba, Nobuhisa Fujita, and Hidetatsu Sato, *Dai ni ji taisen to Mitsubishi zaibatsu* [The Second World War and Mitsbushi zaibatsu] (Tokyo: Nihon keizai shinbun, 1987), 93.

[26] Wataru Sasaki, 'Kokuki no taryo seisan' ['High-Volume Production of Aircraft'], in Nainen kikan henshubu [Internal Combustion Machinery Editorial Division] (ed.), *Kokuki no taryo seisan hoshiki* [Aircraft High-Volume Production Methods] (Tokyo: Sankai-do, 1944), 66–9.

[27] Ibid.

assembly line techniques'.[28] However, this proved extremely difficult to accomplish. Although Nakajima Aircraft and Mitsubishi Heavy Industries, then the two largest Japanese aircraft producers, can be seen as important examples of assembly-line production, they both required prodigious amounts of time and effort to achieve results.

Japanese aircraft companies, as well as the military authorities, were eager to learn foreign production practices. In January 1939, for example, the Military Air Service Board invited two engineers, K. E. Sutton and E. B. Parke, from the Curtiss-Wright Corporation in the USA to present a week-long course of lectures on mass production of aircraft engines. This covered a wide range of subjects from design and production to organization, cost accounting, and the layout of machine tools. In February and March 1942, moreover, visiting engineers from the German firm Junkers lectured to Japanese aircraft engineers on high-volume manufacture of fuselage and engines. Around this time, Nakajima and Mitsubishi were struggling to improve their process controls, while experiments also began with an assembly-line style of fuselage manufacture.

By 1943, engineers from Mitsubishi's Nagoya Aircraft Works had successfully implemented an assembly-line method of fuselage manufacturing called 'zenshin-shiki', similar to the German 'Takt system'.[29] The Japanese Army as well as the aircraft engineers closely studied the 'Takt system', which had been adopted at major German aircraft companies and tried to imitate the system. 'Takt' in German has musical associations and means 'rhythm' or 'measure'.[30] In 1943 one observer described the 'zenshin-shiki' method as follows:

The Mitsubishi Nagoya Works, known as the best aircraft manufacturer in Japan, implemented the zenshin-shiki method ('Takt system' [original author's note]) last autumn, and the achievement was remarkable . . . Originally, the Takt system was devised in Germany . . . as the name 'Takt' suggests the music-loving German adopted this . . . At each station, workers worked on the sections of fuselage under the direction of their squad leaders . . . Then the noisy plant suddenly turned quiet. The allocated time for the operation was over . . . then the workers pushed the section of fuselage forward to the next position. Next, the rising-sun flag was hoisted, and all the workers came together in the centre of plant and lined up . . . Then they returned to their original locations in order to continue working.[31]

With this method, manufacturing activities were conducted within a given time period at each production station after which all items moved forward in unison to the next station.

[28] Ibid.

[29] The Army officially commended this method. See Haruyoshi Aikawa, *Gijutsu oyobi gino kanri* [Management of Technology and Skills] (Tokyo: Toyo Shokan, 1944), 193.

[30] Jonathan Zeitlin, 'Flexibility and Mass Production at War: Aircraft Manufacturing in Britain, the United States, and Germany, 1939–1945', *Technology and Culture* 36, 1 (Jan. 1995), 71.

[31] Nihon Sangyo Keizai Shinbun Sha Seikei Bu [Political and Economic Section of the Japanese Newspaper for Industry and Economy] (ed.), *Zenkoku Mohan Kojo Shisatsu-ki* [The Observation Record of Model Factories across the Nation] (Tokyo: Kasumigaseki Shobo, 1943), 7–10.

Before implementing the '*Takt* system', Mitsubishi's Nagoya Aircraft Works made many changes in order to reform its production process. First, the company decided to build an entirely new greenfield plant, which was completed in 1938. The seven shops at the plant were deliberately arranged according to the planned flow of production. Second, the company sent engineers to 'process analysis seminars' sponsored by the Japan Management Association (JMA).[32] In those seminars, Japanese Taylorist engineers taught the principles of time and motion studies, while also offering practical training on how to organize a plant most efficiently. On the basis of this training, Mitsubishi engineers began to analyse and improve the process of production more systematically. This facilitated the introduction of 'sectional production' as well as the '*Takt* system'.

'Sectional production' meant that the fuselage and wings were divided into sections, each of which was produced; fittings were installed in each section, without waiting for final assembly; and then the entire sections were joined together.[33] The 'most advanced' form of this sectional production was eventually to be found at Mitsubishi's Nagoya Aircraft Works, which introduced 'completely sectional methods' from the design stage of its '*Hiryu*' (Flying Dragon) bomber. The design and manufacturing engineers met regularly from the early stages of the design process in order to decide on the best way to sectionalize the plane. Their recommendations were used to produce blueprints for each section, including the placement of all necessary fittings. It was for its role in the development of the rudimentary practice of product and process design for manufacturing that the Japanese Association of Mechanical Engineers praised the '*Hiryu*' bomber so highly in its fiftieth-anniversary history published in 1949.[34]

After the '*Takt* system' was implemented, the engineers tried to synchronize the diverse operations of all the sections. As the war developed, moreover, Mitsubishi's Nagoya Aircraft Works had to move its production facilities to various plants in order to avoid aerial attacks. Even under such conditions, the works managed to continue production with the sectional method. But when a great earthquake occurred in this area in December 1944, the production facilities were severely damaged, and output was virtually halted.

After Mitsubishi's Nagoya Works implemented the '*Takt* system' for the final assembly of fuselages, it turned to parts deliveries. As one engineer put it, 'The

---

[32] The Japanese name of the Japan Management Association is Nihon Noritsu Kyokai, which literally means 'Japan Efficiency Association'. The Association now officially calls itself 'Japan Management Association'. In this chapter, we follow the latter convention.

[33] Mitsubishi Jukogyo Kabushiki-kaisha Nagoya Kokuki Seisakusho [Mitsubishi Heavy Industries Nagoya Aircraft Works], *Meiko kosaku-bu no senzen-sengo-shi: watashi to kokuki seisan—Moriya Sodan-yaku* [The Pre- and Post-war History of the Nagoya Aircraft Works Machining Division: Aircraft Production and Me—Adviser Moriya] 1988; S. Mizota, *Zosen jukikai sangyo no kigyo shisutemu* [Enterprise Systems in the Shipbuilding and Heavy Machinery Industries] (Tokyo: Moriyama Shoten, 1994).

[34] Nihon Kikai Gakkai [Japan Society of Mechanical Engineers] (ed.), *Nihon kikai kogyo 50 nen* [Mechanical Engineering in Japan: 50 Years] (Tokyo: Nihon Kikai Gakkai, 1949), 977.

greatest enemy of the modified *Takt* system was late [delivery of] parts. The system worked well if the parts were there, and it was easier to line up the parts if the system was working well.'[35] Indeed, the plant produced a detailed report in March 1944, saying explicitly that:

We, Mitsubishi's Nagoya Aircraft Works, implemented the '*Takt* system' in the assembly plant; we completed flow production in the assembly operations [for the fuselage]. But the problem now remaining is how to establish the planned flow production for parts-making . . . This is a very difficult task to accomplish . . . but in order to utilize [the virtues of the '*Takt* system'] fully we prepared to implement flow production in parts-making.[36]

This lengthy report shows that the works recognized the necessity of smooth delivery of parts for assembly once the '*Takt* system' had been established to some extent in fuselage production. But this idea was not fully implemented at the Nagoya Works, probably because of the difficulties of acquiring materials in time and controlling suppliers as the war developed.

Nevertheless, as the production of aircraft increased, some parts such as pistons were required in large quantities and their manufacture was converted almost completely to moving assembly lines. But it proved difficult to implement this method with parts required only in low volumes. In response, Nakajima's Musashino Works came up with a compromise in the form of the 'semi-assembly line production system'.[37] This arrangement grouped together low-volume parts having similar production processes and sought to apply flow-production methods to each group. For example, although gears have differing sizes and shapes, the processes involved in their production are basically the same, allowing a variety of gears to be brought together for processing in a single manufacturing operation. This was similar to so-called group production employed in the UK. Such 'semi-flow production' involved 'analysing and classifying the jobs to be run and then grouping the components so as to allow the machines to be laid out in operation sequence', just as in British 'group production'.[38] In this 'semi-flow production' method, the production area was subdivided into very small operational sections, each with its own foreman and throughput manager responsible for controlling the production schedule. As a result, many lower administrative staff employees were placed on the shop-floor. This labour-intensive control of scheduling was, in a sense, substituted for more capital-intensive controls such as mechanical conveyors and so on.

---

[35] Gakuji Moriya, 'High-Volume Production of Aircraft', 45–6.

[36] Sadaaki Takagi, 'Keikaku sekinin nagare seisan ni okeru sagyo kanri ni tsuite' ['On the shop-floor control in the responsible planned flow production'] (mimeo., 1943), 1–2. This was a 'secret' document written and circulated in the Mitsubishi Nagoya Aircraft Works.

[37] Ichiro Sakuma, 'Seisan-ryoku to nagare sagyo' ['Productive Power and Flow Production Methods'], in Nihon Keizai Renmei Chosa-ka [Japan Economic Federation Research Dept.] (ed.), *Taryo seisan hoshiki jitsugen no gutai-saku* [Practical Approaches to the Realization of High-volume Production] (Tokyo: Sankai-do, 1943), 128.

[38] Frank G. Woollard, *Principles of Mass and Flow Production* (London: Iliffe, 1954), 42.

Changes in design and production methods were constantly being introduced, and each time the production manager would have to race around the factory to look for the parts affected and adjust all the accompanying documentation. The military authorities asked munitions factories to present many accounting documents in an effort to maintain control over costs. Whenever changes in design were made, every document also had to be changed accordingly. Those on the shop-floor were often unfamiliar with administrative procedures, and individual documents were frequently lost or misplaced. As a result, documents often became separated from the parts to which they should have been attached. The document-based control of production volumes and daily plans became virtually meaningless. Having examined this situation, the JMA towards the end of the war advocated a production method, which was later refined and developed into the 'work-centre' method after 1945. This method sought to control the production schedule as well as the volume of work in progress, without relying upon production slips and other documents.[39] The whole shop-floor was divided into many work areas. The whole production flow was now organized as a chain of work areas. Each work area functioned, in a sense, as a small depot. A foreman at each depot dealt with all the necessary paperwork, reducing the administrative burden on the other workers. Two other employees worked at the depot: one mainly checked the volume and quality of parts received from the previous production process, and the other was in charge of keeping up the production schedule. By maintaining close surveillance of these depots, the administrative staff of the factory central office in turn sought to control the shop-floor more tightly and minutely. The true effectiveness of this method remains unknown, but one document suggests that the Army was astonished by its performance. Following factory efficiency assessments, the work-centre system was implemented on an experimental basis at the Tachikawa Aircraft Company around the end of 1943. The Army officers were amazed that production increased by 90 per cent using the same facility and personnel.[40] The promoters of this method thought that it could be applied to other manufacturing situations beyond parts production, but the war came to an end just as several aircraft makers began to experiment with the system.

The German 'Takt sytem', a flow-type production method, resulted in increased production in fuselage assembly, but this also clearly showed that the aircraft production process as a whole, from parts manufacture to assembly, remained unbalanced. The production schedule was not maintained. The

---

[39] See Kuninobu Niizaki, 'Kotei kanri no ichi-hoshiki' ['One Method of Process Control'], *Nihon noritsu* [Japan Efficiency] 3, 5 (1948).

[40] Gunju-sho kouku-heiki-so-kyoku norituka [Efficiency Section of the Air-Force Arms Dept. at the Ministry of Munitions], *Maru-hi: Tachikawa Hikoki Kabushikigaisha Tachikawa kojo oyobi Sunagawa kojo ni taisuru Noritsu Shindan Seika Hokoku (Bassui)* [Secret Document: Abridged Report on the Results of the Efficiency Diagnosis for the Tachikawa and the Sunagawa Plants at the Tachikawa Aircraft Company](1944), in *Kokusaku kenkyu-kai bunsho* [Documentation of the State Policy Research Council], Ca-2-5, no pagination.

so-called end-of-month problem occurred: the greater part of the volume was produced near the end of month, ensuring that the total volume of production ordered was ready by the closing day of the accounting period, but it was done in great haste. Consequently, most plants had quite large production capacity, in order to accommodate urgent work near the end of the month. Japanese engineers became aware of this, and placed a greater emphasis on levelling out the volume of production, in order to avoid the concentration of production near the end of month. If this objective were realized, the volume of production could be increased with the same production facilities. Even at this time, most factories were mainly equipped with general-purpose machines: special-purpose machines remained uncommon.[41] The sequential arrangement of special-purpose machines was rarely implemented, and the workers still set the pace of work. So, the engineers concentrated their efforts on tight control of the production schedule: by dividing the plant into small sections and also by deliberately placing work distribution depots on the shop-floor.

11.4 Implementing and Refining Wartime Production Methods after 1945

For a time after the war, the work-centre method was eagerly introduced in Japan's manufacturing sector. The method had been developed by the aircraft companies towards the close of the war, and further experimentally implemented on a small scale by auto-makers soon after the onset of peace. This method was refined with a daily scheduling component so that it had become fairly well defined by the early 1950s.[42] The work-centre method represented the culmination of wartime production innovations, and the JMA promoted its diffusion. This technique had been implemented in many companies by the 1950s,[43] notably Canon, JVC, Japan Air Brake, Isuzu, the Nagoya Machinery

---

[41] Even at the Kure Naval Aresenal, one of the most advanced plants, general-purpose machines comprised about 40 per cent of all machine tools in the mid-1930s. See Konosuke Odaka, 'Senkan-ki ni okeru shitauke-sei kikai buhin kougyo hattatsu no sho-yoin' [Factors in the Development of the Subcontracting Parts Industry during the Wars], in Takafusa Nakamura (ed.), *Senkan-ki no Nihon keizai bunseki* [Analysis of the Japanese Economy during the Wars] (Tokyo: Yamakawa-shuppan, 1981), 177. Worst of all, the number of machine tools was far too small for the projected increase in the production of aircraft and other goods. Therefore, in 1943 the government established the so-called war standard machine-tools, low-quality model machines, in order to induce tool-builders to increase the production of machines. Recognizing the pressing need for increased production of aircraft, moreover, the government ordered machine-tool builders to stop the production of machines and to produce instead aircraft engines. But the number of machines produced declined after 1944, although the total floor-space devoted to machine-tool factories soared. See Okuda, *People and Management*, 506.

[42] See the May 1949 issue of *Seisan Noritsu*. This method was a major topic when the Committee of Inquiry of the Ministry of International Trade and Industry (MITI) published a book on production control in 1953. See Tsusho sangyo sho sangyo gorika shingikai kanri-bukai [The Section on Production Control of MITI's Committee of Inquiry on the Rationalization of Industries] (ed.), *Kotei-kanri* [Production control], (Tokyo: Nikkan kogyo shinbun, 1953).

[43] See Tetsuro Nakaoka, 'Senchu-sengo no kagakuteki kanri undo (chu)' ['The Wartime/Post-War Scientific Management Movement (Part 2 of 3)'], *Keizai-gaku Zasshi* (Osaka Shiritsu Daigaku) [Economics Magazine (Osaka Municipal University)] 82, 3 (1981).

Works of New Mitsubishi Heavy Industries, Aichi Clock Electric, Kayaba Industries, Toyota Auto Body, and Okuma Iron Works.

In 1957, for example, Canon received the third Okochi Memorial Production Prize in recognition of its 'establishment of a mass-production system for high-performance cameras', and its adoption of the work-centre method was cited as one of the reasons for its award.[44] The company originally employed a voucher system which kept track of parts by production lot, but this proved inadequate for thorough control; as the scale of production increased, managers were 'unable to determine exactly how many [parts] there were and where the precise parts were'.[45] This was a similar situation to that of the aircraft companies during the war. Then, in January 1952, when the company consolidated its manufacturing facilities in a single location, Canon adopted the work-centre method under the guidance of the JMA. The company history indicates that the decision dramatically improved production management by allowing the implementation of 'visual control'.[46]

Because a number of companies which introduced the system came up with major improvements in production control on the factory floor, it was temporarily perceived as a panacea for industry throughout the country. However, the system did not always work well. Implementation required a thorough survey of manufacturing conditions, as well as detailed analytical knowledge of each process; hence, for companies which lacked such information, the preparation period for implementation was excessively long. Diesel Kiki attempted to introduce the system with the assistance of the JMA in July 1953, but, due to a 'short period of preparation . . . numerous changes had to be made in the material collected to facilitate planning . . . Orders for required items had to be modified, and the arrangement of machinery had to be changed in order to simplify progress management, and the abilities of subcontractor factories had to be reinforced.'[47] Recognizing this difficulty, the company decided to abandon its plans to introduce the method. Even in the case of Canon, which is seen as having been comparatively successful, an engineer recalled four years after the project that 'there was problem after problem from the time we started until the method really got going . . . The work-centre method is certainly not the solution for every problem.'[48]

Nevertheless, the complaints about the work-centre method suggest that after the war even major companies did not have detailed information about specific production processes, and also lacked sufficient knowledge to change the production processes by themselves. While the JMA promoted this method, they also diffused industrial engineering knowledge and know-how such as the pro-

[44] Canon, Inc., *Canon-shi: Gijutsu to seihin no 50 nen* [Canon History: 50 Years of Technology and Products] (Tokyo: privately published, 1987), 70.

[45] Ibid., 44.   [46] Ibid., 47.

[47] Diesel Kiki [Diesel Equipment Co., Ltd.], *Diesel Kiki 40 nen-shi* [A 40-Year History of Diesel Equipment] (Tokyo: privately published, 1981), 110–11.

[48] *Kojo Kanri* (1956), 32.

cedures for carrying out time-and-motion studies to many plants. Such know-how was now widely spread across many manufacturing sectors. After acquiring this knowledge, some companies were able to change the layout of production facilities by themselves.

The work-centre method itself also had undeniably significant effects on a large number of Japanese industries in the period immediately following the war. One of these was shipbuilding, which went on to become an important pillar of the post-war Japanese economy. As is quite well known, one of the primary factors that propelled Japan's post-war shipbuilding industry forward was the block-construction method. Post-war block construction of ships drew on both the wartime aircraft industry's sectional production methods, and on wartime American technology. The first location to employ the new method successfully was the Kure Shipyard run by National Bulk Carriers Ltd. (NBC), an American shipping firm. As other shipyards sought to implement the new system, it became an industry-wide standard.[49] While the United States had fully developed section building and structural welding for the fabrication of Liberty ships during the war, Japan had not: her series-built standard ships were still mainly riveted together. But once Hisashi Shinto was invited to become NBC's chief engineer, he changed the production method from the 'what to build' principle to a flowline method.[50] According to Shinto, he was impressed with the production control procedures used for aircraft manufacture during the war, and tried to introduce such procedures into the shipbuilding process at NBC.[51] The availability of high-quality welding technology made it possible to realize his idea.

But just copying the aircraft production procedure in a shipbuilding setting did not result in simultaneously higher output and faster production times; the block method was not particularly effective without control of the production schedule of all processes. 'Imprecise production control often led to the chaotic situation in which some necessary materials were not delivered to designated locations and some were delivered in more than the required amount. Nevertheless, things would still work out',[52] as long as output remained below a certain level. This happy state of affairs was not, however, sustainable. For example, as output at Mitsubishi's Nagasaki Shipyards began to increase from around 1956, conditions no longer allowed the continuation of the 'things

---

[49] Takeaki Teratani, *Kindai Nihon zosen-shi josetsu* [An Introduction to the History of Modern Japanese Shipbuilding], (Tokyo: Gannan-do Shoten, 1979), 162–6. See also Hisashi Shinto, *Zosen seisan gijutsu no hatten to watashi* [The Development of Shipbuilding Technology and I] (Tokyo: Kaiji Press, 1980), 20–2.

[50] Tomohei Chida and Peter Davies, *The Japanese Shipping and Shipbuilding Industries: A History of their Modern Growth* (London: Athlone Press, 1990), 111–12.

[51] Shinto, *Zosen seisan gijutsu no hatten to watashi*.

[52] Kikuichi Kita, 'Kotei kanri no kaizen de kenzo kikan wo ohaba ni shukusho: Mitsubishi Zosen, Nagasaki Zosen-sho ni okeru ogata yusosen kenzo no jitsurei' ['Dramatically Decreasing Construction Time through Process Control Improvement: The Case of Oil Tanker Construction at the Nagasaki Shipyards of Mitsubishi Shipbuilding'], *Manejimento* [Management] 16, 4 (1957), 34.

would still work out' approach. The facility thus began a programme aimed at ensuring that blocks were 'produced at the right time, in the right order, in the right quantity' and reworked the entire production process.[53] Evaluation was entrusted to the JMA in the interest of speeding the process along, and five JMA engineers conducted the evaluation at the facility from 1956. On their advice, the Nagasaki Shipyard was reorganized around flow-production methods.[54] That is, Mitsubishi's Nagasaki Shipyards received the model for block assembly from the NBC Kure Shipyard, 'but achieved a faster departure than any other Japanese shipyards from the level of basic research'.[55] The work-centre method promoted by the JMA contributed to this transformation of the yards.

The work-centre method, then, had a widespread influence for a time following the war, including on the shipbuilding industry. But it began to fade in importance from the late 1950s. According to Kakuzo Morikawa of the JMA, this method was particularly suitable for pre-mechanized conditions, where Detroit-style mass-production methods were not implemented and the production time at each process was neither standardized nor stable.[56] The progress of mechanization might explain the declining influence of this method. But we have to consider another factor, the growing influence of American ideas.

11.5   The Reception and Influence of American Management Methods in Post-War Japan

During the war and the immediate post-war years, Japanese engineers and management scholars often wrote articles using German terms such as '*Takt*' or '*Fliessarbeit*'.[57] Indeed, indigenous techniques strongly influenced by German ideas remained widely diffused in Japan even after the war. But this rapidly changed in the 1950s and 1960s as a flood of American management and production techniques filled the Japanese journals.

After 1946, the Civil Communications Section (CCS) of the Allied Powers' General Headquarters (GHQ) sought to advise and improve the management and quality control of some Japanese communication equipment manufacturers. As a final stage of its activity, the section convened a 'CCS Management Seminar' in 1949 and 1950. Then in 1950 Dr William E. Deming presented an eight-day series of lectures. Thus, it is often said, did Japan realize the importance of Statistical Quality Control (SQC) and begin to refine it. In 1950 the Ministry of Trade and Industry also carried out the Management Training Programme (MTP), which the Far East Air Force had drawn up for Japanese

[53] *Manejimento* [Management] 16, 4 (1957), 34.   [54] Ibid.

[55] Takehei Miyashita, 'Zosengyo no hatten to kozo' ['The Development and Structure of the Shipbuilding Industry'] in Hiromi Arisawa (ed.), *Gendai Nihon sangyo koza* [Lectures on Modern Japanese Industries], vol. IV (Tokyo: Iwanami Shoten, 1960), 183.

[56] Kakuzo Morikawa, *Keiei gori-ka no chishiki* [Knowledge for Rationalizing Management] (Tokyo: Daiyamondo-sha, 1950), 172.

[57] See Shigetaka Mori, *Nagaresagyo no riron* [The Theory of Flow Production] (Tokyo: Akagi shobo, 1947).

supervisors. In the same year, the Ministry of Labour, backed by GHQ, also held the first seminar on Training Within Industry (TWI). Thus, after the war, many American management and production techniques, such as TWI, MTP, SQC, human relations, work factor analysis, marketing, and so on, were introduced in Japan. This process gained momentum when the Japan Productivity Council (JPC), set up in 1955, sent many Japanese businessmen to the United States. By 1965 the council had sent about 660 groups and about 6,600 people, on 'productivity missions' to the USA. After they came back to Japan, these businessmen gave lectures about their own findings in some major cities and also published their reports. These reports ran to 166 volumes, about 40,000 pages in total.[58]

Thus, many American management practices and ideas were introduced in Japan, and attracted great attention. But Japanese engineers and managers did not think that every element of American methods could be applied to domestic manufacturing industry. Therefore, they selectively chose certain elements and adapted them to the Japanese situation. The case of mixed-model assembly encapsulates this approach.

Mixed-model assembly was often praised, especially in the 1970s and 1980s, as the key element responsible for productive flexibility in Japanese manufacturing plants. This practice was often believed to have originated in Japan. But in fact this practice probably originated in the USA, and some Japanese found it applicable to domestic plants. During the war, as we have already seen, the Japanese engineers did not place a strong emphasis on the introduction of the mechanized conveyor system. Unless Japanese plants had implemented Detroit-style mass production, it was widely believed, they could not fully utilize mechanized conveyors. But this notion changed in the mid-1950s. The productivity mission on the electrical appliance industry, which visited the USA in 1955, was amazed at the mechanized conveyors used for the production of cars in Ford's Chicago assembly plant, as well as those for electric refrigerators at the Hotpoint Company, saying:

Unless we implemented mass production, we had thought that the implementation of the mechanized conveyor system was impossible. As Japanese manufacturers had to produce a diverse range of products in smaller quantities, therefore, we could not adopt the mechanized conveyor system. But we had to reconsider this ... At the Ford plant, several models of cars with different colors were produced with just one conveyor line ... Also at the Hotpoint Company, different models of refrigerators were made with one conveyor ... Even in Japan, if we could produce a wide variety of products using the same parts or standardized parts, there would be plenty of scope for adopting the mechanized conveyor system.[59]

[58] Nihon Seisanse Honbu [Japan Productivity Council] (ed.), *Sesansei Undo no 10 nen no ayumi* [The Ten Years' Progress of the Productivity Movement] (Tokyo: JPC, 1965), 39–40.

[59] Japan Productivity Council (ed.), *Denki kogyo: Nihon denki kogyo seisan-sei shisatsu-dan hokoku-sho* [Electrical Appliance Industry: Report of the Japan Productivity Mission on the Electrical Appliance Industry] (Tokyo: JPC, 1956), 138–40; Satoshi Sasaki, *Kagaku-teki kanri ho no Nihon-teki tenkai* [The Development of Scientific Management in Japan] (Tokyo: Yuhikaku, 1998), 288–9 also mentions this.

The productivity mission on the car parts industry also mentions mixed-model assembly, a rather neglected American practice:

A wide variety of products was moving on the same conveyor, and we saw this practice, especially at assembly plants . . . If we had to use the conveyor system in order to move a wide variety of products, we set up separate conveyors for each product . . . in order to avoid the mixing up of products and so on. But after watching the American practices, we thought that we should utilize the conveyors fully following the American way even if this would make it difficult for the production control.[60]

Japanese participants in the productivity missions to the USA thus sought out American practices which could potentially be applied in domestic manufacturing plants. As in the case of mixed-model assembly, Japanese engineers and managers found certain American practices applicable even in domestic factories which did not completely adopt Detroit-style mass-production methods.

Even when some American techniques were adopted, moreover, their content or the implications for management often differed greatly between the two countries. In some cases, Japanese companies which adopted the American ideas encountered difficulties in implementing them effectively. The case of Toyota illustrates this well. In 1950, just after the strike which almost bankrupted the company, Toyota sent two managing directors to the USA in order to investigate the American automobile industry. Both visited Ford's River Rouge Plant, and found the suggestion system at the plant useful, even for Toyota. When they came back, they advocated implementing the suggestion scheme at Toyota. On their recommendation, the 'Toyota Creative Ideas and Suggestion System' was introduced 'as a means of supporting its company-wide improvement activities' in 1951. But within a couple of years of its implementation, the number of suggestions made by workers dropped, and it was only a small group of workers that ever made suggestions. Then the management changed its approach, praising personally, in front of all workers, those who submitted good suggestions, and paying them a monetary reward. But ordinary workers did not respond to this approach either. Temporary workers, by contrast, were eager to get permanent jobs. So, they wanted to show their ability and loyalty to the company, hoping to be offered permanent positions. Once competition between workers began to develop within the framework of long-term employment, the suggestion system was used to identify loyal employees, and they were selected for promotion.

This suggestion system was not, however, enough to change workers' attitudes towards full involvement with job modification and continuous improvement activities. In the early 1950s, Toyota was short of senior workers, because over 2,000 operatives had left the company after the strike. In these circumstances, Toyota reviewed the training of supervisors as 'one of the ways that the

---

[60] Japan Productivity Council (ed.), *Jidosha buhin kogyo: Jidosha buhin kogyo seisan-sei shisatsu-dan hokoku-sho* [Car Parts Industry: The Report of the Japan Productivity Mission on the Car Parts Industry] (Tokyo: JPC, 1956), 95.

company would carry out its aims [on the shop floor] through the management hierarchy'.[61] In order to train supervisors, Toyota seized on TWI. In December 1951, Toyota produced 10 instructors. Thereafter the company continued this programme. In 1953, the company also decided to hold a JMA training seminar. Shigeo Shingo, who contributed much to the formulation of the so-called Toyota Production System, was the trainer and taught time and motion studies and other basic methods of industrial engineering, as well as offering practical instruction in the modification of production processes.[62] Then in 1955 Shingo 'took charge of the industrial engineering training course. By 1981 he had convened this training course over 80 times and the number of his trainees had reached about 3,000.'[63] The approach of these seminars was quite similar to that of the process analysis seminars, which the JMA had offered before and during the war. Thus, by the mid-1950s, Toyota had incorporated TWI and the suggestion system into their own training system, and transformed them into a means of diffusing job modification and continuous improvement activities from foremen and supervisors to the shop-floor.[64] TWI as well as the suggestion system had originated in, or at least were directly imported from, the USA. But Japanese management modified and adjusted them to make them acceptable to their own workers. As a result of such reworking, 'many Japanese companies introduced suggestion systems to follow up on the job modification movement begun by TWI'.[65]

11.6   Development of 'Automation'

In the mid-1950s, the Japanese economy began to pick up. When the popular magazine *Asahi Weekly* did a special feature on 'Detroit Automation' in 1955, it created a sensation in Japan. Yoichi Ueno, well known as an activist in the efficiency movement before the war, introduced the idea of automation as early as 1954. After the mid-1950s, 'automation' became a catchword, and it was believed to show the future direction of technological development. At about the same time, the Japanese government introduced several laws aimed at facilitating the replacement of older machines by new models. 'Automation' did not only mean the introduction of transfer machinery in Japan; rather it involved the enhancement of mechanical equipment or the introduction of new machine tools. 'Automation' or technological development in the Japanese automobile industry during this period actually meant 'to change over to mass production',

---

[61] Rodo horei kyokai [Labour Law Association] (ed.), Kanri/kantoku-sha kunren no jissai [The Practical Side of Training for Supervisors] (Tokyo: Rodo horei kyokai, 1964), 105.

[62] Ibid., 115.

[63] See Shigeo Shingo, *Study of the 'Toyota' Production System from an Industrial Engineering Viewpoint* (Tokyo: JMA, 1981), 2.

[64] Alan G. Robinson and Dean M. Schroeder, 'Training, Continuous Improvement, and Human Relations: TWI Programs and the Japanese Management Style', *California Management Review* (Winter 1993), 51. A large number of small groups scattered around the company reinforced this transformation.

[65] Ibid.

as one observer succinctly pointed out in the early 1960s.[66] In spite of the enthusiasm expressed in the mass media, however, the progress of mechanical automation was quite slow: just 16.5 per cent of all engineering firms employing over 30 workers had introduced it by 1959, whereas 27 per cent had introduced process automation.[67]

Canon was one of the companies which aggressively increased mechanization after the mid-1950s. By the end of the 1950s, the company was putting its efforts into the mechanization and automation of its parts-manufacturing operations, and was proceeding with the changeover to conveyor-type assembly, moving away from its award-winning 'work-centre' method. By 1960, the company 'did away with its re-work section and prohibited the use of files and hammers to fit parts', meaning that a very high degree of precision was required for the interchangeable parts of its cameras.[68] Thus, companies like Canon, which had the in-house capability to analyse their own production processes and overall manufacturing operations and to undertake the introduction of conveyor systems with extensive use of specialized automated equipment, had no further need for the training and experience offered by the work-centre method.[69]

From about 1955, other companies joined Canon in setting up mass-production systems using specialized manufacturing machinery. The conversion to flow production using large numbers of high-performance machine tools brought about huge increases in productivity at each step of the manufacturing process, and dictated the establishment of an improved balance among individual production processes. As a result, Japanese production engineers at such companies quickly began to show interest in a heretofore virtually unused 'approach to statistical analysis of mass production'.[70]

But most companies were more hesitant about accepting such 'automation'. In particular, automobile assemblers such as Toyota and Nissan approached with caution the idea of 'Detroit automation'. Even when Nissan installed transfer machines at its plant in 1956, for example, the machine was re-designed to accommodate two different engine models on a single line.[71] Thus, the idea of mixed-model assembly was applied.

Toyota placed much emphasis on 'humanized automation', rejecting the idea of 'Detroit automation'. But Toyota mechanized its production facility by following a five-year long-term investment policy planned when the company introduced an automated press line in the early 1960s. In June 1960, Toyota and Ford nearly reached agreement on setting up a joint venture for manufacturing

[66] Nihon Jinbun Kagaku-kai [Japan Association of Human Science] (ed.), *Gijutsu kakushin no shakai-teki eikyo* [Social Influences on Technical Innovation] (Tokyo: Tokyo daigaku shuppan-kai, 1963), 7.

[67] Tsusho sangyo sho sangyo kigyo kyoku [MITI's Company Bureau] (ed.), *Waga kuni sangyo no ohtomehshon no genjo to shorai* [The Present and Future of Automation in our Country's Industries] (Tokyo: Hokuetsu bunka kogyo, 1953), 126–7.

[68] Canon, Inc., *Canon-shi*, 68–9.   [69] Ibid.

[70] Kenichi Okuda, *People and Management*, 518.

[71] See Daito, 'Automation and Production Organization'.

and selling a car to be called the 'Publica,' but this failed. During the negotiations, Toyota realized how Ford produced bodies and doors cheaply. Therefore, when Toyota established its new Motomachi plant, the company introduced Danely press machines with rapid die-change features.[72] Using these machines, Toyota pursued quick die-changing, which was crucial for accommodating mixed-model assembly.

Even in the early 1960s, the Japanese market for cars remained small and assemblers wanted to produce a variety of cars. The companies preferred to accommodate many types of cars on a single line, and consequently emphasized the need to equalize the cycle times of all production processes. This was rather similar to the situation at aircraft companies during the war: a rather low level of mechanization combined with the need to produce a variety of products. Hence it is not surprising that certain control methods developed during the war were implemented. According to the company's official history, Nissan introduced the [kit] marshalling method in 1955.[73] This was probably the method originally adopted in UK aircraft and aero-engine factories during the war and then refined by British automotive manufacturers after 1945. Nissan had access to this through its close link with Austin through technology licensing.[74] By introducing this method, Nissan was able to 'improve labor management, and it became easier and more reliable to control the process of production as well as the parts. Furthermore, the company could assemble a variety of cars on just one line.'[75] This achievement was nearly the same as that which Japanese engineers had tried to realize during the war. We might say that Japanese manufacturers' wartime experience anticipated and prepared them for the subsequent borrowing and assimilation of Western production techniques in the 1950s.

Much has been written about the Total Quality Control (TQC) movement. There is no need in this chapter to say more about the enthusiasm with which Japanese managers responded to Dr Deming's lectures and the successful diffusion by the Japanese Union of Scientists and Engineers (JUSE) of the ideas of TQC and QC circles. It is more relevant to reconsider what the QC circles did, and in this context, Cusumano's observation is of interest:

---

[72] Hanji Umehara, 'Kita-michi yuku-michi' ['Past and Coming Years in the Automotive Industry'], *Tetsu to Ko* [Iron and Steel] 59, 8 (1973), 1186–9. According to Michael A. Cusumano, Taiichi Ohno 'first saw Danley stamping presses with rapid die-change features on a trip to the United States during the mid-1950s'. See Cusumano, *Japanese Automobile Industry*, 285.

[73] The Investigation Section of General Affairs Department of Nissan Motor Corporation (ed.), *Nissan Jidosha 30 nen-shi* [The First Thirty Years of Nissan Motor Corporation] (Yokohama: privately published, 1958), 330.

[74] Cusumano described this as an 'American method', but Nissan's official history, upon which he relied, does not mention this as an 'American method': Cusumano, *Japanese Automobile Industry*, 315. On 'kit marshalling' in wartime Britain, see Sebastian Ritchie, 'A New Audit of War: The Productivity of Britain's Wartime Aircraft Industry Reconsidered', *War & Society* 12, 1 (1994), 134; and for its adaptation to production control and component supply in the post-war British automobile industry, see Jonathan Zeitlin, 'Reconciling Automation and Flexibility? Technology and Production in the Postwar British Motor Vehicle Industry,' *Enterprise and Society* 1 (2000).

[75] Nissan Motor Corporation, *The First Thirty Years*, 331.

Despite the reputation of QC circles in Japan and elsewhere, and his role in inventing them, Ishikawa admitted that they did little to improve quality without the support of a comprehensive 'quality assurance' program throughout a firm and its supply network. Nissan and Toyota officials also conceded that QC circles functioned primarily to increase employee participation in company operations and to boost morale by allowing employees to work in groups to solve problems.[76]

Thus we may rightly ask why QC circles were so popular at the time, especially among Japanese manufacturers.

Japanese companies accepted relatively easily the idea of QC circles at each point in the production process. Dividing the shop-floor into a large number of small groups was again quite similar to Japanese wartime practice, and the post-war adoption of the work-centre method by domestic manufacturing companies had important ramifications for the subsequent absorption of QC circles. Also, lower levels of mechanization required careful and delicate labour management. Especially after the strike-torn years immediately after the war and in the early 1950s, labour relations in Japanese companies were tense. Toyota, after the 1950 strike, began to set up large numbers of small groups across the company based on hobby interests, schooling, and district of origin. Nearly all employees joined at least one of these groups. The establishment of such groups created close personal relationships among employees, and greatly contributed toward easing the tense labour relations, as one person in charge of personnel management recollected.[77] This practice probably represented another way of organizing the workers, inducing them to participate more directly in the operations of production. As Statistical Quality Control evolved into the Total Quality Control with QC circles, the company found the circles useful for achieving shop-floor peace while emphasizing the quality of their product.

11.7 Conclusion

After the late 1950s, the Japanese manufacturing sector seems rapidly to have incorporated production control techniques which originated in the United States. These American methods seemed to represent an approach that had not been actively considered by Japanese production engineers, and the rapid spread of the new methods seemed to be supported by an appreciation of their novelty on the part of management. Of course, novelty was not the only factor involved; the most significant reason for the enthusiastic welcome given to American production control methods was the widespread introduction of high-performance machine tools and the changeover to automated flow production. Thus there was finally a critical mass of Japanese companies engaged

---

[76] Cusumano, *Japanese Automobile Industry*, 334.

[77] See Hirohide Tanaka, 'Nihon-teki kanko wo kizuita hito-tachi: Yamamoto Keimei shi ni kiku' ['Persons who created the Japanese Labor Practices: Inteview with Mr Keimei Yamamoto'], *Nihon Rodokyo-kai Zasshi* [Monthly Journal of the Japanese Institute of Labour], 280–2 (1982).

in introducing production systems which required a new 'approach to statistical understanding and analysis of mass production'.

As the new methods took hold, Japanese production engineers quickly forgot about their wartime production efforts and the development of the work-centre method. This method was the result of wholly indigenous efforts during and immediately following the war under the strong influence of German practices. But it was also a highly significant step in the Japanese drive towards scientific management. Tetsuro Nakaoka points out that although this was a 'rudimentary' system, 'it corresponded to the needs of the Japanese machinery industry of the time', also observing that, 'One cannot overlook the importance of such a system having been developed independently prior to the adoption of American methods'.[78]

Moreover, even when American methods seemed to have been implemented, Japanese companies reworked and adjusted them to their own environment. In the 1970s and 1980s, the so-called Japanese production system was 'discovered' and often praised as the most advanced method at that time. In particular, Toyota's methods were praised highly by many international scholars, and attracted world-wide attention as the 'lean production system'. At that point some observers regarded Toyota as having the most advanced production system in the post-war period. Our view is totally different. In the 1950s, and also even in the 1960s, Toyota lagged far behind the then world standard of mass-production techniques. The company did not directly imitate American methods, but tried to emulate the efficiency of US companies both by utilizing indigenous ideas and by re-working foreign ones, adapting them to domestic market conditions. Hence when the 'Toyota production system' was 'discovered' during the 1970s and 1980s, it seemed to be quite different from that of the American automobile industry.

[78] Nakaoka, 'The Wartime/Post-War Scientific Management Movement', 48.

Chapter 12

American Occupation, Market Order, and Democracy: Reconfiguring the Steel Industry in Japan and Germany after the Second World War

GARY HERRIGEL

12.1 Introduction

Engagement with American practices and ideas in the period after the Second World War was different in Germany and Japan than in many of the other political economies considered in this volume because both were militarily occupied countries. Thus, in addition to the diffusion (or in any case, incursion) of American industrial ideas, principles of organization, and technologies by way of markets, scholarly and technical writings, and élite circulation, American ideals were also imposed on Germany and Japan by military governors during the first decade after the war. By analysing the process of restructuring in the steel industries in both occupied countries, this chapter examines the complexity of the notion of 'imposition' in the context of this military occupation. The main argument is that the American occupation dramatically changed both societies by forcing them to grapple with American ideas of social, industrial, and political order. By this I do not mean that the Americans got exactly what they wanted in either case, much less that the two industries (or societies) were 'Americanized'. Rather, the interaction between occupiers and occupied transformed the way that both understood their own interests and what each under-

Thanks to Jonathan Zeitlin, Carol Horton, Steve Tolliday, Kazuo Wada, Dick Samuels, George Steinmetz, Lisa Wedeen, Christian Kleinschmidt, Peter Kjaer, Sei Yonekura, Gerald Feldman, Beverly Crawford, Michael Storper, Robert Salais, Gregory Jackson, Karl Lauschke, the participants of the conference that produced this volume and the members of the Workshop on Organization Theory and State-building at the University of Chicago, the Center for German and European Studies at the University of California Berkeley, the Workshop in Historical Sociology at the University of Michigan, and the Franco-American Seminar on Rationality, Conventions and Institutions for critical readings and discussions of this chapter. Thanks as well to Laura Kjeraulf for valuable research assistance. The usual disclaimers should be strongly enforced in this case, for this is my initial foray into the very fascinating yet complicated history of Japanese industrial development. My hermeneutically challenged condition, I fear, has led me on occasion not always to be able to appreciate the good sense of my more knowledgeable colleagues.

stood to be industrial practice consistent with American ideals. In the steel industry, the result was the emergence of genuinely hybrid firms, industrial structures, and market strategies which were all markedly different from their pre-Occupation ancestors and far more innovative and efficient than the American models that were used to guide their reconstruction.

In making this argument it is important to emphasize immediately that in my view the terrain upon which contestation occurred between American imposers and German and Japanese receivers/resisters *was not* in the first instance organizational or technological, though ultimately the effects of the encounter were felt in these areas. American military authorities did not insist that Japanese or German industrial actors adopt specifically American internal organizational procedures or implement American production techniques or technologies or industrial relations practices. Indeed, they left considerable latitude for local practice in these areas and allowed substantially different practices and forms of organization within the corporation and in industrial relations to emerge than either existed or were then developing in the United States. In any case, both the Germans and the Japanese had long been familiar with and even implemented a wide variety of American techniques before they ever confronted American military reformers.

Rather, the crucial area of engagement concerned the definition of the political, economic, and social terrain upon which industrial institutions and technologies were to be constructed and deployed in each society. Indeed, it was the capacity of the occupying power to establish this terrain of conflict that indicates the enormity of the power it wielded over these two vanquished nations and it was certainly the source of its greatest influence. I will emphasize in particular two forms of Allied action on this terrain, one destructive, the other creative. First, the occupation forces systematically deconstructed the central institutions of market and political order in Germany and Japan: among other things, cartels were outlawed, firms were broken up, leadership structures were attacked, and key associations were either redefined to play a different role or simply outlawed. This destructive impulse was defensive: in the political-economic imaginary of the occupying powers, the industrial institutions that they targeted had previously contributed directly to authoritarianism and the growth of military aggression in those countries.

Second, the presence and position of the Allies as military occupiers of defeated powers enabled them to structure debate about broad normative guidelines for social, industrial, and political reorganization. American insistence on the ground rules for public deliberation about the reconstruction of each society brought principles of order into debate that diverged markedly from pre-war indigenous ideas about the character and interrelationship of social, economic, and political practices and institutions in Germany and Japan. American occupiers believed strongly, for example, that stable democratic societies contained plural, cross-cutting organizations, multiple and countervailing instances of social and economic power, a strong middle class, trade unions, market competition, and efficient, oligopolistically structured industries. As it turned out, the USA allowed the Germans and the Japanese to create these conditions in their societies in their own way—or in any case to make it seem as if they were. But the source of American power was that deliberating actors were not permitted to deny that a society with such features was desirable.[1]

The two sides of the Allied strategy of occupation in both countries are nicely expressed in a summary memorandum by the Supreme Command of the Allied Powers (SCAP) in Japan: 'In the first two years of the Allied Occupation of Japan, SCAP's activities in economic matters [were] directed toward eradicating the old imperialistic, non-democratic economic pattern of life and replacing it with a new framework which would lead Japan into democracy and rightful membership among the community of nations.'[2] In large part these two factors of US power successfully forced actors in Japan and Germany to engage with the categories and logic of the US conception of social order and, as a consequence, to establish new rules for market competition and democratic order. Of course, what was actually achieved did not correspond in either case to what the occupying authorities imagined market order, democracy, or their interdependence to be. German and Japanese actors created the mutual limitations between state and economy that the Allies desired, but in ways that were inescapably and insidiously informed by their own peculiar understandings of the categories and relations that the Americans imposed. Enforced, unavoidable confrontation with American ideas of political and market order, in other words, did not force the Germans and Japanese on to entirely virgin terrain. Rather, it forced them to reflect on their traditional practices in light of American ideals and allowed them to recompose elements from both into a strikingly new style of practice.

In my view, this kind of argument cannot be made by focusing narrowly on the ten year period following the end of the Second World War. A bit of historical background is needed in order persuasively to convey the character of the conflicts that emerged as US occupiers set out to restructure the steel industries in the two occupied countries. Thus, section 12.2 outlines the pre-war and wartime structures of the German and Japanese steel industries, highlighting the political boundaries of the industry in each society and the distinctive character both of property forms and of market organization in both industries. Section 12.3 turns to the lenses through which the occupying military governments interpreted these industries after the war and to the destructive and creative policies they promulgated to change them. The emphasis here is on the categories of American economic and political understanding and on the way in which American authorities understood market order to be related to democracy. Section 12.4 looks at the

---

[1] I have been influenced in my efforts to describe the character of American power as profoundly shaping and yet accommodating of broad experimentation, creativity, and strategizing by the work of the Comaroffs on the colonial experience in South Africa. See Jean Comaroff and John Comaroff, *Of Revelation and Revolution. Christianity, Colonialism and Consciousness in South Africa, Vol. I* (Chicago: University of Chicago Press, 1991), especially their theoretical discussion in ch. 1, 1–48.

[2] SCAP document excerpted in Wolfgang Benz (ed.), 'Amerikanische Besatzungsherrschaft in Japan 1945–1947', in *Vierteljahresschrift für Zeitgeschichte*, 26 (1978), 331.

reception of these ideas and policies and their effects on the development of the industries in both countries during the initial period of growth in the first two decades after the war. In this section German and Japanese understandings of the relationship between market and political order are contrasted with the Allied conceptions outlined in section 12.3. This section narrates how German and Japanese attempts to appropriate American ideas and practices (and American efforts to impose the same) resulted in possibility-creating effects which transformed the principles of market and political order of all the principals.

12.2 The Steel Industry in Germany and Japan in the Period Prior to Occupation

Though the two steel industries differed in size and technological sophistication in this period—the German industry was capable of far larger output than the Japanese industry and was a world leader in technology—there were many similarities between the two in terms of company forms and industry structure. Indeed, by the 1930s in both countries, the steel industry had developed a distinctive bifurcated structure. In each case a large, deeply integrated, diversified producer of relatively high-volume, standardized steel goods accounted for nearly half of the total output in the industry, while a larger number of smaller, more specialized companies, which were often part of much larger diversified combines, accounted for the rest of the industries' domestic output. Interestingly, these similarities in industry structure were achieved with dramatically different positions of the government in the industry. The Japanese industry will be described first, the German second.

THE PRE-1945 JAPANESE STEEL INDUSTRY

There are two distinct yet intertwined strands to the story of the development of the Japanese steel industry up until the end of the Second World War. The first is a story of relations between private industry and government in the coordination of industrial development and the second involves the development of particular forms of governance and industrial structure in the production of steel. The two strands can be introduced separately, but they then begin to intertwine in a series of cartelization laws beginning in the early 1930s and then converge almost completely in the wartime innovations in industrial governance known as Control Associations.

Government–Business Relations

As was the case with an array of early industries in nineteenth-century Japan, the emergence and early development of the steel industry was strongly shaped by the intervention of the state. This early state involvement, as many have noted, was decidedly *not* because the Japanese government was committed to state-controlled development. Rather, most agree that it was because private actors could not be enticed into carrying the initial start-up risks in the indus-try.[3] Indeed, for nearly two decades at the end of the nineteenth century, the Japanese state tried to encourage private actors, in particular large diversified *Zaibatsu* holding companies, to enter the steel industry because it believed that indigenous Japanese steelmaking capacity was crucial for national development and security. Despite these efforts, the *Zaibatsu* were very reluctant to play along. They believed that the start-up costs in steel would be far too high and the profitability and growth prospects for the industry too low to justify mov-ing into the industry—even with state support.[4]

Frustrated by this lack of private interest and convinced of the value of indig-enous production, in 1901 the Japanese state took the bull by the horns and founded the Yahata Steel Works. The new facilities were state of the art for the time (modelled very closely after German steelmaking technology and practice) and had an initial total production capacity of 90,000 tons of steel a year. In all, the investment totalled over 120 million yen.[5] All of this was very impressive by the standards of the time—but in a way that vindicated the reticence of the *Zaibatsu*, for none of it worked very well. The operation of the blast furnaces and the coke ovens in the new venture suffered continual mechanical problems in the early stages and ultimately had to be shut down for two years to have their German designs adapted to the particular kinds of raw material inputs and skills available in Japan. Further, even when everything was up and running, the Yahata Works suffered tremendous operating losses for the first nine years of its existence. When it finally did earn a profit in 1910, it was with state assistance.[6] Growth in the domestic market for steel, moreover, was slow, at least until the beginning of the First World War, and competition from imports was very intense: as late as 1913, imported steel accounted for 64 per cent of home con-sumption and 52 per cent of pig iron consumption.[7]

---

[3] For a discussion of the phenomenon of the state creating the initial firm in the industry and then using its position to lure in private actors, see Richard Samuels, *The Business of the Japanese State: Energy Markets in Comparative and Historical Perspective* (Ithaca: Cornell University Press, 1987), esp. ch. 3, 68–134. The Japanese state, particularly in the decade after the Meiji restoration, pursued this policy of starting and then spinning off or luring in private actors in an array of other industries (beyond coal and steel) as well. On the historical period of industrial promotion and ownership on the part of the Meiji state, see, among many others, E. Sydney Crawcour, 'Economic Change in the Nineteenth Century' and Kozo Yamamura, 'Entrepreneurship, Ownership, and Management in Japan', both in Yamamura (ed.), *The Economic Emergence of Modern Japan* (New York: Cambridge University Press, 1997), 1–49; 294–352; Tessa Morris-Suzuki, *The Technological Transformation of Japan: From the Seventeenth to the Twenty-first Century* (New York: Cambridge University Press, 1994), 71–105; William W. Lockwood, *The Economic Development of Japan* (Princeton: Princeton University Press, 1968), ch. 1.

[4] Hidemasa Morikawa, 'The *Zaibatsu* in the Japanese Iron and Steel Industry', in Hans Pohl (ed.), *Innovation, Know-how, Rationalization and Investment in the German and Japanese Economies, Beiheft 22, Zeitschrift für Unternehmensgeschichte* (1982), 134–50, esp. 136.

[5] Morikawa, 'The *Zaibatsu* in the Japanese Iron and Steel Industry', 140; on the facilities, see Table 12.1 in the text.

[6] Morikawa, 'The *Zaibatsu* in the Japanese Iron and Steel Industry', 141–3.

[7] G. C. Allen, *A Short Economic History of Modern Japan, 1867–1937* (London: George Allen & Unwin, 1946), 74.

None the less, Yahata did succeed in gradually establishing a position for itself in the domestic market (after all, prior to its establishment, imports accounted for virtually all consumption of steel and pig iron). And this success was used by the government to lure other private domestic firms, many from *Zaibatsu* interests, to enter the industry. On the one hand, the Japanese government used attractive munitions, shipbuilding, or railway contracts to entice *Zaibatsu* investments in steel production. This created a demand for pig iron that led to further investment, particularly in Japan's colonial possessions in the initial stage of steelmaking (see Table 12.1). And, once private firms had been lured into the industry, the government encouraged and protected them with numerous subsidies, tax incentives, tariffs, and the like.[8] On the other hand, Yahata itself also helped these novice, less-integrated, and much smaller-scale private operations develop. It supplied the private steel units with domestic pig iron and, more significantly, spun off to the private firms specialized steelmaking and rolling technologies that Yahata was not willing itself to develop because they distracted from the large-scale steel production strategy the state firm was pursuing.[9] Itself a very significant market actor, for much of the first thirty years of its existence Yahata none the less behaved like an incubator for newer, smaller, and above all private entrants into the industry.

With imported steel as the enemy and the technological development of the Japanese economy as their shared goal, public actors and private producers pursued a co-operative, yet none the less competitive and market-based strategy of industry expansion. Bureaucrats, bankers, executives from the *Zaibatsu* holdings, and managers from Yahata engaged in continuous discussions concerning the problems, opportunities, needs, and possibilities of steel production in Japan. One has to be careful to emphasize that this was not planning in any sense. Instead it was élite dialogue on how to structure market-based economic activity in the interest of the nation's development.[10]

Bifurcated Market of Steel Producers

By the 1920s, these business–government relations had fostered the development of a bifurcated industrial structure in the Japanese steel industry, with a large, state-owned, integrated, high-volume producer on the one side and a number of smaller, more specialized, less integrated, private producers on the

---

[8] E. B. Schumpeter (ed.), *The Industrialization of Japan and Manchukuo, 1930–1940* (New York: Macmillan, 1940), 596 ff.

[9] Seiichiro Yonekura, *The Japanese Iron and Steel Industry, 1850–1990* (New York: St. Martins Press, 1994), 57–77.

[10] This distinctive kind of close and co-ordinated relationship between private business (in this case *Zaibatsu*) and government has frequently been noted by scholars of Japanese industrialization—though they disagree about the balance of power between the state and the private economy. A classic account of the Japanese state as a 'developmental state' in which the balance of power is placed in the state's favour is given in Chalmers Johnson's *MITI and the Japanese Miracle* (Palo Alto: Stanford University Press, 1981). Samuels, *The Business of the Japanese State*, and John Owen Haley, *Authority Without Power. Law and the Japanese Paradox* (New York: Oxford University Press, 1991) are good examples of scholars who place the balance of power towards the other end.

Table 12.1. Overview of Initial Private (mainly *Zaibatsu*) Moves into Steel Production in Japan

Name of enterprise (and affiliation)	Date	Initial capacity	Main product area and market
Yahata (State Enterprise)	1901	Crude steel, 90,000 tons/year 2 blast furnaces (160 tons each) 4 open hearths (25 tons each) 2 converters (10 tons each)	Pig iron, steel, rolled products
Sumitomo Chukojo (Sumitomo)	1902 (1912)	Crude steel, 2,000 tons/year	Cast iron (later: cold drawn steel tube for Imperial Navy)
Kobe Seikojo (Suzuki Shoten)	1905	Open hearth (3.5 tons)	Cast and forged steel
Kawasaki Zosen Hyogo (Kawasaki)	1906	2 open hearths (10 tons each)	Cast and forged steel for railway equipment and later shipbuilding
Nihon Seikojo (Mitsui)	1911	2 open hearths (50 tons) 4 open hearths (25 tons) 2 open hearths (5 tons)	Cast and forged steel and weapons for the navy
Nihon Kokan (NKK) (Asano)	1912	2 open hearths (20 tons)	Mannesmann seamless steel tube
Hokuton Wanishi (Mitsui)	1913	1 blast furnace 50 tons	Pig iron
Honeiko Baitan Konsu (Okura Gumi & Chinese National Iron Works)	1915	1 blast furnace (130 tons)	Pig iron (side business from colonial mining operations)

Name of enterprise (and affiliation)	Date	Initial capacity	Main product area and market
Mitsubishi Seitetsu Kenjiho (Mitsubishi-Korea)	1918	1 blast furnace (150 tons)	Pig iron (side business from colonial mining operations)

Source: Adapted from Hidemasa Morikawa, 'The *Zaibatsu* in the Japanese Iron and Steel Industry', in Hans Pohl (ed.), *Innovation, Know-how, Rationalization and Investment in the German and Japanese Economies, Beiheft 22, Zeitschrift für Unternehmensgeschichte* (1982), 142, Table 1 and accompanying text.

other. In order to understand the strategic and competitive dynamics that produced this bifurcated structure in the industry, the interaction of two sets of relations are important: relations between the new *Zaibatsu* market entrants and Yahata and relations between *Zaibatsu* steel production and the rest of the *Zaibatsu* operations.

Plainly, in their initial forays into steel production, the *Zaibatsu* were hedging their bets on the growth of the industry. They wanted to have some position in the industry if it finally did begin to grow, but they did not want to carry the burden of large investments on the scale of Yahata. Consequently, the *Zaibatsu* invested in areas of specialized production in which Yahata was not engaged, or in which import competition was weak. Also, as noted above, these investments were encouraged by special state contracts from downstream operations within the *Zaibatsu* themselves such as in shipbuilding and weapons production. On the whole, then, the scale of these private investments was considerably smaller, less integrated, and far more specialized than Yahata. The *Zaibatsu* did not try to take over the steel market either from imports or from the government's firm. Rather, they entered steel production as niche players, gaining experience and technological know-how, while avoiding direct competition with larger competitors.[11]

Such apparent opportunism was actually in line with what the Japanese government wanted to achieve, for it contributed to the expansion of indigenous steel production capacity and also systematically enhanced the technological sophistication of private firms in areas that the state viewed as progressive. But the strategy was also very congenial to the broader combine-wide strategies of development that the *Zaibatsu* were pursuing during the first decades of the twentieth century. Knowledge of particular steelmaking technologies and

[11] Morikawa, 'The *Zaibatsu* in the Japanese Iron and Steel Industry'.

markets was beneficial to other firms in related markets for machinery, ship-building, engineering, and trade *elsewhere within the conglomerate*. Information circulated in and out of the steel arms of *Zaibatsu* in ways that encouraged learning and innovation in the complex of firms within the *Zaibatsu*'s scope of operations. The co-operative synergy between the steel and/or rolling operations of a *Zaibatsu* steel unit with the particular down-line operations of the conglomerate were the main thing, not vertical technical economies of integration or the lowering of unit costs in the steel unit itself.[12] Often *Zaibatsu* steel units were not integrated (backward or forward beyond rolling operations) and purchased pig iron or scrap (and even steel ingots) from third parties (in particular, Yahata).

As the industry grew, the bifurcation between the high-volume standardized producer, Yahata, and the specialized niche players became more entrenched. In a period of (albeit slow) growth like that in the first decades of the century, this was a very stable structure, beneficial to both parties in the industry. It was not a 'dualist' or core-periphery structure in which the large core firm was the 'most advanced' and the peripheral players less so. The players in the industry were all technologically sophisticated—indeed increasingly so as the new century wore on. They were of different sizes and differently integrated because they were pursuing alternative kinds of profit-making strategies in the same industry.

Governance Problems

The stories of business–government co-operation and industrial bifurcation merge when turbulence and market disorder start to affect action in the industry during the late 1920s and 1930s. The stable reproduction of the bifurcated market was challenged during this period due to the following factors: large investments in new plant and capacity during the First World War resulted in overcapacity after the war; the Japanese economy experienced a post-war recession and then later (along with the world economy) a very severe depression by the beginning of the 1930s; finally, during the 1930s militarist forces within the Japanese state began to attempt to accelerate the growth of domestic steel production in order to ensure domestic supplies for military industries.[13] The first

---

[12] *Zaibatsu* were very distinctive horizontal business organizations: neither hierarchically-controlled multidivisional companies nor holding companies of loose ownership ties, they were complex agglomerations of specialities, whose boundaries were given by family property rather than technology, related markets or even contiguous space. For good discussions of their organization, see Hidemasa Morikawa, *Mitsubishi* (Tokyo: University of Tokyo Press, 1992) and Norbert Voack, *Die japanische 'Zaibatsu' und die Konzentration wirtschaftlicher Macht in ihren Händen*, Ph.D dissertation (University of Erlangen-Nürnberg, 1962).

[13] These factors are generally emphasized in the standard accounts of the move toward cartelization and Control Associations in the inter-war and wartime period. Unless otherwise cited, I have relied primarily on the following sources for my information: Eleanor Hadley, *Anti-trust in Japan* (Princeton: Princeton University Press, 1970), 357–89; Schumpeter, *The Industrialization of Japan and Manchukuo*, esp. 596–604, 680–740, 789–865; T. A. Bisson, *Japan's War Economy* (New York: Macmillan, 1945); Jerome Cohen, *Japan's Economy in War and Reconstruction* (Minneapolis: University of Minnesota Press, 1949); Secretariat, Institute of Pacific Relations (ed.), *Industrial Japan. Aspects of Recent Economic Changes as Viewed by Japanese Writers* (New York: Institute

two factors (overcapacity and recession/depression) made it extremely difficult for producers to find markets for the steel they were *able* to produce, while the third introduced a set of constraints on the range and quantity of steel products they were *allowed* to produce. These pressures gave rise to three different governance solutions: state-enforced private cartelization, merger, and market-supplanting Control Associations.

Cartelization   Overcapacity and recession/depression in the decade immediately after the First World War introduced disorder into markets because it encouraged market poaching. In the face of capacity well in excess of existing demand, firms that had previously left one another's specialized markets alone began not to. Seeking desperately to utilize expensive idle equipment, private speciality producers attempted to enter other specialists markets as well as the standardized markets dominated by Yahata and imports. Price cutting, special purchase and delivery conditions, and other pernicious competitive devices were the techniques deployed to achieve market entry. Unsurprisingly, with their deployment market order in the industry deteriorated: prices became erratic, profits suffered, and employment became unsteady. The classic response to this kind of situation is the formation of a cartel: producers agree to co-operate with one another on price setting and the fixing of terms and conditions of sale in order to bring stability back to the market.

During the 1920s, the formation of cartels was a completely private matter of agreements between consenting firms on a wide array of elements within their competitive industrial environment. Such agreements were typically satisfactory for short periods to stabilize markets, but proved to be relatively fragile unless they were followed by sustained upturn in demand for the product concerned. The Achilles heel of these agreements was that they were never able to cover all producers in particular markets. Without a third party to enforce participation, short-term calculations of advantage tended to outweigh longer-term considerations of stability through co-operation. When the onset of depression at the end of the decade made this irritation in the nature of cartel organization into a serious crisis of market governance, the fragility of these private cartel agreements became a topic of intense debate among both private industrialists and government bureaucrats. Ultimately, debate gave rise to political intervention and, in 1931, the passage of the Major Industries Control Law which brought the authority of the state into the organization of cartels in the industry.[14]

of Pacific Relations, 1941), chs. 1, 3, 8, 13; Johnson, *MITI and the Japanese Miracle*, 116–97; Lockwood, *The Economic Development of Japan*; Johannes Hirschmeier and Tsunehiko Yui, *The Development of Japanese Business, 1600–1980* (2nd edn., London: George Allen & Unwin, 1981), esp. 236–51; Mitsubishi Economic Research Bureau (MERB), *Japanese Industry and Trade. Present and Future* (London: Macmillan, 1936), 40–52, 114–30, 195–210; Leonard H. Lynn and Timothy J. McKeown, *Organizing Business. Trade Associations in America and Japan* (Washington, DC: American Enterprise Institute, 1988), 15–28.

[14] A concise description of the purposes of the law are given in MERB, *Japanese Industry and Trade*, 117.

The Major Industries Control Law was significant because it permitted the state to force non-member producers to comply with the agreements struck by the majority of cartelized producers in a particular product market. This enabled producers in the steel industry to stabilize the specialization relations in the bifurcated structure by establishing cartels that defined the boundaries among product markets as well as the prices in the industry through production and so called joint sales agreements (that is, selling quotas, the collection and allotment of orders, etc.). Such agreements eliminated damaging poaching, maintained the integrity of individual producers' speciality niches, and allowed producers to concentrate on innovation rather than on covering their costs. Such arrangements were attractive to the *Zaibatsu* because they facilitated productive interaction between their steel interests and the rest of their manufacturing business—something that poaching and competitive chaos in steel markets had made more difficult.[15]

Merger   State support of cartels was not enough to introduce stability into the volume sector of the industry. Yahata had been suffering throughout the 1920s and early 1930s by poaching from private firms which sought to unload their surplus production into the volume markets. In addition, several private firms, in particular private independents not part of *Zaibatsu*, had invested in volume capacity during the First World War because it seemed at that time that Yahata would not be able to meet the entire market demand. Ten years later these producers along with Yahata lived in an unstable merciless market of organized overcapacity, under-production, and low profitability. Cartelization in this market was not enough: merger was called for. But because Yahata was a state-owned firm, the creation of a merger had to be a political act.

In 1933–4 this is precisely what occurred: the Japan Steel Manufacturing Company Law was promulgated in 1933 and all of the volume producers in the industry, including Yahata, were consolidated into a single firm the following year.[16] Though the consolidation was encouraged and organized by the Japanese state and it involved a very large state enterprise, the consolidation actually took place through the *privatization* of the Yahata works and the incorporation of the operations of five private iron and steelmaking companies into the newly privatized firm.[17] The new industry structure concentrated virtually all of Japan's pig iron production (97 per cent) into the new consolidated firm. Importantly, the market positions of Japan Steel in both the raw steel and finished-product segments of the industry were significantly less: Japan Steel accounted for 56 per cent of steel ingot production and 52 per cent of finished-steel output. The rest of this, more specialized, output was supplied by the

[15] Yonekura, *Japanese Iron and Steel Industry*, 117–18. MERB, *Japanese Industry and Trade*, 118, lists steel sectors belonging as of 1936; see also Lynn and McKeown, *Organizing Business*.
[16] See MERB, *Japanese Industry and Trade*, 119; Lynn and McKeown, *Organizing Business*.
[17] Yonekura, *Japanese Iron and Steel Industry*, 150.

remaining non-integrated players in the industry, above all by NKK, Kawasaki, Kobe, Sumitomo, and Asano.[18]

As these market-share statistics suggest, this round of industrial stabilization in the early 1930s, driven by co-ordinated public and private action, successfully stabilized the bifurcated structure of specialists and volume producers. This was a co-ordinated, market-oriented steel industry. The final set of governance changes introduced during the early years of the Second World War, however, were of a different character entirely.

Control Associations   With the coming of war, the militarists within the Japanese state began to become increasingly concerned that the nation's industrial capacity be able to be used in the service of the nation's military needs. In their view, the structures that were created during the 1930s were unreliable and unwieldy in this regard, especially the cartels. The Major Industries Control Law had stabilized markets, but it also allowed for the continuous fragmentation of markets and proliferation of cartel organizations, all of which made it very difficult for resources to be consolidated and marshalled for state ends.[19] In September 1941, the government introduced the Major Industries Association Ordinance, the purpose of which was to create extra-firm 'Control Associations' with the power and authority to direct production and the distribution of products in the industry toward the satisfaction of the state's military needs.[20] If the earlier Major Industries Control Law attempted to bring state authority to bear to help private interests govern themselves in a market context, this later law sought to have state interests completely trump the market interests of private firms.[21]

There was considerable debate at the time as to who would staff these associations, with the military preferring its own and private industry favouring their own managers.[22] Ultimately, it was decided that staffing would be by members of the industry itself, but in an interesting way: appointees were taken from among top managers of the operating units of steel companies—rather than, for example, from the further removed but more broadly strategically oriented holding companies of the *Zaibatsu*—and were made to resign their

---

[18] Ibid., 150.   [19] Bisson, *Japan's War Economy*, 25 ff.

[20] Ibid.; J. Cohen, *Japan's Economy in War and Reconstruction*, 1–48, (25–8 explicitly on the Steel Industry); Seiichiro Yonekura, 'Industrial Associations as Interactive Knowledge Creation. The Essence of Japanese Industrial Policy Creation', in *Japanese Yearbook on Business History* 13 (1996), 27–51; Johnson, *MITI and the Japanese Miracle*, 153, 162–3. For an entertaining debate on the post-war influence of these initiatives, see, on the one side (lots of influence), Tesuji Okazaki, 'The Japanese Firm under Wartime Planned Economy', in Masahiko Aoki and Ronald Dore (eds.), *The Japanese Firm. Sources of Competitive Strength* (New York: Oxford University Press, 1994), 350–78; and on the other side (not much influence), Juro Hashimoto, 'How and Why Japanese Economic and Enterprise Systems Were Formed', in *Japanese Yearbook on Business History* 13 (1996), 5–27. As an outsider, I am largely persuaded by Hashimoto's side, but as will be clear below, the entire debate is a bit orthogonal to my concerns.

[21] Lynn and McKeown, *Organizing Business*, 21–4.

[22] Bisson, *Japan's War Economy*; J. Cohen, *Japan's Economy in War and Reconstruction*; Yonekura, 'Industrial Associations'; Johnson, *MITI and the Japanese Miracle*.

positions within their original firms. In effect, the aim of this arrangement was to ensure that the decisions of the Control Associations would not be made with the profits of the steel companies (much less the broader technological synergies of parent *Zaibatsu* operations) uppermost in their minds, but rather with the needs of the industry's immediate military customers. At the same time, since the designers would have intimate knowledge of the industry and its technology, they would be able to set production targets in ways that would be technologically and organizationally possible to achieve without bankrupting the companies.[23]

In the end, however, it seems that the *dirigisme* of the Control Associations was not very successful. They created a coterie of managers with an understanding of the steel industry that was markedly different from the stable bifurcated grouping of specialists and volume producers which had characterized the development of the industry to that point. Rather than volume production as a speciality among an array of specialities, these actors wanted to create an oligopolistic structure in the industry to increase the total volume output of all producers. We will see that these managers and this conception of the industry proved to be very significant in the reconstitution of the Japanese steel industry after the war.[24]

THE PRE-1945 GERMAN STEEL INDUSTRY

The German steel industry was much older and larger than the Japanese industry.[25] None the less, by the middle of the 1930s it had developed a remarkably similar bifurcated industry structure. Just as in Japan, output in the industry was dominated by one large, integrated, and widely diversified producer, the Vereinigte Stahlwerke AG (hereafter, Vestag). Unlike the Japanese firm Yahata Steel, however, Vestag was a private concern formed out of companies which traditionally had no ownership participation of public bodies of any kind.[26] The

---

[23] Yonekura, 'Industrial Associations'; id., *Japanese Iron and Steel Industry*; Lynn and McKeown, *Organizing Business*, agree with Yonekura on the predominance of business leaders in the Associations, though they also suggest that the identity of these figures was blurred since most had spent significant portions of their careers inside the regulating state agencies.

[24] Even Hashimoto and Okasaki agree on this, see Okasaki, 'Japanese Firm' and Hashimoto, 'Japanese Economic and Enterprise Systems'. Also please note that by '*dirigisme*' in this context I mean merely the supplanting of market mechanisms by the directives of the Control Associations. I do not mean to suggest with this word that the Control Associations were direct organs of the state. Control Associations were corporatist bodies staffed by private industrialists acting against market mechanisms to organize industry according to state interests.

[25] The early history of the industry is summarized in this author's book, *Industrial Constructions: The Sources of German Industrial Power* (New York: Cambridge University Press, 1996), ch. 3. See also Wilfried Feldenkirchen, *Die Eisen- und Stahlindustrie des Ruhrgebiets, 1879–1914* (Stuttgart: Franz Steiner Wiesbaden, 1982); and Ulrich Wengenroth, *Unternehmensstrategien und technischer Forschritt. Die deutsche und die britische Stahlindustrie, 1865–1895* (Göttingen: Vandenhoek & Ruprecht, 1986).

[26] For the most part this was true, though there was brief state participation in the firm in the early 1930s. For detail on the changing and complicated ownership structure of Vestag, see Gerhard Mollin, *Montankonzerne und 'Drittes Reich'* (Göttingen: Vandenhoeck & Ruprecht, 1988).

outcome of a merger in 1926 among four of the largest firms in the coal and steel industry, this company accounted for over 50 per cent of all steel and rolled products in Germany soon after it was first established (1927). It continued to be the dominant firm in the industry during the 1930s and into the war, though its output was nearly matched after 1937 by the newly established Nazi state company, the Hermann Göring Werke (hereafter, HGW, more on which below). Vestag produced a broad range of rolled or finished-steel products, but it had only a relatively minimal set of downstream manufacturing subsidiaries (machinery, shipbuilding, etc.), and these were largely hold-over properties from the firms that had entered into the Vestag structure (that is, the firm made no effort actively to expand its downstream non-steel operations after its formation).[27]

As in Japan, most of the rest of the inter-war output in the German industry was, until the late 1930s, produced by relatively smaller, more specialized operations of broadly diversified conglomerate firms, known in Germany as *Konzerne*. These firms (the main ones being Hoesch, Krupp, Klöckner, Mannesmann [after 1932], and the Gutehoffnungshütte) were sprawling combines with far-flung operations throughout the steel-finishing and engineering industries. As in Japan, the output of these firms' steel units tended to be more specialized than the steel operations of the dominant Vestag. Where Vestag concentrated on more standardized fare and larger volumes across finished-steel markets, the steel operations of the *Konzerne* produced speciality and customized finished products. Often, the technical capacity to provide these products was aided by close exchanges of information and expertise with downstream units in the *Konzerne*.

These large conglomerates resembled their Japanese counterpart *Zaibatsu* in that they encouraged intra-*Konzern* exchanges in the interest of furthering the knowledge and innovative capacity of the internal exchanging parties—all to the benefit of the *Konzern* as a whole.[28] Unlike the *Zaibatsu* in Japan, however, the *Konzerne* were not ancient organizations in the country nor did they dominate the German industrial economy: the diversified *Konzern* form only emerged in the first decades of the twentieth century in the steel industry and it stabilized as a form of corporate organization only in the inter-war period.[29] Moreover, the *Konzerne* were largely regional entities and had a narrower palette of companies in a more restricted range of sectors than the *Zaibatsu*. Most German *Konzerne* had operations in steel, coal-mining, engineering (especially heavy engineering), and shipbuilding. Most also had their own trading companies and several had traditionally significant weapons and military-related operations (for example Krupp). Though the so-called *Montankonzerne* tended to have close relationships with major German banks, such as the Deutsche or Dresdner banks, these broadly diversified industrial entities did not

---

[27] For a more detailed description of Vestag, see Herrigel, *Industrial Constructions*, and, for the company during the Third Reich, Mollin, *Montankonzerne und 'Drittes Reich'*.

[28] Herrigel, *Industrial Constructions*, ch. 3.   [29] Ibid.

have the kind of intimate family-esque relations with a *Konzern*-controlled bank as was the case with their *Zaibatsu* counterparts.[30] Finally, in the steel segment itself, unlike the *Zaibatsu* steel producers, for the most part the steel producers within German *Konzerne* were integrated producers—that is, they produced their own pig iron, raw steel, and rolled products.

On the level of the governance of market relations, the Germans also had many similarities with the Japanese: cartels were the mechanism of choice both to maintain market stability and to uphold the integrity of boundaries between specialities. Stable bifurcation in the industry's structure, then, was maintained by a mixture of merger and cartelization in ways that intricately intertwined markets and co-ordination. Though remarkably similar to the Japanese, the Germans arrived at this governance equilibrium in a very different manner. The key difference was that in Japan, bifurcation in the industry structure was a point of departure in the development of the industry, whereas in Germany it was the outcome of many years of turbulence caused by a mid-life crisis in a relatively competitive nineteenth-century industry.

For most of the early decades of the twentieth century, German steel industry markets were extremely volatile, despite the fact that these years were marked on the whole by growth. The problem was that most of the major firms in the industry had emerged during the nineteenth century as suppliers of railway expansion and the large-scale construction attendant on rapid industrialization. And where in Japan only Yahata Steel was fully integrated, *all* of the major German producers had integrated production facilities, often ranging from coal reserves to finished bar and rail. As demand for rails and structural shapes began to level off toward the end of the century, all the players in the industry simultaneously attempted to diversify into newer and more specialized finished-goods markets, giving rise to tremendous and very turbulent competition. Cartels were created continuously, but, though legally binding, they were for the most part ineffective: either new entrants continually entered the market or competition outside the cartel created new finished-goods markets that then attenuated the commitments of customers in existing ones. Merger on a grand scale to consolidate industrial capacity into a small number of oligopolistic producers (the so-called US Steel Strategy) was also discussed, but was never seriously attempted due to the obstinacy of the primarily family-owned and operated steel corporations in the industry.

Finally, in the general economic turmoil and recession of the first few years after the First World War, a solution that split the difference emerged. Volume production was consolidated into a new firm, the Vestag, through a grand merger of production capacity. The rest of the more specialized steel and rolling capacity was then ceded to newly created *Konzerne*, which were able to link their steelmaking operations with extensive, recently acquired, downstream

---

[30] An excellent source on this relationship—which I ignore in this discussion—is Volker Wellhöner, *Grossbanken und Grossindustrie im Kaiserreich* (Göttingen: Vandenhoeck & Ruprecht, 1989).

manufacturing interests. These moves together both reduced the number of producers in the industry and clarified product-market boundaries among producers. And this transformed the cartel into a durable and effective mechanism for
the stabilization of market order in the industry.[31] Indeed, the correspondence
between industrial structure and corporate strategy was so harmonious that,
unlike their Japanese counterparts, German producers never appealed to the
state to intervene in the maintenance of commitment or participation in cartels.

This picture of a strong family resemblance between the bifurcated industrial
structures in the German and Japanese steel industries must be slightly diluted,
however, by two additional aspects of the German case: German producers had
a very different relationship to the state than their Japanese counterparts; and,
largely because of that difference, two additional groups of producers existed in
the later inter-war German industry—both of which grew enormously after
1937 in association with the Nazi regime's efforts to encourage steel production
for war.

German steel producers had a highly ambivalent relationship to the various
governments that existed in Germany prior to the end of the Second World
War.[32] On the one hand, large-scale German industrialists, in the steel industry
as well as in other sectors, were very firm and articulate advocates of the self-
government of industry and of society and correspondingly, of the limitation of
state power. They consistently advocated private control over the market
(though not market freedom), and limited (though not necessarily democratic)
government. Unlike the Japanese case where the government was reconstituted
in the mid-nineteenth century in a way that allowed it to be viewed as an agent
of the interests of the nation, in Germany the Imperial government after unification and then later the Weimar and Nazi regimes never succeeded in erasing
the feeling among non-aristocratic social classes that the government was the
affair of a particular social estate. Throughout the pre-1945 period, as a result,
large-scale industrialists found their own social and ultimately national legitimacy as a social estate (*Stand*) in the existence of strict limits on the state's
capacity to act in the economy in general and in the affairs of private firms and
industrial sectors in particular. Modern industry in a powerful nation, German
industrialists believed, should organize itself.

That said, German industrialists, particularly in heavy industry, were not
beyond engaging in commerce with the state if it proved profitable—as it did
with the massive growth of the German military, particularly its navy, before
the First World War. Nor did they object to the state taking over responsibility
for collective-good services that proved to be too cumbersome to maintain
co-operatively and privately, such as technical training and other forms of public infrastructure. As long as the state did not insinuate its own designs on the

[31] These paragraphs condense a much longer account in Herrigel, *Industrial Constructions*, ch. 3.
[32] Note that this relates to large firms and the German state, and is not a general claim about all
industry and its relationship to government in Germany. For alternative relations between industry
and the state see Herrigel, *Industrial Constructions*, chs. 2 and 5.

operations of private firms, large corporate private industry appreciated the role that a limited state could play in a market economy. If in Japan industry and government co-operated in the development of a private economy in the interest of the nation, in Germany large industry and government coexisted with mutual respect and dependence, yet without a superordinate purpose or interest to bring them together in sustained co-operation.

Evidence of the salience of this distinction can be seen in the way in which the structure of the steel industry shifted in the late 1930s when the Nazi regime began to attempt to change the rules of the game and become more directive in the affairs of private steel production. The established private enterprises in the industry baulked at Nazi efforts to channel their steel output away from their market strategies and toward the satisfaction of state needs. Even the formation of their own Control Associations (*Wirtschaftsgruppen*) in the context of the development of the first Nazi Four-Year plan in the mid-1930s, staffed very often by trusted former trade association officials, did not attenuate large firms' mistrust of the state or temper their resistance to its projects.[33] The German steel producers were willing to incorporate the state's demand for steel into their output (as they had traditionally done in the past), but they were not willing to turn completely away from their private commercial operations or strategies in the process.[34]

This resistance on the part of the industry gave rise to two developments that changed the composition and structure of the industry. On the one hand, a number of private firms emerged which were willing to exploit the resistance of the major producers to the state and cater exclusively to the steel needs of the Nazi government. These new firms cobbled together smaller, non-integrated pig iron, steel, and rolling mills that were broadly dispersed throughout Germany (and after the war began, throughout the expanded Reich) and organized their output completely around the specialized demands of the military or military-related sectors. The largest and most significant firm of this type was Friedrich Flick AG, which operated as a somewhat loose *Konzern*-like structure, but largely within the steel branch. That is, instead of seeking synergy between sectors (for example steel and engineering), Flick encouraged synergy among specialized and non-integrated steel operations.[35]

---

[33] Ironically, the Control Associations in Japan were modelled after these *Wirtschaftsgruppen*, though they seem to have been far more successful in controlling the affairs of the steel industry in Japan than their German counterparts. On the *Wirtschaftsgruppen* as role model for the Control Associations, see J. Cohen, *Japan's Economy in War and Reconstruction*, 1–48, and Yonekura, 'Industrial Associations'.

[34] Mollin, *Montankonzerne und 'Drittes Reich'*; see also R. J. Overy, 'Heavy Industry in the Third Reich: The Reichswerke Crisis', ' "Primacy Always Belongs to Politics": Gustav Krupp and the Third Reich', and 'The Reichswerke "Herman Göring": A Study in German Economic Imperialism', in id., *War and Economy in the Third Reich* (New York: Oxford University Press, 1994), 91–174.

[35] Other producers of this character were Otto Wolf AG, Röchling (from the Saar region), and Ballestrem (from Silesia). These firms (or their steel holdings) were all spun out of the Vestag after the latter firm's 1932 restructuring.

The other development was the creation by the Nazi regime of its own massive firm, the Hermann Göring Werke (HGW). Established in 1936, this wholly owned state firm soon became the second largest producer of steel in the Third Reich.[36] The HGW was a sprawling agglomeration of cobbled-together old firms and new greenfield sites. It was established to disrupt the private power of the *Montankonzerne* in the procurement of raw materials for steel production and in the delivery of steel for the military (and also to enhance the bureaucratic power of Herman Göring and his henchmen within the Nazi state apparatus).[37] By garnering a privileged position in the administered distribution system (the *Wirtschaftsgruppen*), the mere presence of the state firm could force the recalcitrant private companies to shift their attentions toward Nazi economic ends. As a producer of steel, however, though it had high levels of mostly standardized output primarily for military-related ends, HGW was far less efficient along a whole array of measures than the major privates were, in particular the Vestag. (See Table 12.2.)

Table 12.2. Comparative Data on Vestag and Hermann Göring Werke, Germany, 1942–1943

Index	Vestag	HGW (1943)
Ratio of coal to pig iron production	4.7:1	10.5:1
Ratio of coke to pig iron production	0.5:1	2:1
Crude steel production/employee (in tons)	25.5	10.1
Percentage foreign workers in total work-force	32	58
Sales/employee (Reichs Marks)	11,558	6,333
State credits received (millions of RM in 1945)	0.18	3.2
Depreciation (RM/ton of pig iron)	9	12

Source: G. Mollin, *Montankonzerne und 'Drittes Reich'* (Göttingen: Vandenhoeck & Ruprecht, 1988), 256, Table 28.

From a comparative point of view, the significance of the HGW and the newer specialized steel *Konzerne*, such as Flick and Otto Wolff, was to make the German steel sector slightly more fragmented than the Japanese industry on the eve of war, though the general bifurcation between specialists and standard-volume producers was maintained. That said, it is remarkable how generically similar the industries were: both the Japanese and German industries had large centralized steel companies which produced in high volumes, though in the German case there were two such firms, together accounting for market shares

[36] On the foundation and development of this firm, see Mollin, *Montankonzerne und 'Drittes Reich'*, and Overy, 'Reichswerke "Hermann Göring"', and 'Heavy Industry in the Third Reich', 144–74, 93–118.

[37] Mollin, *Montankonzerne und 'Drittes Reich'*, has a very extensive discussion of the political ties and interest of the state bureaucrats involved in the formation and operation of the HGW and their political interests in the *Konzerne*.

comparable to that controlled by Japan Steel alone (at least in steel and finished-goods markets). Likewise, both industries had significant portions of market share accounted for by steel-producing subsidiaries of much more broadly diversified industrial combines. These firms, in both countries, were much stronger in specialized and lower-volume markets than was the case with the large dominant firms.

The major disturbance in the analogy between the two industries in the end has to do with the different relationship between steel producers and the state in Germany as compared with that in Japan. In Germany, there was greater industrial fragmentation as the economy mobilized for war because the private firms were unwilling to allow their sovereignty to be compromised for state interests. There was no analogy in Japan to the upstart catfish-type war production firms such as Flick which came into being to exploit opportunities for war contracts created by the reluctance of the private sector. In Japan, private firms were more able to reconcile their interests with those of the state (and vice versa) and hence there was no opening for opportunistic profiteering.[38]

12.3 Occupation and the Imposition of New American Rules of Political-Economic Order, 1945–1957

With the defeat of Germany and Japan in the Second World War by the Allied Powers, the steel industries in both defeated countries came under very critical scrutiny. Top executives with close ties to the former enemy regimes were arrested and assets in the industry appropriated by the occupying powers. The personalities in the industry and their culpability in the aims and crimes of the defeated regimes, however, were not the main factor that drove the Allies to seek reform.[39] Rather, it was the very structure of the industry itself that became

[38] Mollin, *Montankonzerne und 'Drittes Reich'*, provides interesting insights into these producers, although they are not the primary focus of his concern, see esp. 257–70.

[39] For two discussions from the early post-war years of the general belief that German Big Business, especially the steel and chemical industries, helped the Nazis to power and underwrote the Nazi war effort, see Gustav Stolper, *German Realities* (New York: Reynal and Hitchcock, 1948), 172–96; and Joachim Jösten, *Germany: What Now?* (Chicago: Ziff-Davis Publishing Co., 1948), pt. 3. Graham D. Taylor discusses this phenomenon as it affected American antitrust policies in occupied Germany in his 'The Rise and Fall of Anti-Trust in Occupied Germany, 1945–48', *Prologue* 11, 1 (1979): 22–39, esp. 27–8. These popular understandings of the implication of German business in Nazism did not, as much other work has shown, correspond to the actual relationship between the two. Stolper makes this point, in a very impatient way, in the book mentioned above. But for the most recent scholarship and the complexities of the relationship, see Henry Turner, *German Big Business and the Rise of Hitler* (New York: Oxford University Press, 1985); Peter Hayes, *Industry and Ideology. IG Farben in the Nazi Era* (New York: Cambridge University Press, 1987); Reinhard Neebe, *Grossindustrie, Staat und NSDAP, 1930–1933* (Göttingen: Vandenhoeck & Ruprecht, 1981); Gottfried Plumpe, *Die IG Farbenindustrie AG. Wirtschaft, Technik und Politik, 1904–1945* (Berlin: Duncker & Humblot, 1990); Peter Wolfram Schreiber, *IG Farben. Die unschuldige Kriegsplaner* (Stuttgart: Neuer Weg, 1978); Raymond Stokes, *Divide and Prosper: The Heirs of I. G. Farben under Allied Authority* (Berkeley: University of California Press, 1988), ch. 1; Neil Gregor, *Daimler Benz in the Third Reich* (New Haven: Yale University Press, 1998); and for a general discussion, Dick Geary, 'The Industrial Elite and the Nazis', in Peter D. Stachura (ed.), *The Nazi Machtergreifung* (London: Allen & Unwin, 1983), 85–100.

a target: the Allies attacked both the bifurcated structure of the steel producer market and the character of the companies that constituted it as crucial obstacles to democratic order in Japanese and German society.

The attack on the steel industry was consistent with the way in which the Allies dealt with industrial forms that resembled those of the steel industry in all other sectors. The dominant self-understanding of the American and Allied occupying powers in Germany and Japan is easy to detect in the contemporary writings and memoirs of participants in the occupation. One especially concise formulation is given by Eleanor Hadley in her 1970 book which draws extensively on her experiences as an antitrust specialist in Japan during the occupation. I present this lengthy quotation because it underscores the linkage that the victorious powers made between economic organization and political democracy and their sense of legitimacy in intervening in the restructuring of the fundamental rules of economic behaviour that structured practice in both societies:

World War II differed radically from previous wars in the terms imposed by the victors on the defeated. Previously exactions had been limited to territorial changes, restrictions on the military establishment, and reparations. World War II, representing for the first time 'total' warfare, extended to the peace conditions 'total' peace, with demands for change in the political and economic structure of the defeated powers. In both Germany and Japan the victors attempted to revamp the social structure, to establish democracy. In the words of one descriptive title, the Germans and the Japanese were 'forced to be free' . . . Allied leaders saw the expansionist foreign policy of Germany and Japan as the product of their undemocratic governments, and believed that the future security interests of their own countries required nothing less than the social reconstruction of these two nations. By themselves, proscription of army-navy-airforce, along with territorial adjustments were insufficient. Nothing less than basic social reconstruction was needed if democracy, which would be peaceable, were to take root . . .

The programs for democratization in Germany and Japan were essentially similar. In both instances they called for a new constitution, new leadership, and change in the structure of the economy. Economic change was demanded for political reasons. The Allies believed that a democratic constitution would be meaningless unless the key pressure groups of the nation supported its ideology. In both Germany and Japan concentrated business was seen as one of the most powerful pressure groups, and because German and Japanese concentrated business was not considered to support democracy but rather oligarchy, it became a target of occupation reform.[40]

Why did the US occupiers consider the structures of the 'concentrated' businesses in Germany and Japan to be 'oligarchic' as opposed to democratic? It is not sufficient to classify this orientation as a simple free-market, anti-monopoly position, although advocates of smaller business and 'free' markets were important minority voices in both occupation governments.[41] On the contrary, the

[40] Hadley, *Anti-trust in Japan*, 3. Reference to a 'descriptive title' is to John D. Montgomery, *Forced to be Free: The Artificial Revolution in Germany and Japan* (Chicago: University of Chicago Press, 1957).

[41] In both Germany and Japan there was a battle within the American camp between radical progressive trust-busters and more conservative so-called New Deal advocates of American-style

most influential elements of the occupation governments were quite sympathetic to big business and were not at all opposed to some limits on market
competition. They were, in the words of Theodore Cohen, 'New Dealers' who
believed deeply in the progressive and democratic potential of large scale, mass
production industry.[42] But their vision of large-scale industry was one in which
the power of large market actors was checked, as they believed it to be in the
United States, by other strong and organized actors, both in their own markets
and in the society at large: rather than monopoly, they supported oligopoly;
rather than cartelization and co-operation among producers, they favoured vertical integration, competitive (oligopolistic) price setting, and the stimulation of
stable consumer demand; instead of paternalism and employer discretion in
wage setting and labour markets, they favoured trade unionism and collective
bargaining; instead of multi-sectoral private conglomerate empires, they preferred to have corporate actors confine their growth to the vertically related
processes of individual industries.[43]

In each instance, the American ideal was to create the economic and social
conditions for the emergence of plural sources of power which would have to
compete with and respect one another. It was precisely the absence of such conditions, as the quotation above from Hadley indicates, that the Allies believed
made possible the authoritarianism and demagoguery of the two enemy
regimes. In the context of industry, this pluralist understanding of a democratic

oligopolistic big business and mass production. On these divisions within US policy in Germany, see
Volker Berghahn, *Unternehmer und Politik in der Bundesrepublik* (Frankfurt: Suhrkamp, 1985),
84–111; and Taylor, 'Anti-Trust in Occupied Germany', 22–39. For an account of the deconcentra-
tion and decartelization process in Germany by a radical trust-buster, see James D. Martin, *All
Honorable Men* (Boston: Little Brown, 1950). For debates within the occupation of Japan, see
Theodore Cohen, *Remaking Japan. The American Occupation as New Deal* (New York: The Free
Press, 1987), 3–48, 154–86, 301–98; Hadley, *Anti-trust in Japan*, pt. 1; T. A. Bisson, *Zaibatsu
Dissolution in Japan* (Berkeley: University of California Press, 1954), *passim*. New Dealers were
more concerned with the consumer than with size of enterprise: as long as prices could be estab-
lished competitively, they believed that this would have salutary macroeconomic (and political)
effects. For the arguments of one of the seminal developers of this kind of thinking in the USA, see
Thurman Arnold, *The Bottlenecks of Business* (New York: Reynal and Hitchcock, 1940). In general
on the divisions in American antitrust thinking during this period, see, for example, Rudolph Peritz,
Competition Policy in America, 1888–1992 (New York: Oxford University Press, 1996), 111–229.
For a concise general overview of New Deal conceptions of the state and democracy in relation to
the market see Alan Brinkley, 'The New Deal and the Idea of the State' in Steven Fraser and Gary
Gerstle (eds.), *The Rise and Fall of the New Deal Order, 1930–1980* (Princeton: Princeton University
Press, 1989), 85–121.

[42] The New Deal is invoked even in the title of his memoir: T. Cohen, *Remaking Japan: The
American Occupation as New Deal*.

[43] Arnold characteristically observed: 'The inevitable result of the destruction of competitive
domestic markets by private combinations, cartels and trade associations is illustrated by Germany
today . . . Industrial Germany became an army with a place for everyone, and everyone was required
to keep his place in a trade association or cartel. Here was arbitrary power without public control
and regimentation without public leadership. That power, exercised without public responsibility
was constantly squeezing the consumer. *There was only one answer*. Germany was organized to
such an extent that it needed a general and Hitler leaped into power. Had it not been Hitler it would
have been someone else. When a free market was destroyed, state control of distribution had to follow' (*The Bottlenecks of Business*, 15–16) [italics added].

economy was connected to an emerging American conception of industrial mass production and its social and political preconditions. The most significant connection had to do with the possibilities for market restructuring and the achievement of scale economies that would be made possible by the elimination of monopolies and cartels. Inefficient large producers protecting costly specialist production processes and cartels of specialized small producers would no longer be able to survive in an environment that insisted on competition and frowned upon collaboration.[44]

American reformers believed that the creation of this kind of alternative environment in the economy of the occupied societies would have significant positive employment and income effects and, ultimately, social and political ones as well. Traditional middle classes, key villains in the American understanding of the rise of fascism, would decay without the collusive mechanisms they had used to protect their social and economic position.[45] The emergence of efficient, large-scale oligopolistically structured industries would replace the old middle classes with a less differentiated industrial and service middle class. Without cartels and multi-sectoral vertical monopolies to enhance their social, economic, and political positions, 'feudal' industrialists would be forced to concentrate their energies and resources on the achievement of efficiency and to compete for profits in markets and influence in politics with other firms and associations.[46]

In order to achieve these goals in the occupied political economies, the Allies developed a strategy that involved both *destructive* and *creative* elements. The most notable destructive strategy involved direct intervention in the property-rights structure of industrial holdings in order to 'deconcentrate' industries deemed by the Allies to have 'unhealthy' or 'undemocratic' monopoly structures. The same logic was also applied to cartels: the Allied powers immediately advocated the elimination of all cartels in both economies.[47] The creative

[44] See, for example, the testimony of Paul Hoffmann, head of the Economic Cooperation Administration, before the Senate Foreign Relations Committee, *Executive Sessions of the Senate Foreign Relations Committee (Historical Series) Volume II, Eighty-First Congress, First and Second Sessions, 1949–1950*, 184 and *passim*.

[45] For the linkage between traditional middle classes and fascism, see Seymour Martin Lipset's seminal work, *Political Man: The Social Bases of Politics* (New York: Anchor/Doubleday, 1963).

[46] In addition to the above references, this characterization of American views is drawn from a reading of T. Cohen, *Remaking Japan*; Bisson, Zaibatsu *Dissolution in Japan*; Hadley, *Anti-trust in Japan*; Montgomery, *Forced to be Free*; Volker Berghahn, 'West German Reconstruction and American Industrial Culture, 1945–1960', in Reiner Pommerin (ed.), *The American Impact on Postwar Germany* (Oxford: Berghahn Books, 1995), 65–82; Richard Pells, *The Liberal Mind in a Conservative Age. American Intellectuals in the 1940s and 1950s* (Middletown: Wesleyan University Press, 1989). On feudal elements in industry, see Reinhard Bendix, *Work and Authority in Industry* (Berkeley: University of California Press, 1956); David Landes, 'Japan and Europe: Contrasts in Industrialization', in Lockwood, *State and Economic Enterprise in Japan*, 93–182; Gustav Ranis, 'The Community Centered Entrepreneur in Japanese Development', *Explorations in Entrepreneurial History* 8, 2 (1955), 80–98; Johannes Hirschmeier, *The Origins of Entrepreneurship in Meiji Japan* (Cambridge, MA: Harvard University Press, 1964); and Ralf Dahrendorf, *Society and Democracy in Germany* (New York: Norton, 1964) (on the notion of 'feudal remnants').

[47] According to Article 12, Economic Section (Section B), *Potsdam Agreement. Joint Report on Results of the Anglo-Soviet-American Conference* (Berlin, 1945), released 2 August 1945: 'At the

element of occupation strategy came with efforts to create rules for practice in the economy that would prevent the re-emergence of the structures that the Allied governments destroyed and which would encourage instead the development of strong, countervailing organizations in markets and in society. In both occupied countries, this resulted in a very aggressive emphasis on antitrust legislation as well as a positive disposition toward the emergence of an organized labour movement.[48] Both kinds of shifts, American reformers believed, would foster the democratization of the occupied economies. Significantly, though the American occupiers viewed the formation of trade unions and employers' organizations favourably and understood their competition and co-operation to be constitutive of stable democratic order, they disapproved of direct joint governance by these groups: bargaining over wages among independent organizations was acceptable and democratic; parity participation in single organizations to determine and set wages and prices was not. The USA wanted plural sources of social, economic, and political power.[49]

In both occupied countries, the steel industry became a primary target of both the destructive and creative aspects of Allied occupation reform. The huge market positions of Japan Steel and Vestag and HGW were considered by the occupiers to be excessive: they were concentrations of economic power that

earliest practicable date, the German economy shall be decentralized for the purpose of eliminating the present excessive concentration of economic power as exemplified in particular by cartels, syndicates, trusts and other monopolistic arrangements' (excerpts reprinted as Appendix B in Gustav Stolper, German Realities, 264–72, quoted matter in text from 267).

Similarly, in Japan, as early as 6 Sept. 1945, the President of the United States issued to the Strategic Command of the Allied Powers (SCAP) an executive order, No. 244 for the Liquidation of Holding Companies. This edict charged the SCAP with:

(1) breaking up existing Zaibatsu companies and other holding company arrangements;
(2) elimination of various mechanisms for the creation of private monopolies;
(3) the foundation of a system of free competition.

The first two measures were understood to be transitional measures; the third was to be a fundamental principle of order in the new Japan. See Bisson, Zaibatsu Dissolution in Japan, 234–44; and Joseph Hiroshi Iyori, Das japanische Kartelrecht. Entwicklungsgeschichte, Grundprinzipien und Praxis (Cologne: Carl Heymans Verlag, 1967), 21.

[48] 'United States policy was unequivocal in demanding legal support for Trade Unions for the first time in Japanese history. In September 1945, President Truman ordered that "encouragement shall be given and favor shown to the development of organizations in labor, industry, and agriculture, organized on a democratic basis"', in Andrew Gordon, The Evolution of Labor Relations in Japan. Heavy Industry, 1853–1955 (Cambridge, MA: Harvard University Press, 1985), 331.

[49] Diethelm Prowe, 'German Democratization as Conservative Restabilization: The Impact of American Policy', in Jeffrey M. Diefendorf, Axel Frohn, and Hermann-Josef Rupieper (eds.), American Policy and the Reconstruction of West Germany, 1945–1955 (New York: Cambridge University Press, 1993), 307–30. Prowe, however, suggests that American authorities opposed more corporatist arrangements because they involved a different kind of democratic representation than that which the Americans wanted. He opposes these corporatist groups to political parties. I do not believe that this is the appropriate distinction to make: rather, Americans did not approve of the corporate arrangements Germans proposed because they preferred representative parties. The occupiers disapproved of the proposed groups because they understood them to be monopolistic and hence likely to undermine social competition and pluralism. US occupiers did not have a conception of democracy narrowly confined to party competition and elections.

circumvented efficiency-inducing competition and gave producers the potential to hold society hostage to their interests. Likewise, the companies that were part of diversified combines (*Konzerne* and *Zaibatsu*) were attacked as unreasonable concentrations of power: co-operation was viewed as collusion that undermined developments toward greater efficiency and larger scale within individual market segments. Both types of firm, and the bifurcated market structures that they created in the steel industry, became the targets for dismantling by military authorities in both Germany and Japan. Naturally there was resistance in both societies to these reforms. The remainder of this section details the reforms that were pursued and implemented in both countries. Section 12.4 addresses the reality of resistance and the transformation of indigenous and American ideals it ultimately produced.

ALLIED POLICIES FOR THE DECONCENTRATION, DECARTELIZATION, AND DEMOCRATIZATION OF GERMAN STEEL

The American occupiers believed that the extreme integration and cartelization that had traditionally characterized the structure of the German steel industry both blocked the diffusion of efficient technologies and industrial practices in this industry and created dangerous monopoly power in the economy. Outsiders viewed the relatively large numbers of small and flexible rolling mills attached to individual companies as a sign of steel mill monopoly power and its inefficiency, rather than as an explicit strategy on the part of producers to expand their range of products. The Allies agreed that the industry needed to be deconstructed and rearranged in a way that would allow more 'modern' and 'healthy'—that is, larger-scale and less horizontally diversified—forms of industrial production to take root.[50]

To this end, all of the assets of the iron and steel industry were seized by the military governments in 1946 and early 1947.[51] All six of the major Ruhr steel producers (Vestag, Krupp, Hoesch, Klöckner, GHH, and Mannesmann), which alone during the 1930s (before the formation of HGW) had controlled over 78 per cent of German crude steel production, were completely broken up, as were the sprawling HGW and Flick operations (though in these cases there was the additional factor of capacity and units lost in liberated non-German territories). Bargaining between the Allies, the German government, and the iron and steel

---

[50] Lucius Clay, ultimately only a moderate enthusiast for deconcentration, was very clear on the unacceptability of the traditional organization of the iron and steel industry. Clay was confident that the break-up of the industry would lead in an efficient direction because it would follow a plan drawn up by a committee organized by George Wolf of the United States Steel Corporation, see Clay, *Decision in Germany* (Garden City, NY: Doubleday & Co., 1950), 329–30.

[51] Isabel Warner, 'Allied–German Negotiation on the Deconcentration of the West German Steel Industry', in Ian Turner (ed.), *Reconstruction in Post-War Germany. British Occupation Policy and the Western Zones* (Oxford: Berg, 1989), 155–85, and more generally, Isabel Warner, *Steel and Sovereignty. The Deconcentration of the West German Steel Industry, 1949–1954* (Mainz: Verlag Philipp von Zabern, 1996) *passim*.

industry in the last years of the 1940s resulted in a constant game of strategic recombination in the industry. Properties and plants were torn apart and repositioned both on paper and in fact.[52] Ultimately, 23 new companies were created, 13 from the old Vestag alone.[53] (See Table 12.3.) The new structure of the industry was organized around steelworks: each new company was dominated by one works. But, significantly, the degree of diversification in the finished rolling-mill product palettes in each of the newly created steel firms was disrupted. Though the Deconcentration Authority for the Steel industry (*Stahltreuhändervereinigung*) attempted to retain as much integration between existing steel and rolling mills as 'was technically necessary', much capacity was deemed inessential and allocated to other steel-production units. In this way, parts of the industry's rolling-mill capacity was spread across the industry, rather than within firms. This kind of distribution of plant linkages was often extremely awkward: in some cases rolling mills were realigned with steelworks that could not supply them with the proper kind of steel and separated from ones that previously had.[54] But this kind of allocation of capacity was desirable from the point of view of the reformers both because it was difficult otherwise to achieve the goal of creating more companies to enlarge the arena of competition and because it raised the costs of diversification and created an incentive for firms to grow by increasing the scale of rolling mills they possessed. A political conception of the proper relationship between market order and democratic order, in other words, trumped narrower concerns of technical and economic efficiency.[55]

This intentional reallocation of finishing capacity among steel works was most dramatic in the Vestag successor companies, but a narrowing of rolling-mill capacity occurred within the steelmaking operations of the old *Konzerne* as

[52] Warner, 'Allied-German Negotiation'.

[53] See Stahltreuhändervereinigung, *Die Neuordnung der Eisen und Stahlindustrie im Gebiet der Bundesrepublik Deutschland* (Munich: C. H. Beck'sche Verlagsbuchhandlung, 1954); Ernst Schröder, 'Die Westdeutsche Montanindustrie heute', *Der Volkswirt* 44 (1952), 27–32; K. H. Herchenröder, Johan Schäfer, and Manfred Zapp, *Die Nachfolger der Ruhrkonzerne. Die Neuordnung der Montanindustrie* (Düsseldorf: Econ Verlag, 1953); Gerd Baare, *Ausmass und Ursachen der Unternehmungskonzentration der deutschen Stahlindustrie im Rahmen der Montanunion, ein internationaler Vergleich*, Ph.D. dissertation (Tübingen, 1965); Franz Lammert, *Das Verhältnis zwischen der Eisen schaffenden und der Eisen verarbeitenden Industrie seit dem ersten Weltkrieg*, Ph.D. dissertation (University of Cologne, 1960), 140–55; and Paul Weil, *Wirtschaftsgeschichte des Ruhrgebietes* (Essen: Siedlungsverband Ruhrkohlenbezirk Essen, 1970). The Stahltreuhändervereinigung work, the Schroeder article, and the Herchenroeder *et al.* volume all have charts outlining lineages of firms.

[54] Stahltreuhändervereinigung, *Neuordnung*, 129 ff., 301–4. For complaints about disruptions to rolling mills, see Schroeder, 'Westdeutsche Montanindustrie'. Schroeder lists nine newly created companies with uneconomic combinations of steelmaking and rolling capacity.

[55] Stahltreuhändervereinigung, *Neuordnung*, 129–31 and 191–3. Norman Pounds in 1953 noted that the recomposition of the industry had not followed the guidelines that technical efficiency would have dictated: 'It is too early to suggest the shape which the future organization of the heavy industry of the Ruhr is likely to take. It is clear, however, that the unwisdom of too great a fragmentation has been realized', in Norman J. G. Pounds, *The Ruhr. A Study in Historical and Economic Geography* (Bloomington: Indiana University Press, 1952), 259.

Table 12.3. The 23 'New' German Steel Enterprises Created after the Second World War

New enterprise	Old parent
Deutsche Edelstahlwerke AG, Krefeld	Vestag
Rheinisch-Westfälische Eisen- und Stahlwerke AG, Mülheim/Ruhr	Vestag
Bergbau- und Industriewerte GmbH, Düsseldorf	Vestag
Hüttenwerke Phönix AG, Duisburg	Vestag
Dortmund-Hörder Hüttenunion, AG, Dortmund	Vestag
Gußtahlwerk Bochumer Verein AG, Bochum	Vestag
August Thyssen-Hütte AG, Duisberg-Hamborn	Vestag
Hüttenwerke Siegerland AG, Siegen	Vestag
Gußtahlwerk Witten AG, Witten	Vestag
Rheinische Röhrenwerke AG, Mülheim	Vestag
Niederrheinische Hütte AG, Duisburg	Vestag
Stahlwerke Südwestfalen AG, Geisweid	Vestag
Ruhrstahl AG, Hattingen	Vestag
Mannesmann AG, Düsseldorf	Mannesmanröhren-Werke
Hansche Werke AG, Duisburg-Großenbaum	Mannesmanröhren-Werke
Stahl- und Walzwerke Rasselstein/Andernach AG, Neuweid	Otto Wolf
Stahlwerke Bochum AG, Bochum	Otto Wolf
Nordwestdeutscher Hütten- und Bergwerksverein	Klöckner
Hoesch-Werke AG, Dortmund (Hoesch)	Hoesch
Hüttenwerk Rheinhausen AG, Rheinhausen (Krupp)	Krupp
Eisenwerk-Gesellschaft Maximilianshütte AG, Sulzbach-Rosenberg Hütte (Friedrich Flick KG)	Friedrich Flick KG
Luitpoldhütte AG, Amberg	Reichswerke AG für Erzbergbau und Eisenhütten
Hüttenwerk Oberhausen AG, Oberhausen	Gutehoffnungshütte Aktienverein für Bergbau und Hüttenbetrieb

Note: Ilseder Hütte, the twenty-fourth enterprise, was left unchanged by the rearrangements.

Source: Stahltreuhändervereinigung, *Die Neuordnung der Eisen und Stahlindustrie im Gebiet der Bundesrepublik Deutschland* (Munich: C. H. Beck'sche Verlagsbuchhandlung, 1954).

well. Often this was less the result of explicit deconcentration efforts than the outcome of war damage or dismantling losses in rolling-mill plant which the deconcentration authorities did not attempt to redress. But, whatever the cause, the key outcome of the deconcentration for all firms' steel operations was the need seriously to rethink their strategy in rolled production. In particular, the old pre-war strategy of the *Konzerne* steel units which emphasized product proliferation, speciality production, and broad horizontal synergy and diversification not only was now seriously undermined, that is, physically and technologically, but became only one of a number of alternative strategies available to producers as they set about reconstructing themselves.

In addition to this specialization of core steel production units (blast furnaces, steel mills, rolling mills), deconcentration also aimed at severing all but the most essential non-steel ties within the old *Konzerne* enterprises.[56] The logic here was the same as within steel more narrowly defined. The broad cross-sectoral diversification of the *Montankonzerne* was viewed as both inefficient and monopolistic. Such enterprise structures were undesirable from an Allied point of view in two ways: they allowed steel operations to run inefficiently by providing them with guaranteed markets; while at the same time they effectively subsidized the downstream operations, thereby perpetuating inefficiency and market fragmentation there as well. The Allies did not view the *Konzerne* as dynamic entities pursuing strategies of continuous innovation and product proliferation; they understood them instead to be archaic and pernicious economic forms that constituted obstacles to efficiency, democracy, and progress.[57] The effect of breaking up cross-sectoral linkages, however, was none the less to undermine the dynamic economies of technological variety that had sustained the *Konzerne* in the pre-war period.

All of Vestag's machinery and manufacturing interests were collected together in the Rheinische Stahlwerke AG, its coal holdings into three independent companies, and its trading businesses into the Handelsunion AG.[58] The same vertical deconstruction occurred in most of the *Konzerne*. Mannesmann, Klöckner, and Hoesch, for example, were stripped of crucial steel plants, all but their most-needed coal-mines, and most of their downstream engineering interests.[59] The GHH had its networks of coal-mines and steel mills expropriated but retained its extensive holdings in manufacturing, primarily in heavy engineering and shipbuilding. Its sizeable former steel operations in Oberhausen were con-

---

[56] Norman J. G. Pounds, *The Ruhr A Study in Historical and Economic Geography* (Bloomington: Indiana University Press, 1952), 129–31 and 133–41.

[57] For this interpretation, see Kurt Pritzkoleit, *Männer, Mächte, Monopole. Hinter den Türen er westdeutschen Wirtschaft* (2nd edn., Düsseldorf: Karl Rauch, 1960); William Manchester, *The Arms of Krupp* (New York: Bantam Books, 1968); Joseph Borkin and Charles Welsh, *Germany's Master Plan: The Story of an Industrial Offensive* (New York: Duell, Sloane and Pearce, 1943).

[58] Herchenroeder *et al.*, *Nachfolger der Ruhr Konzerne*, 57–118; Kurt Pritzkoleit, *Männer, Mächte, Monopole*, 131–41.

[59] On the break-up of Hoesch, see Herchenroeder *et al.*, *Nachfolger der Ruhr Konzerne*, 230–41. On Mannesmann, 199–218, and on Klöckner, 171–94.

stituted as a fully independent steel company, the Hüttenwerk Oberhausen AG (HOAG).[60] Krupp was forced to sign a commitment to separate legally all of the family's steelmaking facilities (concentrated primarily in the works at Rheinhausen) and attendant coal-mines from the rest of its business, collect them into a holding company, and sell them off. The major non-steel lines of Krupp's business included some trading companies and a wide variety of specialized heavy engineering workshops located around Essen and in the Ruhr Valley more generally.[61]

Deconcentration thus significantly reorganized and decentralized both the production of steel and rolled products and the broad, cross-sectorally diversified companies that produced them. The bifurcated industry structure of giant mass producers surrounded by smaller specialist producers was eliminated. Individual producers were smaller, less integrated (especially with forward non-steel sectors), and less diversified producers of rolled-steel products. These developments totally destroyed the coherence of both the industry and the companies in it as they had existed prior to the war.

The strategic intent behind all of these deconcentration efforts (namely, to create the economic conditions for the formation of oligopolies and integrated mass producers and the social conditions for pluralist democratic order) was reinforced through the simultaneous implementation of very strict decartelization measures in the industry. An outright ban on all cartelization in the German economy was announced by all three Allied governments in the West in 1947.[62] The ban was enforced by the Allied occupying governments until 1955 and continued to be enforced by the Germans thereafter until it was replaced by a new law governing competition in 1957. This ban on cartels was complemented on a European level in the coal and steel industry where the European Coal and Steel Community Treaty and Organization (ECSC) made the rejection of all forms of cartel a condition of membership.[63]

The intention of this change of rule regarding cartels was to alter the environment in which producers in the steel industry competed. In the pre-war

[60] Dietrich Wilhelm von Menges, *Unternehmens-Entscheide. Ein Leben für die Wirtschaft* (Düsseldorf: Econ Verlag, 1976) (former GHH manager); Herchenroeder *et al.*, *Nachfolger der Ruhr Konzerne*, 119–40.

[61] For complicated reasons, the steelmaking facilities were never, in fact, sold off but were kept in a separate holding, apart from the rest of its business in the manner dictated by the Allied settlement, until the late 1960s—when the firm went bankrupt and had to be completely reorganized. Descriptions of the reconstitution and development of the Krupp empire are given in Gert von Klass, *Krupps* (London: Sidgwick & Jackson, 1954), 431–3; Herchenroeder *et al.*, *Nachfolger der Ruhr Konzerne*, 145–70; Dietrich Weder, *Die 200 Grössten deutschen Aktiengesellschaften 1913–1962 Beziehungen zwischen Grösse, Lebensdauer und Wettbewerbschancen von Unternehmen*, Ph.D. dissertation (Frankfurt, 1968), 323; and *Wirtschaftswoche* 2 (14 Jan. 1972), 62–3; *Manager Magazine* 12 (1973), 30–4; and 11 (1975), 27–34.

[62] British Zone Law No. 56; American Zone Law No. 78; and French Zone Law No. 96. Eberhard Günther, 'Entwurf eines Gesetzes gegen Wettbewerbsbeschränkungen', in *Wirtschaft und Wettbewerb*, 1, 1 (Nov. 1951), 17–40. A contemporary overview, the article is also good on contrasting American thinking about antitrust and countervailing power with German views.

[63] Duncan Burn, *The Steel Industry 1939–1959* (Cambridge: Cambridge University Press, 1961), 407–16; and Lammert, *Das Verhältnis*, 201–4.

industry structure, concentration, speciality production, horizontal diversification, and cartelization were integrally related. Volume production of standard
products was concentrated in the large Vestag and HGW combines, while production of lower-volume and more specialized products was undertaken by the
smaller, horizontally diversified *Konzerne*. The large-volume producers strove
for scale economies, while the *Konzerne* looked for horizontal synergies and
economies of scope across their many manufacturing businesses. An anomaly of
this structure is that it gave the smaller *Konzern*-based specialist producers an
interest in the formation of cartels to stabilize competition. To these producers,
with their wide variety of smaller-sized rolling units, oriented toward specialization, yet with high fixed costs despite their small size, the threat of ruinous
price and terms-based competition during downturns in the business cycle was
very real. By forming cartels, they could stabilize the structure of specialization.[64]

The Allied and ECSC reform measures undermined the integral logic of this
industry structure. In effect, the reforms discouraged strategies of specialization
and customization in finished products (by taking away the governance mechanisms and business forms that made them stable and profitable) and created
the possibility for the radically fragmented pieces of the industry to be recombined, both within firms and across the industry as a whole, in ways that would
facilitate larger-scale, more 'efficient', production. Through specialized investment in select finished-product markets, greater capacity in each product-market segment could be spread across a more limited number of firms than had
been possible in the past. Bifurcation would be replaced by oligopoly; monopoly and specialization by mass production. In any case, the absence of *Konzerne*
strategies of scope and synergy across sectors and the stabilization mechanism
of the cartel meant that the possibility of making profits in a broad variety of
rolled products, with only a modest scale commitment in each, was significantly
reduced.

Finally, the occupying powers were extremely active on the labour side in the
industry. The pre-war German steel industry was notably hostile to trade unionism. This not only was true of the industry during the Third Reich but had actually characterized the industry's outlook from its very inception. While, by 1913,
Germany had the largest social democratic movement and one of the world's
largest trade union movements, there was no union representation in the steel
industry at that time.[65] The High Commissioner for Germany within the occupying government created an Office of Labour Affairs which attempted, systematically, to encourage the rejuvenation of the German labour movement

---

[64] On the logic behind this claim that cartels were used in the stabilization of specialization, see
the more extensive discussion in Herrigel, *Industrial Constructions*, ch. 2.

[65] Elizabeth Domansky-Davidsohn, 'Der Grossbetrieb als Organizationsproblem des Deutschen
Metallarbeiter-Verbandes vor dem Ersten Weltkrieg', in Hans Mommsen (ed.), *Arbeiterbewegung
und industrieller Wandel. Studien zu gewerkschaftlichen Organisationsproblemen im Reich und an
der Ruhr* (Wuppertal: Peter Hammer, 1980), 95–116.

after its period of enforced dormancy within the Third Reich.[66] According to a US Policy Directive for the High Commissioner issued in November 1949, cited by Michael Fichter, the goal of the United States in the labour and industrial relations field was 'to encourage the development of free, democratic trade unions and the negotiation of agreements and cooperative settlement of problems between them and employer associations'.[67] The Office of Labour Affairs sponsored hundreds of trips by trade unionists to the United States to learn about 'the American Way of Life'. It also sponsored a Training Within Industry (TWI) programme which established educational committees on the shop-floor level designed to undermine what were viewed as the 'authoritarian or paternalistic' shop-floor relations in German firms and impart more democratic, 'American' habits. Production units in the steel industry figured very prominently in these programmes. Unions began to assert themselves very aggressively in the industry almost immediately after the occupation began.[68]

DECONCENTRATION, ZAIBATSU DISSOLUTION, AND ALLIED EFFORTS TO CREATE A DEMOCRATIC ECONOMY IN JAPAN

The Allied occupying force in Japan, Strategic Command of the Allied Powers (SCAP), attacked the bifurcated structure of the Japanese steel industry and the diversified multisectoral Zaibatsu holding companies with at least as much alacrity as they had the analogous structures in Germany. The procedure, timing of the reforms, and official position of the authorities differed significantly in Japan, however. In particular, unlike Germany, there was not actually a Military Government in Japan. Because there was always a clear division between the government and the military in Japan, and because the government did not collapse as it had in Germany (and perhaps because there were not enough people knowledgeable about the country and capable of speaking the language), the occupation forces dissolved only the military and its institutions, but decided to run all of its reforms through the existing civilian government structure, even as that structure was to be changed. All reforms in politics— including a new constitution and the reconstitution of parties, changing the suffrage laws, etc.—as well as policies of deconcentration and promulgation of new rules regarding market behaviour were formally legislated by the Japanese government and carried out by the bureaucracy—though always at the bidding

---

[66] See Michael Fichter, 'Hicog and the Unions in West Germany. A Study of Hicog's Labor Policy toward the Deutscher Gewerkschaftsbund, 1949–1952' in Diefendorf et al., American Policy and the Reconstruction of West Germany, 257–80; for contemporary accounts by a sympathetic American which describe US measures, see Matthew A. Kelly, 'The Reconstitution of the German Trade Union Movement', Political Science Quarterly 64 (1949), 24–49; and M. A. Kelly, 'Labor Relations in American-Occupied Germany', in Colston E. Warne (ed.), Labor in Postwar America, (Brooklyn: Remsen Press, 1949), 607–21.

[67] Fichter, 'Hicog and the Unions in West Germany', 260.

[68] Ibid., 262.

of the SCAP.[69] None the less, Allied antipathy toward 'monopoly' power and multisectoral conglomerates came through very clearly in the policies that were ultimately put in place. As in Germany, the SCAP pursued policies that were both destructive and creative.

The earliest and most economically significant targets of SCAP destructive policy were the *Zaibatsu* firms.[70] These were attacked in a number of ways, all of which effectively destroyed the coherent, centrally-controlled, and diversified *Zaibatsu* as a governance structure in the Japanese economy.[71] First, a Holding Company Liquidation Commission (HCLC) was established which designated 83 holding companies for immediate dissolution. This list included all of the major *Zaibatsu* property interests (Mitsui, Mitsubishi, Sumitomo, Yasuda, Asano, etc.) and all of those which had significant interests in the steel industry. Then in 1947 the ten major families who controlled these 83 designated companies and other companies were declared to be '*Zaibatsu*' families. The ownership shares of both the families and the holding companies were transferred to the HCLC. Further, a 1947 anti-monopoly law, also designed by the SCAP but implemented by the Japanese government, completely outlawed holding companies (among other things). Then, the resale of the assets of the dissolved holdings were limited in crucial ways: In particular, no one was permitted to purchase more than 1 percent of the shares of any given company.[72] Taken

[69] For general overviews of US occupation and its structure see, in addition to T. Cohen, *Remaking Japan*, Kazuo Kawai, *Japan's American Interlude* (Chicago: University of Chicago Press, 1960); Richard B. Finn, *Winners in Peace. MacArthur, Yoshida and Postwar Japan* (Berkeley: University of California Press, 1992); Hadley, *Anti-trust in Japan*; and Robert E. Ward and Sakamoto Yoshikazu (eds.), *Democratizing Japan. The Allied Occupation* (Honolulu: University of Hawaii Press, 1987).

[70] J. Cohen, *Japan's Economy in War and Reconstruction*, 427, quotes Edwin Pauley as representative of SCAP feelings regarding the *Zaibatsu*: 'Japan's *Zaibatsu* . . . are the comparatively small group of persons, closely integrated both as families and in their corporate organizations, who throughout the modern history of Japan have controlled not only finance, industry and commerce, but also the government. They are the greatest war potential of Japan. It was they who made possible all Japan's conquests and aggressions. Not only were the *Zaibatsu* as responsible for Japan's militarism as the militarists themselves, but they profited immensely by it. Even now, in defeat, they have actually strengthened their monopoly position . . . Unless the *Zaibatsu* are broken up, the Japanese have little prospect of ever being able to govern themselves as free men. As long as the *Zaibatsu* survive, Japan will be their Japan.'

[71] Excellent comparisons of the pre- and post-*Zaibatsu* dissolution character of large Japanese corporations can be found in Hadley, *Anti-trust in Japan*, 205–315, and Kozo Yamamura, *Economic Policy in Postwar Japan. Growth versus Economic Democracy* (Berkeley: University of California Press, 1967).

[72] On the American influences on this law and Japanese modifications, see Masahiko Aoki, 'The Japanese Firm in Transition', in Kazo Yamamura and Yasukichi Yasuba (eds.), *The Political Economy of Japan. Volume 1. The Domestic Transformation* (Stanford: Stanford University Press, 1987), 263–88; Hadley, *Anti-trust in Japan*, *passim*; and Hirosi Acino, 'Zehn Jahre Antimonopolgesetz in Japan', in Georg Jahn and Kurt Junckerstorff (eds.), *Internationales Handbuch der Kartelpolitik* (Berlin: Duncker & Humblot, 1958), 307–26; Iyori, *Das japanische Kartelrecht*; Yoshio Kanazawa, 'The Regulation of Corporate Enterprise: The Law of Unfair Competition and the Control of Monopoly Power', in Arthur Taylor von Mehren (ed.), *Law in Japan. The Legal Order in a Changing Society* (Cambridge, MA: Harvard University Press, 1963), 480–507.

together these policies effectively eliminated the traditional structure of family control through holdings of broadly diversified, multisectoral operations.

Next, there was a purge of leading managers in the largest companies (not only *Zaibatsu*): the SCAP mandated the removal of all the wartime chief officers of 200 important companies in the economy. Moreover, participation in the new management of the restructured companies by *Zaibatsu* family members or any of the high-ranking directors of 240 *Zaibatsu*-related companies was banned for a period of ten years. There are disagreements in the literature as to how many managers were actually purged as a result of these measures: Hadley estimates around 2,500 while Aoki places the figure at 3,500. In any case, it seems fairly clear that an entire generation of top leadership in firms was wiped away by SCAP efforts to cleanse the positions of power in the economy of 'militarists and ultra-nationalists'.[73]

So much for the diversified, multisectoral combines. These reforms undermined the old system of central co-ordination of diversified operating units and created essentially autonomous and isolated units, weakly related by interpenetrated property ties (and, crucially, without any controlling property interest). In the context of the steel industry, as in the German case, this had the consequence of cutting loose the individual producers from the security and technological and market resources provided by their association with up- and downstream conglomerate units. The most dramatic example of this was the case of Kawasaki Steel which, due to the personnel and property changes in the Kawasaki Group, severed itself completely from its long-time parent, Kawasaki Heavy Industries. As in the German case, this new autonomy would profoundly affect the strategies that individual producers believed that they could then pursue (particularly after the break-up of Japan Steel, as will be discussed below).

A final destructive intervention was the abolition of the Control Association infrastructure that had been created during the war at the bidding of the military. These organizations were viewed as monopolistic, market-hostile forces, and as sources of cartelization—all of which undermined the emergence of a plurality of organizational forms in the economy. The Iron and Steel Control Association was among the first to be abolished (in February 1946 by SCAP decree) but by the end of the year all remaining Control Associations had been ordered to be abolished.[74] The USA was hostile to the control function of the Control Associations, that is, the central control and allocation of investment resources within an industry. But the SCAP was not opposed to the existence of trade associations in an industry, *per se*. The Trade Association Act of July 1948, while banning all control functions and all other activities that were leading or could lead to cartel, monopolistic, or unfair trade practices, explicitly allowed trade associations to 'receive voluntary submissions of statistical data and to publish such data in summary form without disclosing business information or

---

[73] Aoki, 'The Japanese Firm in Transition', 268–9; Hadley, *Anti-trust in Japan*.
[74] J. Cohen, *Japan's Economy in War and Reconstruction*, 431.

conditions of any particular entrepreneur'.[75] The law also allowed the dissemination of technical information and co-operation in research. In other words, as long as trade associations acted like trade associations in the United States, representing the interests of their members and providing non-market related services and publicity, they were acceptable and fully compatible with a pluralist economic order.[76]

As in Germany, the SCAP's actions were not all destructive. The military authority also made an effort to create a set of rules in the industry that would foster the development of a pluralist political and economic order. American-style antitrust legislation was passed on 14 April 1947 by the unreformed Japanese parliament. As in Germany, it completely outlawed all forms of cartel arrangements and severely curtailed the possibilities for merger and holding company arrangements as well. In the spring of 1948, US authorities allowed the law to be reformed to reduce the severity of obstacles to merger. They also established at that time a Deconcentration Review Board, staffed by American business and government officials, to review each individual market arrangement to determine precisely whether or not it constituted unfair competition. This innovation slowed the initial intense anti-monopolist fervour of US deconcentration policy, but it established an effective mechanism to ensure that only organizations consistent with American pluralist conceptions of market order would find the space to reconstitute themselves.[77]

The central achievement of this new system was the break-up of Japan Steel. Like the Vestag (and the HGW), this large firm was taken to be an 'unreasonable' concentration of economic power by the Allied authorities. Japan Steel was not fragmented into as many pieces as Vestag: instead of the 13 that came out of the Vestag, Japan Steel was broken into two new and independent firms, Yahata Steel and Fuji Steel.[78] This move had a significant impact on the industry, in two ways. First, it increased the number of integrated producers in the industry from two to three (in addition to Yahata and Fuji, NKK was also integrated). Second, and more significantly, by making the formerly state-controlled Yahata and Fuji private firms with a responsibility to sell their production on the market at competitive prices, the break-up of the firm introduced significant

[75] Yamamura, *Economic Policy in Postwar Japan*, 25.

[76] For a discussion of US and Japanese trade association similarities and differences, before, during, and after the occupation period, see Lynn and McKeown, *Organizing Business*, 24–32 and *passim*.

[77] Shen-Chang Hwang, *Das Japanische Antimonopolgesetz im Lichte des deutschen Kartellrechts*, Ph.D. dissertation (Department of Law, Ruprecht-Karl University, Heidelberg, 1968), 10 f.; Iyori, *Das japanische Kartelrecht;* Hadley, *Antitrust in Japan*; and T. Cohen, *Remaking Japan*. Cohen in particular gives step-by-step detail on the formation of the Deconcentration Review Board, 353–77.

[78] Yahata Steel took over the Yahata Works while Fuji took over the Hirohata, Kamaishi, Wanishi, and Fuji Works: Yonekura, *Japanese Iron and Steel Industry*, 189–211; id., 'The Post-War Japanese Iron and Steel Industry: Continuity and Discontinuity', in Etsuo Abe and Yoshaka Suzuki (eds.), *Changing Patterns of International Rivalry* (Tokyo: University of Tokyo Press, 1991), 193–241; and Walter Scheppach, *Die japanische Stahlindustrie* (Hamburg: Mitteilungen des Instituts für Asienkunde, No. 48, 1972).

uncertainty in the supply of pig iron to non-integrated producers. In the past, non-integrated firms such as Kawasaki, Sumitomo, and Kobe had been content to purchase pig iron from Japan Steel at subsidized prices. Now this was no longer possible—and, moreover, since Fuji and Yahata had to be competitive in finished-product markets as well, they posed a significant competitive threat to the non-integrated steel producers.[79] The SCAP's hope was that introducing greater competition into the integrated steel sector would ultimately foster the adoption of more efficient, vertically integrated structures by all remaining producers and ideally, would result in the emergence of a new, stable, oligopolistic market structure.

Indeed, it is clear that the combination of destructive and creative policies completely ruined the set of institutions and practices that had created the distinctive, bifurcated structure in the pre-war steel industry. Instead of one there were several integrated, volume producers and the speciality producers no longer had dynamic relations with conglomerate partners nor the ability to arrange and stabilize the boundaries of their specialities in the market through cartelization. The new environment that the SCAP reforms created, moreover, structured incentives in such as way as to make the pursuit of volume and integration more attractive than attempting to recast the strategy of specialization.

Finally, again as in Germany, moves were made to foster the emergence of an organized labour presence both in Japanese industry generally and in the steel industry in particular. As in Germany, during the entire period of industrialization, labour organizations had suffered from considerable repression in the steel industry (though, unlike Germany, in Japan they had made few inroads elsewhere).[80] In October 1945, General Douglas MacArthur (the head of the SCAP) instructed the Japanese Diet (parliament) to enact labour legislation that would protect the rights of labour, including the right to organize. Ultimately three crucial laws were enacted in the ensuing year which established a very favourable environment for the formation of trade unions and collective bargaining.[81] The first law, the Labour Union Law, guaranteed workers the right to organize, collectively bargain, and strike. Discrimination by employers for union activity was also outlawed. This law further established a national Central Labour Relations Commission as well as regional commissions in each prefecture. The second law, The Labour Relations Adjustment Act, explicated

[79] Yonekura's article, 'Postwar Reform in Management and Labor: The Case of the Steel Industry' in Juro Teranishi and Yutaka Kosai (eds.), *The Japanese Experience of Economic Reforms* (London: St. Martins Press, 1993), 205–40 addresses precisely this issue of oligopoly creation.

[80] This is not to say that there were not efforts to organize or that there were no union-like organizations prior to the occupation. There were. But they were very embattled and persecuted by both the state and employers. See the outstanding account of pre-war developments in heavy industry in Gordon, *Evolution of Labor Relations in Japan*, 1–326. Theodore Cohen notes that the Japanese labour movement reached its pre-war high for union organization in 1936 with 420,000 members—approximately 7 per cent of the non-agricultural workforce: T. Cohen, *Remaking Japan*, 191.

[81] Gordon, *Evolution of Labor Relations in Japan*, 330–9; and id., *The Wages of Affluence. Labor and Management in Postwar Japan* (Cambridge, MA: Harvard University Press, 1998), ch. 3.

more precisely the role and purview of these boards. Essentially they were constituted as arbitration boards in which representatives from labour, employers, and government sat in deliberation. Finally, a third piece of legislation, the Labour Standards Law, set minimum hours, wages, insurance, injury compensation, and unemployment benefits for all workers, unionized and non-unionized.[82]

The American authorities conceived of these moves as essential for the creation of a social infrastructure in Japan capable of sustaining democracy. Crucially, however, though supporters of union organization in markets for political reasons, the US authorities were strong opponents of political activism on the part of unions. As Cohen writes of the view of the occupation labour department that he headed, 'To the Americans, political unionism, that is unions as partners of a party, was an anti-democratic and un-American concept'.[83] Japanese unions and workers did not agree with this idea, as we shall see. Indeed, in the following section we will see that many of the framework interventions undertaken by the occupying governments encountered resistance from the occupied societies, giving rise to a significant social, economic, and political transformation in each society.

12.4   The Appropriation and Transformation of American Ideals in Germany and Japan

Most treatments of the American occupation of Japan and Germany have narratives that begin, as this one has, with aggressive American efforts to reform the indigenous systems. But typically the narratives then turn to internal conflicts or incoherence within the occupation governments, intransigence on the part of powerful local leaders, and finally to a weakening of American resolve for reform, especially in the economy, with the onset of the cold war and in particular the outbreak of the Korean war—all of which are held to undermine American reform ideals and lead to a 'reverse course' in policy or the reconstitution of central institutions and actors from the pre-war societies.[84]

Without in any way wanting to cast doubt upon the *existence* of factors disrupting American resolve and distracting attention, the discussion here will depart significantly from this traditional narrative of American failure. Its claim, as indicated at the outset, is both that American occupation profoundly altered the German and the Japanese political economies and that indigenous institutions, actors, and ideas transformed the American ideas in the process of appropriating them. This mutual transformation of the occupied and the occupiers occurred not by the Allies imposing particular institutional forms on the

---

[82] Gordon, *Evolution of Labor Relations in Japan*, 330–9.

[83] T. Cohen, *Remaking Japan*, 204.

[84] Examples of this kind of narrative with respect to the steel industry are Gloria Müller, *Mitbestimmung in der Nachkriegszeit. Britische Besatzungsmacht-Unternehmer-Gewerkschaften* (Düsseldorf: Schwann, 1987); ead., *Strukturwandel und Arbeitnehmerrechte. Die wirtschaftliche Mitbestimmung in der Eisen- und Stahlindustrie, 1945–1975* (Essen: Klartext Verlag, 1991).

occupied lands and the occupied populations blocking some and not blocking others and/or deconstructing those imposed structures. Rather, the mutual transformation occurred because of the way in which contestation itself was constituted. American power successfully established a range of normative social, economic, and political background rules according to which deliberation about the construction of new institutions had to take place. Yet, these background rules (pluralism, cross-cutting power, anti-monopolism, limited state power, etc.) were neither unambiguous nor directly reducible to a finite set of clearly 'American' organizational forms or practices. The deliberative process of identifying a practice or institution that could be legitimately recognized as consistent with higher normative goals resulted in the mutual transformation of all participants' understanding of the object of deliberation. We will look at the German case first and the Japanese second.

ENGAGEMENT WITH AMERICAN REFORM EFFORTS IN GERMANY

The defeat of the Nazis and occupation by foreign anti-National-Socialist powers utterly delegitimized the understandings of market order and state power that the Nazis had attempted to institutionalize during the Third Reich. At the same time, a space was created by the occupation for the reassertion of ideas about social and economic order that had been suppressed by the Nazis. Holders of these views were forced by the situation, however, to articulate their positions anew in ways that addressed the Allied critique of the authoritarian elements within the German political economy and of the Allies' priorities for a pluralist democratic order in the economy and the polity. This joint, interpenetrated, recomposition of imposed and traditional categories occurred in three areas in the context of the steel industry debate: (a) in the area of property rights in industry and the sovereignty of private property ownership; (b) in the area of codetermination, workplace democracy, and the scope of trade unionism; and (c) regarding the reconstitution of market competition, antitrust, and cartelization.

Property in Industry

The property issue was central to all discussions of industrial restructuring in the occupation period and it was of decisive importance for restructuring in the steel industry.[85] While the US government and German industrialists both opposed the elimination of private control of the steel industry (unlike the British[86]), the two did not share the same conception of the political significance

---

[85] Isabel Warner, *Steel and Sovereignty, passim*, emphasizes this aspect of industrial interest above all others (such as industrial structure, cartels, etc.) on the domestic front in the negotiations with the Allies.

[86] British interests in socializing the steel industry, their ultimate inability to assert their interests over American ones, and the lost opportunities of this fact have received inordinate attention in the cold war-influenced literature on the deconcentration process in the steel industry after the war. I overlook this dimension of the occupation because it was brief and its effects, such as they were, are

of private property or of its role in political order. In the American view, private property was, at its most basic level, a constitutive feature of a market economy, without which there could be no exchange. By protecting its sovereignty in market competition, and yet at the same time opposing monopoly, Americans believed that an ensuing healthy competition among private capital would drive innovation in the economy, expand the spectrum of opportunity for individual private actors, and create the social power of organization to limit the unhealthy growth of state power. This view of property assumed an equality among property holders and understood social order to be a competitive equilibrium among plural sources of social and political power. Private property was constitutive of the American conception of liberal-democratic pluralism.

German industrialists had a conception of private property that had little to do with equality, liberal freedoms, democracy, or competitively constructed social order. For them, private industrial property entailed both a particular status in society and a whole range of mutual obligations with other social groups, the nation, and the state. Whether as individual private property holder or as managerial representative of diffused joint-stock capital, the status associated with private industry came from an idea that society was divided into specialized roles—skilled workers, farmers, artisans, shopkeepers, bankers, lawyers, etc.—and that private industrialists played the esteemed role of steering industrial production, providing employment and income, housing, education, and social provision for their employees, and driving industrial progress. It was significant for the private industrialists that their ownership of property entailed authority: if they were obligated to provide their employees with the benefits just noted, they fully believed that their employees owed them unconditional respect for their authority, and all matters relating to it, both within and outside the factory walls.[87]

German private industrialists, in other words, understood themselves to be playing a crucial role, *as a corporate group*, in the maintenance of the traditional social and political order in Germany. As mentioned in section 12.2, this role was to be performed in support of and in conjunction with, but always independently of, the state. Large industrialists understood their role as contributing to the greatness of the nation, not to the greatness of the state. For them, the state, as a higher authority, had complementary ends and reciprocal obligations to its citizenry to maintain public order, respect the order of status and entitlements in society, and provide for the developmental needs of the nation. Industrialists believed that the power of the state should be limited *vis-à-vis* the rights of property, just as they believed their own power over their workers was

orthogonal to the points I am making about the process. For characteristic examples of writing that place the emphasis differently, see Eberhard Schmidt, *Die Verhinderte Neuordnung* (Frankfurt: EVA, 1970); and Turner, *Reconstruction in Post-War Germany*, esp. 37–154.

[87] On the paternalistic views of German large-scale private industrialists, see the discussion of Gustav Krupp in excellent essay by Richard Overy, ' "Primacy Always Belongs to Politics" '; Turner, *German Big Business and the Rise of Hitler*; Berghahn, *Unternehmer und Politik*.

tempered by obligations. Such mutual recognition and limitation contributed to good order. Though perhaps obvious, it is important to underscore that there is nothing in this view of the rights and status of private property that held that the state should be democratic or even equally limited in its power relative to all social groups—hence German industrialists' tolerance of Nazi labour measures and labour repression but resistance to Nazi efforts to influence managerial decision making.[88]

In the context of Allied occupation and debates about the deconcentration of the steel industry, then, private property was not private property. And, in the face of Allied power (all steel assets were held by the Allied Military Government in trust), German steel industrialists found it prudent in the struggles over restructuring to formulate their arguments for the social and political value of private property in industry in a way that was consistent with American understanding. Crucially, they did this in a way that did not involve the simple appropriation of the American view. Rather they highlighted elements within their traditional view that resonated with the American one. In particular, an important industrial argument regarding private property throughout the occupation period (and well into the mature Federal Republic) involved the valuable role that private capital and its organizations could play in the limitation of state power and as a bulwark against the return of authoritarianism. Weak property rights and poorly organized industrial associations, they argued, made society vulnerable to unjust incursions of state power and prey for demagogic political actors.[89] Further, this idea of the central significance of sovereign private power being capable of limiting state power was linked to the traditional idea of self-government (*Selbstverwaltung*), in which private industry as well as its associations were understood to have the social obligation and privilege of being able to govern their own affairs without outside interference.[90] These very traditional ideas about the significance and rights of property could be made to appear consistent with American desires to establish a regime of private property that would foster social pluralism, limited government, and economic progress.

What the new German positions did not do, however, was abandon the idea of society being composed of deeply entrenched, functional groups, whose location and identity involved complex notions of status, entitlement, and mutual obligation. By emphasizing the key limiting role of private property on the

---

[88] Overy, ' "Primacy always belongs to Politics" '; Turner, *Big Business and the Rise of Hitler*, *passim*. For conflicts between the state and industry that reveal private industry's norms regarding the limits of state power in their domain, see especially Overy, 'Heavy Industry in the Third Reich'.

[89] Various articles from *Die Volkswirt*, most by anonymous industrialists in the context of the codetermination debate in the late 1940s, make this argument.

[90] See essays by Berg on self-government in Fritz Berg, *Die Westdeutsche Wirtschaft in der Bewährung. Ausgewählte Reden aus den Jahren 1950 bis 1965* (Hagen: Linnepe Verlagsgesellschaft KG, 1966).

growth of state power, German industrialists were simply highlighting the part of their understanding of property that was compatible with the American notion of property. There was nothing manipulative or dishonest about this: it was simply a way in which the American discourse could be understood within the traditional frames of German understanding. Indeed, the American insistence on democratic pluralism encouraged this kind of recomposition in the German view.

By the same token, it is important to insist on the idea that this was not a process of the Americans being somehow duped. When the military authority decided to make the restructured steel assets exchangeable for shares in the *Altkonzerne* in May 1951, the Americans did so because they understood German arguments as acceptance of their own understanding of private property as a vehicle for social and political pluralism. According to James Stewart Martin, who did not believe this, the view was: 'They were not Nazis; they are businessmen.'[91] The American occupiers understood the category of 'businessman' or private industrialist to have a transparent meaning—as did the Germans. Yet, in both cases the Germans and Americans were nodding their heads in agreement when the content of what they agreed upon differed quite radically.

For the German actors, their own position merely involved a recalibration of the kinds of entitlements and mutual obligations that were to be publicly associated with private property: private property in industry was still understood to be crucial for the maintenance of social order and hence deserving of respect and recognition. The American view denied that distinctions of status and entitlement could be politically drawn among private actors, while the adapted German view assumed this to be a foundational dimension of what was meant by private control of industry. In both forms of understanding, however, private property constituted a countervailing power against the authority of the state— and this was crucial for Allied approval.

An additional arena in which a recalibration of indigenous understandings of social categories and political status took place in confrontation with American ideas was in the context of the governance of the workplace and the labour market. As in the case of property in industry, German and American understandings of markets, social order, and democracy diverged, despite formal harmonies and the appearance of consistency in social and political understanding. The interesting aspect of this process of recomposition is that it occurred in interaction with the developments in the reconstitution of property just outlined.

---

[91] Martin, *All Honorable Men*, 91. The reference to Nazis in this quote may distract: the view conveyed is that the Americans were persuaded that a businessman was a businessman was a businessman, while I am pointing out that this was not the case. Where Martin would probably associate the corporate dimensions of the German industrialists' understanding of property immediately with Nazism, I would not. The corporate idea of private property is much older and industrialists were forced to recompose the idea to accommodate the Nazi regime as well.

Codetermination, Workplace Democracy, and the Scope of Trade Unionism

As noted earlier, the Allied Military Government was favourably disposed to the formation of trade unionism in occupied Germany and very much encouraged the development of collective bargaining practices. This was in line with their idea of pluralist democracy as a system of countervailing powers. German workers embraced this idea enthusiastically. But they also extended it in a way that went far beyond the American understanding of a trade union or of the proper boundaries of countervailing power in the labour market. The reason for this stems from the fact that, unlike the American occupiers, the German workers understood the traditional roots of the social and political lines that German industrialists were drawing around their property. In response, the labour movement sought to apply American notions of the limitation of power (or at least that language) to what it viewed as unwarranted concentration of corporate social power in the form of independent managerial control over industrial enterprises. In the German labour movement's view, the only way to obtain countervailing power was to have it inside industrial enterprises.[92]

To be sure, this idea had a wide variety of indigenous precursors: social catholics and social democrats had both advocated variants of industrial democracy during the Weimar Republic (and even before) and syndicalist ideas had informed the spontaneous takeovers of factories after both the First and Second World Wars.[93] But the particular variant that emerged after the Second World War in the context of occupation and restructuring in the steel industry was distinguished from these by the way in which its defenders indicated its consistency with American understandings of pluralist democratic order. Previously, codetermination had been argued for very much in corporate social terms, that is, as an argument for the social entitlement of the working class to be able to co-influence the organization of work and the direction of the production unit's investment. In the post-war variant, the language was not about social entitlement for the working class, but about the need to limit concentrations of private power with countervailing social organizational power in order to create secure democratic order.

Though they were wary of the threat to their own understanding of private property that codetermination seemed to pose, the American governors, and other American observers, seemed to be persuaded that, though unconventional, codetermination was at least consistent with their own democratic concerns. Clark Kerr, an early observer of the German labour movement from the United States, made sense of codetermination in this revealing way:

[92] Müller, *Strukturwandel und Arbeitnehmerrechte*; Wolfgang Streeck, 'Codetermination: After Four Decades', in id., *Social Institutions and Economic Performance. Studies of Industrial Relations in Advanced Capitalist Economies* (London: Sage, 1992), 137–68.

[93] On the history of codetermination efforts in Germany prior to the 1950s, see Hans Jürgen Teuteberg, *Geschichte der industriellen Mitbestimmung in Deutschland* (Tübingen: JCB Mohr [Paul Siebeck], 1961).

The program defies labeling for it is not social democracy, nor catholic corporatism, nor socialism, nor capitalism, nor syndicalism, nor voluntarism, although it bears some similarity to each. It might, perhaps, be termed 'joint economic pluralism.' 'Pluralistic' [sic] because it envisages many loci of power, 'economic' because it emphasizes the role of functional interest groups; 'joint' because power is shared by capital and labor. It is a sort of meeting ground between liberalism and socialism, for it has elements of both private enterprise and social control. It is also a meeting ground for capital and organized labor since each, in Germany, favors a privately controlled economy. The battle between them is not an ideological one of capitalism versus socialism; rather it is more a practical but nonetheless quite intense one over how joint shall self-administration be.[94]

The German unions sold codetermination as an institution for the limitation of private industrial power and the American occupiers understood it that way. Yet codetermination was not only that. It was also clearly understood by both trade unionists and employers in Germany as an institutional move to secure and improve the social and political status of the working class. It challenged the social and political entitlements involved in the industrialists' understanding of their social position and asserted the rights and entitlements of workers. This latter dimension of the struggle over codetermination was very far removed from the liberal pluralist concerns of the Americans. It involved the recalibration of corporate social rights. Or, as the eminent legal theorist Ernst-Joachim Mestmäcker stated in criticism in the 1970s, when revision of the law was being widely debated, the legal justification of codetermination by the trade unions has a liberal pluralist heritage, but this heritage has always paradoxically supported a system of political entitlement for the participating social groups.[95]

The system of codetermination was thus neither the direct product of Allied occupation nor thinkable in its precise form without it. Consistent enough with American pluralist ideals to lead the American authorities to keep at arm's length over the German debate in the passage of the law, calling it a 'German affair', the system of codetermination was also an arrangement that never spawned any imitators in the United States. By all accounts, however, the system was a tremendous advantage for the steel producers during the great postwar economic boom. It fostered co-operation between labour and management in the labour market and on the shop-floor and thereby gave steel producers remarkable flexibility in work and production.[96]

Reconstitution of Market Competition, Antitrust, and Cartelization

As we saw in the previous section, Allied intervention into the order of markets in the steel industry was extremely aggressive: firms were broken up and recon-

---

[94] Clark Kerr, 'The Trade Union Movement and the Redistribution of Power in Postwar Germany', *Quarterly Journal of Economics* 68, 4 (1954), 553–4.

[95] Ernst-Joachim Mestmäcker, 'Mitbestimmung und Vermögensverteilung. Alternativen zur Umverteilung von Besitzständen', in id., *Recht und ökonomisches Gesetz. Über die Grenzen von Staat, Gesellschaft und Privatautonomie* (Baden Baden: Nomos Verlag, 1978), 135–54, esp. 143 ff.

[96] On the post-war benefits of the system, see Streeck, 'Codetermination'; Kathleen Thelen, *Union of Parts* (Ithaca: Cornell University Press, 1991), *passim*; Herrigel, *Industrial Constructions*, ch. 6.

stituted as much more specialized units, downstream linkages to machinery and other product markets were severed, and all manner of cartelization was banned. It was plain to the steel managers that the Allies were attempting to block the reconstitution of the old pre-war industrial structure and encourage a more mass-production-oriented, oligopolistic industrial structure to develop.

German steel managers were divided on the attractiveness of this situation. Some, such as Hermann Reusch of the GHH *Konzern*, found the entire Allied programme unacceptable and abandoned GHH's commitment to the steel industry and concentrated its business in the machinery sector. Others, however, recognized the kind of strategy the Allies were encouraging as reminiscent of a strategy that earlier actors in the industry, such as August Thyssen, had advocated, unsuccessfully, earlier in the century.[97] Thyssen and his allies failed in the earlier period both because of the intransigence of existing property relations in the industry, which blocked consolidation efforts, and because the bifurcated industrial structure of a large-volume producer and numerous more specialized *Konzerne* proved to be a profitable alternative strategy.[98] With the old firms physically and technologically broken up into distinct property units and Allied hostility toward the *Konzerne* strategy, the mass-production and oligopoly strategy appeared both attractive and possible. Here, producers found that a strategy for development was being made available to them that was both consistent with a dimension of their past and in line with the American understanding of 'modern' industrial structure.

Seen in this light, it should be of little surprise that the industry developed in precisely this way during the 1950s and 1960s: in response to strong demand during the Korean war, and then subsequently with the growth of the automobile and other consumer goods industries, producers invested heavily in new, larger-scale, modern rolling-mill equipment, in particular wide-strip mills, that enabled them to achieve greater production efficiencies at higher volumes. At the same time, a process of concentration occurred in the industry over the course of the 1950s as the smaller units created by the Allied deconcentration reforms were recombined to consolidate industry capacity and accommodate the large-scale technologies.

This process of recomposition in the industry can be clearly seen in the way that the companies of the old Vestag were recombined during the initial period of expansion. Of the 13 steel companies originally created out of the break-up of the Vestag in 1953, only four were left by the beginning of the 1960s: the August Thyssen Hütte AG (ATH), the Phoenix-Rheinrohr AG, Rheinische Stahlwerk AG (Rheinstahl) and the Dortmund-Hörde Hütten Union

---

[97] On Thyssen's (essentially American) image of the industry, see Alfred D. Chandler, Jr., *Scale and Scope. The Dynamics of Industrial Capitalism* (Cambridge, MA: Harvard University Press, 1990), 493, and the discussion in Herrigel, *Industrial Constructions*, ch. 3.

[98] Ibid. See also Christian Kleinschmidt, *Rationalisierung als Unternehmenstrategie. Die Eisen- und Stahlindustrie des Ruhrgebiets zwischen Jahrhundertwende und Weltwirtschaftskrise* (Essen: Klartext, 1993) for an excellent discussion of the inter-war period.

(DHHU).[99] For the most part, the consolidation and expansion of these companies did not result in direct competition between them. Instead, each expanded and consolidated its capacity in areas that the others ignored. The vast majority of the steelmaking capacity of the old Vestag was reincorporated within the operations of either ATH or Phoenix-Rheinrohr and their product offerings and specializations in rolled-product markets were for the most part not overlapping. ATH specialized in the production of lighter, semi-finished and finished rolled sheet and coils, and wire and speciality steels, whereas Phoenix-Rheinrohr specialized in the production of steel pipe, heavy plate, semi-finished steels, and raw iron.[100] The co-ordinated growth and reconsolidation of these two companies was facilitated by the fact that both enterprises were controlled by different members of the Thyssen family.

Rheinstahl, which had very broad ownership, was less interested in getting involved in the reintegration of the Vestag steelmaking interests. This company received all of the non-steel manufacturing interests of the old Vestag as a result of deconcentration. It held an interest in only a small number of relatively small and specialized steelworks from the old Vestag, and, though it continued to produce steel during the 1950s, it was a very minor player in the industry.[101] The DHHU, for its part, was an important producer of crude steel. But unlike its larger Vestag cousins, the company was not as broadly diversified across steel industry markets. By the beginning of the 1960s, the company was concentrated on two general areas: it was the Federal Republic's largest producer of heavy plate and an important producer of steel bars and structural steel.[102]

Together, these four firms, with their co-ordinated capacities, accounted for nearly 35 per cent of the crude steel output of West Germany in 1960–1—ATH and Phoenix-Rheinrohr together accounted for 20.91 per cent. Another 41 per cent of total industry output was taken up by the steel operations of the former *Konzerne* Hoesch, Klöckner, Mannesmann, Hüttenwerk Oberhausen AG (HOAG),[103] and Krupp.[104] On the whole these producers followed the DHHU pattern of specialization, each attempting to organize its steel and rolled-

---

[99] The attentive reader will note that three of these companies—ATH, Phoenix, and Rheinstahl—correspond, at least in name, to companies that existed prior to the formation of the Vestag, and which were instrumental in its creation. DHHU, as noted above, was a creation of the allies, brought about by the merger of two plants from the old Vestag. Prior to the creation of the Vestag, the Dortmund plant of the DHHU was part of Hugo Stinnes' giant Rhein-Elbe Union, while the Hörde plant belonged to the Phoenix group. The two plants were formally linked together by the Vestag during the 1930s, but both continued to operate independently. All of these companies, especially the former three, differed substantially, in holdings and specializations, from these earlier incarnations.

[100] Bernd Huffschmid, *Das Stahlzeitalter beginnt erst* (Munich: Verlag Moderne Industrie, 1965), 110–15; on Thyssen family ownership, see Lammert, *Das Verhältnis*, 207–8; Müller, *Strukturwandel*, 300–3.

[101] On Rheinstahl, see Huffschmid, *Stahlzeitalter*, 185–91; Müller, *Strukturwandel*, 304–5; and Lammert, *Verhältnis*, 206–7.

[102] Huffschmid, *Stahlzeitalter*, 147–55.   [103] HOAG belonged to GHH until 1945.

[104] Krupp's steel interests were grouped in a separate holding which included the Hütten- und Bergwerke Rheinhausen AG.

product output in a way that would give the company a strong position in a limited number of markets. Thus, for example, Mannesmann concentrated its production on steel pipe and a variety of high-quality, fine, and zinc-treated plate steels, Krupp on speciality steels, structural steels, bar, and semis, Hoesch on sheet steels and fine plate, HOAG on heavy plate, structural steels, and speciality wire, and Klöckner on sheet steels, fine plates, wire, and bar.[105] The producers accounting for the remaining 15 per cent of the industry's output in 1960, who were located both inside and outside the Ruhr, followed a similar pattern of specialization.

The emergence of oligopolized sub-markets and more specialized firms oriented toward mass rather than speciality production ultimately resulted in a distinctly different industry structure than that which had existed before the Second World War. Before the war, only six companies controlled nearly three-quarters of the industry's total output, whereas by 1961 that same share was divided among nine enterprises. More significantly, the successor companies of the old Vestag did not recapture the same dominant share of industry output that had belonged to their ancestor: the Vestag had accounted for roughly half the industry's total output during the 1930s, whereas the successor companies were only able to achieve 35 per cent by 1961. Decartelization and deconcentration in the 1940s and early 1950s had succeeded in creating a more internally specialized and more competitive German steel industry—much as the Allies had intended when they broke up the old structure at the end of the war.

None the less, it would be misleading to characterize this newly structured German steel industry as a complete triumph of Americanization. Its new structure and practices differed in significant ways from the American pluralist conception of mass production and oligopoly. First, as indicated above, the actors in the industry, both in management and labour, understood themselves to be corporate groups with social and political status in the broader society and with an understanding of mutual obligation and responsibility. By all accounts, this mutually accepting, and through codetermination, institutionalized, arrangement created significant flexibility within high-volume production. Labour and management were able to co-operate in ways that allowed firms to take on special orders, produce in varied lot sizes, and reallocate labour and materials within plants in ways that the more rigid, bargained-out arrangements of equal contracting citizens in American plants did not allow.

Indeed, second, this reserve of flexibility within the new industry and its structure led the actors in the industry to favour a modified rather than the strongly American version of cartel law in the debates over the reform of German antitrust law that raged during the 1950s.[106] This modified version accepted the

---

[105] Huffschmid, *Stahlzeitalter* provides information of the product profile and output of each of these companies, 122–46, 156–84, and 192–7.

[106] On this debate, see Rüdiger Robert, *Konzentrationspolitik in der Bundesrepublik—Das Beispiel der Entstehung des Gesetzes gegen Wettbewerbsbeschränkung* (Berlin: Duncker & Humblot, 1976); Viola Gräfin von Bethusy-Huc, *Demokratie und Interessenpolitik* (Wiesbaden: Franz Steiner, 1962); and Berghahn, *Industrie und Politik*.

general American injunction to condemn cartels and monopoly of any kind. It insisted, however, on a concession to flexibility in that it rejected the proposal for a strict ban on cartels in favour of negotiated and constructive solutions to problems of market order among the industry players and the Cartel Office of the Federal government. Moreover, several forms of cartels, in particular rationalization cartels, were permitted under the antitrust law that was accepted by parliament in 1957. Both of these characteristics of the new rules of market order were sympathetically received (if not enthusiastically lobbied for) by the steel industry because they created space among firms for the flexible orientation to the market that the situation inside the plants was making possible. These provisions in the law proved to be significant for the industry in the late 1960s when overcapacity became a problem and producers avoided ruthless and destructive competition by forming distribution syndicates for the allocation of their finished steel. These syndicates both removed pricing from competition and facilitated the recombination of capacity and specialization within the industry.[107]

In conclusion, it is clear that though the German steel industry adopted or was forced to adopt American principles of market order and production and was profoundly changed by this encounter, this in no way resulted in an erasure of distinctively German features in the production of steel. Indeed, German producers embraced the vocabulary and practices of Americanism and pluralism, but in doing so transformed them in ways that either were consistent with their own prior understandings and practices or extended the received principles in ways that were not in evidence in the USA nor foreseen by the Allied reformers. We will now see that, on this level, the case was very similar in Japan.

ENGAGING ALLIED REFORMS IN OCCUPIED JAPAN

Allied reforms engendered debate and transformation in different areas in Japan than in Germany, in part because the character of the occupation was different, and in part because the two were very different political economies in different geopolitical locations. In Japan, for example, the centrality of private property for stable political order was never debated in the same way as it was in Germany. This is not necessarily because Americans and Japanese held the same political understanding of private property in a democratic market order, but more likely because there were neither British socialists nor threatening Communist neighbours to push the alternative case in Japan.[108] None the less, at a different level, there were many similarities between German and Japanese experiences. In particular, in Japan as in Germany, defeat and reform created spaces in the debate for the re-articulation of abandoned, defeated, or unreal-

---

[107] Burkhardt Röper (ed.), *Rationaisierungseffekte der Walzstahlkontore und der Rationalisierungsgruppen* (Berlin: Duncker & Humblot, 1974).

[108] And the SCAP simply refused to allow the minority Communist views in Japan to push the issue onto the agenda.

ized conceptions of social and industrial order from the past that the wartime regime had suppressed. Moreover, as in Germany, those advancing such ideas, though liberated, were also constrained by American insistence on the creation of a pluralistic democratic economy. The central areas of contestation relevant to the reform of the Japanese steel industry were: (a) the reconstitution of mutually limiting relations between the state and industry; (b) the recomposition of firms, associations, and industrial structure; and (c) the recasting of authority in production. The outcome of these encounters was a completely new way of understanding the identity, boundaries, and organization of the steel industry.

Government–Business Relations

American reformers took it as their goal to establish a system of countervailing powers between the Japanese state and economy in order to prevent actors in either realm from gaining unchecked monopoly power. The abolition of the Japanese military, the break-up of the *Zaibatsu* holding companies, and the dissolution of the old Control Associations, for example, were all guided by the conviction that these institutions had been either monopolistic or unaccountable (or both) and in their basic structures were incapable of being reformed in a way that would allow them to be checked. As noted earlier, however, the central institutions of the civil bureaucracy were not targeted by the SCAP either for dissolution or even for serious reform.[109] Unlike the other institutions, the SCAP believed that it was possible to check the power of the existing state bureaucracy through the construction of countervailing forces and institutions both in the state and in society. The creation of parliamentary institutions that monitored the bureaucracy, the creation of conditions that favoured the formation of strong oligopolies in the economy, and the construction of a trade association law that both forbade monopolization and encouraged the formation of interest groups, for example, all were directed at the construction of a set of rules and a population of social organizations that could fragment social, economic, and political power.[110] Rather than destroy the bureaucracy as it did the other 'authoritarian' institutions, the SCAP attempted to make it one organization among many in a plural system of social and political power. Though it is possible to say that this is what the reforms did indeed achieve, the self-understanding of the Japanese actors in the change and the quality of state relations with industry differed very significantly from what the Americans had in mind.

Some Japan scholars view the above strategy on the part of the SCAP with great irony because they point out that by removing the *Zaibatsu* and the military from the field of play, the occupiers actually enhanced the power of the civil bureaucracy in the post-war environment by removing the traditional indigenous checks on its power without introducing effective countervailing

---

[109] On the feeble efforts to reform the state bureaucracy on the part of the SCAP, see T. Cohen, *Remaking Japan*, 378–97.

[110] Haley, *Authority Without Power*, 139–68.

institutional forces.[111] This view, however, significantly underestimates the constraints the new reforms placed on bureaucratic power, and overestimates the power of the elements within the Japanese bureaucracy that believed in unilateral state control over resource allocation. After the reforms, bureaucratic ministries were newly constrained by the constellation of power in the parliamentary arena and by the competing interests and agendas of other minstries.[112] Moreover, Bai Gao argues that with the coming of the market-oriented Dodge Plan and the foundation of MITI, the elements within the bureaucracy which favoured a heavily *dirigiste* planned economy lost influence in the construction of state intervention to those who favoured a more co-operative and 'managed' approach to private actors.[113] In other words, the civil bureaucracy that led Japan through its post-war miracle had an interest in limiting itself as an organizational actor in the economy and encouraging the development of private organizations and markets.

Even though seen in this way the reformed bureaucracy contains features that are compatible with American concerns for the limitation of power, it seems clear that the bureaucrats themselves understood their role and relation to private industry to be consistent with the long-standing developmentalist traditions of the Japanese state. But for the brief period of the ascendancy of the military in the 1930s and early 1940s, the bureaucracy was never interested in directly controlling the allocation of resources in the economy and preferred instead to assist private firms and to use the market to achieve goals that all agreed were significant for the nation. The SCAP believed it had modified the structural location of the bureaucracy in a way that made it compatible with its own New Deal-informed conception of a market-friendly interventionist state, while the Japanese believed that they were returning to more traditional mechanisms for the pursuit of national economic greatness. Neither was entirely wrong.

Thus, even in the immediate aftermath of the reforms, when it was in fact the case that the traditional checks on the bureaucracy had been removed, there were clear limits on its capacity and interest in intervening in the economy—some of which were unknown to and unanticipated by the reformers themselves. The subsequent development of the private economy and the

---

[111] Chalmers Johnson acidly observes about this: 'Ironically, it was during the Occupation that one of the fondest dreams of the wartime "control bureaucrats" [was] finally realized. With the militarists gone, the *Zaibatsu* facing dissolution and SCAP's decision to try to set the economy on its feet, the bureaucracy found itself working for the *tenno* [MacArthur] who really possessed the attributes of "absolutism".' See his 'Japan: Who Governs? An Essay on Official Bureaucracy', *Journal of Japanese Studies* 2, 1 (1975), 16, quoted in Haley, *Authority Without Power*, 141.

[112] Haley, *Authority Without Power*, ch. 7.

[113] Bai Gao, 'Arisawa Hiromi and His Theory for a Managed Economy', *Journal of Japanese Studies* 20, 1 (1994), 115–53; id., *Economic Ideology and Japanese Industrial Policy. Developmentalism 1931–1965* (New York: Cambridge University Press, 1997). The key shift that occurred with the Dodge Plan and the foundation of MITI, according to Gao, was that 'from then on state intervention in the Japanese economy changed from control over resource allocation to control over credit': ibid., 150.

associations that engaged with it created the kind of societal checks against bureaucratic power for which the Americans hoped. But as is clear from the steel industry example, this new development resulted in a new constellation of relations between the three relevant actors (state, firms, associations) which produced not only mutual limitation, but also a co-operative orientation toward industrial improvement and technological transformation that ultimately far surpassed the competitive capacity of the American industry, embedded in its own less co-operative system of countervailing power.

Business, Associations, and the State

The dissolution of the Iron and Steel Control Association and the break-up of Japan Steel and of the *Zaibatsu* destabilized the steel industry in a whole variety of ways. The smaller and more specialized steel producers were simultaneously confronted with direct competition from larger, integrated firms and deprived of the dynamic and nurturing downstream exchanges with diverse *Zaibatsu* units that had previously sustained their strategies of specialization. The large-volume producer Japan Steel was broken up into two new units that found themselves in the unaccustomed position of having to compete with one another. The top leadership of all the major firms in the industry was purged and the central association linking firms and fostering intra-industry informational exchange was banned. Neither the structure of the industry, nor the identity of the firms that would constitute it, nor the identity or interests of the managers that governed the firms could be taken for granted, nor could their strategies for reconstitution be clearly defined.

The only aspect of the situation that was very clearly defined for all the actors was that the Americans found the old bifurcated structure of volume producers and specialists to be unacceptable and that they wanted to encourage instead the formation of a competitive oligopoly structure with some number of volume-producing, integrated firms. If clarity in the midst of disarray grabs the attention, it should not be surprising that the actors in the industry attempted to recompose themselves along the American lines. There was no debate about reconstituting the old structure of bifurcation—because the conditions that had made it possible no longer existed.[114] Ex-*Zaibatsu* firms such as Kawasaki, Kobe, and Sumitomo found that they either had to abandon steel production

---

[114] It may strike some readers that I am overestimating the difference in the conditions facing the steel producers because the reconstitution of Japanese corporations formerly affiliated with *Zaibatsu* into *Keiretsu* was in the end simply the re-creation of the old *Zaibatsu* conditions under a new name. I disagree with this view—and moreover, so does much of the historical literature. Rather, from the perspective of interest here, the recasting of groups as *Keiretsu* is, in fact, a further indication of the way in which the steel firms were cut off in the new structure. Though *Keiretsu* involved extensive inter-unit cross-holdings and co-operation, the centralized control of the *Zaibatsu Honsha* (central holding office) was eliminated. The salience of inward co-ordination and co-operation by steel units with filial downstream consumers relative to non-filial downstream consumers was weakened, while the pressure to reorganize along mass-production lines was considerably strengthened, as we shall see. On the crucial differences between pre-war *Zaibatsu* and post-war *Keiretsu*, see, among others, Yamamura, *Economic Policy in Postwar Japan*, 110–28.

entirely, or commit themselves to strategies of higher-volume production and integration. In the end, the latter option was the one that they took.[115]

This compelling structuralism belies a great deal of local innovation on the part of Japanese actors as they attempted to define precisely what the American constraints actually allowed and how much of their own knowledge of steel-making and organization could be utilized in the pursuit of the new kind of strategy in production. Destabilization and constraint forced actors to draw on their knowledge of steel manufacture and organization in new ways and thereby led them to possibilities in the organization of integrated volume production and in the nature of oligopolistic competition that were not part of the American programme. At the same time, their success in doing so dramatically changed the face of the industrial structure of steel production in Japan.

The move toward integration and volume production was very explicitly conceived of as an effort to adopt an American strategy and move away from the old strategy of specialization. Kawasaki led the way in the integration of the non-integrated works with the construction of its dramatic Chiba Works in 1953. The firm's new chairman, Yataro Nishiyama, decided to construct this plant because he was convinced that the only way for his firm, and for the entire industry, to survive, was to adopt an American approach to production:

The Japanese iron and steel industry must cut costs and develop the ability to compete internationally by switching to the American mode of production and away from the European mode of small-lot production. It is necessary to construct an integrated works with a blast furnace.[116]

Rather than purchase or build upstream operations to complement Kawasaki's existing facilities, Nishiyama sought to gain a competitive advantage by constructing a giant new integrated facility on a greenfield site located on the sea coast to facilitate easy and massive in-and-out transport through deep harbours. The construction of the Chiba Works was announced in late 1950 and was completed in 1953. According to Yonekura, at the time its construction was announced, it was forecast to have a capacity equal to nearly a fifth of Japan's entire 1950 iron and steel output. The plant had 'two 500 tons per charge (tpc) blast furnaces, six 100 tpc open-hearth furnaces, matching slab mills and hot and cold strip mills with annual capacities of 350,00 tons of pig iron and 500,000 tons of crude steel'.[117] The Chiba Works also had an extremely efficient and consolidated plant layout: the plant had only 60 kilometres of rail track linking its various operations. This was 440 kilometres less than the Yahata works at the time.[118] In this and other ways, the Chiba Works took American principles of large-scale production and the technologies to achieve it and constructed them in ways that outstripped the best forms of organization of its domestic

---

[115] Yonekura, *Japanese Iron and Steel Industry*; Scheppach, *Die japanische Stahlindustrie*, 27–31.
[116] Yonekura, 'Postwar Reform in Management and Labor', 221.
[117] Id., *Japanese Iron and Steel Industry*, 213.
[118] Ibid., 213 ff.; Scheppach, *Die japanische Stahlindustrie*, 88 ff.

competitors. Moreover, in the efficiency of organization and the completeness and newness of its conception, the works exceeded the ambition of most American producers of the time as well. In many ways, Chiba took Americanism and perfected it beyond what American producers themselves had achieved.[119]

The success of the Chiba Works emboldened the other significant non-integrated producers to follow suit. Soon after the construction of the Chiba Works, both Sumitomo and Kobe decided to integrate their blast furnace operations, first in 1953 and 1954 by buying smaller integrated works and then, like Kawasaki, by building their own greenfield plants near deep harbours. Others followed these firms like dominos. By 1961, there were ten fully integrated steel producers in Japan: Yahata, Fuji, NKK, Kawasaki, Sumitomo, Kobe, Amagasaki, Nakayama, Osaka, and Nisshin.[120] The top six firms (Yahata, Fuji, NKK, Kawasaki, Sumitomo, Kobe) accounted for about 90 per cent of pig iron production and 80 per cent of crude steel, but the market structure of producers was much more oligopolistic (being evenly distributed among all producers) (see Table 12.4).

As in the German case, the effect of the allied interventions in Japan was to create a set of competitive conditions in industry that gave rise to a very different, less monopolistic, more pluralistic industrial structure of high-volume

Table 12.4. Concentration in Pig Iron and Rolled Steel in Japan, 1950–1967

No. of firms	Percentage of total production					
	1950	1955	1960	1965	1966	1967
Pig iron						
3	89	81	69	58	57	60
6	91	94	87	89	88	93
8	92	97	94	95	96	98
10	93	98	95	98	98	99
Hot-rolled steel						
3	50	51	51	45	46	49
6	68	69	69	70	70	76
8	74	75	75	75	77	81
10	77	77	79	78	80	83

Source: Walter Scheppach, *Die japanische Stahlindustrie* (Hamburg: Mitteilungen des Instituts für Asienkunde, No. 48, 1972), 94.

[119] Indeed, it was only in the 1960s that the balance of US production began to shift toward facilities located near deep harbour ports and even then significant portions of capacity were still located inland in the Pittsburgh–Youngstown area. See the discussion in Burn, *Steel Industry*, 518–36, esp. 528–31.

[120] Scheppach, *Die Japansiche Stahlindustrie*, 88 ff.

producers. Indeed, if you compare the industry structures of the Japanese, German, and US steel industries by the beginning of the 1960s, they all look very similar (see Table 12.5). The difference, of course, was that the Japanese (and the Germans) by this time were using forms of organization (and as we shall see shortly, technologies as well) in production that were superior to those in place in the American industry. The imitators had taken American principles and made them better by making them their own. What made oligopolistic competition in Japan so much more dynamic than the same form of competitive industry structure in the USA?

Table 12.5. Comparison of Market Shares of the Largest Firms in the USA, the UK, West Germany, and Japan, 1965 (percentage of total industry output)

USA	UK	FRG	Japan
US Steel 25	US 12.5	ATH 23.3	Yahata 18.6
Bethlehem 16	RTB 12.2	Hoesch 15.6	Fuji 17.4
Republic 7.5	Wales 10.1	Krupp 10.5	Kokan 10.4
National 6.5	Colvilles 9.8	Klöckner 8.3	Kawatetsu 10.4
Armco 5.9	S&L 7.4	Mannesmann 7.1	Sumikin 10.1
J&L 5.6	GKN 7.3	Oberhausen 6.1	Kobe 6
Inland 4.9	Dorman 7.2	Salzgitter 5.1	Nisshin 2.2
Youngstown 4.6	John 6.2	Röchling 3.4	Nakayama 2
Kaiser 2.1	S. Durham 5.5	Dilinger 3.3	Otani 1.5
Wheeling 1.6	Consett 3.7	Ilseder 2.9	Daido 1.4

Source: Adapted from Walter Scheppach, *Die japanische Stahlindustrie* (Hamburg: Mitteilungen des Instituts für Asienkunde, No 48, 1972), 94.

The answer is that, in embracing competitive oligopoly, the Japanese reinterpreted and transformed the American understanding of oligopolistic competition in a way that produced both extremely rapid growth and striking leaps in technology and innovation. The key was the union of competition and co-operation among actors in the industry. Actors in the industry did not abandon the co-operative exchanges among firms and between firms and the bureaucracy that had been a hallmark of development in the old pre-war structures. Instead, they recast the operation of co-operation within the industry away from the old model of co-operation between a state monopoly and broadly diversified holding companies to co-operation among relatively equal rival steel firms—a strategy, as we saw earlier, that was foreshadowed by the industry reform efforts of steel managers who had been active in the Control Associations during the war.

The advocates of co-operative oligopoly within the industry were eagerly encouraged and guided by bureaucrats within MITI and by players within the newly reconstituted trade associations (many of whom had previously been

members of the Control Association as well) who saw it as their mission to foster the reconstruction and modernization of private industry in the interest of
the recovery of the nation. These bureaucrats and trade association figures, as
we saw, abandoned their market-supplanting or *dirigiste* orientation toward
industrial co-ordination and turned toward an older market-preserving
approach of discussion, collective priority setting, and problem solving, with
private operating units in the industry. Under the new conditions of pluralistic
oligopoly and competing institutional actors, this new system was congenial (if
not identical) to the original political-economic norms established by the
Americans for democratic industrial practice.[121]

This dynamic three-way interaction between MITI bureaucrats, trade associations, and newly competitive firms led to the development of two sequential
'Rationalization Plans' for the steel industry which drove its rapid technological
development in the 1950s (and then, with subsequent plans, beyond).[122] In the
development and execution of these plans, MITI engaged steel producers in constant discussions and exchanges about their capacity requirements and technological needs, while the trade associations played a crucial role in bringing
managers from the various firms together to discuss mutual technological, production, and market concerns with one another and with representatives from
the bureaucracy (in particular MITI). All of the discussions involved the
identification of competitive organization and technology from abroad and
developing strategies for the adaptation and appropriation of those practices
in Japan.[123] MITI used its resources to affirm and co-ordinate decisions on

[121] And in part, as Lynn and McKeown, *Organizing Business*, 55–119 emphasize throughout
their account, this was facilitated by the fact that the bureaucrats and association leaders had personal experience within steel companies or had dealt with the *Zaibatsu* holdings directly. Chalmers
Johnson also emphasizes the centrality of personal contacts between bureaucrats, upper-level steel
managers, and trade association officials, and their revolving affiliations, as central for facilitating
tripartite co-ordination in steel—and also, paradoxically, for maintaining the autonomy of each:
'All of the big six [steel producers] had high-ranking former MITI officials on their boards of directors . . . But even so it was hard to give direct orders to the industry because so many of its executives were top leaders in business organizations such as Keidanren and Keisai Doyukai': see
Johnson, *MITI and the Japanese Miracle*, ch. 7, *passim*, quotation, 268. This emphasis on Johnson's
part is ironic because after pointing out the autonomy of the actors, he suggests that, 'From the era
of priority production just after the war down to approximately 1960, MITI had exercised detailed
control over investments in the steel industry': ibid., 269. On his own evidence, it seems more accurate to conclude that MITI engaged in extensive dialogue over the direction of investment with the
other two interlocutors.
[122] A concise outline of these plans is provided in James E. Vestal, *Planning for Change.
Industrial Policy and Japanese Economic Development, 1945–1990* (New York: Oxford University
Press, 1993), 115–44, esp. 118–32. An overview of the first plan is given in the English publication of
the Japanese Iron and Steel Industry Federation, *The Iron and Steel Industry and Fabricated
Products, 1952* (Tokyo: Tokyo Liaison and Translation Service).
[123] On state steel policy: Vestal, *Planning for Change*; Patricia O'Brien, 'Industry Structure as a
Competitive Advantage: The History of Japan's Post-war Steel Industry', *Business History* 34
(1992), 128–59; ead., 'Governance Systems in Steel: The American and Japanese Experience', in J.
Rogers Hollingsworth, Phillipp Schmitter, and Wolfgang Streeck (eds.), *Governing Capitalist
Economies. Performance and Control of Economic Sectors* (New York: Oxford University Press,
1994), 43–71. On the co-operative and collectively deliberative nature of trade association, government, and industry interaction, see Yonekura, 'Industrial Associations', 27–51; Lynn and

capacity expansion that the three-way discussion produced and it also subsidized co-operative research efforts, co-ordinated by the trade associations and involving all the major firms, on the development of key technologies which had been identified as crucial for the industry's competitive evolution. Most crucial in this regard were, of course, basic oxygen furnace (BOF) technology, the development of high-quality domestic refractory brick for use in blast furnaces, and, later, continuous casting technology.[124]

These new technologies, of course, were systematically incorporated into the physical plant in virtually all of the new greenfield facilities built alongside deep harbour ports following the opening of the Chiba Works, described above. In fact, the decision to integrate production and to adopt the BOF were linked. Japan's steel producers were plagued by extremely high materials costs—especially due to the extremely high cost of imported scrap—and sought integration and BOF because they minimized these costs. Integration backward into pig iron production radically reduced producers' need for scrap inputs and BOF was an attractive technology because it could make high-quality steel with little or no scrap.[125] Crucially, MITI and the Japan Iron and Steel Federation worked hard in the mid-1950s to ensure that the technologies necessary for this lower-cost production strategy would diffuse among all the producers in the industry.[126]

This co-operative investment strategy was enormously successful. At the beginning of the 1960s, Japanese steel producers had more BOF capacity than any other steel-producing nation in the world. By the beginning of the 1970s, the same was true of continuous casting. The enormity of these achievements should not go unappreciated. In effect, co-operation among the state, associations, and leading firms, recomposed and reconstituted through engagement with American reforms, produced a radical revolution in the mass production of steel. The American model had been appropriated and transformed in a way that ultimately led the industry to surpass US producers in technological sophistication, plant layout, organization, capacity, and quality of output.[127]

McKeown, *Organizing Business*, 90–119, 140–71; Kozo Yamamura, 'The Role of Government in Japan's "Catch-up" Industrialization: A Neoinstitutionalist Perspective', in Hyung Ki Kim, Michio Muramatsu, T. J. Pempel, and Kozo Yamamura (eds.), *The Japanese Civil Service and Economic Development* (New York: Oxford University Press, 1995), 103–32, esp. 117–22.

[124] The first rationalization plan, which ran from 1950 to 1954, focused on the replacement of capacity and the modernization of technologies, while the second rationalization plan (1955–60) targeted capacity expansion and the dramatic adoption of new technology—11 new blast furnaces for the production of pig iron were constructed in the period and a phenomenal 13 BOFs for steel production were introduced. See Vestal, *Planning for Change*, 122 ff.; and Lynn and McKeown, *Organized Business*, 90–119, 140–71.

[125] This argument is laid out in Gordon, *Wages of Affluence*, 63. The problem of high scrap costs had actually shaped the way that numerous producers adopted technology in the pre-war period as well: see, for example, the interesting article by Yoischi Kobayakawa, 'Problems of Technology Choice Faced by the Private-Sector Steel Industry in Prewar Japan. Nippon Steel Pipe's Steel Manufacturing Integration and the Introduction of Converters', *Japanese Yearbook of Business History* 13 (1996), 53–71.

[126] Gordon, *Wages of Affluence*, 62–3.   [127] Ibid., 60–1.

So, there you have it: the Japanese industry was dramatically transformed by the SCAP reforms and the direction of that transformation was, on the level of industrial structure (oligopolistic competition), and in the strategy of production (high-volume mass production), very consistent with what the Americans wanted and attempted to encourage. None the less, the outcome diverged markedly from what Americans understood to be pluralist oligopoly or integrated steel production. Rather than apply the American understanding of pluralist competition and the limitation of power through the construction of comparable adversaries, the Japanese organized oligopoly, with the eager involvement of the state and of newly reconstituted trade associations, as a process of collaborative discussion and co-operative learning. Rather than attempt to replicate the organization of American integrated steel production, the Japanese took American principles and applied them in a way that was more efficient and elegant. Developments in the workplace, as the next section will show, which were also profoundly influenced by American intervention and example, only enhanced the dynamism of Japanese steel firms.

Recasting Authority in Production: Pluralism and Status

As in the German case, Japanese workers seized the opportunity presented to them by Allied support of trade unionism and worker organization. Only in Japan, where the extent of trade union organization prior to occupation had always been relatively muted and suppressed, the encouragement of organization had the effect of opening the floodgates to massive pent-up interest in worker organization: in October 1945, there were only 5,000 people in trade unions in Japan; by December 1946, that number had increased astronomically to nearly five million.[128] Workers took the American political values of democracy and the need for countervailing organizational power in society very seriously, but as was the case with their German counterparts, they interpreted these political values through the lens of their own traditions and of their understanding of the social and political position of workers in Japanese society. The result was a profoundly different set of industrial relations institutions than Americans typically associated with voluntary labour organization.

As in Germany, the structures of authority at the level of the enterprise emerged immediately as a constitutive arena in the Japanese worker's conception of democratic order. For the Japanese worker, the creation of a countervailing organization in the labour market was not sufficient, by itself, to check the power of management in society at large, or within the enterprise. This was because the enterprises themselves were structured in ways that presupposed a hierarchy of status in society in which managers enjoyed ascribed social privilege and workers a caste-like social denigration. The creation of organizational power for labour without explicitly attacking these traditional differences in

---

[128] Id., 'Contests for the Workplace' in id. (ed.), *Postwar Japan as History* (Berkeley: University of California Press, 1993), 378.

status would not have created equality or plural power. It would merely have created a social organization for a subordinate estate in society.[129]

Thus, in the initial years of the occupation when the expansion of worker organization was most dramatic, organized workers in numerous industries, including steel, repeatedly called the role of managerial authority into question, and attacked status distinctions in plants between blue-collar and white-collar workers—all in the name of democracy.[130] In response to worker demands labour-management councils were formed which gave workers joint control with management over the workplace, personnel management, and corporate strategy.[131] And, according to Andrew Gordon, this made possible a whole cascade of profound firm-level reforms:

> through council deliberations or collective bargaining, workers eliminated many petty and substantive status divisions between white-collar staff and blue-collar staff that they found pervasive and repugnant throughout the prewar era. Under union pressure managers did away with separate gates, dining halls, and toilets as well as distinctions in dress and terminology (some companies replaced the terms *worker* and *staff* with the single term *employee*). Workers also gained a new equality in wages and bonuses. Some enterprises replaced a distinction between workers, who were paid by the day, and staff, who were paid by the month, with a common calculation in terms of monthly wages and paid bonuses to all employees as multiples of this monthly amount.[132]

These targets of democratic struggle were not anticipated by the Americans when they advocated labour organization as a key component of democratization, but American encouragement made the struggles possible. Moreover, American advocacy of democratic order also provided them with a vocabulary (democratization) in which acceptably to express traditional desires about status change against the claims of other positions (in particular, management) in the social order. Moreover, (initial) SCAP toleration of the gains won allowed them to become institutionalized in Japanese factories.[133] It is also true that many of these early gains were possible because management itself was

---

[129] On the significance of social status in working-class action in pre-war Japan, see Thomas C. Smith, 'The Right to Benevolence: Dignity and Japanese Workers, 1890–1920', in id., *Native Sources of Japanese Industrialization, 1750–1920* (Berkeley: University of California Press, 1988), 236–71; for a discussion of how these sentiments played out at the NKK steel mill in the early postwar years, see Gordon, *Wages of Affluence*, ch. 2.

[130] Otake Hideo writes: 'The term "democratization" thus acquired in some quarters an extremely radical content, quite contrary to the intentions of the Occupation. Especially with respect to the economic order, it was variously used to legitimize "enterprise democratization", management participation by labor, and even management exclusively by labor': 'The Zaikai under the Occupation: The Formation and Transformation of Managerial Councils', in Ward and Yoshikazu, *Democratizing Japan*, 366–91 (quotation, 366–7).

[131] On these councils, see ibid.; Ikuo Kume, *Disparaged Success. Labor Politics in Postwar Japan* (Ithaca: Cornell University Press, 1998), 59–67; Gordon, *Evolution of Labor Relations in Japan*, 339–49; and id., *Wages of Affluence*, chs. 1–3, esp. 36 ff.

[132] Id., 'Contests for the Workplace', 379; id., *Wages of Affluence*, chs. 1–3 for discussion of these events at NKK.

[133] For a contemporary observer's interesting effort to make sense of labour gains, see T. Cohen, *Remaking Japan*, 206.

extremely weakened, both economically and ideologically, by the defeat, occupation, purges, etc. and could not but cede to labour's transformative demands.

All of this naturally was very short-lived in the occupation: the SCAP's enthusiasm for labour power shifted abruptly when labour began to cross the line into politics in January 1947 and threatened to topple the government with a general strike. As in the German case, the American conception of democratic order did not include 'politicized' labour organizations; only countervailing organizations in the labour market. American resistance to labour's moves into politics emboldened the Japanese employers to counterattack and roll back many of the reforms that the workers had won in the initial years.[134] But the rollback, though significant, was not by any means a return to the social order in the factory or in society that antedated the initial years of labour mobilization. The initial post-war achievements created a new balance of power and new poles in debate in the struggle between labour and capital. In future struggles, employers argued away from the gains workers had made; but they could not and did not argue for the complete abandonment of the workplace and status change that the early gains had implemented. To that extent, mobilization under American encouragement and in the name of democracy brought lasting change to working life in Japan.

There was, however, significant rollback. Struggle over authority and power in the workplace continued for another fifteen years or so in Japan before a stable equilibrium was reached, and much of this struggle ultimately produced outcomes favourable to employers and against the most aggressive factions within the unions. Of the early enterprise-level gains, unions were forced to compromise most significantly on the degree of their formal influence on labour-management councils. Their role was shifted from direct participation in decision making with management to the lesser one of a body to be consulted by management in its decisions and planning. This was a setback, to be sure; but not a defeat. According to Kume, 'management retained the power to control the management system in the company, but labor maintained its right to be consulted in the case of personnel as well as managerial decision making . . . Labor continued to be a legitimate participant within the company rather than devolve into a mere production factor.'[135]

Gains in the form of wage payment and production control were also rolled back over the course of the 1950s and 1960s, and in these cases, the rollback was driven by management's desire to implement what they considered to be more advanced American practices. The result, however, was not more Americanism or even familiar forms of managerial control. Rather, given the organizational strength of labour on the shop-floor, managers were forced to recompose the

---

[134] The change in American priorities is a cornerstone of the historical narrative of the Japanese occupation. For the story of its impact on labour see the various discussions in Kume, *Disparaged Success*; Gordon, *Wages of Affluence*, ch. 3; and Joe Moore, *Japanese Workers and the Struggle for Power, 1945–1947* (Madison: University of Wisconsin Press, 1983).

[135] Kume, *Disparaged Success*, 61.

American techniques in the interest of achieving their desired goals in production. The result in both cases was the creation of the very distinctive Japanese hybrid systems of combined seniority and merit-based wages and decentralized, shop-floor-based, cross-functional, quality control.

The homogeneous seniority-based payment system actually began to diffuse in Japanese industry during the final years of the war before it spread widely during the period of worker radicalism in the initial post-war years.[136] Employers disliked the pure seniority system, calling it 'evil egalitarianism',[137] because it offered them no way to link pay to performance on the job. They began to attack it as soon as the balance of power began to shift in their favour during the course of the 1950s. In the steel industry an effort was made to introduce American 'job wages', in which different jobs would be compensated at different rates—the idea was not to supplant seniority completely as a method of payment (the newly mobilized workers had far too much invested in the old system for that), but to factor in the differential contribution of different jobs to the overall value output of the company.

Though it made some initial headway, the solution soon proved inadequate. First, the technological changes in the steel industry over the course of the 1950s created a more automated industry and as a result undermined the rough correspondence that had existed between seniority and level of skill in the early post-war years. Increasingly, young and old were engaged in comparable production tasks—a fact that made younger workers resentful of the seniority principle. Second, management itself became dissatisfied with the job-wage system because, when combined with the seniority principle, it allowed workers within a given category to have their wages increased over time regardless of performance. These problems led at the beginning of the 1960s to a shift away from seniority and job wages to pay based on a combination of seniority and merit— a solution that satisfied both older workers and younger workers, as well as managers interested in maintaining a tight relationship between pay and performance. In this case, an American idea introduced by management (linking payment to the contribution of the specific job) was modified beyond recognition (pay for individual performance) in an effort to adapt it to Japanese circumstances.[138]

This was similarly the case with the emergence of quality control circles. In this case the American idea was the creation of centralized bureaux for production engineers responsible, in good Taylorist fashion, for instructing shop-floor workers how best to maintain quality in production. Centralized quality control engineering bureaux were thus established in numerous steel production facilities during the early 1950s. But it soon became apparent that the production engineers from the Quality Control bureaux were greeted with significant distrust on the shop-floor from both production worker and foreman, both of

---

[136] Gordon, *Evolution of Labor Relations in Japan*, 257–98, 374–86.
[137] Id., *Wages of Affluence*, 66.   [138] Ibid., 66–70, 164–7.

whom regarded their own knowledge of production as far superior to that of the distant and élite engineering interlopers. Given this resistance, steel management began to reverse the centralizing impulse in the quality control initiative. Instead of separating engineering from the shop-floor, they began to create committees—or 'circles'—that systematically brought them together. As Gordon notes, these reversals of the Taylorist American logic of separating engineering and planning from production and execution marked 'the first stirrings of quality control as a system of widespread small group activities, foreshadowing a shift in the meaning of the abbreviation QC from "quality control" to "quality circles". This was the start of a crucial breakthrough to "total quality control" (TQC) involving technicians, foreman and the rank and file.'[139]

Once again, in the context of a new balance of power between management and labour, American techniques were modified and transformed in improbable ways. The hybrid institutional forms that emerged in this case as well as in that of the wage-payment system, however, were hailed as cornerstones of Japanese competitive advantage in international manufacturing markets during much of the post-war period.[140]

Though very significant for the long-term competitiveness of Japanese production, these workplace-level conflicts were at the time dwarfed in intensity by industry-level conflicts over the institutional role and position of organized labour in the Japanese political economy. Literally epochal struggles during the 1950s focused on the issue of the scope of collective bargaining and the extent to which labour would be able to act as a countervailing power in post-war Japanese society: should collective bargaining concentrate on the level of the enterprise or should it be extended to the level of the entire industry: that is, enterprise versus industrial unionism? This was a matter of particularly intense conflict in the steel industry throughout the 1950s, with the leadership factions within the unions at all the major producers solidaristically fighting for industry-wide bargaining and the employers at these firms just as solidaristically opposing this.

The conflict resulted in three major strikes in 1956, 1957, and 1959. Had the unions been able to win one of these strikes and force the employers into accepting industrial unionism, it would have positioned them as sovereign institutional rivals of capital in debates on the distribution of the social surplus. It would also have placed them in a position to re-enhance their power within the enterprises.[141] But this was not to be. In the heat of each of the strikes, the

---

[139] Ibid., 71.

[140] The two most notable authors bringing this dimension of Japanese success to the forefront are, of course, J. C. Abegglen, *The Japanese Factory* (Glencoe, IL: Free Press, 1958), and *Management and Worker: The Japanese Solution* (Tokyo: Sophia University Press, 1973); and Ronald Dore, *British Factory/Japanese Factory. The Origins of National Diversity in Industrial Relations* (Berkeley: University of California Press, 1973).

[141] And hence they could have blocked entirely the Americanization efforts noted above, in favour of an internationally unprecedented control of production by labour. The Japan of today could have been dramatically different from what it became; this possibility existed until at least as

employers were able to get one or more enterprise unions to break from the national coalition (frequently the more conservative Yahata union) and thereby insist on local rather than industrial wage deals. After the defeat in 1959, factionalism within the unions, already a source of instability during the 1950s, intensified. Eventually those elements within the unions which did not want to further jeopardize their enterprise-level gains by continued conflict at the industry level gained control. They abandoned the idea of industrial unionism and re-focused union energy on the enterprise level.[142]

A victory in the struggle for industrial unions might have launched the trade-union movement in Japan on to a European trajectory, where strong centralized unions bargained with peak-level employers associations over the distribution of much of society's surplus. So these losses at the end of the 1950s were significant for Japanese labour. But they did not by any means erase the significant gains in organizational, economic, and social power *relative to their pre-war position* that the unions had gained in the early post-war years. With American encouragement and support, Japanese workers were able to redefine their status within enterprises—and in the society at large—by abolishing the symbolic markers of difference and asserting institutional reforms that insisted upon equal treatment (and equitable and just reward) among employees within the enterprise, and respectful and informed relations between all employees and management. Here again, a very distinctively Japanese reality was created with the vocabulary and at crucial moments encouragement of American occupiers.

12.5   Conclusion

This chapter has attempted to show the way in which political ideas about the proper relationship between democracy and market order were constitutive in the reconstruction of core institutions and practices in both the German and Japanese political economies after the Second World War. The power of the American occupation was its ability to establish the discursive and conceptual terrain upon which debate and struggle for the reconstruction of industrial institutions and governance mechanisms would take place. My claim here is neither that the US occupiers had a unitary conception of the kind of economy they wanted to see emerge in Germany and Japan, nor that they had specific institutional arrangements or governance practices that they viewed as indispensable

late as 1959. Gordon and others, in emphasizing the significance of the pivotal decade of the 1950s, have rightly criticized those who try to portray the Japanese system as somehow entrenched in trans-historical institutional and cultural features of the Japanese people. See Gordon, *Wages of Affluence, passim.*

[142] This account follows Andrew Gordon, 'Conditions for the Disappearance of the Japanese Working Class Movement', in Elizabeth Perry (ed.), *Putting Class in its Place. Worker Identities in East Asia* (Berkeley: University of California Press, 1996), 11–52; Gordon, *Wages of Affluence, passim*; as well as Kume, *Disparaged Success.* For the very important point about factionalism within the steel workers unions during the 1950s, I have relied on the excellent Ph.D. dissertation by Akira Suzuki, 'The Polarization of the Union Movement in Post War Japan: Politics in the Unions of Steel and Railway Workers' (University of Wisconsin-Madison, 1997), ch. 2.

for the construction of a pluralist economy. Indeed, both cases of steel-industry reconstruction have shown in a very rich way how the occupation tolerated and even encouraged institutional forms that corresponded neither to their original conceptions nor to the institutional arrangements that then existed in the United States itself. Rather, the occupation first destroyed those institutional arrangements in the economy that it deemed morally and politically incompatible with a pluralist democratic economy and then established normative guidelines and pressures intended to encourage the creation of countervailing institutional powers within the economy that German and Japanese actors were compelled to take into account. This was a profound form of domination and the chapter makes plain that it led to the permanent transformation of both societies.[143]

The irony in the story, of course, is that the transformation of the German and Japanese steel industries created forms of practice and institutional design in production and at the level of the industry as a whole that were far more competitive than their American competitors for much of the post-war period. Destabilization and political constraint imposed by the occupation gave rise to a remarkable process of collective reflection among actors in both societies about the plasticity and recomposability of their own practices and institutions. As we saw in both cases, experimentation with technologies and organizational practices from elsewhere (especially America) as well as struggles for control were part and parcel of this process of collective reflexivity and redefinition. The new structures that emerged and became so competitive were in many cases distinctive hybrids that combined traditional indigenous practices with American ideas in ways that resembled neither but which often established entirely new international standards of performance.

---

[143] Again, for their reflections on power in this way see Comaroff and Comaroff, *Of Revelation and Revolution.*

Index